## EXPONENTIAL AND LOGARITHMIC FUNCTIONS

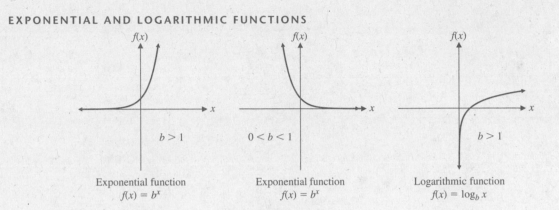

Exponential function
$f(x) = b^x$
$b > 1$

Exponential function
$f(x) = b^x$
$0 < b < 1$

Logarithmic function
$f(x) = \log_b x$
$b > 1$

## REPRESENTATIVE POLYNOMIAL FUNCTIONS (DEGREE > 2)

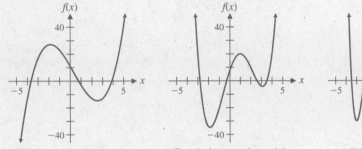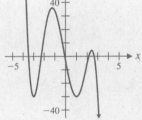

Third-degree polynomial
$f(x) = x^3 - x^2 - 14x + 11$

Fourth-degree polynomial
$f(x) = x^4 - 3x^3 - 9x^2 + 23x + 8$

Fifth-degree polynomial
$f(x) = -x^5 - x^4 + 14x^3 + 6x^2 - 45x - 3$

## REPRESENTATIVE RATIONAL FUNCTIONS

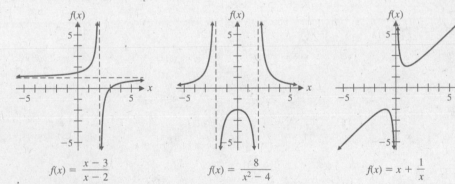

$f(x) = \dfrac{x - 3}{x - 2}$

$f(x) = \dfrac{8}{x^2 - 4}$

$f(x) = x + \dfrac{1}{x}$

## GRAPH TRANSFORMATIONS

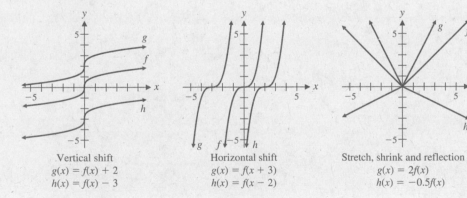

Vertical shift
$g(x) = f(x) + 2$
$h(x) = f(x) - 3$

Horizontal shift
$g(x) = f(x + 3)$
$h(x) = f(x - 2)$

Stretch, shrink and reflection
$g(x) = 2f(x)$
$h(x) = -0.5f(x)$

## ADDITIONAL CALCULUS TOPICS

# CALCULUS
## AND
## COLLEGE MATHEMATICS FOR BUSINESS, ECONOMICS, LIFE SCIENCES
## &
## SOCIAL SCIENCES
### TWELFTH EDITION

## Raymond Barnett
*Merritt College*

## Michael Ziegler
*Marquette University*

## Karl Byleen
*Marquette University*

**Prentice Hall**
is an imprint of

Copyright © 2011, 2008, 2006 Pearson Education, Inc.

Publishing as Pearson Prentice Hall, 75 Arlington Street, Boston, MA  02116.

ISBN-13: 978-0-321-65509-7
ISBN-10: 0-321-65509-5

3 4 5 6 V069 14 13 12

**Prentice Hall**
is an imprint of

www.pearsonhighered.com

# CONTENTS

PREFACE   ix

## 1 Differential Equations   1

1-1 Basic Concepts   2
1-2 Separation of Variables   12
1-3 First-Order Linear Differential Equations   24
*Chapter 1 Review   35*
*Review Exercise   36*

## 2 Taylor Polynomials and Infinite Series   39

2-1 Taylor Polynomials   40
2-2 Taylor Series   53
2-3 Operations on Taylor Series   64
2-4 Approximations Using Taylor Series   73
*Chapter 2 Review   83*
*Review Exercise   84*

## 3 Probability and Calculus   87

3-1 Improper Integrals   88
3-2 Continuous Random Variables   95
3-3 Expected Value, Standard Deviation, and Median   108
3-4 Special Probability Distributions   117
*Chapter 3 Review   128*
*Review Exercise   129*

Appendices A and B are found in the following publications: Calculus for Business, Economics, Life Sciences and Social Sciences 11E (0-13-232818-6) and College Mathematics for Business, Economics, Life Sciences and Social Sciences 11E (0-13-157225-3).

## C Tables   133

**Table III** Area Under the Standard Normal Curve   133

## D Special Calculus Topic   135

D-1 Interpolating Polynomials and Divided Differences   135

Answers   A-1

Solutions to Odd-Numbered Exercises   S-1

Index   I-1

Applications Index   Inside Back Cover

# PREFACE

We have prepared this supplement to support *Calculus for Business, Economics, Life Sciences, and Social Sciences* 11/e, and *College Mathematics for Business, Economics, Life Sciences, and Social Sciences* 11/e.

Chapter 1, Differential Equations, begins by introducing some of the basic concepts and terminology used in the study of differential equations. Separable differential equations, first-order linear differential equations, and related applications are covered thoroughly. All of the growth models introduced in Section 6-3 of *Calculus* 11/e or Section 13-3 of *College Mathematics* 11/e are covered again, this time with increased emphasis on student recognition of the relevant model.

Chapter 2, Taylor Polynomials and Infinite Series, begins by discussing the approximation of functions by Taylor polynomials. Infinite series follow naturally from this presentation, eliminating the need for a lengthy treatment of series of constants and all the associated convergence tests. The operations that can be performed on Taylor series and the effects of these operations on the interval of convergence are carefully discussed, since this is the way Taylor series are used in most real-world applications.

Chapter 3, Probability and Calculus, begins by discussing improper integrals. The remainder of the chapter discusses properties of continuous probability density functions, including the uniform, exponential, and normal probability distributions. The presentation, though comprehensive, assumes no previous experience with probability.

Appendix D contains a section that can be covered at appropriate points in the text.

Section D-1, Interpolating Polynomials and Divided Differences, can be covered at any point in the book.

The pedagogy and design mirrors that of *Calculus for Business, Economics, Life Sciences, and Social Sciences* 11/e and *College Mathematics for Business, Economics, Life Sciences, and Social Sciences* 11/e. This allows easy and seamless integration of this supplement into any course.

*R. A. Barnett*
*M. R. Ziegler*
*K. E. Byleen*

# Differential Equations

1-1   Basic Concepts

1-2   Separation of Variables

1-3   First-Order Linear
      Differential Equations

Chapter 1 Review

Review Exercise

# INTRODUCTION

Mathematical models for many processes in nature and society are best developed by understanding the *rate* at which the process occurs. For example, the rate of growth of a population may be proportional to the size of the population, proportional to the difference between the size and a fixed limit, or proportional to both the size and the difference between the size and a fixed limit. These descriptions of the rate may be expressed succinctly by *differential equations* such as

$$\frac{dy}{dt} = ky \qquad \frac{dy}{dt} = k(M - y) \qquad \frac{dy}{dt} = ky(M - y)$$

Solutions to these differential equations provide models for several types of exponential growth—unlimited, limited, and logistic (see Table 1, page 20). Unfortunately, there is no single method that will solve all the differential equations that may be encountered—even in very simple applications. In this chapter, after discussing some basic concepts in the first section, we will consider methods for solving several types of differential equations that have significant applications.

1

## Section 1-1 BASIC CONCEPTS

- Solutions of Differential Equations
- Implicit Solutions
- Application

In this section we review the basic concepts introduced in Section 6-3 of *Calculus* 11/e or Section 13-3 of *College Mathematics* 11/e and discuss in more detail what is meant by a solution to a differential equation, including explicit and implicit representations. Specific techniques for solving differential equations will be discussed in the next two sections.

### ■ Solutions of Differential Equations

A **differential equation** is an equation involving an unknown function, usually denoted by $y$, and one or more of its derivatives. For example,

$$y' = 0.2xy \tag{1}$$

is a differential equation. Since only the first derivative of the unknown function $y$ appears in this equation, it is called a **first-order** differential equation. In general, the **order** of a differential equation is the highest derivative of the unknown function present in the equation.

Notice that we used $y'$ rather than $dy/dx$ to represent the derivative of $y$ with respect to $x$ in equation (1). This is customary practice in the study of differential equations. Unless indicated otherwise, we will assume that $y$ is a function of the independent variable $x$ and that $y'$ refers to the derivative of $y$ with respect to $x$.

*Explore & Discuss* **1**

Consider the differential equation

$$y' = 2x \tag{2}$$

(A) Which of the following functions satisfy this equation?

$$y = x^2 \qquad y = 2x^2 \qquad y = x^2 + 2$$

(B) Can you find other functions that satisfy equation (2)? What form do all these functions have?

(C) Discuss possible solution methods for any differential equation of the form

$$y' = f(x)$$

Figure 1A shows a *slope field* for equation (1). Recall from Section 6-3 of *Calculus* 11/e or Section 13-3 of *College Mathematics* 11/e that the **slope field** for a differential equation is obtained by drawing tangent line segments determined by the equation at each point in a grid. This provides a geometric interpretation of the equation that indicates the general shape of solutions to the equation. Due to the large number of line segments, slope fields are usually generated by a graphing utility. Figure 1B shows the slope field for $y' = 0.2xy$ on a TI-83 Plus.

Now consider the function

$$y = 2e^{0.1x^2}$$

whose derivative is

$$y' = 2e^{0.1x^2}(0.2x) = 0.4xe^{0.1x^2}$$

Substituting for $y$ and $y'$ in equation (1) gives

$$y' = 0.2xy \tag{1}$$
$$0.4xe^{0.1x^2} = 0.2x(2e^{0.1x^2})$$
$$0.4xe^{0.1x^2} = 0.4xe^{0.1x^2}$$

which is certainly true for all values of $x$. This shows that the function $y = 2e^{0.1x^2}$ is a **solution** of equation (1). But this function is not the only solution.

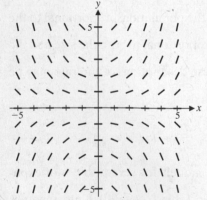

(A)

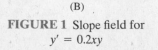

(B)

**FIGURE 1** Slope field for $y' = 0.2xy$

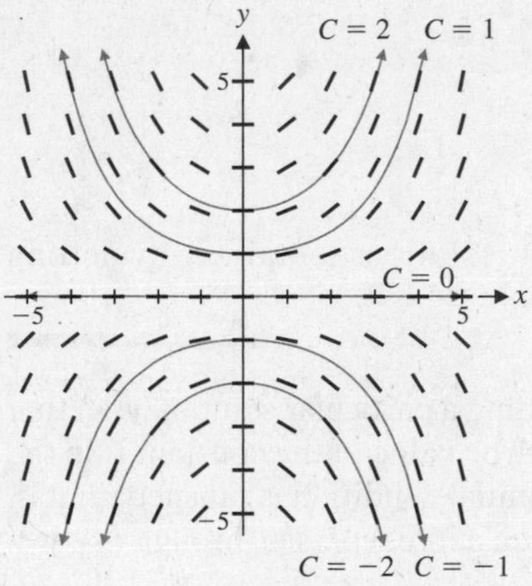

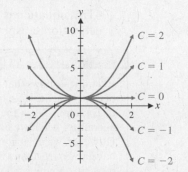

FIGURE 2 Particular solutions of
$y' = 0.2xy$

In fact, if $C$ is any constant, substituting $y = Ce^{0.1x^2}$ and $y' = 0.2xCe^{0.1x^2}$ in equation (1) yields the identity

$$y' = 0.2xy$$
$$0.2xCe^{0.1x^2} = 0.2xCe^{0.1x^2}$$

It turns out that all solutions of $y' = 0.2xy$ can be obtained from $y = Ce^{0.1x^2}$ by assigning $C$ appropriate values; hence, $y = Ce^{0.1x^2}$ is called the **general solution** of equation (1). The collection of all functions of the form $y = Ce^{0.1x^2}$ is called the **family of solutions** of equation (1). The function $y = 2e^{0.1x^2}$, obtained by letting $C = 2$ in the general solution, is called a **particular solution** of equation (1). Other particular solutions are

$$y = -2e^{0.1x^2} \qquad C = -2$$
$$y = -e^{0.1x^2} \qquad C = -1$$
$$y = 0 \qquad C = 0$$
$$y = e^{0.1x^2} \qquad C = 1$$

These particular solutions are graphed in Figure 2, along with the slope field for equation (1). Notice how the particular solutions conform to the shapes indicated by the slope field.

## INSIGHT

Most, but not all, first-order differential equations have general solutions that involve one arbitrary constant. Whenever we find a function, such as $y = Ce^{0.1x^2}$, that satisfies a given differential equation for any value of the constant $C$, we will assume this function is the general solution.

## EXAMPLE 1

**Verifying the Solution of a Differential Equation**  Show that

$$y = Cx^2 + 1$$

is the general solution of the differential equation

$$xy' = 2y - 2$$

On the same set of axes, graph the particular solutions obtained by letting $C = -2, -1, 0, 1$, and 2.

## SOLUTION

Substituting $y = Cx^2 + 1$ and $y' = 2Cx$ in the differential equation, we have

$$xy' = 2y - 2$$
$$x(2Cx) = 2(Cx^2 + 1) - 2$$
$$2Cx^2 = 2Cx^2 + 2 - 2$$
$$2Cx^2 = 2Cx^2$$

which shows that $y = Cx^2 + 1$ is the general solution. The particular solutions corresponding to $C = -2, -1, 0, 1$, and 2 are graphed in Figure 3.

FIGURE 3 Particular solutions of $xy' = 2y - 2$

**MATCHED PROBLEM 1** Show that

$$y = Cx + 1$$

is the general solution of the differential equation

$$xy' = y - 1$$

On the same set of axes, graph the particular solutions obtained by letting $C = -2, -1, 0, 1,$ and 2.

---

In many applications, we will be interested in finding a particular solution $y(x)$ that satisfies an **initial condition** of the form $y(x_0) = y_0$. The value of the constant $C$ in the general solution must then be selected so that this initial condition is satisfied; that is, so that the solution curve (from the family of solution curves) passes through $(x_0, y_0)$.

**EXAMPLE 2** **Finding a Particular Solution** Using the general solution in Example 1, find a particular solution of the differential equation $xy' = 2y - 2$ that satisfies the indicated initial condition, if such a solution exists.

(A) $y(1) = 3$    (B) $y(0) = 3$    (C) $y(0) = 1$

**SOLUTION** (A) *Initial condition $y(1) = 3$.* From Example 1, the general solution of the differential equation is

$$y = Cx^2 + 1$$

Substituting $x = 1$ and $y = 3$ in this general solution yields

$$3 = C(1)^2 + 1$$
$$= C + 1$$
$$C = 2$$

Thus, the particular solution satisfying the initial condition $y(1) = 3$ is

$$y = 2x^2 + 1$$

See the graph labeled $C = 2$ in Figure 3.

(B) *Initial condition $y(0) = 3$.* Substituting $x = 0$ and $y = 3$ in the general solution yields

$$3 = C(0)^2 + 1$$

No matter what value of $C$ we select, $C(0)^2 = 0$ and this equation reduces to

$$3 = 1$$

which is *never* valid. Thus, we must conclude that there is no particular solution of this differential equation that will satisfy the initial condition $y(0) = 3$.

(C) *Initial condition $y(0) = 1$.* Substituting $x = 0$ and $y = 1$ in the general solution, we have

$$1 = C(0)^2 + 1$$
$$1 = 1$$

This equation is valid for *all* values of $C$. Thus, all solutions of this differential equation satisfy the initial condition $y(0) = 1$ (see Fig. 3). ∎

## INSIGHT

Example 2 shows that there may be exactly one particular solution, no particular solution, or many particular solutions that satisfy a given initial condition. The situation illustrated in Example 2A is the most common. Most first-order differential equations you will encounter will have exactly one particular solution that satisfies a given initial condition. However, we must always be aware of the possibility that a differential equation may not have any particular solutions that satisfy a given initial condition, or the possibility that it may have many particular solutions satisfying a given initial condition.

**MATCHED PROBLEM 2** Using the general solution in Matched Problem 1, find a particular solution of the differential equation $xy' = y - 1$ that satisfies the indicated initial condition, if such a solution exists.

(A) $y(1) = 2$    (B) $y(0) = 2$    (C) $y(0) = 1$

## ■ Implicit Solutions

The solution method we will discuss in the next section always produces an implicitly defined solution of a differential equation. In most of the differential equations we will consider, an explicit form of the solution can then be found by solving this implicit equation for the dependent variable in terms of the independent variable.

**Explore & Discuss 2**    Let $y$ be a function defined implicitly by the equation

$$xy = C \quad \text{where } C \text{ is a constant} \tag{3}$$

(A) Differentiate equation (3) implicitly to obtain a differential equation involving $y$ and $y'$.

(B) Solve equation (3) for $y$, and verify that the resulting function satisfies the differential equation you found in part (A).

**EXAMPLE 3**    **Verifying an Implicit Solution** If $y$ is defined implicitly by the equation

$$y^3 + e^y - x^4 = C \tag{4}$$

show that $y$ satisfies the differential equation

$$(3y^2 + e^y)y' = 4x^3$$

**SOLUTION**    We use implicit differentiation to show that $y$ satisfies the given differential equation:

$$y^3 + e^y - x^4 = C$$

$$\frac{d}{dx}(y^3 + e^y - x^4) = \frac{d}{dx}C \quad \text{Remember, } \frac{d}{dx}x^n = nx^{n-1} \text{ and } \frac{d}{dx}e^x = e^x,$$

$$\boxed{\frac{d}{dx}y^3 + \frac{d}{dx}e^y - \frac{d}{dx}x^4 = 0} \quad \text{but } \frac{d}{dx}y^n = ny^{n-1}y' \text{ and } \frac{d}{dx}e^y = e^y y'$$

$$\text{since } y \text{ is a function of } x.$$

$$3y^2y' + e^y y' - 4x^3 = 0$$

$$(3y^2 + e^y)y' = 4x^3$$

Since the last equation is the given differential equation, our calculations show that any function $y$ defined implicitly by equation (4) is a solution of this differential equation. We cannot find an explicit formula for $y$, since none exists in terms of finite combinations of elementary functions.

**MATCHED PROBLEM 3**    If $y$ is defined implicitly by the equation

$$y + e^{y^2} - x^2 = C$$

show that $y$ satisfies the differential equation

$$(1 + 2ye^{y^2})y' = 2x$$

| EXAMPLE 4 | **Finding an Explicit Solution** If $y$ is defined implicitly by the equation |

$$y^2 - x^2 = C$$

show that $y$ satisfies the differential equation

$$yy' = x$$

Find an explicit expression for the particular solution that satisfies the initial condition $y(0) = 2$.

**SOLUTION**  Using implicit differentiation, we have

$$y^2 - x^2 = C$$
$$\frac{d}{dx}(y^2 - x^2) = \frac{d}{dx}C$$
$$2yy' - 2x = 0$$
$$2yy' = 2x$$
$$yy' = x$$

which shows that $y$ satisfies the given differential equation. Substituting $x = 0$ and $y = 2$ in $y^2 - x^2 = C$, we have

$$y^2 - x^2 = C$$
$$(2)^2 - (0)^2 = C$$
$$4 = C$$

Thus, the particular solution satisfying $y(0) = 2$ is a solution of the equation

$$y^2 - x^2 = 4$$

or

$$y^2 = 4 + x^2$$

Solving the last equation for $y$ yields two explicit solutions,

$$y_1(x) = \sqrt{4 + x^2} \qquad \text{and} \qquad y_2(x) = -\sqrt{4 + x^2}$$

The first of these two solutions satisfies

$$y_1(0) = \sqrt{4 + (0)^2} = \sqrt{4} = 2$$

whereas the second satisfies

$$y_2(0) = -\sqrt{4 + (0)^2} = -\sqrt{4} = -2$$

Thus, the particular solution of the differential equation $yy' = x$ that satisfies the initial condition $y(0) = 2$ is

$$y(x) = \sqrt{4 + x^2}$$

Check:

$$y'(x) = \frac{1}{2}(4 + x^2)^{-1/2}2x \qquad\qquad\qquad yy' = x$$

$$= \frac{x}{\sqrt{4 + x^2}} \qquad\qquad \sqrt{4 + x^2}\left(\frac{x}{\sqrt{4 + x^2}}\right) \stackrel{?}{=} x$$

$$x \stackrel{\checkmark}{=} x$$

MATCHED PROBLEM 4 | If $y$ is defined implicitly by the equation

$$y^2 - x = C$$

show that $y$ satisfies the differential equation

$$2yy' = 1$$

Find an explicit expression for the particular solution that satisfies the initial condition $y(0) = 3$.

## APPLICATION

In economics, the price of a product is often studied over time, and so it is natural to view price as a function of time. Let $p(t)$ be the price of a particular product at time $t$. If $p(t)$ approaches a limiting value $\bar{p}$ as $t$ approaches infinity, then the price for this product is said to be **dynamically stable** and $\bar{p}$ is referred to as the **equilibrium price.** (Later in this chapter, this definition of equilibrium price will be related to the one given in Section 7-2 of *Calculus* 11/e or Section 14-2 of *College Mathematics* 11/e.) In order to study the behavior of price as a function of time, economists often assume that the price satisfies a differential equation. This approach is illustrated in the next example.

EXAMPLE 5 | **Dynamic Price Stability** The price $p(t)$ of a product is assumed to satisfy the differential equation

$$\frac{dp}{dt} = 10 - 0.5p \qquad t \geq 0$$

(A)  Show that

$$p(t) = 20 - Ce^{-0.5t}$$

is the general solution of this differential equation, and evaluate

$$\bar{p} = \lim_{t \to \infty} p(t)$$

(B)  On the same set of axes, graph the three particular solutions that satisfy the initial conditions $p(0) = 40, p(0) = 10$, and $p(0) = 20$.

(C)  Discuss the long-term behavior of the price of this product.

SOLUTION | (A)  $p(t) = 20 - Ce^{-0.5t}$

$p'(t) = 0.5Ce^{-0.5t}$

Substituting in the given differential equation, we have

$$\frac{dp}{dt} = 10 - 0.5p$$

$$0.5Ce^{-0.5t} = 10 - 0.5(20 - Ce^{-0.5t})$$

$$= 10 - 10 + 0.5Ce^{-0.5t}$$

$$= 0.5Ce^{-0.5t}$$

which shows that $p(t) = 20 - Ce^{-0.5t}$ is the general solution of this differential equation. To find the equilibrium price, we must evaluate $\lim_{t \to \infty} p(t)$.

$$\bar{p} = \lim_{t \to \infty} p(t)$$

$$= \lim_{t \to \infty} (20 - Ce^{-0.5t})$$

$$= 20 - C \lim_{t \to \infty} e^{-0.5t} \qquad \lim_{t \to \infty} e^{-0.5t} = 0$$

$$= 20 - C \cdot 0$$

$$= 20$$

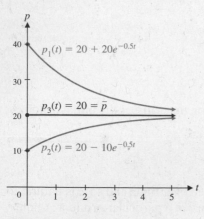

**FIGURE 4** Particular solutions of

$$\frac{dp}{dt} = 10 - 0.5p$$

(B) Before we can graph the three particular solutions, we must evaluate the constant $C$ for each of the indicated initial conditions. In each case, we will make use of the equation

$$p(0) = 20 - Ce^0 = 20 - C$$

| Case 1 | Case 2 | Case 3 |
|---|---|---|
| $p(0) = 40$ | $p(0) = 10$ | $p(0) = 20$ |
| $20 - C = 40$ | $20 - C = 10$ | $20 - C = 20$ |
| $C = -20$ | $C = 10$ | $C = 0$ |
| $p_1(t) = 20 + 20e^{-0.5t}$ | $p_2(t) = 20 - 10e^{-0.5t}$ | $p_3(t) = 20$ |

The graphs are shown in Figure 4.

(C) From part (A), we know that $p'(t) = 0.5Ce^{-0.5t}$ and that the equilibrium price is $\bar{p} = 20$. If

$$p(0) = 20 - C > \bar{p} = 20 \qquad \boxed{\text{Add } -20 \text{ to both sides.}}$$
$$-C > 0 \qquad \text{or} \qquad C < 0$$

then $p'(t)$ is negative and the price decreases and approaches $\bar{p}$ as $t \to \infty$ [See the graph of $p_1(t)$ in Fig. 4]. On the other hand, if

$$p(0) = 20 - C < \bar{p} = 20 \qquad \boxed{\text{Add } -20 \text{ to both sides.}}$$
$$-C < 0 \qquad \text{or} \qquad C > 0$$

then $p'(t)$ is positive and the price increases and approaches $\bar{p}$ as $t \to \infty$ [See the graph of $p_2(t)$ in Fig. 4]. Finally, if

$$p(0) = 20 - C = \bar{p} = 20 \qquad \boxed{\text{Add } -20 \text{ to both sides.}}$$
$$-C = 0 \qquad \text{or} \qquad C = 0$$

then $p'(t) = 0$ and the price is constant for all $t$ [See the graph of $p_3(t)$ in Fig. 4]. ■

**MATCHED PROBLEM 5** (A) Show that the price function

$$p(t) = 25 - Ce^{-0.2t}$$

is the general solution of the differential equation

$$\frac{dp}{dt} = 5 - 0.2p \qquad t \geq 0$$

Find the equilibrium price $\bar{p}$.

(B) On the same set of axes, graph the three particular solutions that satisfy the initial conditions $p(0) = 40, p(0) = 5,$ and $p(0) = 25$.

(C) Discuss the long-term behavior of the price of this product.

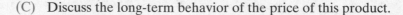

*Answers to Matched Problems*   **1.**

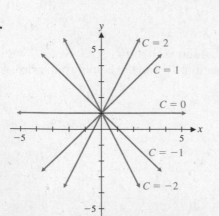

**2.** (A) $y = x + 1$

(B) No particular solution exists.

(C) $y = Cx + 1$ for any $C$

**3.** $\frac{d}{dx}y + \frac{d}{dx}e^{y^2} - \frac{d}{dx}x^2 = \frac{d}{dx}C$

$y' + e^{y^2}2yy' - 2x = 0$

$(1 + 2ye^{y^2})y' = 2x$

**4.** $y(x) = \sqrt{9 + x}$

**5.** (A) $\bar{p} = 25$

(B)

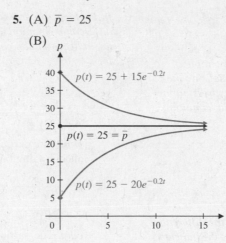

(C) If $p(0) > \bar{p}$, then the price decreases and approaches $\bar{p}$ as a limit. If $p(0) < \bar{p}$, then the price increases and approaches $\bar{p}$ as a limit. If $p(0) = \bar{p}$, then the price remains constant for all $t$.

# Exercise 1-1

**A**  *In Problems 1–10, show that the given function is the general solution of the indicated differential equation.*

**1.** $y = Cx^2$; $xy' = 2y$

**2.** $y = Cx^3$; $xy' = 3y$

**3.** $y = \dfrac{C}{x}$; $xy' = -y$

**4.** $y = \dfrac{C}{x^2}$; $xy' = -2y$

**5.** $y = C(x - 1)^2$; $(x - 1)y' = -y$

**6.** $y = \dfrac{C}{x + 1}$; $(x + 1)y' = -y$

**7.** $y = Ce^{x^2}$; $y' = 2xy$

**8.** $y = Ce^{-x^3}$; $y' = -3x^2 y$

**9.** $y = 4 + Ce^x$; $y' = y - 4$

**10.** $y = 1.5 + Ce^{-2x}$; $y' = 3 - 2y$

*In Problems 11–14, determine which of the slope fields (A)–(D) is associated with the indicated differential equation. Briefly justify your answer.*

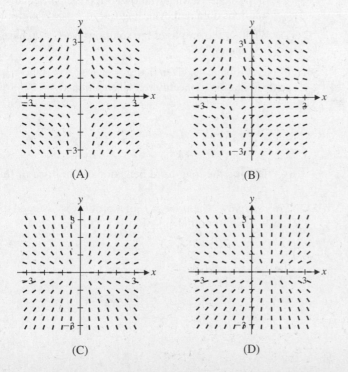

(A)  (B)

(C)  (D)

**11.** $xy' = 2y$ (From Problem 1)

**12.** $(x - 1)y' = -y$ (From Problem 5)

**13.** $xy' = -y$ (From Problem 3)

**14.** $(x + 1)y' = -y$ (From Problem 6)

*In Problems 15–18, use the appropriate slope field—(A), (B), (C), or (D), given for Problems 11–14 (or a copy)—to graph the particular solutions obtained by letting $C = -2, -1, 0, 1,$ and $2$.*

**15.** $y = Cx^2$ (From Problem 1)

**16.** $y = C(x - 1)^2$ (From Problem 5)

**17.** $y = \dfrac{C}{x}$ (From Problem 3)

**18.** $y = \dfrac{C}{x + 1}$ (From Problem 6)

**B**  *In Problems 19–28, show that the given function y is the general solution of the indicated differential equation. Find a particular solution satisfying the given initial condition.*

**19.** $y = 3 + 3x + Ce^x$; $y' = y - 3x$; $y(0) = 4$

**20.** $y = -4 - 4x + Ce^x$; $y' = y + 4x$; $y(0) = 1$

**21.** $y = xe^{-x} + Ce^{-x}$; $y' = -y + e^{-x}$; $y(0) = 0$

**22.** $y = xe^{2x} + Ce^{2x}$; $y' = 2y + e^{2x}$; $y(0) = 0$

**23.** $y = 2x + \dfrac{C}{x}$; $xy' = 4x - y$; $y(1) = 1$

**24.** $y = -3x + \dfrac{C}{x}$; $xy' + y = -6x$; $y(1) = 2$

**25.** $y = 2 + Cx^{-3}$; $xy' + 3y = 6$; $y(1) = 3$

**26.** $y = 2 + Cx^{-4}$; $xy' + 4y = 8$; $y(1) = -2$

**27.** $y = x + Cx^{1/2}$; $2xy' - y = x$; $y(1) = 0$

**28.** $y = x^2 + Cx^{1/2}$; $2xy' - y = 3x^2$; $y(1) = 1$

*If y is defined implicitly by the given equation, use implicit differentiation in Problems 29–32 to show that y satisfies the indicated differential equation.*

**29.** $y^3 + xy - x^3 = C$; $(3y^2 + x)y' = 3x^2 - y$

**30.** $y + x^3 y^3 - 2x = C$; $(1 + 3x^3 y^2)y' = 2 - 3x^2 y^3$

**31.** $xy + e^{y^2} - x^2 = C; (x + 2ye^{y^2})y' = 2x - y$

**32.** $y + e^{xy} - x = C; (1 + xe^{xy})y' = 1 - ye^{xy}$

*If y is defined implicitly by the given equation, use implicit differentiation in Problems 33–36 to show that y satisfies the indicated differential equation. Find an explicit expression for the particular solution that satisfies the given initial condition.*

**33.** $y^2 + x^2 = C; yy' = -x; y(0) = 3$

**34.** $y^2 + 2x^2 = C; yy' = -2x; y(0) = -4$

**35.** $\ln(2 - y) = x + C; y' = y - 2; y(0) = 1$

**36.** $\ln(5 - y) = 2x + C; y' = 2(y - 5); y(0) = 2$

*In Problems 37–40, use the general solution y of the differential equation to find a particular solution that satisfies the indicated initial condition. Graph the particular solutions for x ≥ 0.*

**37.** $y = 2 + Ce^{-x}; y' = 2 - y$

    (A) $y(0) = 1$     (B) $y(0) = 2$     (C) $y(0) = 3$

**38.** $y = 4 + Ce^{-x}; y' = 4 - y$

    (A) $y(0) = 2$     (B) $y(0) = 4$     (C) $y(0) = 5$

**39.** $y = 2 + Ce^x; y' = y - 2$

    (A) $y(0) = 1$     (B) $y(0) = 2$     (C) $y(0) = 3$

**40.** $y = 4 + Ce^x; y' = y - 4$

    (A) $y(0) = 2$     (B) $y(0) = 4$     (C) $y(0) = 5$

*In Problems 41–44, write a differential equation that expresses the given description of a rate. Explain what each symbol in your differential equation represents.*

**41.** The number of employees of a corporation increases at a rate of 650 employees per year.

**42.** The blood pressure in the aorta decreases between heartbeats at a rate proportional to the blood pressure.

**43.** When a pizza is taken out of a hot oven, it cools off at a rate proportional to the difference between the temperature of the pizza and the room temperature of 72°F.

**44.** When a gallon of milk is taken out of a refrigerator, it warms up at a rate proportional to the difference between the room temperature of 75°F and the temperature of the milk.

C *In Problems 45 and 46, use the general solution y of the differential equation to find a particular solution that satisfies the indicated initial condition. Graph the particular solutions for x ≥ 0.*

**45.** $y = \dfrac{10}{1 + Ce^{-x}}; y' = 0.1y(10 - y)$

    (A) $y(0) = 1$     (B) $y(0) = 10$     (C) $y(0) = 20$

    [*Hint:* The particular solution in part (A) has an inflection point at $x = \ln 9$.]

**46.** $y = \dfrac{5}{1 + Ce^{-x}}; y' = 0.2y(5 - y)$

    (A) $y(0) = 1$     (B) $y(0) = 5$     (C) $y(0) = 10$

    [*Hint:* The particular solution in part (A) has an inflection point at $x = \ln 4$.]

*In Problems 47 and 48, use the general solution y of the differential equation to find a particular solution that satisfies the indicated initial condition.*

**47.** $y = Cx^3 + 2; xy' = 3y - 6$

    (A) $y(0) = 2$     (B) $y(0) = 0$     (C) $y(1) = 1$

**48.** $y = Cx^4 + 1; xy' = 4y - 4$

    (A) $y(0) = 1$     (B) $y(0) = 0$     (C) $y(1) = 2$

*In Problems 49 and 50, use window dimensions Xmin = −5, Xmax = 5, Ymin = −5, and Ymax = 5.*

**49.** Given that $y = x + Ce^{-x}$ is the general solution of the differential equation $y' + y = 1 + x$:

    (A) In the same viewing window, graph the particular solutions obtained by letting $C = 0, 1, 2$, and 3.

    (B) What do the graphs of the solutions for $C > 0$ have in common?

    (C) In the same viewing window, graph the particular solutions obtained by letting $C = 0, -1, -2$, and $-3$.

    (D) What do the graphs of the solutions for $C < 0$ have in common?

**50.** Repeat Problem 49, given that $y = x + (C/x)$ is the general solution of the differential equation $xy' = 2x - y$.

## Applications

**51.** *Price stability.* The price $p(t)$ of a product is assumed to satisfy the differential equation

$$\frac{dp}{dt} = 0.5 - 0.1p$$

    (A) Show that

$$p(t) = 5 - Ce^{-0.1t}$$

    is the general solution of this differential equation, and evaluate

$$\bar{p} = \lim_{t \to \infty} p(t)$$

    (B) Graph the particular solutions that satisfy the initial conditions $p(0) = 1$ and $p(0) = 10$.

    (C) Discuss the long-term behavior of the price of this product.

**52.** *Price stability.* The price $p(t)$ of a product is assumed to satisfy the differential equation

$$\frac{dp}{dt} = 0.8 - 0.2p$$

    (A) Show that

$$p(t) = 4 - Ce^{-0.2t}$$

is the general solution of this differential equation, and evaluate

$$\overline{p} = \lim_{t \to \infty} p(t)$$

(B) Graph the particular solutions that satisfy the initial conditions $p(0) = 2$ and $p(0) = 8$.

(C) Discuss the long-term behavior of the price of this product.

**53.** *Continuous compound interest.* If money is deposited at the continuous rate of \$200 per year into an account earning 8% compounded continuously, then the amount $A$ in the account after $t$ years satisfies the differential equation

$$\frac{dA}{dt} = 0.08A + 200$$

(A) Show that

$$A = Ce^{0.08t} - 2{,}500$$

is the general solution of this differential equation.

(B) Graph the particular solutions satisfying $A(0) = 0$ and $A(0) = 1{,}000$.

(C) Compare the long-term behavior of the two particular solutions in part (B). Discuss the effect of the value of $A(0)$ on the amount in the account.

**54.** *Continuous compound interest.* If money is withdrawn at the continuous rate of \$500 per year from an account earning 10% compounded continuously, then the amount $A$ in the account after $t$ years satisfies the differential equation

$$\frac{dA}{dt} = 0.1A - 500$$

(A) Show that

$$A = 5{,}000 + Ce^{0.1t}$$

is the general solution of this differential equation.

(B) Graph the particular solutions satisfying $A(0) = 4{,}000$ and $A(0) = 6{,}000$.

(C) Compare the long-term behavior of the two particular solutions in part (B). Discuss the effect of the value of $A(0)$ on the amount in the account.

**55.** *Population growth—Verhulst growth model.* The number $N(t)$ of bacteria in a culture at time $t$ is assumed to satisfy the Verhulst growth model

$$\frac{dN}{dt} = 100 - 0.5N$$

(A) Show that

$$N(t) = 200 - Ce^{-0.5t}$$

is the general solution of this differential equation, and evaluate

$$\overline{N} = \lim_{t \to \infty} N(t)$$

where $\overline{N}$ is the equilibrium size of the population.

(B) Graph the particular solutions that satisfy $N(0) = 50$ and $N(0) = 300$.

(C) Discuss the long-term behavior of this population.

**56.** *Population growth—logistic growth model.* The population $N(t)$ of a certain species of animal in a controlled habitat at time $t$ is assumed to satisfy the logistic growth model

$$\frac{dN}{dt} = \frac{1}{500}N(1{,}000 - N)$$

(A) Show that

$$N(t) = \frac{1{,}000}{1 + Ce^{-2t}}$$

is the general solution of this differential equation, and evaluate $\overline{N} - \lim_{t \to \infty} N(t)$.

(B) Graph the particular solutions that satisfy $N(0) = 200$ and $N(0) = 2{,}000$.

(C) Discuss the long-term behavior of this population.

**57.** *Rumor spread—Gompertz growth model.* The rate of propagation of a rumor is assumed to satisfy the Gompertz growth model

$$\frac{dN}{dt} = Ne^{-0.5t}$$

where $N(t)$ is the number of individuals who have heard the rumor at time $t$.

(A) Show that $N(t) = Ce^{-2e^{-0.5t}}$ is the general solution of this differential equation, and evaluate $\overline{N} = \lim_{t \to \infty} N(t)$.

(B) Graph the particular solutions that satisfy $N(0) = 100$ and $N(0) = 200$.

(C) Discuss the effect of the value of $N(0)$ on the long-term propagation of this rumor.

## Section 1-2 SEPARATION OF VARIABLES

- Separation of Variables
- Exponential Growth
- Limited Growth
- Logistic Growth
- Comparison of Exponential Growth Phenomena

### ■ Separation of Variables

In this section we will develop a technique called *separation of variables*, which can be used to solve differential equations that can be expressed in the form

$$f(y)y' = g(x) \tag{1}$$

We have already used this technique informally in Section 6-3 of *Calculus* 11/e or in Section 13-3 of *College Mathematics* 11/e to derive some important exponential growth laws. We will now present a more formal development of this method.

*Explore & Discuss 1*

Use algebraic manipulation to express each of the following equations in the form $f(y)y' = g(x)$, or explain why it is not possible to do so.

(A) $y' = x^2 y$

(B) $y' = x^2 + xy$

(C) $y' = x^2 y + xy$

The solution of differential equations by the method of separating the variables is based on the substitution formula for indefinite integrals. If $y = y(x)$ is a differentiable function of $x$, then

$$\int f(y)\, dy = \int f[y(x)]y'(x)\, dx \quad \text{Substitution formula for indefinite integrals} \tag{2}$$

If $y(x)$ is also the solution of (1), then $y(x)$ and $y'(x)$ must satisfy (1). That is,

$$f[y(x)]y'(x) = g(x)$$

Substituting $g(x)$ for $f[y(x)]y'(x)$ in (2), we have

$$\int f(y)\, dy = \int g(x)\, dx \tag{3}$$

Thus, the solution $y(x)$ of (1) is given implicitly by this equation. If both indefinite integrals can be evaluated, and if the resulting equation can be solved for $y$ in terms of $x$, then we have an explicit solution of the differential equation (1).

This discussion is summarized in Theorem 1:

**THEOREM 1  SEPARATION OF VARIABLES**

The solution of the differential equation

$$f(y)y' = g(x) \tag{1}$$

is given implicitly by the equation

$$\int f(y)\, dy = \int g(x)\, dx \tag{3}$$

> ## INSIGHT
>
> Differentials can be used to make the connection between equations (1) and (3) in Theorem 1:
>
> $$f(y)y' = g(x) \qquad \text{Substitute } y' = \frac{dy}{dx} \text{ in equation (1).}$$
>
> $$f(y)\frac{dy}{dx} = g(x) \qquad \text{Multiply both sides by } dx.$$
>
> $$f(y)\,dy = g(x)\,dx \qquad \text{Integrate both sides.}$$
>
> $$\int f(y)\,dy = \int g(x)\,dx \qquad \text{Equation (3)}$$

Some prefer to go directly from equation (1) to equation (3). Others may want to switch to differential notation and multiply both sides by $dx$. The choice is up to you. We will illustrate both approaches in the following examples.

---

### EXAMPLE 1    Using Separation of Variables, Solve: $y' = 2xy^2$.

**SOLUTION**  We choose to introduce differentials:

$$y' = 2xy^2 \qquad \text{Multiply by } 1/y^2 \text{ to separate the variables.}$$

$$\frac{1}{y^2}y' = 2x \qquad \text{Substitute } y' = \frac{dy}{dx}.$$

$$\frac{1}{y^2}\frac{dy}{dx} = 2x \qquad \text{Multiply both sides by } dx \text{ and write } 1/y^2 \text{ as a power form.}$$

$$y^{-2}\,dy = 2x\,dx \qquad \text{Integrate both sides.}$$

$$\int y^{-2}\,dy = \int 2x\,dx \qquad \text{Use the power rule to evaluate each integral.}$$

$$\frac{y^{-1}}{-1} + C_1 = 2\frac{x^2}{2} + C_2 \qquad \text{Simplify each side algebraically and combine } C_1 \text{ and } C_2 \text{ into a single arbitrary constant.}$$

$$-\frac{1}{y} = x^2 + C \qquad \text{Solve for } y.$$

$$y = -\frac{1}{x^2 + C} \qquad \text{General solution of } y' = 2xy^2.$$

Check:

$$y' = \frac{d}{dx}\left(-\frac{1}{x^2 + C}\right) = \frac{2x}{(x^2 + C)^2}$$

$$y' = 2xy^2 \qquad \text{Substitute } y' = 2x/(x^2 + C)^2 \text{ and } y = -1/(x^2 + C).$$

$$\frac{2x}{(x^2 + C)^2} \overset{?}{=} 2x\left(-\frac{1}{x^2 + C}\right)^2$$

$$\frac{2x}{(x^2 + C)^2} \overset{\checkmark}{=} \frac{2x}{(x^2 + C)^2}$$

This verifies that our general solution is correct. (You should develop the habit of checking the solution of each differential equation you solve, as we have done here. From now on, we leave it to you to check most of the examples worked in the text.)

MATCHED PROBLEM 1    Solve: $y' = 4x^3y^2$.

> INSIGHT
>
> In some cases the general solution obtained by the technique of separation of variables does not include all the solutions to a differential equation. For example, the constant function $y = 0$ also satisfies the differential equation in Example 1. (Verify this.) Yet the solution $y = 0$ cannot be obtained from the expression
>
> $$y = -\frac{1}{x^2 + C}$$
>
> for any choice of the constant $C$. Solutions of this type are referred to as **singular solutions** and are usually discussed in more advanced courses. We will not attempt to find the singular solutions of any of the differential equations we consider.

**EXAMPLE 2**  **Using Separation of Variables**  Find the general solution of

$$(1 + x^2)y' = 2x(y - 1)$$

Then find the particular solution that satisfies the initial condition $y(0) = 3$.

**SOLUTION**  First, we find the general solution:

$$(1 + x^2)y' = 2x(y - 1) \qquad \text{Multiply by } 1/(1 + x^2) \text{ and } 1/(y - 1) \text{ to separate the variables.}$$

$$\frac{y'}{y - 1} = \frac{2x}{1 + x^2} \qquad \text{Convert to an equation involving indefinite integrals (Theorem 1).}$$

$$\int \frac{dy}{y - 1} = \int \frac{2x\,dx}{1 + x^2} \qquad \text{Evaluate each indefinite integral.}$$

$$\ln|y - 1| = \ln(1 + x^2) + C$$

where $C$ is an arbitrary constant. Notice that the two constants of integration always can be combined to form a single arbitrary constant. Also note that we can use $1 + x^2$ instead of $|1 + x^2|$, since $1 + x^2$ is always positive. In order to solve for $y$, we convert this last equation to exponential form:

$$|y - 1| = e^{\ln(1+x^2)+C}$$
$$= e^C e^{\ln(1+x^2)} \qquad \text{Use the property } e^{\ln r} = r.$$
$$= e^C(1 + x^2)$$

It can be shown that if we replace $e^C$ with an arbitrary constant $K$, then we can omit the absolute value signs on the left side of the last equation.* The resulting equation can then be used to find the general solution to the original differential equation. Thus,

$$y - 1 = K(1 + x^2)$$
$$y = 1 + K(1 + x^2) \qquad \textit{General solution}$$

To find the particular solution that satisfies $y(0) = 3$, we substitute $x = 0$ and $y = 3$ in the general solution and solve for $K$:

$$3 = 1 + K(1 + 0)$$
$$K = 2$$
$$y = 1 + 2(1 + x^2) = 3 + 2x^2 \qquad \textit{Particular solution}$$

---

*In any problem involving separation of variables, you may assume that the equations

$$e^{\ln|f(y)|} = e^C g(x) \qquad \text{and} \qquad f(y) = Kg(x)$$

are equivalent (both $C$ and $K$ are arbitrary constants). Justification of this assumption involves properties of the absolute value and exponential functions.

**MATCHED PROBLEM 2**   Find the general solution of

$$(2 + x^4)y' = 4x^3(y - 3)$$

Then find the particular solution that satisfies the initial condition $y(0) = 5$.

*Explore & Discuss* **2**   Let $y$ be a positive quantity that is changing with respect to time $t$, such as the price of a product, the amount of a drug in the bloodstream, or the population of a country. If $y$ satisfies the differential equation

$$\frac{dy}{dt} = ky \qquad k \text{ a constant}$$

then we say that the rate of change of $y$ with respect to time $t$ is proportional to $y$. If $k > 0$, $y$ is increasing (growing); if $k = 0$, $y$ is constant; and if $k < 0$, $y$ is decreasing (declining or decaying). For each of the following differential equations, write a brief verbal description of the rate of change of $y$ and discuss the possible increasing/decreasing properties of $y$.

(A) $\dfrac{dy}{dt} = k$    (B) $\dfrac{dy}{dt} = k(100 - y)$    (C) $\dfrac{dy}{dt} = ky(100 - y)$

## ■ Exponential Growth

We now return to the study of exponential growth models first begun in Section 6-3 of *Calculus* 11/e or Section 13-3 of *College Mathematics* 11/e. This time we will place more emphasis on determining the relevant model for a particular application and on using separation of variables to solve the corresponding differential equation in each problem. We begin with the familiar exponential growth model.

**DEFINITION**   **Exponential Growth Model**

If the rate of change with respect to time $t$ of a quantity $y$ is proportional to the amount present, then $y$ satisfies the differential equation

$$\frac{dy}{dt} = ky$$

Exponential growth includes both the case where $y$ is increasing and the case where $y$ is decreasing (or decaying).

**EXAMPLE 3**   **Product Analysis**   Mothballs of a certain brand evaporate at a rate proportional to their volume, losing half their volume every 4 weeks. If the volume of each mothball is initially 15 cubic centimeters and a mothball becomes ineffective when its volume reaches 1 cubic centimeter, how long will these mothballs be effective?

**SOLUTION**   The volume of each mothball is decaying at a rate proportional to its volume. This indicates that an exponential growth model is appropriate for this problem. If $V$ is the volume of a mothball after $t$ weeks, then

$$\frac{dV}{dt} = kV \qquad \text{Exponential growth model}$$

Since the initial volume is 15 cubic centimeters, we know that $V(0) = 15$. After 4 weeks, the volume will be half the original volume, so $V(4) = 7.5$. Summarizing these requirements, we have the following exponential decay model:

$$\frac{dV}{dt} = kV$$

$$V(0) = 15 \qquad V(4) = 7.5$$

We want to determine the value of $t$ that satisfies the equation $V(t) = 1$. First, we use separation of variables to find the general solution of the differential equation:

$$\frac{dV}{dt} = kV$$

$$\frac{1}{V}\frac{dV}{dt} = k$$

$$\int \frac{dV}{V} = \int k\,dt$$

$$\ln V = kt + C \qquad \text{We can write ln V in place of ln |V|, since V > 0.}$$

$$V = e^{kt+C} = e^C e^{kt} = Ae^{kt} \qquad \text{General solution}$$

where $A = e^C$ is a positive constant. Now we use the initial condition to determine the value of the constant $A$:

$$V(0) = Ae^0 = A = 15$$

$$V(t) = 15e^{kt}$$

Next, we apply the condition $V(4) = 7.5$ to determine the constant $k$:

$$V(4) = 15e^{4k} = 7.5$$

$$e^{4k} = \frac{7.5}{15} = 0.5$$

$$4k = \ln 0.5$$

$$k = \frac{\ln 0.5}{4} \approx -0.1733$$

$$V(t) = 15e^{(t/4)\ln 0.5} \qquad \text{Particular solution}$$

The graph of $V(t)$ is shown in Figure 1. To determine how long the mothballs will be effective, we find $t$ when $V = 1$:

$$V(t) = 1$$

$$15e^{(t/4)\ln 0.5} = 1$$

$$e^{(t/4)\ln 0.5} = \frac{1}{15}$$

$$\frac{t}{4}\ln 0.5 = \ln \frac{1}{15}$$

$$t = \frac{4\ln \frac{1}{15}}{\ln 0.5} \approx 15.6 \text{ weeks}$$

$V$

15

$V(t) = 15e^{(t/4)\ln 0.5}$

7.5

1

0    4         15.6    $t$

**FIGURE 1** Exponential decay

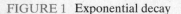

**MATCHED PROBLEM 3** Repeat Example 3 if the mothballs lose half their volume every 5 weeks.

## ■ Limited Growth

In certain situations there is an upper limit (or a lower limit), say $M$, on the values a variable can assume. This limiting value leads to the *limited growth model*.

**DEFINITION** **Limited Growth Model**

If the rate of change with respect to time $t$ of a quantity $y$ is proportional to the difference between $y$ and a limiting value $M$, then $y$ satisfies the differential equation

$$\frac{dy}{dt} = k(M - y)$$

When we speak of limited growth, we will include both the case where $y$ increases and approaches $M$ from below and the case where $y$ decreases and approaches $M$ from above.

**EXAMPLE 4**  **Sales Growth**  The annual sales $S$ of a new company are expected to grow at a rate proportional to the difference between the sales and an upper limit of \$20 million. The sales are 0 initially and \$4 million for the second year of operation.

(A)  Find the sales $S$ during year $t$.

(B)  Find $S(10)$ and $S'(10)$ and interpret.

(C)  In what year should the sales be expected to reach \$15 million?

**SOLUTION**  (A)  The description of the sales growth indicates that a limited growth model is appropriate for this problem. That is,

$$\frac{dS}{dt} = k(20 - S) \qquad \text{Limited growth model}$$

$$S(0) = 0 \qquad S(2) = 4$$

Separating the variables in the limited growth model, we have

$$\frac{dS}{dt} = k(20 - S) \qquad \text{Multiply both sides by } dt \text{ and } 1/(20 - S).$$

$$\frac{1}{20 - S} \, dS = k \, dt \qquad \text{Integrate both sides.}$$

$$\int \frac{1}{20 - S} \, dS = \int k \, dt \qquad \text{Evaluate both integrals.}$$

$$-\ln(20 - S) = kt + C \qquad \text{We can write } -\ln(20 - S) \text{ in place}$$

$$\ln(20 - S) = -kt - C \qquad \text{of } -\ln |20 - S|, \text{ since } 0 < S < 20.$$

$$20 - S = e^{-kt - C} = e^{-C} e^{-kt} = A e^{-kt} \qquad A = e^{-C}$$

$$S = 20 - A e^{-kt} \qquad \text{General solution}$$

Now we use the conditions $S(0) = 0$ and $S(2) = 4$ to determine the constants $A$ and $k$:

$$S(0) = 20 - A e^0$$

$$= 20 - A = 0$$

$$A = 20$$

$$S(t) = 20 - 20 e^{-kt}$$

$$S(2) = 20 - 20 e^{-2k} = 4$$

$$20 e^{-2k} = 16$$

$$e^{-2k} = \frac{16}{20} = 0.8$$

$$-2k = \ln 0.8$$

$$k = -\frac{\ln 0.8}{2} \approx 0.1116$$

$$S(t) = 20 - 20 e^{(t/2)\ln 0.8} \qquad \text{Particular solution}$$

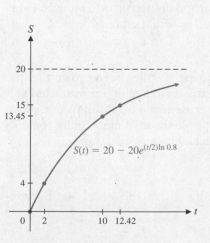

$S(t) = 20 - 20e^{(t/2)\ln 0.8}$

**FIGURE 2**  Limited growth

The graph of $S(t)$ is shown in Figure 2. Notice that the upper limit of \$20 million is a horizontal asymptote.

(B) $\quad S'(t) = -20\left(\dfrac{1}{2}\ln 0.8\right)e^{(t/2)\ln 0.8}$

$\qquad S(10) = 20 - 20e^{5\ln 0.8} \approx \$13.45$ million

$\qquad S'(10) = -20\left(\dfrac{1}{2}\ln 0.8\right)e^{5\ln 0.8} \approx \$0.7312$ million

During the tenth year, the sales are approximately \$13.45 million and are increasing at the rate of \$0.7312 million per year.

(C) $\quad S(t) = 20 - 20e^{(t/2)\ln 0.8} = 15$

$\qquad\qquad 20e^{(t/2)\ln 0.8} = 5$

$\qquad\qquad e^{(t/2)\ln 0.8} = \dfrac{5}{20} = 0.25$

$\qquad\qquad \dfrac{t}{2}\ln 0.8 = \ln 0.25$

$\qquad\qquad t = \dfrac{2\ln 0.25}{\ln 0.8} \approx 12.43$ years

The annual sales will exceed \$15 million in the thirteenth year.

**MATCHED PROBLEM 4**  Repeat Example 4 if the sales during the second year are \$3 million.

## ■ Logistic Growth

If a quantity first begins to grow exponentially but then starts to approach a limiting value, it is said to exhibit *logistic growth*. More formally, we have:

**DEFINITION  Logistic Growth Model**

If the rate of change with respect to time $t$ of a quantity $y$ is proportional to both the amount present and the difference between $y$ and a limiting value $M$, then $y$ satisfies the differential equation

$$\frac{dy}{dt} = ky(M - y)$$

Theoretically, functions satisfying logistic growth models can be increasing or decreasing, just as was the case for the exponential and limited growth models. However, decreasing functions are seldom encountered in actual practice.

**EXAMPLE 5**  **Population Growth**  In a study of ciliate protozoans, it has been shown that the rate of growth of the number of *Paramecium caudatum* in a medium with fixed volume is proportional to the product of the number present and the difference between an upper limit of 375 and the number present. Suppose the medium initially contains 25 paramecia. After 1 hour there are 125 paramecia. How many paramecia are present after 2 hours?

**SOLUTION**  If $P$ is the number of paramecia in the medium at time $t$, then the model for this problem is the following logistic growth model:

$$\frac{dP}{dt} = kP(375 - P)$$

$$P(0) = 25 \quad P(1) = 125$$

We want to find $P(2)$. First, we separate the variables and convert to an equation involving indefinite integrals:

$$\frac{1}{P(375 - P)}\frac{dP}{dt} = k$$

$$\int \frac{1}{P(375 - P)}\,dP = \int k\,dt \qquad (4)$$

The integral on the left side of (4) can be evaluated either by using an algebraic identity* or by using formula 9 in Table II in *Calculus* 11/e or in *College Mathematics* 11/e. We will use formula 9 with $u = P$, $a = 375$, and $b = -1$:

$$\int \frac{1}{u(a + bu)}\,du = \frac{1}{a}\ln\left|\frac{u}{a + bu}\right| \qquad \text{Formula 9}$$

$$\int \frac{1}{P(375 - P)}\,dP = \frac{1}{375}\ln\left|\frac{P}{375 - P}\right| \qquad \text{Since } 0 < P < 375, \text{ the absolute value signs can be omitted.}$$

$$= \frac{1}{375}\ln\left(\frac{P}{375 - P}\right)$$

Returning to equation (4), we have

$$\frac{1}{375}\ln\left(\frac{P}{375 - P}\right) = \int k\,dt = kt + D \qquad D \text{ is a constant.}$$

$$\ln\left(\frac{P}{375 - P}\right) = 375kt + 375D$$

$$\frac{P}{375 - P} = e^{375kt + 375D} = e^{Bt}e^{C} \qquad B = 375k, C = 375D$$

$$P = 375e^{Bt}e^{C} - Pe^{Bt}e^{C}$$

$$P(e^{Bt}e^{C} + 1) = 375e^{Bt}e^{C}$$

$$P = \frac{375e^{Bt}e^{C}}{e^{Bt}e^{C} + 1} \qquad \text{Multiply numerator and denominator by } e^{-Bt}e^{-C} \text{ and let } A = e^{-C}.$$

$$= \frac{375}{1 + Ae^{-Bt}} \qquad \text{General solution}$$

Now we use the conditions $P(0) = 25$ and $P(1) = 125$ to evaluate the constants $A$ and $B$:

$$P(0) = \frac{375}{1 + A} = 25$$

$$375 = 25 + 25A$$

$$A = 14$$

$$P(t) = \frac{375}{1 + 14e^{-Bt}}$$

$$P(1) = \frac{375}{1 + 14e^{-B}} = 125$$

$$375 = 125 + 1{,}750e^{-B}$$

$$e^{-B} = \frac{250}{1{,}750} = \frac{1}{7}$$

$$-B = \ln\frac{1}{7} = -\ln 7$$

$$B = \ln 7$$

$$P(t) = \frac{375}{1 + 14e^{-t\ln 7}} \qquad \text{Particular solution}$$

---

*If you do not wish to use Table II to evaluate integrals, you can use the algebraic identity

$$\frac{1}{x(a - x)} = \frac{1}{a}\left[\frac{(a - x) + x}{x(a - x)}\right] = \frac{1}{a}\left(\frac{1}{x} + \frac{1}{a - x}\right)$$

to evaluate this integral and in all logistic growth problems in Exercise 1-2.

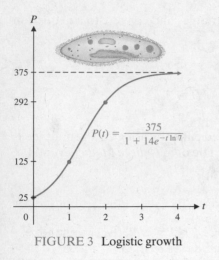

FIGURE 3  Logistic growth

To determine the population after 2 hours, we evaluate $P(2)$:

$$P(2) = \frac{375}{1 + 14e^{-2\ln 7}} \approx 292 \text{ paramecia}$$

The graph of $P(t)$ is shown in Figure 3. This S-shaped curve is typical of logistic growth functions. Notice that the upper limit of 375 is a horizontal asymptote. ■

## ■ Comparison of Exponential Growth Phenomena

The graphs and equations given in Table 1 compare several widely used growth models. These are divided basically into two groups: unlimited growth and limited growth. Following each equation and graph is a short (and necessarily incomplete) list of areas in which each model is used.

MATCHED PROBLEM 5    Repeat Example 5 for 15 paramecia initially and 150 paramecia after 1 hour.

*Answers to Matched Problems*

1. $y = -1/(x^4 + C)$
2. General solution: $y = 3 + K(2 + x^4)$; particular solution: $y = 5 + x^4$
3. 19.5 weeks
4. (A) $S(t) = 20 - 20e^{(t/2)\ln 0.8}$
   (B) $S(10) = \$11.13$ million; $S'(10) = \$0.7211$ million; during the tenth year, the sales are approximately $11.13 million and are increasing at the rate of $0.72211 million per year.
   (C) $t \approx 17.06$ yr; annual sales will exceed $15 million near the beginning of the eighteenth year.
5. 343 paramecia

| TABLE 1  Exponential Growth | | | | |
|---|---|---|---|---|
| **Description** | **Model** | **Solution** | **Graph** | **Uses** |
| **Unlimited growth:** Rate of growth is proportional to the amount present | $\dfrac{dy}{dt} = ky$ <br> $k, t > 0$ <br> $y(0) = c$ | $y = ce^{kt}$ | | • Short-term population growth (people, bacteria, etc.) <br> • Growth of money at continuous compound interest <br> • Price–supply curves |
| **Exponential decay:** Rate of growth is proportional to the amount present | $\dfrac{dy}{dt} = -ky$ <br> $k, t > 0$ <br> $y(0) = c$ | $y = ce^{-kt}$ | | • Depletion of natural resources <br> • Radioactive decay <br> • Light absorption in water <br> • Price–demand curves <br> • Atmospheric pressure ($t$ is altitude) |
| **Limited growth:** Rate of growth is proportional to the difference between the amount present and a fixed limit | $\dfrac{dy}{dt} = k(M - y)$ <br> $k, t > 0$ <br> $y(0) = 0$ | $y = M(1 - e^{-kt})$ | | • Sales fads (for example, skateboards) <br> • Depreciation of equipment <br> • Company growth <br> • Learning |
| **Logistic growth:** Rate of growth is proportional to the amount present and to the difference between the amount present and a fixed limit | $\dfrac{dy}{dt} = ky(M - y)$ <br> $k, t > 0$ <br> $y(0) = \dfrac{M}{1 + c}$ | $y = \dfrac{M}{1 + ce^{-kMt}}$ | | • Long-term population growth <br> • Epidemics <br> • Sales of new products <br> • Rumor spread <br> • Company growth |

## Exercise 1-2

**A**  *In Problems 1–4, write a differential equation that describes the rate of change of the indicated quantity.*

1. The annual sales $y$ of a company are increasing at the rate of \$100,000 per year.

2. The annual sales $y$ of a company are increasing at a rate proportional to the annual sales.

3. The fish population $y$ in a lake is growing at a rate proportional to the difference between the population and an upper limit of 10,000 fish.

4. In a community of 100,000, the number of people $y$ who have contracted an infectious disease is growing at a rate proportional to the product of the number of people who have contracted the disease and the number who have not.

*In Problems 5–8, write a verbal description of the rate of change of the given quantity satisfying the indicated differential equation, and discuss the increasing/decreasing properties of the quantity.*

5. The annual sales $y$ (in millions of dollars) of a company satisfy $dy/dt = 0.1y$ and $y(0) = 2$.

6. The annual sales $y$ (in millions of dollars) of a company satisfy $dy/dt = 0.1(8 - y)$ and $y(0) = 2$.

7. The number of people $y$ who have heard a rumor satisfies $dy/dt = 0.2y(5{,}000 - y)$ and $y(0) = 1$.

8. The amount of a drug $y$ (in milliliters) in a patient's bloodstream satisfies $dy/dt = -0.2y$ and $y(0) = 5$.

*In Problems 9–12, show that the technique of separation of variables is applicable by writing each differential equation in the form $f(y)y' = g(x)$.*

9. $xy' = y^2$

10. $y' + xy = 2x$

11. $xy' + xy = 3y$

12. $y' + \dfrac{x}{y} = \dfrac{2}{y}$

*In Problems 13–22, find the general solution for each differential equation. Then find the particular solution satisfying the initial condition.*

13. $y' = 3x^2$; $y(0) = -1$

14. $y' = 2e^{2x}$; $y(0) = 5$

15. $y' = \dfrac{2}{x}$; $y(1) = 2$

16. $y' = \dfrac{2}{\sqrt[3]{x}}$; $y(1) = 2$

17. $y' = y$; $y(0) = 10$

18. $y' = y - 10$; $y(0) = 15$

19. $y' = 25 - y$; $y(0) = 5$

20. $y' = 3x^2 y$; $y(0) = \dfrac{1}{2}$

21. $y' = \dfrac{y}{x}$; $y(1) = 5$; $x > 0$

22. $y' = \dfrac{y}{x^2}$; $y(-1) = 2e$

**B**  *In Problems 23–32, find the general solution for each differential equation. Then find the particular solution satisfying the initial condition.*

23. $y' = \dfrac{1}{y^2}$; $y(1) = 3$

24. $y' = \dfrac{x^2}{y^2}$; $y(0) = 2$

25. $y' = ye^x$; $y(0) = 3e$

26. $y' = -y^2 e^x$; $y(0) = \dfrac{1}{2}$

27. $y' = \dfrac{e^x}{e^y}$; $y(0) = \ln 2$

28. $y' = y^2(2x + 1)$; $y(0) = -\dfrac{1}{5}$

29. $y' = xy + x$; $y(0) = 2$

30. $y' = (2x + 4)(y - 3)$; $y(0) = 2$

31. $y' = (2 - y)^2 e^x$; $y(0) = 1$

32. $y' = \dfrac{x^{1/2}}{(y - 5)^2}$; $y(1) = 7$

*In Problems 33–38, find the general solution for each differential equation. Do not attempt to find an explicit expression for the solution.*

33. $y' = \dfrac{1 + x^2}{1 + y^2}$

34. $y' = \dfrac{6x - 9x^2}{2y + 4}$

35. $xyy' = (1 + x^2)(1 + y^2)$

36. $(xy^2 - x)y' = x^2 y - x^2$

37. $x^2 e^y y' = x^3 + x^3 e^y$

38. $y' = \dfrac{xe^x}{\ln y}$

**C**  *In Problems 39–44, find an explicit expression for the particular solution for each differential equation.*

39. $xyy' = \ln x$; $y(1) = 1$

40. $y' = \dfrac{xe^{x^2}}{y}$; $y(0) = 2$

41. $xy' = x\sqrt{y} + 2\sqrt{y}$; $y(1) = 4$

42. $y' = x(x - 1)^{1/2}(y - 1)^{1/2}$; $y(1) = 1$

43. $yy' = xe^{-y^2}$; $y(0) = 1$

44. $yy' = x(1 + y^2)$; $y(0) = 1$

*In Problems 45–48, discuss why the technique of separation of variables does not produce an explicit solution $y = h(x)$ of the differential equation.*

45. $y' + xy = 2$

46. $y' - 5x^2 y = e^x$

47. $3yy' = e^{x^2} + y$

48. $y' = \dfrac{9x^2 + 2}{5y^4 - 6}$

49. In many applications involving limited growth, the limiting value $M$ is not known in advance and must be determined from data. Suppose that the rate of change with respect to time $t$ of a quantity $y$ is proportional to the difference between $y$ and an unknown upper limit $M$, and that $y = 0$ when $t = 0$. Using the limited growth model, it follows that

$$y = M(1 - e^{-kt}) \qquad M > 0, k > 0$$

(A)  Using the data given in the table, find two equations that $M$ and $k$ must satisfy, and solve each equation for $M$.

| $t$ | 1 | 1.5 |
|-----|---|-----|
| $y$ | 3 | 4 |

(B)  Use a graphing utility and the equations from part (A) to approximate $M$ to one decimal place.

**50.**  Refer to Problem 49. Suppose that the rate of change with respect to time of a quantity $y$ is proportional to the difference between $y$ and an unknown lower limit $M$,

and that $y = 10$ when $t = 0$. Using the limited growth model, it follows that

$$y = M + (10 - M)e^{-kt} \qquad M > 0, k > 0$$

(A)  Using the data given in the table, find two equations that $M$ and $k$ must satisfy, and solve each equation for $M$.

| $t$ | 1 | 5 |
|-----|---|---|
| $y$ | 9 | 7 |

(B)  Use a graphing utility and the equations from part (A) to approximate $M$ to one decimal place.

## Applications

**51.**  *Continuous compound interest.*  Provident Bank offers a CD (certificate of deposit) that earns 5.5% compounded continuously. How much will a $5,000 investment be worth in 5 years?

**52.**  *Continuous compound interest.*  Equitable Federal offers a CD that earns 4.9% compounded continuously. How much will a $10,000 investment be worth in 4 years?

**53.**  *Advertising.*  A company is using radio advertising to introduce a new product to a community of 100,000 people. Suppose the rate at which people learn about the new product is proportional to the number who have not yet heard of it.

(A)  If no one is aware of the product at the start of the advertising campaign and after 7 days 20,000 people are aware of the product, how long will it take for 50,000 people to become aware of the product?

(B)  Suppose the company is dissatisfied with the result in part (A) and wants to decrease to 14 days the amount of time it takes for 50,000 people to become aware of their product. For this to happen, how many people must become aware of the product during the first 7 days?

**54.**  *Advertising.*  Prior to the beginning of an advertising campaign, 10% of the potential users of a certain brand are aware of the brand name. After the first week of the campaign, 20% of the consumers are aware of the brand name. Assume that the percentage of informed consumers is growing at a rate proportional to the product of the percentage of informed consumers and the percentage of uninformed consumers.

(A)  What percentage of consumers will be aware of the brand name after 5 weeks of advertising?

(B)  Suppose the company is dissatisfied with the result in part (A) and increases the intensity of the advertising campaign so that 25% of the consumers are aware of the brand name after the first week. What effect does this have on the percentage of consumers that will be aware of the brand name after 5 weeks of advertising?

**55.**  *Product analysis.*  A company wishes to analyze a new room deodorizer. The active ingredient evaporates at a

rate proportional to the amount present. Half of the ingredient evaporates in the first 30 days after the deodorizer is installed. If the deodorizer becomes ineffective after 90% of the active ingredient has evaporated, how long will one of these deodorizers remain effective?

**56.**  *Product analysis.*  A chlorine-based pool disinfectant contains an active ingredient with a half-life of 4 days. How often should this disinfectant be added to a pool to ensure that at least 30% of the active ingredient is present?

**57.**  *Personal income.*  According to the U.S. Census Bureau, total personal income was $6,201 billion in 1995 and $8,407 billion in 2000. If personal income is increasing at a rate proportional to total personal income, find the total personal income and the rate of change of total personal income in 2010 and interpret.

**58.**  *Corporate profits.*  According to the U.S. Bureau of Economic Analysis, total corporate profits in domestic industries were $576.8 billion in 1995 and $946.2 billion in 2000. If corporate profits are increasing at a rate proportional to total corporate profits, find the total corporate profits and the rate of change of total corporate profits in 2010 and interpret.

**59.**  *Sales analysis.*  The annual sales of a new company are expected to grow at a rate proportional to the difference between the sales and an upper limit of $5 million. If the sales are 0 initially and $1 million during the fourth year of operation, find the sales and the rate of change of the sales during the fifteenth year and interpret. How long (to the nearest year) will it take for the sales to grow to $4 million?

**60.**  *Sales analysis.*  The annual sales of a company have declined from $8 million 2 years ago to $6 million today. If the annual sales continue to decline at a rate proportional to the difference between the annual sales and a lower limit of $3 million, find the sales and the rate of change of the sales 3 years from now and interpret. How long (to the nearest year) will it take for the sales to decline to $3.5 million?

**61.**  *Sales analysis.*  A new company has 0 sales initially, sales of $2 million during the first year, and sales of $5 million during the third year. If the annual sales $S$ are assumed to be growing at a rate proportional to the difference between the sales and an unknown upper limit $M$, then by the limited growth model,

$$S(t) = M(1 - e^{-kt}) \qquad M > 0, k > 0$$

where $t$ is time (in years) and $S(t)$ represents sales (in millions of dollars). Use approximation techniques to find $k$ to one decimal place and $M$ to the nearest million. (See Problem 49.)

62. *Sales analysis.* Refer to Problem 61. Approximate $k$ to one decimal place and $M$ to the nearest million if the sales during the third year are \$4 million and all other information is unchanged.

*Newton's law of cooling states that the rate of change of the temperature of an object is proportional to the difference between the temperature of the object and the temperature of the surrounding medium. Use this law to formulate models for Problems 63–66, and then solve using the techniques discussed in this section.*

63. *Manufacturing.* As part of a manufacturing process, a metal bar is to be heated in an oven until its temperature reaches 500°F. The oven is maintained at a constant temperature of 800°F. The temperature of the bar before it is placed in the oven is 80°F. After 2 minutes in the oven, the temperature of the bar is 200°F. How long should the bar be left in the oven?

64. *Manufacturing.* The next step in the manufacturing process described in Problem 63 calls for the heated bar to be cooled in a vat of water until its temperature reaches 100°F. The water in the vat is maintained at a constant temperature of 50°F. If the temperature of the bar is 500°F when it is first placed in the water and the bar has cooled to 400°F after 5 minutes in the water, how long should the bar be left in the water?

65. *Food preparation.* A pie is removed from an oven where the temperature is 325°F and placed in a freezer with a constant temperature of 25°F. After 1 hour in the freezer, the temperature of the pie is 225°F. What is the temperature of the pie after 4 hours in the freezer?

66. *Food preparation.* A roast is taken from a freezer where the temperature is 20°F and placed in an oven with a constant temperature of 350°F. After 1 hour in the oven, the temperature of the roast is 185°F. What is the temperature of the roast after 3 hours in the oven?

67. *Population growth.* A culture of bacteria is growing at a rate proportional to the number present. The culture initially contains 100 bacteria. After 1 hour there are 140 bacteria in the culture.

    (A) How many bacteria will be present after 5 hours?
    (B) When will the culture contain 1,000 bacteria?

68. *Population growth.* A culture of bacteria is growing in a medium that can support a maximum of 1,100 bacteria. The rate of change of the number of bacteria is proportional to the product of the number present and the difference between 1,100 and the number present. The culture initially contains 100 bacteria. After 1 hour there are 140 bacteria.

(A) How many bacteria are present after 5 hours?
(B) When will the culture contain 1,000 bacteria?

69. *Simple epidemic.* An influenza epidemic has spread throughout a community of 50,000 people at a rate proportional to the product of the number of people who have been infected and the number who have not been infected. If 100 individuals were infected initially and 500 were infected 10 days later:

    (A) How many people will be infected after 20 days?
    (B) When will half the community be infected?

70. *Ecology.* A fish population in a large lake is declining at a rate proportional to the difference between the population and a lower limit of 5,000 fish.

    (A) If the population has declined from 15,000 fish 3 years ago to 10,000 today, find the population 6 years from now.
    (B) Suppose the fish population 6 years from now turns out to be 8,000, indicating that the lake might be able to support more than the original lower limit of 5,000 fish. Assuming that the rate of change of the population is still proportional to the difference between the population and an unknown lower limit $M$, approximate $M$ to the nearest hundred.

*Body temperature is one factor that crime scene investigators use to estimate time of death. Use Newton's law of cooling in Problems 71 and 72 and assume that the body temperature at the time of death is 98.6°F.*

71. *Crime scene investigation.* A body was found in an alley at 1:00 A.M. The temperature that night was a constant 50°F and the temperature of the body at the time of discovery was 75°F. The investigator left the body in the alley while she looked for other evidence. She checked the body temperature again at 2:00 A.M. and found that the temperature of the body had dropped to 70°F. What was the time of death (to the nearest hour)?

72. *Crime scene investigation.* A body was found stashed in a freezer at 3:00 P.M. The temperature in the freezer was a constant 0°F. A crime scene investigator determined that the temperature of the body at the time of discovery was 55°F. Evidence suggested that death occurred in the freezer. Leaving the body in the freezer, the investigator checked the body temperature again at 4:00 P.M. and found that the temperature had dropped to 50°F. What was the time of death (to the nearest hour)?

73. *Sensory perception.* A person is subjected to a physical stimulus that has a measurable magnitude, but the intensity of the resulting sensation is difficult to measure. If $s$ is the magnitude of the stimulus and $I(s)$ is the intensity of sensation, experimental evidence suggests that

$$\frac{dI}{ds} = k\frac{I}{s}$$

for some constant $k$. Express $I$ as a function of $s$.

74. *Learning.* The number of words per minute, $N$, a person can type increases with practice. Suppose the rate of change of $N$ is proportional to the difference between $N$ and an upper limit of 140. It is reasonable to assume that

a beginner cannot type at all. Thus, $N = 0$ when $t = 0$. If a person can type 35 words per minute after 10 hours of practice:

(A) How many words per minute can that individual type after 20 hours of practice?

(B) How many hours must that individual practice to be able to type 105 words per minute?

75. *Rumor spread.* A rumor spreads through a population of 1,000 people at a rate proportional to the product of

the number who have heard it and the number who have not heard it. If 5 people initiated a rumor and 10 people had heard it after 1 day:

(A) How many people will have heard the rumor after 7 days?

(B) How long will it take for 850 people to hear the rumor?

## Section 1-3 FIRST-ORDER LINEAR DIFFERENTIAL EQUATIONS

- The Product Rule Revisited
- Solution of First-Order Linear Differential Equations
- Applications

### The Product Rule Revisited

The solution method we will discuss in this section is based in part on the product rule. Recall that if $F$ and $S$ are two differentiable functions, then

$$(FS)' = FS' + F'S$$

In Chapter 4 of *Calculus* 11/e or Chapter 10 of *College Mathematics* 11/e, we started with $F$ and $S$ and used the product rule to find $(FS)'$. Now we want to use the rule in the reverse direction. That is, we want to be able to recognize when a sum of two terms is the derivative of a product.

*Explore & Discuss 1*

Match each sum on the left with the derivative of a product on the right, where $y$ is an unknown differentiable function of $x$.

(A) $x^2 y' + 2xy$       (1) $(e^{x^2} y)'$

(B) $e^{-2x} y' - 2e^{-2x} y$      (2) $\left( \dfrac{1}{x^2} y \right)'$

(C) $\dfrac{1}{x^2} y' - \dfrac{2}{x^3} y$      (3) $(e^{-2x} y)'$

(D) $e^{x^2} y' + 2xe^{x^2} y$      (4) $(x^2 y)'$

In general, if $m(x)y' + n(x)y$ is the derivative of a product, how must $m(x)$ and $n(x)$ be related?

### Solution of First-Order Linear Differential Equations

A differential equation that can be expressed in the form

$$y' + f(x)y = g(x)$$

is called a **first-order linear differential equation.**

For example,

$$y' + \frac{2}{x} y = x \tag{1}$$

is a first-order linear differential equation with $f(x) = 2/x$ and $g(x) = x$. This equation cannot be solved by the method of separation of variables. (Try to separate the variables to convince yourself that this is true.) Instead, we will change the form of the equation by multiplying both sides by $x^2$:

$$x^2 y' + 2xy = x^3 \tag{2}$$

How was $x^2$ chosen? We will discuss that shortly. Let us first see how this choice leads to a solution of the problem.

Since $(x^2)' = 2x$, the left side of (2) is the derivative of a product:

$$x^2 y' + 2xy = (x^2 y)'$$

Thus, we can write equation (2) as

$$(x^2 y)' = x^3$$

Now we can integrate both sides:

$$\int (x^2 y)' \, dx = \int x^3 \, dx$$

$$x^2 y = \frac{x^4}{4} + C$$

Solving for $y$, we obtain the general solution

$$y = \frac{x^2}{4} + \frac{C}{x^2}$$

The function $x^2$, which we used to transform the original equation into one we could solve as illustrated, is called an *integrating factor*. It turns out that there is a specific formula for determining the integrating factor for any first-order linear differential equation. Furthermore, this integrating factor can then be used to find the solution of the differential equation, just as we used $x^2$ to find the solution to (1). The formula for the integrating factor and a step-by-step summary of the solution process are given in the box that follows.

---

**PROCEDURE**

**Solving First-Order Linear Differential Equations**

**Step 1.** Write the equation in the **standard form:**

$$y' + f(x)y = g(x)$$

**Step 2.** Compute the **integrating factor:**

$$\cdot \; I(x) = e^{\int f(x) \, dx}$$

[When evaluating $\int f(x) \, dx$, choose 0 for the constant of integration.]

**Step 3.** Multiply both sides of the standard form by the integrating factor $I(x)$. The left side should now be in the form $[I(x)y]'$:

$$[I(x)y]' = I(x)g(x)$$

**Step 4.** Integrate both sides:

$$I(x)y = \int I(x)g(x) \, dx$$

[When evaluating $\int I(x)g(x) \, dx$, include an arbitrary constant of integration.]

**Step 5.** Solve for $y$ to obtain the **general solution:**

$$y = \frac{1}{I(x)} \int I(x)g(x) \, dx$$

---

**EXAMPLE 1**   **Using an Integrating Factor** Solve: $2xy' + y = 10x^2$.

**SOLUTION**   **Step 1.** Multiply both sides by $1/(2x)$ to obtain the standard form:

$$y' + \frac{1}{2x}y = 5x \qquad f(x) = \frac{1}{2x} \text{ and } g(x) = 5x$$

**Step 2.** Find the integrating factor:

$$I(x) = e^{\int f(x)\,dx}$$

$$= e^{\int [1/(2x)]\,dx} \qquad \text{Assume } x > 0 \text{ and choose } 0$$
$$\qquad\qquad\qquad \text{for the constant of integration.}$$

$$= e^{(1/2)\ln x} \qquad r \ln t = \ln t^r$$

$$= e^{\ln x^{1/2}} \qquad e^{\ln r} = r, r > 0$$

$$= x^{1/2} \qquad \text{Integrating factor}$$

**Step 3.** Multiply both sides of the standard form by the integrating factor:

$$x^{1/2}\left(y' + \frac{1}{2x}\,y\right) = x^{1/2}(5x)$$

$$x^{1/2}y' + \tfrac{1}{2}x^{-1/2}y = 5x^{3/2} \qquad \text{The left side should have}$$
$$\qquad\qquad\qquad\qquad\qquad \text{the form } [I(x)y]'.$$

$$(x^{1/2}y)' = 5x^{3/2}$$

**Step 4.** Integrate both sides:

$$\int (x^{1/2}y)'\,dx = \int 5x^{3/2}\,dx \qquad \text{Include an arbitrary constant}$$
$$\qquad\qquad\qquad\qquad\qquad\qquad \text{of integration on the right side.}$$

$$x^{1/2}y = 2x^{5/2} + C$$

**Step 5.** Solve for $y$:

$$y = \frac{1}{x^{1/2}}(2x^{5/2} + C)$$

$$= 2x^2 + \frac{C}{x^{1/2}} \qquad \text{General solution}$$

> **MATCHED PROBLEM 1**  Solve: $xy' + 3y = 4x$.

## INSIGHT

The requirement that a differential equation be written in standard form is essential. To see why, consider the following equation, which is not written in standard form:

$$2y' + 4y = 8 \qquad\qquad (3)$$

If we proceed without first converting (3) to standard form, the integrating factor formula gives us

$$I(x) = e^{\int 4\,dx} = e^{4x}$$

Notice that we chose 0 for the constant of integration when computing $I(x)$. This is not a requirement but a recommendation. Ignoring this recommendation will not result in an error, but will complicate the subsequent calculations.

Returning to the differential equation in (3) and multiplying both sides by $I(x)$, we have:

$$2e^{4x}y' + 4e^{4x}y = 8e^{4x} \qquad\qquad (4)$$

The left side of (4) is supposed to be $[e^{4x}y]'$. But

$$[e^{4x}y]' = e^{4x}y' + 4e^{4x}y \qquad\qquad (5)$$

Since the left side of (4) does not equal $[e^{4x}y]'$ (the coefficients of the $y'$ terms are different), we cannot integrate both sides of (4) to find the solution to (3). In other words, $I(x) = e^{4x}$ is *not an integrating factor for* (3).

Whenever you use an integrating factor, you should check that

$$[I(x)y]' = I(x)y' + I'(x)y$$

as we did in (5), before proceeding to solve for $y$.

As an exercise, you should provide a correct solution to equation (3).

In Example 1, notice that we assumed $x > 0$ to avoid introducing absolute value signs in the integrating factor. Many of the problems in this section will require the evaluation of expressions of the form $e^{\int [h'(x)/h(x)]dx}$. In order to avoid the complications caused by the introduction of absolute value signs, we will assume that the domain of $h(x)$ has been restricted so that $h(x) > 0$. This will simplify the solution process. We state the following familiar formulas for convenient reference (in the first formula the constant of integration is 0, in accordance with Step 2):

**SUMMARY**  **BASIC FORMULAS INVOLVING THE NATURAL LOGARITHM FUNCTION**

If the domain of $h(x)$ is restricted so that $h(x) > 0$, then

$$\int \frac{h'(x)}{h(x)}\,dx = \ln h(x) \qquad \text{and} \qquad e^{\ln h(x)} = h(x)$$

## INSIGHT

If a first-order linear differential equation is written in standard form, then multiplying both sides of the equation by its integrating factor will always convert the left side of the equation into the derivative of $I(x)y$. Thus, it is possible to omit Steps 3 and 4, and proceed directly to Step 5. This approach is illustrated in the next example. You decide which is easier to use—the step-by-step procedure or the formula in Step 5.

**EXAMPLE 2**  **Using an Integrating Factor** Find the particular solution of the equation

$$y' + 2xy = 4x$$

satisfying the initial condition $y(0) = 5$.

**SOLUTION**  Since the equation is already in standard form, we begin by finding the integrating factor:

$$I(x) = e^{\int 2x\,dx} = e^{x^2} \quad f(x) = 2x$$

Proceeding directly to Step 5, we have

$$y = \frac{1}{I(x)} \int I(x)g(x)\,dx \quad g(x) = 4x$$

$$= \frac{1}{e^{x^2}} \int e^{x^2}(4x)\,dx$$

$$= e^{-x^2} \int 4xe^{x^2}\,dx \qquad \text{Let } u = x^2,\ du = 2x\,dx.$$

$$= e^{-x^2}(2e^{x^2} + C)$$

$$= 2 + Ce^{-x^2} \qquad \text{General solution}$$

Substituting $x = 0$ and $y = 5$ in the general solution, we have

$$5 = 2 + C$$

$$C = 3$$

$$y = 2 + 3e^{-x^2} \quad \text{Particular solution}$$

**MATCHED PROBLEM 2**  Find the particular solution of $y' + 3x^2y = 9x^2$ satisfying the initial condition $y(0) = 7$.

⚠ CAUTION:

**1.** When integrating both sides of an equation such as

$$(x^{1/2}y)' = 5x^{3/2}$$

remember that

$$x^{1/2}y = \int 5x^{3/2}\, dx \neq 2x^{5/2} \quad \text{"+ C" is missing in the antiderivative.}$$

**Remember: A constant of integration must be included when evaluating**

$$\int I(x)g(x)\, dx.$$

If you omit this constant, you will not be able to find the general solution of the differential equation. See Step 4 of Example 1 for the correct procedure.

**2.** $\dfrac{1}{e^{x^2}} \int 4xe^{x^2}\, dx \neq \dfrac{1}{e^{x^2}} \int 4xe^{x^2}\, dx = \int 4x\, dx$

Just as a variable factor cannot be moved across the integral sign, **a variable factor outside the integral sign cannot be used to cancel a factor inside the integral sign.** See Example 2 for the correct procedure.

## APPLICATIONS

If $P$ is the initial amount deposited into an account earning $100r\%$ compounded continuously, and $A$ is the amount in the account after $t$ years, then we saw in Section 6-3 of *Calculus* 11/e or in Section 13-3 of *College Mathematics* 11/e that $A$ satisfies the exponential growth equation

$$\frac{dA}{dt} = rA \qquad A(0) = P$$

Now suppose that money is continuously withdrawn from this account at a rate of \$$m$ per year. Then the amount $A$ in the account at time $t$ must satisfy

$$\begin{pmatrix} \text{Rate of change} \\ \text{of amount } A \end{pmatrix} = \begin{pmatrix} \text{Rate of growth} \\ \text{from continuous} \\ \text{compounding} \end{pmatrix} - \begin{pmatrix} \text{Rate of} \\ \text{withdrawal} \end{pmatrix}$$

$$\frac{dA}{dt} \qquad = \qquad rA \qquad - \qquad m$$

or

$$\frac{dA}{dt} - rA = -m \tag{6}$$

*Explore & Discuss* **2**  How does equation (6) change if instead of making continuous withdrawals, money is continuously deposited into the account at the rate of \$$m$ per year?

---

**EXAMPLE 3**  **Continuous Compound Interest**  An initial deposit of \$10,000 is made into an account earning 8% compounded continuously. Money is then continuously withdrawn at a constant rate of \$1,000 a year until the account is depleted. Find the amount in the account at any time $t$. When will the amount be 0? What is the total amount withdrawn from this account?

**SOLUTION**  The amount $A$ in the account at any time $t$ must satisfy [see equation (6)]

$$\frac{dA}{dt} - 0.08A = -1,000 \qquad A(0) = 10,000$$

The integrating factor for this equation is

$$I(t) = e^{\int -0.08\,dt} = e^{-0.08t} \quad f(t) = -0.08$$

Multiplying both sides of the differential equation by $I(t)$ and following the step-by-step procedure, we have

$$e^{-0.08t}\frac{dA}{dt} - 0.08e^{-0.08t}A = -1{,}000e^{-0.08t}$$

$$(e^{-0.08t}A)' = -1{,}000e^{-0.08t}$$

$$e^{-0.08t}A = \int -1{,}000e^{-0.08t}\,dt$$

$$= 12{,}500e^{-0.08t} + C$$

$$A = 12{,}500 + Ce^{0.08t} \quad \textit{General solution}$$

Applying the initial condition $A(0) = 10{,}000$ yields

$$A(0) = 12{,}500 + C = 10{,}000$$

$$C = -2{,}500$$

$$A(t) = 12{,}500 - 2{,}500e^{0.08t} \quad \textit{Amount in the account at any time t}$$

To determine when the amount in the account is 0, we solve $A(t) = 0$ for $t$:

$$A(t) = 0$$

$$12{,}500 - 2{,}500e^{0.08t} = 0$$

$$12{,}500 = 2{,}500e^{0.08t}$$

$$5 = e^{0.08t}$$

$$t = \frac{\ln 5}{0.08} \approx 20.118 \text{ years}$$

Thus, the account is depleted after 20.118 years.

**MATCHED PROBLEM 3**   Repeat Example 3 if the account earns 5% compounded continuously.

**EXAMPLE 4**   **Equilibrium Price**   In economics, the supply $S$ and the demand $D$ for a commodity often can be considered as functions of both the price, $p(t)$, and the rate of change of the price, $p'(t)$. (Thus, $S$ and $D$ are ultimately functions of time $t$.) The **equilibrium price at time $t$** is the solution of the equation $S = D$. If $p(t)$ is the solution of this equation, then the **long-range equilibrium price** is

$$\bar{p} = \lim_{t \to \infty} p(t)$$

For example, if

$$D = 50 - 2p(t) + 2p'(t) \qquad S = 20 + 4p(t) + 5p'(t)$$

and $p(0) = 15$, then the equilibrium price at time $t$ is the solution of the equation

$$50 - 2p(t) + 2p'(t) = 20 + 4p(t) + 5p'(t)$$

This simplifies to

$$p'(t) + 2p(t) = 10 \quad f(t) = 2 \text{ and } g(t) = 10$$

which is a first-order linear equation with integrating factor

$$I(t) = e^{\int 2\,dt} = e^{2t}$$

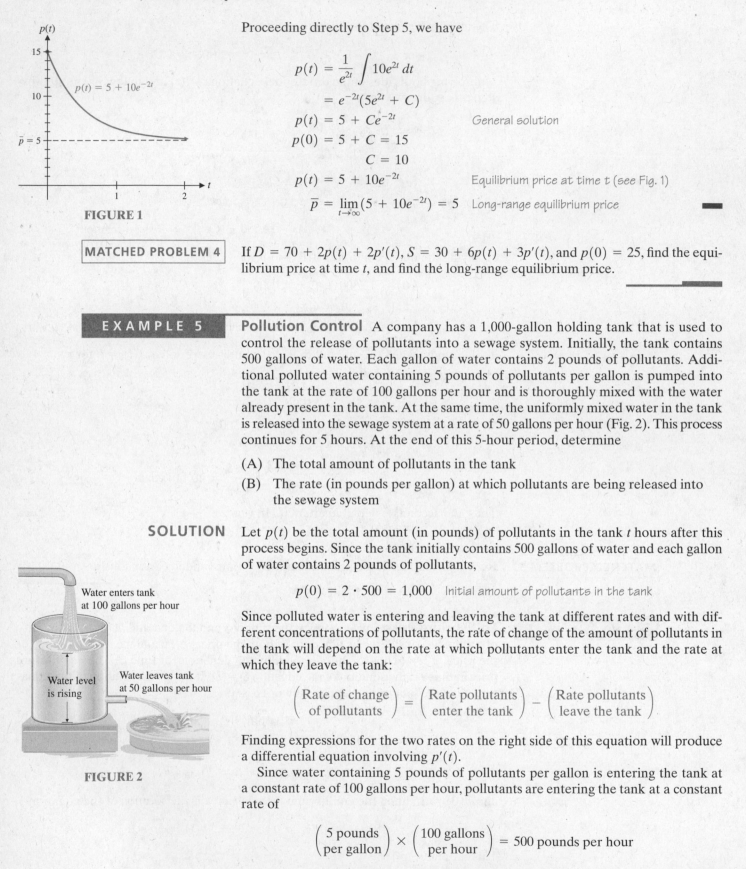

**FIGURE 1**

Proceeding directly to Step 5, we have

$$p(t) = \frac{1}{e^{2t}} \int 10e^{2t}\, dt$$

$$= e^{-2t}(5e^{2t} + C)$$

$$p(t) = 5 + Ce^{-2t} \qquad \text{General solution}$$

$$p(0) = 5 + C = 15$$

$$C = 10$$

$$p(t) = 5 + 10e^{-2t} \qquad \text{Equilibrium price at time } t \text{ (see Fig. 1)}$$

$$\bar{p} = \lim_{t \to \infty}(5 + 10e^{-2t}) = 5 \qquad \text{Long-range equilibrium price}$$

**MATCHED PROBLEM 4** If $D = 70 + 2p(t) + 2p'(t)$, $S = 30 + 6p(t) + 3p'(t)$, and $p(0) = 25$, find the equilibrium price at time $t$, and find the long-range equilibrium price.

**EXAMPLE 5** **Pollution Control** A company has a 1,000-gallon holding tank that is used to control the release of pollutants into a sewage system. Initially, the tank contains 500 gallons of water. Each gallon of water contains 2 pounds of pollutants. Additional polluted water containing 5 pounds of pollutants per gallon is pumped into the tank at the rate of 100 gallons per hour and is thoroughly mixed with the water already present in the tank. At the same time, the uniformly mixed water in the tank is released into the sewage system at a rate of 50 gallons per hour (Fig. 2). This process continues for 5 hours. At the end of this 5-hour period, determine

(A) The total amount of pollutants in the tank

(B) The rate (in pounds per gallon) at which pollutants are being released into the sewage system

**SOLUTION** Let $p(t)$ be the total amount (in pounds) of pollutants in the tank $t$ hours after this process begins. Since the tank initially contains 500 gallons of water and each gallon of water contains 2 pounds of pollutants,

$$p(0) = 2 \cdot 500 = 1{,}000 \qquad \text{Initial amount of pollutants in the tank}$$

Since polluted water is entering and leaving the tank at different rates and with different concentrations of pollutants, the rate of change of the amount of pollutants in the tank will depend on the rate at which pollutants enter the tank and the rate at which they leave the tank:

$$\left(\begin{array}{c}\text{Rate of change}\\ \text{of pollutants}\end{array}\right) = \left(\begin{array}{c}\text{Rate pollutants}\\ \text{enter the tank}\end{array}\right) - \left(\begin{array}{c}\text{Rate pollutants}\\ \text{leave the tank}\end{array}\right)$$

Finding expressions for the two rates on the right side of this equation will produce a differential equation involving $p'(t)$.

Since water containing 5 pounds of pollutants per gallon is entering the tank at a constant rate of 100 gallons per hour, pollutants are entering the tank at a constant rate of

$$\left(\begin{array}{c}5\text{ pounds}\\ \text{per gallon}\end{array}\right) \times \left(\begin{array}{c}100\text{ gallons}\\ \text{per hour}\end{array}\right) = 500\text{ pounds per hour}$$

How fast are the pollutants leaving the tank? Since the total amount of pollutants in the tank is increasing, the rate at which the pollutants leave the tank will depend on the amount of pollutants in the tank at time $t$ and the amount of water in the tank at time $t$. Since 100 gallons of water enter the tank each hour and only 50 gallons

Water enters tank at 100 gallons per hour

Water level is rising

Water leaves tank at 50 gallons per hour

**FIGURE 2**

leave each hour, the amount of water in the tank increases at the rate of 50 gallons per hour. Thus, the amount of water in the tank at time $t$ is

$$\begin{pmatrix} \text{Initial amount} \\ \text{of water} \end{pmatrix} + \begin{pmatrix} \text{Gallons} \\ \text{per hour} \end{pmatrix} \times \begin{pmatrix} \text{Number} \\ \text{of hours} \end{pmatrix}$$

$$500 \qquad + \qquad 50t$$

The amount of pollutants in each gallon of water at any time $t$ is

$$\frac{p(t)}{500 + 50t} \qquad \frac{\begin{pmatrix} \text{Total amount} \\ \text{of pollutants} \end{pmatrix}}{\begin{pmatrix} \text{Total amount} \\ \text{of water} \end{pmatrix}} = \text{Pollutants per gallon}$$

Since water is leaving the tank at the rate of 50 gallons per hour, the rate at which the pollutants are leaving the tank is

$$\frac{50p(t)}{500 + 50t} = \frac{p(t)}{10 + t}$$

Thus, the rate of change of $p(t)$ must satisfy

$$\begin{pmatrix} \text{Rate of change} \\ \text{of pollutants} \end{pmatrix} = \begin{pmatrix} \text{Rate pollutants} \\ \text{enter the tank} \end{pmatrix} - \begin{pmatrix} \text{Rate pollutants} \\ \text{leave the tank} \end{pmatrix}$$

$$p'(t) \qquad = \qquad 500 \qquad - \qquad \frac{p(t)}{10 + t}$$

This gives the following model for this problem:

$$p'(t) = 500 - \frac{p(t)}{10 + t} \qquad p(0) = 1{,}000$$

or

$$p'(t) + \frac{1}{10 + t} p(t) = 500 \qquad \text{First-order linear equation with } f(t) = 1/(10 + t) \text{ and } g(t) = 500$$

$$I(t) = e^{\int dt/(10+t)} = e^{\ln(10+t)} = 10 + t \qquad \text{Integrating factor}$$

$$p(t) = \frac{1}{10 + t} \int 500(10 + t)\, dt \qquad \text{Proceeding directly to Step 5}$$

$$= \frac{1}{10 + t}\left[250(10 + t)^2 + C\right]$$

$$p(t) = 250(10 + t) + \frac{C}{10 + t} \qquad \text{General solution}$$

$$1{,}000 = 250(10) + \frac{C}{10} \qquad \text{Initial condition: } p(0) = 1{,}000$$

$$\frac{C}{10} = -1{,}500$$

$$C = -15{,}000$$

$$p(t) = 250(10 + t) - \frac{15{,}000}{10 + t} \qquad \text{Particular solution (see Fig. 3)}$$

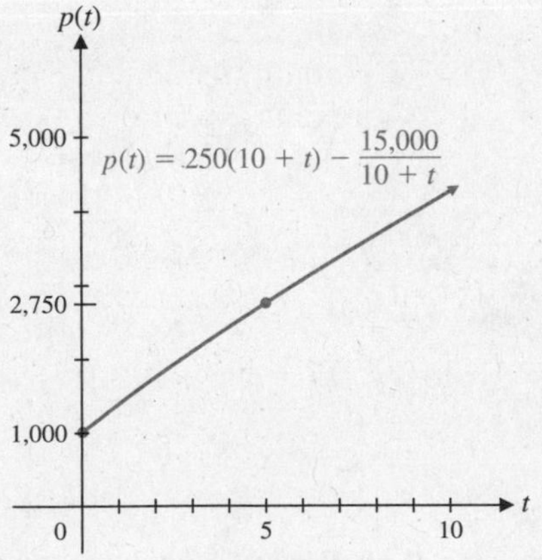

**FIGURE 3**

(A) To find the total amount of pollutants after 5 hours, we evaluate $p(5)$:

$$p(5) = 250(15) - \frac{15{,}000}{15} = 2{,}750 \text{ pounds}$$

(B) After 5 hours, the tank contains 750 gallons of water. The rate at which pollutants are being released into the sewage system is

$$\frac{2{,}750}{750} \approx 3.67 \text{ pounds per gallon} \qquad \blacksquare$$

> **MATCHED PROBLEM 5**  Repeat Example 5 if water is released from the tank at the rate of 75 gallons per hour.

*Answers to Matched Problems*

1. $I(x) = x^3; y = x + (C/x^3)$
2. $I(x) = e^{x^3}; y = 3 + 4e^{-x^3}$
3. $A(t) = 20{,}000 - 10{,}000e^{0.05t}; A = 0$ when $t = (\ln 2)/0.05 \approx 13.863$ yr; total withdrawals $= \$13{,}863$
4. $p(t) = 10 + 15e^{-4t}; \bar{p} = 10$
5. $p(t) = 125(20 + t) - \dfrac{12{,}000{,}000}{(20 + t)^3}$ \quad (A) 2,357 lb \quad (B) Approx. 3.77 lb/gal

# Exercise 1-3

*In all problems, assume $h(x) > 0$ whenever $\ln h(x)$ is involved.*

**A** *In Problems 1–4, is the differential equation first-order linear? If so, find $f(x)$ and $g(x)$ so that the equation can be expressed in the form $y' + f(x)y = g(x)$.*

1. $y' = ky$
2. $y' = k(M - y)$
3. $y' = ky(M - y)$
4. $y' = kx^2(M - y)$

*In Problems 5–12, replace each ? with a function of x that will make the integrand equal to the derivative of a product, and then find the antiderivative. Choose 0 for the constant of integration.*

5. $\int (x^3 y' + ?\, y)\, dx$
6. $\int (e^{2x} y' + ?\, y)\, dx$
7. $\int (e^{-3x} y' + ?\, y)\, dx$
8. $\int (x^{1/2} y' + ?\, y)\, dx$
9. $\int (?\, y' + 4x^3 y)\, dx$
10. $\int (?\, y' + 0.2e^{0.2x} y)\, dx$
11. $\int (?\, y' - 0.5e^{-0.5x} y)\, dx$
12. $\int \left( ?\, y' - \dfrac{1}{x^2}\, y \right) dx$

*In Problems 13–24, find the integrating factor, the general solution, and the particular solution satisfying the given initial condition.*

13. $y' + 2y = 4; y(0) = 1$
14. $y' - 3y = 3; y(0) = -1$
15. $y' + y = e^{-2x}; y(0) = 3$
16. $y' - 2y = e^{3x}; y(0) = 2$
17. $y' - y = 2e^x; y(0) = -4$
18. $y' + 4y = 3e^{-4x}; y(0) = 5$
19. $y' + y = 9x^2 e^{-x}; y(0) = 2$
20. $y' - 3y = 6\sqrt{x}\, e^{3x}; y(0) = -2$
21. $xy' + y = 2x; y(1) = 1$
22. $xy' - 3y = 4x; y(1) = 1$
23. $xy' + 2y = 10x^3; y(2) = 8$
24. $xy' - y = \dfrac{9}{x^2}; y(2) = 3$

**B** *In Problems 25–34, find the integrating factor $I(x)$ for each equation, and then find the general solution.*

25. $y' + xy = 5x$
26. $y' - 2xy = 6x$
27. $y' - 2y = 4x$
28. $y' + y = x^2$
29. $xy' + y = xe^x$
30. $xy' + 2y = xe^{3x}$
31. $xy' + y = x \ln x$
32. $xy' - y = x \ln x$
33. $2xy' + 3y = 20x$
34. $2xy' - y = 6x^2$

*In Problems 35–38, suppose that the indicated "solutions" were given to you by a student whom you are tutoring in this class.*

(A) How would you have the student check each solution?

(B) Is the solution right or wrong? If the solution is wrong, explain what is wrong and how it can be corrected.

(C) Show a correct solution for each incorrect solution, and check the result.

**35.** Equation: $y' + \dfrac{1}{x}y = (x+1)^2$

Solution: $y = \dfrac{1}{x}\displaystyle\int x(x+1)^2\,dx$

$\qquad = \dfrac{1}{3}(x+1)^3 + C$

**36.** Equation: $y' + \dfrac{2}{x}y = x-1$

Solution: $y = \dfrac{1}{x^2}\displaystyle\int x^2(x-1)\,dx$

$\qquad = \dfrac{1}{2}(x-1)^2 + C$

**37.** Equation: $y' + 3y = e^{-x}$

Solution: $y = \dfrac{1}{e^{3x}}\displaystyle\int e^{2x}\,dx$

$\qquad = \dfrac{1}{2}e^{-x}$

**38.** Equation: $y' - 2y = e^x$

Solution: $y = e^{2x}\displaystyle\int e^{-x}\,dx$

$\qquad = -e^x$

C  *In Problems 39–44, find the general solution two ways. First use an integrating factor and then use separation of variables.*

**39.** $y' = \dfrac{1-y}{x}$        **40.** $y' = \dfrac{y+2}{x+1}$

**41.** $y' = \dfrac{2x+2xy}{1+x^2}$      **42.** $y' = \dfrac{4x+2xy}{4+x^2}$

**43.** $y' = 2x(y+1)$      **44.** $y' = 3x^2(y+2)$

**45.** Use an integrating factor to find the general solution of the unlimited growth model,

$$\frac{dy}{dt} = ky$$

[*Hint:* Remember, the antiderivative of the constant function 0 is an arbitrary constant $C$.]

**46.** Use an integrating factor to find the general solution of the limited growth model,

$$\frac{dy}{dt} = k(L-y)$$

**47.** Discuss how the solution process for

$$y' + f(x)y = g(x)$$

simplifies if $f(x)$ and $g(x)$ are constants $a \neq 0$ and $b$, respectively, and find the general solution.

**48.** Discuss how the solution process for

$$y' + f(x)y = g(x)$$

simplifies if $g(x) = 0$ for all $x$, and find the general solution.

## Applications

**49.** *Continuous compound interest.* An initial deposit of $20,000 is made into an account that earns 4.8% compounded continuously. Money is then withdrawn at a constant rate of $3,000 a year until the amount in the account is 0. Find the amount in the account at any time $t$. When is the amount 0? What is the total amount withdrawn from the account?

**50.** *Continuous compound interest.* An initial deposit of $50,000 is made into an account that earns 4.5% compounded continuously. Money is then withdrawn at a constant rate of $5,400 a year until the amount in the account is 0. Find the amount in the account at any time $t$. When is the amount 0? What is the total amount withdrawn from the account?

**51.** *Continuous compound interest.* An initial deposit of $P is made into an account that earns 4.75% compounded continuously. Money is then withdrawn at a constant rate of $1,900 a year. After 10 years of continuous withdrawals, the amount in the account is 0. Find the initial deposit $P$.

**52.** *Continuous compound interest.* An initial deposit of $P is made into an account that earns 4.25% compounded continuously. Money is then withdrawn at a constant rate of $3,400 a year. After 5 years of continuous withdrawals, the amount in the account is 0. Find the initial deposit $P$.

**53.** *Continuous compound interest.* An initial deposit of $5,000 is made into an account earning 7% compounded continuously. Thereafter, money is deposited into the account at a constant rate of $2,100 per year. Find the amount in this account at any time $t$. How much is in this account after 5 years?

**54.** *Continuous compound interest.* An initial deposit of $10,000 is made into an account earning 6.25% compounded continuously. Thereafter, money is deposited into the account at a constant rate of $5,000 per year. Find the amount in this account at any time $t$. How much is in this account after 10 years?

**55.** *Continuous compound interest.* An initial investment of $10,000 is made in a mutual fund. Additional investments are made at the rate of $1,000 per year. After 10 years the fund has a value of $30,000. What continuous compound rate of interest has this fund earned over this 10 year period? Express the answer as a percentage, correct to two decimal places.

**56.** *Continuous compound interest.* A trust fund is established by depositing $100,000 into an account. Withdrawals are made at the rate of $20,000 per year. After 5 years, the trust fund contains $15,000. What continuous compound rate of interest has this fund earned over this 5 year period? Express the answer as a percentage, correct to two decimal places.

57. *Supply–demand.* The supply $S$ and demand $D$ for a certain commodity satisfy the equations

$$S = 35 - 2p(t) + 3p'(t) \quad \text{and}$$

$$D = 95 - 5p(t) + 2p'(t)$$

If $p(0) = 30$, find the equilibrium price at time $t$ and the long-range equilibrium price.

58. *Supply–demand.* The supply $S$ and demand $D$ for a certain commodity satisfy the equations

$$S = 70 - 3p(t) + 2p'(t) \quad \text{and}$$

$$D = 100 - 5p(t) + p'(t)$$

If $p(0) = 5$, find the equilibrium price at time $t$ and the long-range equilibrium price.

59. *Pollution.* A 1,000-gallon holding tank contains 200 gallons of water. Initially, each gallon of water in the tank contains 2 pounds of pollutants. Water containing 3 pounds of pollutants per gallon enters the tank at a rate of 75 gallons per hour, and the uniformly mixed water is released from the tank at a rate of 50 gallons per hour. How many pounds of pollutants are in the tank after 2 hours? At what rate (in pounds per gallon) are the pollutants being released after 2 hours?

Water enters tank at 75 gallons per hour

Water leaves tank at 50 gallons per hour

60. *Pollution.* Rework Problem 59 if water is entering the tank at the rate of 100 gallons per hour.

61. *Pollution.* Rework Problem 59 if water is entering the tank at the rate of 50 gallons per hour.

62. *Pollution.* Rework Problem 59 if water is entering the tank at the rate of 150 gallons per hour.

63. *Pollution.* Refer to Problem 59. When will the tank contain 1,000 pounds of pollutants? Round answer to one decimal place.

64. *Pollution.* Refer to Problem 62. When will the tank contain 1,500 pounds of pollutants? Round answer to one decimal place.

In an article in the College Mathematics Journal *(Jan. 1987, 18:1),* Arthur Segal proposed the following model for weight loss or gain:

$$\frac{dw}{dt} + 0.005w = \frac{1}{3,500}C$$

where $w(t)$ is a person's weight (in pounds) after $t$ days of consuming exactly $C$ calories per day. Use this model to solve Problems 65–68.

65. *Weight loss.* A person weighing 160 pounds goes on a diet of 2,100 calories per day. How much will this person weigh after 30 days on this diet? How long will it take this person to lose 10 pounds? Find $\lim_{t \to \infty} w(t)$ and interpret the results.

66. *Weight loss.* A person weighing 200 pounds goes on a diet of 2,800 calories per day. How much will this person weigh after 90 days on this diet? How long will it take this person to lose 25 pounds? Find $\lim_{t \to \infty} w(t)$ and interpret the results.

67. *Weight loss.* A person weighing 130 pounds would like to lose 5 pounds during a 30 day period. How many calories per day should this person consume to reach this goal?

68. *Weight loss.* A person weighing 175 pounds would like to lose 10 pounds during a 45 day period. How many calories per day should this person consume to reach this goal?

In 1960, William K. Estes proposed the following model for measuring a student's performance in the classroom:

$$\frac{dk}{dt} + \ell k = \lambda \ell$$

where $k(t)$ is the student's knowledge after $t$ weeks (expressed as a percentage and measured by performance on examinations), $\ell$ is a constant called the coefficient of learning and representing the student's ability to learn (expressed as a percentage and determined by IQ or some similar general intelligence predictor), and $\lambda$ is a constant representing the fraction of available time the student spends performing helpful acts that should increase knowledge of the subject (studying, going to class, and so on). Use this model to solve Problems 69 and 70.

69. *Learning theory.* Students enrolled in a beginning Spanish class are given a pretest the first day of class to determine their knowledge of the subject. The results of the pretest, the coefficient of learning, and the fraction of time spent performing helpful acts for two students in the class are given in the table. Use the Estes model to predict the knowledge of each student after 6 weeks in the class.

| | Score on Pretest | Coefficient of Learning | Fraction of Helpful Acts |
|---|---|---|---|
| Student A | 0.1 (10%) | 0.8 | 0.9 |
| Student B | 0.4 (40%) | 0.8 | 0.7 |

70. *Learning theory.* Refer to Problem 69. When will both students have the same level of knowledge? Round answer to one decimal place.

# CHAPTER 1 REVIEW

## Important Terms, Symbols, and Concepts

### 1-1 Basic Concepts

A **differential equation** is an equation involving an unknown function, usually denoted by $y$, and one or more of its derivatives. The **order** of a differential equation is the highest derivative of the unknown function present in the equation. We only consider **first-order** differential equations. The **slope field** for a differential equation is obtained by drawing tangent line segments determined by the equation at each point in a grid. The **solution** of a differential equation is a function $y$ with the property that substituting $y$ and $y'$ into the equation produces a statement that is true for all values of $x$. A solution of a differential equation that involves an arbitrary constant is called the **general solution,** and collectively, is referred to as the **family of solutions.** Each member of the family of solutions is called a **particular solution.** A condition of the form $y(x_0) = y_0$ is called an **initial condition** and is often used to specify a particular solution.

### 1-2 Separation of Variables

The solution of a differential equation that can be written in the form

$$f(y)y' = g(x)$$

is given implicitly by the equation

$$\int f(y)\,dy = \int g(x)\,dx$$

If both integrals can be evaluated and if the resulting equation can be solved for $y$ in terms of $x$, then we have an explicit solution for the differential equation. A solution to a differential equation that cannot be obtained from the general solution is referred to as a singular solution. The basic exponential growth models and their solutions are as follows:

| Exponential Growth | | |
|---|---|---|
| **Description** | **Model** | **Solution** |
| **Unlimited growth:** Rate of growth is proportional to the amount present | $\dfrac{dy}{dt} = ky$ $k, t > 0$ $y(0) = c$ | $y = ce^{kt}$ |
| **Exponential decay:** Rate of growth is proportional to the amount present | $\dfrac{dy}{dt} = -ky$ $k, t > 0$ $y(0) = c$ | $y = ce^{-kt}$ |
| **Limited growth:** Rate of growth is proportional to the difference between the amount present and a fixed limit | $\dfrac{dy}{dt} = k(M - y)$ $k, t > 0$ $y(0) = 0$ | $y = M(1 - e^{-kt})$ |
| **Logistic growth:** Rate of growth is proportional to the amount present and to the difference between the amount present and a fixed limit | $\dfrac{dy}{dt} = ky(M - y)$ $k, t > 0$ $y(0) = \dfrac{M}{1 + c}$ | $y = \dfrac{M}{1 + ce^{-kMt}}$ |

### 1-3 First-Order Linear Differential Equations

A differential equation that can be expressed in the form

$$y' + f(x)y = g(x)$$

is called a **first-order linear differential equation.**

Equations in this form can be solved by the following procedure:

PROCEDURE

**Solving First-Order Linear Differential Equations**

**Step 1.** Write the equation in the **standard form:**

$$y' + f(x)y = g(x)$$

**Step 2.** Compute the **integrating factor:**

$$I(x) = e^{\int f(x)\,dx}$$

[When evaluating $\int f(x)\,dx$, choose 0 for the constant of integration.]

**Step 3.** Multiply both sides of the standard form by the integrating factor $I(x)$. The left side should now be in the form $[I(x)y]'$:

$$[I(x)y]' = I(x)g(x)$$

**Step 4.** Integrate both sides:

$$I(x)y = \int I(x)g(x)\,dx$$

[When evaluating $\int I(x)g(x)\,dx$, include an arbitrary constant of integration.]

**Step 5.** Solve for $y$ to obtain the **general solution:**

$$y = \frac{1}{I(x)}\int I(x)g(x)\,dx$$

## REVIEW EXERCISE

*Work through all the problems in this chapter review and check your answers in the back of this supplement. Answers to all review problems are there along with section numbers in italics to indicate where each type of problem is discussed. Where weaknesses show up, review appropriate sections in the text.*

A  *In Problems 1 and 2, show that the given function is the general solution of the indicated differential equation.*

1.  $y = C\sqrt{x}$; $2xy' = y$

2.  $y = 1 + Ce^{-x/3}$; $3y' + y = 1$

*In Problems 3 and 4, determine which of the following slope fields is associated with the indicated differential equation. Briefly justify your answer.*

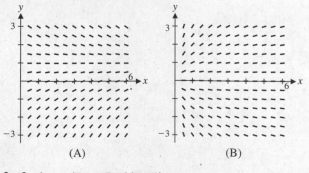

(A)                              (B)

3.  $2xy' = y$ (From Problem 1)

4.  $3y' + y = 1$ (From Problem 2)

*In Problems 5 and 6, use the appropriate slope field—(A) or (B), given for Problems 3 and 4 (or a copy)—to graph the particular solutions obtained by letting $C = -2, -1, 0, 1,$ and 2.*

5.  $y = C\sqrt{x}$  (From Problem 1)

6.  $y = 1 + Ce^{-x/3}$ (From Problem 2)

*In Problems 7 and 8, write a differential equation that describes the rate of change of the indicated quantity.*

7.  The price $y$ of a product is decreasing at a rate proportional to the difference between the price and a lower limit of $5.

8.  The price $y$ of a product is increasing at a rate proportional to the price.

*In Problems 9 and 10, write a verbal description of the rate of change of the given quantity satisfying the indicated differential equation and discuss the increasing/decreasing properties of the quantity.*

9.  The number of people $y$ who have contracted an infectious disease satisfies $dy/dt = 0.2y(50{,}000 - y)$ and $y(0) = 1$.

10.  The amount of radioactive material $y$ (in grams) at the site of a nuclear accident satisfies $dy/dt = -0.001y$ and $y(0) = 100$.

*In Problems 11–14, determine whether the differential equation can be written in the separation of variables form*

$f(y)y' = g(x)$, *the first-order linear form* $y' + f(x)y = g(x)$, *both forms, or neither form.*

**11.** $y' + 3x^2y + 9 = e^x(1 - x + y)$

**12.** $y' + 10y + 25 = 5xy^2$

**13.** $y' + 3y = xy + 2x - 6$

**14.** $y^2y' = 4xe^{x-y}$

*In Problems 15–20, find the general solution.*

**15.** $y' = -\dfrac{4y}{x}$

**16.** $y' = -\dfrac{4y}{x} + x$

**17.** $y' = 3x^2y^2$

**18.** $y' = 2y - e^x$

**19.** $y' = \dfrac{5}{x}y + x^6$

**20.** $y' = \dfrac{3 + y}{2 + x}, \ x > -2$

**B** *In Problems 21–28, find the particular solution that satisfies the given condition.*

**21.** $y' = 10 - y; y(0) = 0$

**22.** $y' + y = x; y(0) = 0$

**23.** $y' = 2ye^{-x}; y(0) = 1$

**24.** $y' = \dfrac{2x - y}{x + 4}; y(0) = 1$

**25.** $y' = \dfrac{x}{y + 4}; y(0) = 0$

**26.** $y' + \dfrac{2}{x}y = \ln x; y(1) = 2$

**27.** $yy' = \dfrac{x(1 + y^2)}{1 + x^2}; y(0) = 1$

**28.** $y' + 2xy = 2e^{-x^2}; y(0) = 1$

**C**

**29.** Solve the following differential equation two ways. First use an integrating factor and then use separation of variables.

$$xy' - 4y = 8$$

**30.** Give an example of an equation that can be solved using an integrating factor but cannot be solved using separation of variables. Solve the equation.

**31.** Give an example of an equation that can be solved using separation of variables but cannot be solved using an integrating factor. Solve the equation.

**32.** Find the general solution of the following differential equation and then find the particular solutions satisfying the indicated initial conditions.

$$xy' - 5y = -10$$

(A) $y(0) = 2$     (B) $y(0) = 0$     (C) $y(1) = 1$

**33.** Find an explicit expression for the particular solution of the differential equation $yy' = x$ satisfying the initial condition $y(0) = 4$.

**34.** Given that $y = x^3 + Cx$ is the general solution of the differential equation $xy' - y = 2x^3$:

(A) In the same viewing window, graph the particular solutions obtained by letting $C = 0, 1, 2$, and 3.

(B) What do the graphs of the solutions for $C > 0$ have in common?

(C) In the same viewing window, graph the particular solutions obtained by letting $C = 0, -1, -2$, and $-3$.

(D) What do the graphs of the solutions for $C < 0$ have in common?

**35.** The rate of change with respect to time $t$ of a quantity $y$ is proportional to the difference between $y$ and an unknown upper limit $M$, and $y = 0$ when $t = 0$. Using the limited growth law, it follows that

$$y = M(1 - e^{-kt}) \qquad M > 0, k > 0$$

(A) Using the data given in the table, find two equations that $M$ and $k$ must satisfy, and solve each equation for $M$.

| $t$ | 2 | 5 |
|---|---|---|
| $y$ | 4 | 7 |

(B) Use a graphing utility and the equations from part (A) to approximate $M$ to one decimal place.

## APPLICATIONS

**36.** *Depreciation.* A refrigerator costs $500 when it is new. The value of the refrigerator depreciates to $25 over a 20-year period. If the rate of change of the value is proportional to the value, find the value of the refrigerator 5 years after it was purchased.

**37.** *Sales growth.* The rate of growth of the annual sales of a new company is proportional to the difference between

the annual sales $S$ and an upper limit of $200,000. Assume that $S(0) = 0$.

(A) If the sales are $50,000 for the first year, how long will it take for the annual sales to reach $150,000? Round answer to the nearest year.

(B) Suppose the company is dissatisfied with the results in part (A) and wants the annual sales to reach $150,000 after just 3 years. In order for this to happen, what must the sales be for the first year? Round answer to the nearest thousand dollars.

**38.** *Price stability.* The supply $S$ and the demand $D$ for a certain commodity satisfy the equations

$$S = 100 + p + p' \quad \text{and} \quad D = 200 - p - p'$$

(A) Find the equilibrium price $p$ at any time $t$.

(B) Find the long-range equilibrium price $\bar{p}$.

(C) Graph the particular solutions that satisfy the initial conditions $p(0) = 75$ and $p(0) = 25$.

(D) Discuss the long-term behavior of the price of this commodity.

**39.** *Continuous compound interest.* An initial deposit of $60,000 is made into an account that earns 5% compounded continuously. Money is then withdrawn at a constant rate of $5,000 per year until the amount in the account is 0. Find the amount in the account at any time $t$. When is the amount 0? What is the total amount withdrawn from the account?

**40.** *Continuous compound interest.* An initial investment of $15,000 is made in a mutual fund. Additional investments are made at the rate of $2,000 per year. After 10 years the fund has a value of $50,000. What continuous compound rate of interest has this fund earned over this 10-year period? Express the answer as a percentage, correct to two decimal places.

**41.** *Crop yield.* The yield $y(t)$ (in bushels per acre) of a corn crop satisfies the equation

$$\frac{dy}{dt} = 100 + e^{-t} - y$$

If $y(0) = 0$, find $y$ at any time $t$.

**42.** *Pollution.* A 1,000-gallon holding tank contains 100 gallons of unpolluted water. Water containing 2 pounds of pollutant per gallon is pumped into the tank at the rate of 75 gallons per hour. The uniformly mixed water is released from the tank at 50 gallons per hour.

(A) Find the total amount of pollutants in the tank after 2 hours. Round answer to one decimal place.

(B) When will the tank contain 700 pounds of pollutants? Round answer to one decimal place.

**43.** *Ecology.* A bird population on an island is declining at a rate proportional to the difference between the population and a lower limit of 200 birds.

(A) If the population has declined from 1,000 birds 5 years ago to 500 today, find the population 4 years from now.

(B) Suppose the bird population 4 years from now turns out to be 400, indicating that the island might be able to support more than the original lower limit of 200 birds. Assuming that the rate of change of the population is still proportional to the difference between the population and an unknown lower limit $M$, approximate $M$ to the nearest integer.

**44.** *Rumor spread.* A single individual starts a rumor in a community of 200 people. The rumor spreads at a rate proportional to the number of people who have not yet heard the rumor. After 2 days, 10 people have heard the rumor.

(A) How many people will have heard the rumor after 5 days?

(B) How long will it take for the rumor to spread to 100 people?

# Taylor Polynomials and Infinite Series

CHAPTER **2**

2-1  Taylor Polynomials

2-2  Taylor Series

2-3  Operations on Taylor Series

2-4  Approximations Using Taylor Series

Chapter 2 Review

Review Exercise

## INTRODUCTION

The circuits inside a calculator are capable only of performing the basic operations of addition, subtraction, multiplication, and division. Yet, many calculators have keys that allow you to evaluate functions such as $e^x$, $\ln x$, and $\sin x$. How is this done? In most cases, the values of these functions are *approximated* by using a carefully selected polynomial, and, of course, polynomials can be evaluated by using the basic arithmetic operations. Thus, the approximating polynomials give the calculator the capability of evaluating nonpolynomial functions. In this chapter we will study one type of approximating polynomial, called a *Taylor polynomial* after the English mathematician Brook Taylor (1685–1731). For a given function $f$, we will determine the coefficients of the Taylor polynomial, the values of $x$ where the Taylor polynomial approximates $f(x)$, and the accuracy of this approximation. We will also consider various applications involving Taylor polynomials, including the approximation of definite integrals.

## Section 2-1 TAYLOR POLYNOMIALS

- Higher-Order Derivatives
- Approximating $e^x$ with Polynomials
- Taylor Polynomials at 0
- Taylor Polynomials at $a$
- Application

### ■ Higher-Order Derivatives

Up to this point, we have considered only the first and second derivatives of a function. In the work that follows, we will need to find higher-order derivatives. For example, if we start with the function $f$ defined by

$$f(x) = x^{-1}$$

then the first derivative of $f$ is

$$f'(x) = -x^{-2}$$

and the second derivative of $f$ is

$$f''(x) = \frac{d}{dx}f'(x) = \frac{d}{dx}(-x^{-2}) = 2x^{-3}$$

Additional higher-order derivatives are found by successive differentiation. Thus, the third derivative is

$$f^{(3)}(x) = \frac{d}{dx}f''(x) = \frac{d}{dx}(2x^{-3}) = -6x^{-4}$$

the fourth derivative is

$$f^{(4)}(x) = \frac{d}{dx}f^{(3)}(x) = \frac{d}{dx}(-6x^{-4}) = 24x^{-5}$$

and so on.

In general, the symbol $f^{(n)}$ is used to represent the **nth derivative of the function f.** When stating formulas involving a function and its higher-order derivatives, it is convenient to let $f^{(0)}$ represent the function $f$ (that is, the zeroth derivative of a function is just the function itself).

The order of the derivative must be enclosed in parentheses, because in most contexts, $f^n(x)$ is interpreted to mean the $n$th power of $f$, not the $n$th derivative. Thus, $f^2(x) = [f(x)]^2 = f(x)f(x)$, while $f^{(2)}(x) = f''(x)$.

Finding a particular higher-order derivative is a routine calculation. However, finding a formula for the $n$th derivative for arbitrary $n$ requires careful observation of the patterns that develop as each successive derivative is found. Study the next example carefully. Many problems in this chapter will involve similar concepts.

EXAMPLE 1    **Finding nth Derivatives**   Find the $n$th derivative of

$$f(x) = \frac{1}{1+x} = (1+x)^{-1}$$

SOLUTION    We begin by finding the first four derivatives of $f$:

$$f(x) = (1+x)^{-1}$$
$$f'(x) = (-1)(1+x)^{-2}$$
$$f''(x) = (-1)(-2)(1+x)^{-3}$$
$$f^{(3)}(x) = (-1)(-2)(-3)(1+x)^{-4}$$
$$f^{(4)}(x) = (-1)(-2)(-3)(-4)(1+x)^{-5}$$

Notice that we did not multiply out the coefficient of $(1 + x)^{-k}$ in each derivative. Our objective is to look for a pattern in the form of each derivative. Multiplying out these coefficients would tend to obscure any pattern that is developing. Instead, we observe that each coefficient is a product of successive negative integers that can be written as follows:

$$(-1) = (-1)^1(1)$$
$$(-1)(-2) = (-1)^2(1)(2)$$
$$(-1)(-2)(-3) = (-1)^3(1)(2)(3)$$
$$(-1)(-2)(-3)(-4) = (-1)^4(1)(2)(3)(4)$$

Next we note that the product of natural numbers in each expression on the right can be written in terms of a factorial:*

$$(-1) = (-1)^1 1! \quad 1! = 1$$
$$(-1)(-2) = (-1)^2 2! \quad 2! = 1 \cdot 2$$
$$(-1)(-2)(-3) = (-1)^3 3! \quad 3! = 1 \cdot 2 \cdot 3$$
$$(-1)(-2)(-3)(-4) = (-1)^4 4! \quad 4! = 1 \cdot 2 \cdot 3 \cdot 4$$

Substituting these last expressions in the derivatives of $f$, we have

$$f'(x) = (-1)^1 1!(1 + x)^{-2} \quad n = 1$$
$$f''(x) = (-1)^2 2!(1 + x)^{-3} \quad n = 2$$
$$f^{(3)}(x) = (-1)^3 3!(1 + x)^{-4} \quad n = 3$$
$$f^{(4)}(x) = (-1)^4 4!(1 + x)^{-5} \quad n = 4$$

This suggests that

$$f^{(n)}(x) = (-1)^n n!(1 + x)^{-(n+1)} \quad \text{Arbitrary } n$$

**MATCHED PROBLEM 1**   Find the $n$th derivative of $f(x) = \ln x$.

*Explore & Discuss* **1**

(A) Compute the first six derivatives of the polynomial function
$p(x) = 7x^3 - 9x^2 + 4x - 15$.

(B) Let $q(x)$ be a polynomial of degree $n$. For which orders are the higher-order derivatives of $q(x)$ equal to the constant function $f(x) = 0$?

## ■ Approximating $e^x$ with Polynomials

We have already seen that the irrational number $e$ and the exponential function $e^x$ play important roles in many applications, including continuous compound interest, population growth, and exponential decay, to name a few. Now, given the function

$$f(x) = e^x$$

we would like to construct a polynomial function $p$ whose values are close to the values of $f$, at least for some values of $x$. If we are successful, then we can use the values of $p$ (which are easily computed) to approximate the values of $f$. We begin by trying to approximate $f$ for values of $x$ near 0 with a first-degree polynomial of the form

$$p_1(x) = a_0 + a_1 x \tag{1}$$

We want to place conditions on $p_1$ that will enable us to determine the unknown coefficients $a_0$ and $a_1$. Since we want to approximate $f$ for values near 0, it is reasonable to require that $f$ and $p_1$ agree at 0. Thus,

$$a_0 = p_1(0) = f(0) = e^0 = 1$$

---

*For $n$ a natural number, $n! = n(n - 1)! = n(n - 1)(n - 2) \cdots \cdot 2 \cdot 1$ and $0! = 1$.

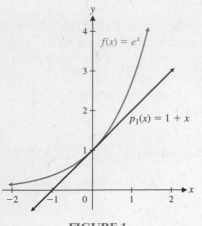

**FIGURE 1**

This determines the value of $a_0$. To determine the value of $a_1$, we require that both functions have the same slope at 0. Since $p_1'(x) = a_1$ and $f'(x) = e^x$, this implies that

$$a_1 = p_1'(0) = f'(0) = e^0 = 1$$

Thus, after substituting $a_0 = 1$ and $a_1 = 1$ into (1), we obtain

$$p_1(x) = 1 + x$$

which is a first-degree polynomial satisfying

$$p_1(0) = f(0) \qquad \text{and} \qquad p_1'(0) = f'(0)$$

How well does $1 + x$ approximate $e^x$? Examining the graph in Figure 1, it appears that $1 + x$ is a good approximation to $e^x$ for $x$ very close to 0. However, as $x$ moves away from 0, in either direction, the distance between the values of $1 + x$ and $e^x$ increases and the accuracy of the approximation decreases.

Now we will try to improve this approximation by using a second-degree polynomial of the form

$$p_2(x) = a_0 + a_1 x + a_2 x^2 \tag{2}$$

We need three conditions to determine the coefficients $a_0$, $a_1$, and $a_2$. We still require that $p_2(0) = f(0)$ and $p_2'(0) = f'(0)$, and add the condition that $p_2''(0) = f''(0)$. This ensures that the graphs of $p_2$ and $f$ have the same concavity at $x = 0$. Proceeding as before, we compute the first and second derivatives of $p_2$ and $f$ and apply these conditions:

$$
\begin{aligned}
p_2(x) &= a_0 + a_1 x + a_2 x^2 & f(x) &= e^x \\
p_2'(x) &= \phantom{a_0 +} a_1 + 2a_2 x & f'(x) &= e^x \\
p_2''(x) &= \phantom{a_0 + a_1 +} 2a_2 & f''(x) &= e^x
\end{aligned}
$$

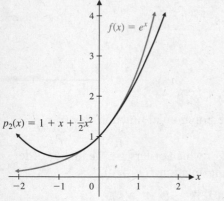

**FIGURE 2**

Thus,

$$
\begin{aligned}
a_0 &= p_2(0) = f(0) = e^0 = 1 & \text{implies} & & a_0 &= 1 \\
a_1 &= p_2'(0) = f'(0) = e^0 = 1 & \text{implies} & & a_1 &= 1 \\
2a_2 &= p_2''(0) = f''(0) = e^0 = 1 & \text{implies} & & a_2 &= \tfrac{1}{2}
\end{aligned}
$$

and, substituting in (2), we obtain

$$p_2(x) = 1 + x + \tfrac{1}{2}x^2$$

The graph is shown in Figure 2.

Comparing Figures 1 and 2, we see that for values of $x$ near 0, the graph of $p_2(x) = 1 + x + \tfrac{1}{2}x^2$ is closer to the graph of $f(x) = e^x$ than was the graph of $p_1(x) = 1 + x$. Thus, the polynomial $1 + x + \tfrac{1}{2}x^2$ can be used to approximate $e^x$ over a larger range of $x$ values than the polynomial $1 + x$.

It seems reasonable to assume that a third-degree polynomial would yield a still better approximation. Using the notation for higher-order derivatives discussed earlier in this section, we can state the required condition as

Find

$$p_3(x) = a_0 + a_1 x + a_2 x^2 + a_3 x^3 \tag{3}$$

Satisfying

$$p_3^{(k)}(0) = f^{(k)}(0) \qquad k = 0, 1, 2, 3$$

As before, we obtain the value of the additional coefficient $a_3$ by adding the requirement that

$$p_3^{(3)}(0) = f^{(3)}(0)$$

Thus,

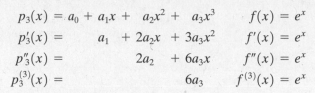

$$
\begin{aligned}
p_3(x) &= a_0 + a_1 x + a_2 x^2 + a_3 x^3 & f(x) &= e^x \\
p_3'(x) &= a_1 + 2a_2 x + 3a_3 x^2 & f'(x) &= e^x \\
p_3''(x) &= 2a_2 + 6a_3 x & f''(x) &= e^x \\
p_3^{(3)}(x) &= 6a_3 & f^{(3)}(x) &= e^x
\end{aligned}
$$

Applying the conditions $p_3^{(k)}(0) = f^{(k)}(0)$, we have

$$
\begin{aligned}
a_0 = p_3(0) &= f(0) = e^0 = 1 & \text{implies} && a_0 &= 1 \\
a_1 = p_3'(0) &= f'(0) = e^0 = 1 & \text{implies} && a_1 &= 1 \\
2a_2 = p_3''(0) &= f''(0) = e^0 = 1 & \text{implies} && a_2 &= \tfrac{1}{2} \\
6a_3 = p_3^{(3)}(0) &= f^{(3)}(0) = e^0 = 1 & \text{implies} && a_3 &= \tfrac{1}{6}
\end{aligned}
$$

Substituting in (3), we obtain

$$
p_3(x) = 1 + x + \tfrac{1}{2}x^2 + \tfrac{1}{6}x^3
$$

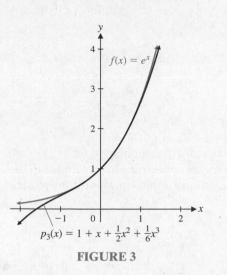

$p_3(x) = 1 + x + \frac{1}{2}x^2 + \frac{1}{6}x^3$

**FIGURE 3**

Figure 3 indicates that the approximations provided by $p_3$ are an improvement over those provided by $p_2$ and $p_1$.

In order to compare all three approximations, Table 1 lists the values of $p_1, p_2, p_3$, and $e^x$ at selected values of $x$, and Table 2 lists the absolute value of the difference between these polynomials and $e^x$ for the same $x$ values. The values of $e^x$ were obtained by using a calculator and are rounded to six decimal places. Comparing the columns in Table 2, we see that increasing the degree of the approximating polynomial decreases the difference between the polynomial and $e^x$. Thus, $p_2$ provides a better approximation than $p_1$, and $p_3$ provides a better approximation than $p_2$. We will have more to say about the accuracy of these approximations later in this chapter.

**TABLE 1**  $p_1(x) = 1 + x$, $p_2(x) = 1 + x + \frac{1}{2}x^2$, $p_3(x) = 1 + x + \frac{1}{2}x^2 + \frac{1}{6}x^3$

| $x$ | $p_1(x)$ | $p_2(x)$ | $p_3(x)$ | $e^x$ |
|---|---|---|---|---|
| −0.2 | 0.8 | 0.820 | 0.818 667 | 0.818 731 |
| −0.1 | 0.9 | 0.905 | 0.904 833 | 0.904 837 |
| 0 | 1 | 1 | 1 | 1 |
| 0.1 | 1.1 | 1.105 | 1.105 167 | 1.105 171 |
| 0.2 | 1.2 | 1.220 | 1.221 333 | 1.221 403 |

**TABLE 2**

| $x$ | $\lvert p_1(x) - e^x \rvert$ | $\lvert p_2(x) - e^x \rvert$ | $\lvert p_3(x) - e^x \rvert$ |
|---|---|---|---|
| −0.2 | 0.018 731 | 0.001 269 | 0.000 064 |
| −0.1 | 0.004 837 | 0.000 163 | 0.000 004 |
| 0 | 0 | 0 | 0 |
| 0.1 | 0.005 171 | 0.000 171 | 0.000 004 |
| 0.2 | 0.021 403 | 0.001 403 | 0.000 069 |

*Explore & Discuss* **2**

(A) Let $p(x)$ be a polynomial of degree $n \geq 1$. Explain why each of the following is true:

    (1) The graph of $p(x)$ intersects any vertical line in exactly one point.

    (2) The graph of $p(x)$ intersects any horizontal line in at most $n$ points.

    (3) The graph of $p(x)$ has neither horizontal nor vertical asymptotes.

(B) Discuss the graphs of

$$
f(x) = e^x \qquad g(x) = \frac{5x^2 - 3x + 1}{2x^2 + 3} \qquad h(x) = \frac{1}{x - 2} \qquad k(x) = e^{-x^2}
$$

Use part (A) to explain why none of the graphs of $f$, $g$, $h$, or $k$ is the graph of a polynomial function.

## ■ Taylor Polynomials at 0

The process we have used to determine $p_1, p_2,$ and $p_3$ can be continued. Given any positive integer $n$, we define

$$p_n(x) = a_0 + a_1 x + a_2 x^2 + \cdots + a_n x^n$$

and require that

$$p_n^{(k)}(0) = f^{(k)}(0) \qquad k = 0, 1, 2, \ldots, n$$

The polynomial $p_n$ is called a *Taylor polynomial*. Before determining $p_n$ for $f(x) = e^x$, it will be convenient to make some general statements concerning the relationship between $a_k$, $p_n^{(k)}(0)$, and $f^{(k)}(0)$ for an arbitrary function $f$. First, $p_n(x)$ is differentiated $n$ times to obtain the following relationships:

$$
\begin{aligned}
p_n(x) &= a_0 + a_1 x + a_2 x^2 + \quad\; a_3 x^3 + \cdots + a_n x^n \\
p_n'(x) &= \qquad a_1 + 2a_2 x + \quad\; 3a_3 x^2 + \cdots + n a_n x^{n-1} \\
p_n''(x) &= \qquad\qquad 2a_2 \;+ (3)(2)a_3 x + \cdots + n(n-1)a_n x^{n-2} \\
p_n^{(3)}(x) &= \qquad\qquad\qquad\quad (3)(2)a_3 \;+ \cdots + n(n-1)(n-2)a_n x^{n-3} \\
&\;\;\vdots \\
p_n^{(n)}(x) &= n(n-1)(n-2) \cdots (1)a_n = n!a_n
\end{aligned}
$$

Evaluating each derivative at 0 and applying the requirement that $p_n^{(k)}(0) = f^{(k)}(0)$ leads to the following equations:

$$
\begin{aligned}
a_0 &= p_n(0) &&= f(0) \\
a_1 &= p_n'(0) &&= f'(0) \\
2a_2 &= p_n''(0) &&= f''(0) \\
(3)(2)a_3 &= p_n^{(3)}(0) &&= f^{(3)}(0) \\
&\;\;\vdots &&\;\;\vdots \\
n!a_n &= p_n^{(n)}(0) &&= f^{(n)}(0)
\end{aligned}
$$

Solving each equation for $a_k$, we have

$$a_k = \frac{p_n^{(k)}(0)}{k!} = \frac{f^{(k)}(0)}{k!}$$

This relationship enables us to state the general definition of a Taylor polynomial.

---

**DEFINITION** | **Taylor Polynomial at 0\***

If $f$ has $n$ derivatives at 0, then the ***n*th-degree Taylor polynomial for $f$ at 0** is

$$p_n(x) = a_0 + a_1 x + a_2 x^2 + \cdots + a_n x^n = \sum_{k=0}^{n} a_k x^k$$

where

$$p_n^{(k)}(0) = f^{(k)}(0) \qquad \text{and} \qquad a_k = \frac{f^{(k)}(0)}{k!} \qquad k = 0, 1, 2, \ldots, n$$

\* Taylor polynomials at 0 are also often referred to as *Maclaurin polynomials*, but we will not use this terminology.

This result can be stated in a form that is more readily remembered as follows:

| DEFINITION | **Taylor Polynomial at 0: Concise Form** |
|---|---|

The $n$th-degree Taylor polynomial for $f$ at 0 is

$$p_n(x) = f(0) + f'(0)x + \frac{f''(0)}{2!}x^2 + \cdots + \frac{f^{(n)}(0)}{n!}x^n = \sum_{k=0}^{n} \frac{f^{(k)}(0)}{k!}x^k$$

provided $f$ has $n$ derivatives at 0.

In each of the preceding definitions, notice that we have used both the **expanded notation**

$$p_n(x) = a_0 + a_1 x + a_2 x^2 + \cdots + a_n x^n$$

$$= f(0) + f'(0)x + \frac{f''(0)}{2!}x^2 + \cdots + \frac{f^{(n)}(0)}{n!}x^n$$

and the more compact **summation notation**

$$p_n(x) = \sum_{k=0}^{n} a_k x^k = \sum_{k=0}^{n} \frac{f^{(k)}(0)}{k!}x^k$$

to represent a finite sum. Recall that we used summation notation earlier in Chapters 6 and 7 of *Calculus* 11/e or Chapters 13 and 14 of *College Mathematics* 11/e. In this chapter we will place more emphasis on the expanded notation, since it is usually easier to visualize a polynomial written in this notation, but we will continue to include the summation notation where appropriate.

### INSIGHT

The $n$th degree Taylor polynomial for $f$ at 0 is always a polynomial, but it does not necessarily have degree $n$, because it is possible that $f^{(n)}(0) = 0$. For example, if $g(x) = x^3 + 2x + 1$, then $g'(x) = 3x^2 + 2$ and $g''(x) = 6x$, so $g(0) = 1$, $g'(0) = 2$, and $g''(0) = 0$. Therefore, for this function $g$ at 0, the second-degree Taylor polynomial is $p_2(x) = 1 + 2x$, which actually has degree 1, not 2.

Returning to our original function $f(x) = e^x$, it is now an easy matter to find the $n$th-degree Taylor polynomial for this function. Since $(d/dx)e^x = e^x$, it follows that

$$f^{(k)}(x) = e^x \qquad f^{(k)}(0) = e^0 = 1 \qquad a_k = \frac{f^{(k)}(0)}{k!} = \frac{1}{k!}$$

for all values of $k$. Thus, for any $n$, the $n$th-degree Taylor polynomial for $e^x$ is

$$p_n(x) = 1 + x + \frac{1}{2!}x^2 + \frac{1}{3!}x^3 + \cdots + \frac{1}{n!}x^n = \sum_{k=0}^{n} \frac{1}{k!}x^k$$

| EXAMPLE 2 | **Approximation Using a Taylor Polynomial**  Find the third-degree Taylor polynomial at 0 for $f(x) = \sqrt{x + 4}$. Use $p_3$ to approximate $\sqrt{5}$. |
|---|---|

**SOLUTION**   **Step 1.**  *Find the derivatives:*

$$f(x) = (x + 4)^{1/2}$$
$$f'(x) = \tfrac{1}{2}(x + 4)^{-1/2}$$
$$f''(x) = -\tfrac{1}{4}(x + 4)^{-3/2}$$
$$f^{(3)}(x) = \tfrac{3}{8}(x + 4)^{-5/2}$$

**Step 2.** *Evaluate the derivatives at* 0:

$$f(0) = 4^{1/2} = 2$$
$$f'(0) = \tfrac{1}{2}(4^{-1/2}) = \tfrac{1}{4}$$
$$f''(0) = -\tfrac{1}{4}(4^{-3/2}) = -\tfrac{1}{32}$$
$$f^{(3)}(0) = \tfrac{3}{8}(4^{-5/2}) = \tfrac{3}{256}$$

**Step 3.** *Find the coefficients of the Taylor polynomial:*

$$a_0 = \frac{f(0)}{0!} = f(0) = 2$$

$$a_1 = \frac{f'(0)}{1!} = f'(0) = \frac{1}{4}$$

$$a_2 = \frac{f''(0)}{2!} = \frac{-\frac{1}{32}}{2} = -\frac{1}{64}$$

$$a_3 = \frac{f^{(3)}(0)}{3!} = \frac{\frac{3}{256}}{6} = \frac{1}{512}$$

**Step 4.** *Write down the Taylor polynomial:*

$$p_3(x) = 2 + \tfrac{1}{4}x - \tfrac{1}{64}x^2 + \tfrac{1}{512}x^3 \quad \text{Taylor polynomial.}$$

To use $p_3$ to approximate $\sqrt{5}$, we must first determine the appropriate value of $x$:

$$f(x) = \sqrt{x+4} = \sqrt{5} \quad \text{Square both sides.}$$
$$x + 4 = 5 \quad \text{Solve for } x.$$
$$x = 1$$

Thus,

$$\sqrt{5} = f(1) \approx p_3(1) = 2 + \tfrac{1}{4} - \tfrac{1}{64} + \tfrac{1}{512} \approx 2.236\,328\,1$$

[*Note:* The value of $\sqrt{5}$ obtained by using a calculator is, to six decimal places, 2.236 068.]

Compare the evaluation of $p_3(1)$ shown in Figure 4 to the value obtained in Example 2. The figure illustrates graphically that $p_3(x)$ is a good approximation to $f(x)$ for values of $x$ near 0.

**FIGURE 4** Graphs of $f(x) = \sqrt{x+4}$ and $p_3(x)$

MATCHED PROBLEM 2    Find the second-degree Taylor polynomial at 0 for $f(x) = \sqrt{x+9}$. Use $p_2(x)$ to approximate $\sqrt{10}$.

Many of the problems in this section involve approximating the values of functions that also can be evaluated with the use of a calculator. In such problems you should compare the Taylor polynomial approximation with the calculator value (which is also an approximation). This will give you some indication of the accuracy of the Taylor polynomial approximation. Later in this chapter we will discuss methods for determining the accuracy of any Taylor polynomial approximation.

EXAMPLE 3    **Taylor Polynomials at 0** Find the $n$th-degree Taylor polynomial at 0 for $f(x) = e^{2x}$.

SOLUTION

| Step 1 | Step 2 | Step 3 |
|---|---|---|
| $f(x) = e^{2x}$ | $f(0) = 1$ | $a_0 = f(0) = 1$ |
| $f'(x) = 2e^{2x}$ | $f'(0) = 2$ | $a_1 = f'(0) = 2$ |

$$f''(x) = 2^2 e^{2x} \qquad f''(0) = 2^2 \qquad a_2 = \frac{f''(0)}{2!} = \frac{2^2}{2!}$$

$$f^{(3)}(x) = 2^3 e^{2x} \qquad f^{(3)}(0) = 2^3 \qquad a_3 = \frac{f^{(3)}(0)}{3!} = \frac{2^3}{3!}$$

$$\vdots \qquad\qquad \vdots \qquad\qquad \vdots \qquad \vdots$$

$$f^{(n)}(x) = 2^n e^{2x} \qquad f^{(n)}(0) = 2^n \qquad a_n = \frac{f^{(n)}(0)}{n!} = \frac{2^n}{n!}$$

**Step 4.** *Write down the Taylor polynomial:*

$$p_n(x) = 1 + 2x + \frac{2^2}{2!}x^2 + \frac{2^3}{3!}x^3 + \cdots + \frac{2^n}{n!}x^n = \sum_{k=0}^{n} \frac{2^k}{k!}x^k$$

---

**MATCHED PROBLEM 3**   Find the $n$th-degree Taylor polynomial at 0 for $f(x) = e^{x/3}$.

## ■ Taylor Polynomials at *a*

Suppose we want to approximate the function $f(x) = \sqrt[4]{x}$ with a polynomial. Since $f^{(k)}(0)$ is undefined for $k = 1, 2, \ldots$, this function does not have a Taylor polynomial at 0. However, all the derivatives of $f$ exist at any $x > 0$. How can we generalize the definition of the Taylor polynomial to approximate functions such as this one? Before answering this question, we need to review a basic property of polynomials.

We are used to expressing polynomials in powers of $x$. However, it is also possible to express any polynomial in powers of $x - a$ for an arbitrary number $a$. For example, the following three expressions all represent the same polynomial:

$$\begin{aligned}
p(x) &= 3 + 2x \qquad\quad + x^2 & \text{Powers of x.} \\
&= 6 + 4(x - 1) + (x - 1)^2 & \text{Powers of x − 1.} \\
&= 3 - 2(x + 2) + (x + 2)^2 & \text{Powers of x + 2.}
\end{aligned}$$

To verify this statement, we expand the second and third expressions:

$$\begin{aligned}
6 + 4(x - 1) + (x - 1)^2 &= 6 + 4x - 4 + x^2 - 2x + 1 \\
&= 3 + 2x + x^2 \\
3 - 2(x + 2) + (x + 2)^2 &= 3 - 2x - 4 + x^2 + 4x + 4 \\
&= 3 + 2x + x^2
\end{aligned}$$

Now we return to the problem of generalizing the definition of the Taylor polynomial. Proceeding as we did before, given a function $f$ with $n$ derivatives at a number $a$, we want to find an $n$th-degree polynomial $p_n$ with the property that

$$p^{(k)}(a) = f^{(k)}(a) \qquad k = 0, 1, \ldots, n$$

That is, we require that $p_n$ and its first $n$ derivatives agree with $f$ and its first $n$ derivatives at the number $a$. It turns out that it is much easier to find $p_n$ when it is expressed in powers of $x - a$. The general expression for an $n$th-degree polynomial in powers of $x - a$ and its first $n$ derivatives are as follows:

$$\begin{aligned}
p_n(x) &= a_0 + a_1(x - a) + a_2(x - a)^2 + & a_3(x - a)^3 + \cdots + a_n(x - a)^n \\
p_n'(x) &= \qquad\quad a_1 \quad + 2a_2(x - a) + & 3a_3(x - a)^2 + \cdots + na_n(x - a)^{n-1} \\
p_n''(x) &= \qquad\qquad\qquad\quad 2a_2 \quad + & (3)(2)a_3(x - a) + \cdots + n(n - 1)a_n(x - a)^{n-2} \\
p_n^{(3)}(x) &= & (3)(2)(1)a_3 \qquad + \cdots + n(n - 1)(n - 2)a_n(x - a)^{n-3} \\
&\vdots
\end{aligned}$$

$$p_n^{(n)}(x) = n(n - 1) \cdot \cdots \cdot (1)a_n = n!a_n$$

Now we evaluate each function at $a$ and apply the appropriate condition:

$$
\begin{aligned}
a_0 = p_n(a) &= f(a) & k = 0 \\
a_1 = p'_n(a) &= f'(a) & k = 1 \\
2a_2 = p''_n(a) &= f''(a) & k = 2 \\
(3)(2)(1)a_3 = p_n^{(3)}(a) &= f^{(3)}(a) & k = 3 \\
\vdots \qquad\qquad & \quad \vdots & \vdots \\
n!a_n = p_n^{(n)}(a) &= f^{(n)}(a) & k = n
\end{aligned}
$$

Thus, each coefficient of $p_n$ satisfies

$$
a_k = \frac{f^{(k)}(a)}{k!} \qquad k = 0, 1, 2, \ldots, n
$$

---

**DEFINITION**  **Taylor Polynomial at $a$**

The **$n$th-degree Taylor polynomial at $a$** for a function $f$ is

$$
p_n(x) = f(a) + f'(a)(x - a) + \frac{f''(a)}{2!}(x - a)^2 + \cdots + \frac{f^{(n)}(a)}{n!}(x - a)^n
$$

$$
= \sum_{k=0}^{n} \frac{f^{(k)}(a)}{k!}(x - a)^k
$$

provided $f$ has $n$ derivatives at $a$.

---

Theoretically, a function $f$ has an $n$th-degree Taylor polynomial at any value $a$ where it has $n$ derivatives. In practice, we usually choose $a$ so that $f$ and its derivatives are easy to evaluate at $x = a$. For example, if $f(x) = \sqrt{x}$, then good choices for $a$ are 1, 4, 9, 16, and so on.

**EXAMPLE 4**    **Taylor Polynomials at $a$**  Find the third-degree Taylor polynomial at $a = 1$ for $f(x) = \sqrt[4]{x}$. Use $p_3(x)$ to approximate $\sqrt[4]{2}$.

**SOLUTION**    **Step 1.**  *Find the derivatives:*

$$
\begin{aligned}
f(x) &= x^{1/4} \\
f'(x) &= \tfrac{1}{4} x^{-3/4} \\
f''(x) &= -\tfrac{3}{16} x^{-7/4} \\
f^{(3)}(x) &= \tfrac{21}{64} x^{-11/4}
\end{aligned}
$$

**Step 2.**  *Evaluate the derivatives at $a = 1$:*

$$
\begin{aligned}
f(1) &= 1 \\
f'(1) &= \tfrac{1}{4} \\
f''(1) &= -\tfrac{3}{16} \\
f^{(3)}(1) &= \tfrac{21}{64}
\end{aligned}
$$

**Step 3.**  *Find the coefficients of the Taylor polynomial:*

$$
\begin{aligned}
a_0 &= f(1) = 1 \\
a_1 &= f'(1) = \tfrac{1}{4} \\
a_2 &= \frac{f''(1)}{2!} = \frac{-\tfrac{3}{16}}{2} = -\frac{3}{32} \\
a_3 &= \frac{f^{(3)}(1)}{3!} = \frac{\tfrac{21}{64}}{6} = \frac{7}{128}
\end{aligned}
$$

**Step 4.** *Write down the Taylor polynomial:*

$$p_3(x) = 1 + \tfrac{1}{4}(x - 1) - \tfrac{3}{32}(x - 1)^2 + \tfrac{7}{128}(x - 1)^3$$

Now we use the Taylor polynomial to approximate $\sqrt[4]{2}$:

$$\sqrt[4]{2} = f(2) \approx p_3(2) = 1 + \tfrac{1}{4} - \tfrac{3}{32} + \tfrac{7}{128} = 1.210\,937\,5$$

[*Note:* The value obtained by using a calculator is $\sqrt[4]{2} \approx 1.189\,207\,1$.]    ▬

**MATCHED PROBLEM 4**    Find the second-degree Taylor polynomial at $a = 8$ for $f(x) = \sqrt[3]{x}$. Use $p_2(x)$ to approximate $\sqrt[3]{9}$.

## APPLICATIONS

**EXAMPLE 5**    **Average Price**   Given the demand function

$$p = D(x) = \sqrt{2{,}500 - x^2}$$

use the second-degree Taylor polynomial at 0 to approximate the average price (in dollars) over the demand interval $[10, 40]$.

**SOLUTION**    The average price over the demand interval $[10, 40]$ is given by

$$\text{Average price} = \frac{1}{30} \int_{10}^{40} \sqrt{2{,}500 - x^2}\, dx$$

This integral cannot be evaluated by any of the techniques we have discussed. However, we can use a Taylor polynomial to approximate the value of the integral. Omitting the details (which you should supply), the second-degree Taylor polynomial at 0 for $D(x) = \sqrt{2{,}500 - x^2}$ is

$$p_2(x) = 50 - \tfrac{1}{100}x^2$$

Assuming that $D(x) \approx p_2(x)$ for $10 \le x \le 40$ (see Fig. 5), we have

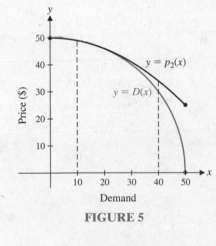

**FIGURE 5**

$$\text{Average price} \approx \frac{1}{30} \int_{10}^{40} \left( 50 - \frac{1}{100}x^2 \right) dx$$

$$= \frac{1}{30} \left( 50x - \frac{1}{300}x^3 \right) \Big|_{10}^{40}$$

$$= \$43 \qquad ▬$$

**MATCHED PROBLEM 5**    Given the demand function

$$p = D(x) = \sqrt{400 - x^2}$$

use the second-degree Taylor polynomial at 0 to approximate the average price (in dollars) over the demand interval $[5, 15]$.

Example 5 illustrates the convenience of using Taylor polynomials to approximate more complicated functions, but it also raises several important questions. First, how can we determine whether it is reasonable to use the Taylor polynomial to approximate a function on a given interval? Second, is it possible to find the Taylor polynomial without resorting to successive differentiation of the function? Finally, can we determine the accuracy of an approximation such as the approximate average price found in Example 5? These questions will be discussed in the next three sections.

1. $f^{(n)}(x) = (-1)^{n-1}(n-1)!\,x^{-n}$

2. $p_2(x) = 3 + \frac{1}{6}x - \frac{1}{216}x^2$; $\sqrt{10} \approx 3.162\,037$

3. $p_n(x) = 1 + \frac{1}{3}x + \frac{1}{3^2}\frac{1}{2!}x^2 + \frac{1}{3^3}\frac{1}{3!}x^3 + \cdots + \frac{1}{3^n}\frac{1}{n!}x^n = \sum_{k=0}^{n} \frac{1}{3^k}\frac{1}{k!}x^k$

4. $p_2(x) = 2 + \frac{1}{12}(x-8) - \frac{1}{288}(x-8)^2$; $\sqrt[3]{9} \approx 2.079\,861\,1$

5. $\dfrac{1}{10}\displaystyle\int_5^{15}\left(20 - \dfrac{x^2}{40}\right)dx = \dfrac{415}{24} \approx \$17.29$

## Exercise 2-1

A  *In Problems 1–4, find $f^{(3)}(x)$.*

1. $f(x) = \dfrac{1}{x}$

2. $f(x) = \ln(1 + x)$

3. $f(x) = e^{-x}$

4. $f(x) = \sqrt{x}$

*In Problems 5–8, find $f^{(4)}(x)$.*

5. $f(x) = \ln(1 + 3x)$

6. $f(x) = e^{5x}$

7. $f(x) = \sqrt{1 + x}$

8. $f(x) = \dfrac{1}{2 + x}$

*In Problems 9–16, find the indicated Taylor polynomial at 0.*

9. $f(x) = e^{-x}; p_4(x)$

10. $f(x) = e^{4x}; p_3(x)$

11. $f(x) = (x + 1)^3; p_4(x)$

12. $f(x) = (1 - x)^4; p_3(x)$

13. $f(x) = \ln(1 + 2x); p_3(x)$

14. $f(x) = \ln(1 + \frac{1}{2}x); p_4(x)$

15. $f(x) = \sqrt[3]{x + 1}; p_3(x)$

16. $f(x) = \sqrt[4]{x + 16}; p_2(x)$

17. (A) Find the third-degree Taylor polynomial $p_3(x)$ for $f(x) = x^4 - 1$ at 0. For which values of $x$ is $|p_3(x) - f(x)| < 0.1$?

   (B) Find the fourth-degree Taylor polynomial $p_4(x)$ for $f(x) = x^4 - 1$ at 0. For which values of $x$ is $|p_4(x) - f(x)| < 0.1$?

18. (A) Find the fourth-degree Taylor polynomial $p_4(x)$ for $f(x) = x^5$ at 0. For which values of $x$ is $|p_4(x) - f(x)| < 0.01$?

   (B) Find the fifth-degree Taylor polynomial $p_5(x)$ for $f(x) = x^5$ at 0. For which values of $x$ is $|p_5(x) - f(x)| < 0.01$?

*In Problems 19–24, find the indicated Taylor polynomial at the given value of a.*

19. $f(x) = e^{x-1}; p_4(x)$ at 1

20. $f(x) = e^{2x}; p_3(x)$ at $\frac{1}{2}$

21. $f(x) = x^3; p_3(x)$ at 1

22. $f(x) = x^2 - 6x + 10; p_2(x)$ at $x = 3$

23. $f(x) = \ln(3x); p_3(x)$ at $\frac{1}{3}$

24. $f(x) = \ln(2 - x); p_4(x)$ at 1

25. Use the third-degree Taylor polynomial at 0 for $f(x) = e^{-2x}$ and $x = 0.25$ to approximate $e^{-0.5}$.

26. Use the fourth-degree Taylor polynomial at 0 for $f(x) = \ln(1 - x)$ and $x = 0.1$ to approximate $\ln 0.9$.

27. Use the second-degree Taylor polynomial at 0 for $f(x) = \sqrt{x + 16}$ and $x = 1$ to approximate $\sqrt{17}$.

28. Use the third-degree Taylor polynomial at 0 for $f(x) = \sqrt{(x + 4)^3}$ and $x = 1$ to approximate $\sqrt{125}$.

29. Use the fourth-degree Taylor polynomial at 1 for $f(x) = \sqrt{x}$ and $x = 1.2$ to approximate $\sqrt{1.2}$.

30. Use the third-degree Taylor polynomial at 4 for $f(x) = \sqrt{x}$ and $x = 3.95$ to approximate $\sqrt{3.95}$.

B  *In Problems 31–36, find $f^{(n)}(x)$.*

31. $f(x) = \dfrac{1}{4 - x}$     32. $f(x) = \dfrac{4}{1 + x}$

33. $f(x) = e^{3x}$     34. $f(x) = \ln(2x + 1)$

35. $f(x) = \ln(6 - x)$     36. $f(x) = e^{x/2}$

*In Problems 37–42, find the nth-degree Taylor polynomial at 0. Write the answer in expanded notation.*

37. $f(x) = \dfrac{1}{4 - x}$     38. $f(x) = \dfrac{4}{1 + x}$

39. $f(x) = e^{3x}$     40. $f(x) = \ln(2x + 1)$

41. $f(x) = \ln(6 - x)$     42. $f(x) = e^{x/2}$

*In Problems 43–50, find the nth-degree Taylor polynomial at the indicated value of a. Write the answer in expanded notation.*

43. $f(x) = \dfrac{1}{x}$; at $-1$     44. $f(x) = \dfrac{2}{x}$; at 1

45. $f(x) = \ln x$; at 1     46. $f(x) = e^x$; at $-2$

47. $f(x) = e^{-x}$; at 2

48. $f(x) = \ln(2 - x)$; at $-1$

49. $f(x) = x \ln x$; at 3

50. $f(x) = xe^x$; at 1

51. Let $f(x) = x^4 + 5x^2 + 1$.

   (A) Find the third-degree Taylor polynomial $p_3(x)$ for $f$ at 0.

   (B) What is the degree of the polynomial $p_3(x)$?

**52.** Let $f(x) = x^5 + 2x^3 + 8x^2 + 1$.

    (A) Find the fourth-degree Taylor polynomial $p_4(x)$ for $f$ at 0.

    (B) What is the degree of the polynomial $p_4(x)$?

**53.** Let $f(x) = x^6 + 2x^3 + 1$. For which values of $n$ does $p_n(x)$, the $n$th-degree Taylor polynomial for $f$ at 0, have degree $n$?

**54.** Let $f(x) = x^4 - 1$. For which values of $n$ does $p_n(x)$, the $n$th-degree Taylor polynomial for $f$ at 0, have degree $n$?

C

**55.** Find the first three Taylor polynomials at 0 for $f(x) = \ln(1 + x)$, and use these polynomials to complete the following two tables. Round all table entries to six decimal places.

| $x$ | $p_1(x)$ | $p_2(x)$ | $p_3(x)$ | $f(x)$ |
|---|---|---|---|---|
| −0.2 | | | | |
| −0.1 | | | | |
| 0 | | | | |
| 0.1 | | | | |
| 0.2 | | | | |

| $x$ | $\lvert p_1(x) - f(x) \rvert$ | $\lvert p_2(x) - f(x) \rvert$ | $\lvert p_3(x) - f(x) \rvert$ |
|---|---|---|---|
| −0.2 | | | |
| −0.1 | | | |
| 0 | | | |
| 0.1 | | | |
| 0.2 | | | |

**56.** Repeat Problem 55 for $f(x) = \sqrt{1 + x}$.

**57.** Refer to Problem 55. Graph $f(x) = \ln(1 + x)$ and its first three Taylor polynomials in the same viewing window. Use $[-3, 3]$ for both the $x$ range and the $y$ range.

**58.** Refer to Problem 56. Graph $f(x) = \sqrt{1 + x}$ and its first three Taylor polynomials in the same viewing window. Use $[-3, 3]$ for both the $x$ range and the $y$ range.

**59.** Consider $f(x) = e^x$ and its third-degree Taylor polynomial $p_3(x)$ at 0. Use graphical approximation techniques to find all values of $x$ such that $\lvert p_3(x) - e^x \rvert < 0.1$.

**60.** Consider $f(x) = \ln(1 + x)$ and its third-degree Taylor polynomial $p_3(x)$ at 0. Use graphical approximation techniques to find all values of $x$ such that

$$\lvert p_3(x) - \ln(1 + x) \rvert < 0.1$$

*In Problems 61–64, write the answer in both expanded notation and summation notation.*

**61.** Find the $n$th-degree Taylor polynomial for $f(x) = e^x$ at any point $a$.

**62.** Find the $n$th-degree Taylor polynomial for $f(x) = \ln x$ at any $a > 0$.

**63.** Find the $n$th-degree Taylor polynomial for $f(x) = 1/x$ at any $a \neq 0$.

**64.** Find the $n$th-degree Taylor polynomial at 0 for $f(x) = (x + c)^n$, where $c$ is a constant.

**65.** Let $f(x)$ be a polynomial. For which values of $n$ is the $n$th-degree Taylor polynomial for $f$ at 0 equal to $f(x)$? Explain.

**66.** Explain how any polynomial expressed in powers of $x$ may be rewritten in powers of $x - 1$ using Taylor polynomials. Illustrate your method for $f(x) = 1 + x^2 + x^4$.

**67.** Let $p_{10}(x)$ be the tenth-degree Taylor polynomial for $f(x) = e^x$ at 0. Do values of $x$ exist for which $\lvert p_{10}(x) - f(x) \rvert \geq 100$? Explain.

**68.** Let $p_{12}(x)$ be the twelfth-degree Taylor polynomial for $f(x) = 1/x$ at 1. Do values of $x \neq 0$ exist for which $\lvert p_{12}(x) - f(x) \rvert \geq 100$? Explain.

## Applications

**69.** *Average price.* Given the demand equation

$$p = D(x) = \tfrac{1}{10}\sqrt{10{,}000 - x^2}$$

use the second-degree Taylor polynomial at 0 to approximate the average price (in dollars) over the demand interval $[0, 30]$.

**70.** *Average price.* Given the demand equation

$$p = D(x) = \tfrac{1}{5}\sqrt{1{,}600 - x^2}$$

use the second-degree Taylor polynomial at 0 to approximate the average price (in dollars) over the demand interval $[0, 15]$.

**71.** *Average price.* Refer to Problem 69. Use the second-degree Taylor polynomial at $a = 60$ to approximate the average price over the demand interval $[60, 80]$.

**72.** *Average price.* Refer to Problem 70. Use the second-degree Taylor polynomial at $a = 24$ to approximate the average price over the demand interval $[24, 32]$.

**73.** *Production.* The rate of production of a mine (in millions of dollars per year) is given by

$$R(t) = 2 + 8e^{-0.1t^2}$$

Use the second-degree Taylor polynomial at 0 to approximate the total production during the first 2 years of operation of the mine.

**74.** *Production.* The rate of production of an oil well (in millions of dollars per year) is given by

$$R(t) = 5 + 10e^{-0.05t^2}$$

Use the second-degree Taylor polynomial at 0 to approximate the total production during the first 3 years of operation of the well.

**75.** *Medicine.* The rate of healing for a skin wound (in square centimeters per day) is given by

$$A'(t) = \frac{-75}{t^2 + 25}$$

The initial wound has an area of 12 square centimeters. Use the second-degree Taylor polynomial at 0 for $A'(t)$ to approximate the area of the wound after 2 days.

**76.** *Medicine.* Rework Problem 75 for $A'(t) = \dfrac{-60}{t^2 + 20}$.

**77.** *Pollution.* On an average summer day in a particular large city the air pollution level (in parts per million) is given by

$$P(x) = 20\sqrt{3x^2 + 25} - 80$$

where $x$ is the number of hours elapsed since 8:00 A.M. Use the second-degree Taylor polynomial for $P(x)$ at $a = 5$ to approximate the average pollution level during the 10-hour period from 8:00 A.M. to 6:00 P.M.

**78.** *Pollution.* On an average summer day in another large city the air pollution level (in parts per million) is given by

$$P(x) = 10\sqrt{4x^2 + 36} - 50$$

where $x$ is the number of hours elapsed since 8:00 A.M. Use the second-degree Taylor polynomial for $P(x)$ at $a = 4$ to approximate the average pollution level during the 8-hour period from 8:00 A.M. to 4:00 P.M.

**79.** *Learning.* In a particular business college, it was found that an average student enrolled in an advanced typing class progresses at a rate of $N'(t) = 6e^{-0.01t^2}$ words per minute per week, $t$ weeks after enrolling in a 15-week course. At the beginning of the course an average student could type 40 words per minute. Use the second-degree Taylor polynomial at 0 for $N'(t)$ to approximate the improvement in typing after 5 weeks in the course.

**80.** *Learning.* In the same business college, it was also found that an average student enrolled in a beginning shorthand class progressed at a rate of $N'(t) = 12e^{-0.005t^2}$ words per minute per week, $t$ weeks after enrolling in a 15-week course. At the beginning of the course none of the students could take any dictation by shorthand. Use the second-degree Taylor polynomial at 0 for $N'(t)$ to approximate the improvement after 5 weeks in the course.

*In Problems 81–88, use a graphing utility to graph the given function and its second-degree Taylor polynomial at a in the indicated viewing window.*

**81.** *Average price.* From Problem 69,

$p = D(x) = \frac{1}{10}\sqrt{10{,}000 - x^2}$

$a = 0$

$x$ range: $[0, 100]$

$p$ range: $[0, 10]$

**82.** *Average price.* From Problem 70,

$p = D(x) = \frac{1}{5}\sqrt{1{,}600 - x^2}$

$a = 0$

$x$ range: $[0, 40]$

$p$ range: $[0, 8]$

**83.** *Production.* From Problem 73,

$y = R(t) = 2 + 8e^{-0.1t^2}$

$a = 0$

$t$ range: $[0, 5]$

$y$ range: $[0, 10]$

**84.** *Production.* From Problem 74,

$y = R(t) = 5 + 10e^{-0.05t^2}$

$a = 0$

$t$ range: $[0, 5]$

$y$ range: $[0, 15]$

**85.** *Pollution.* From Problem 77,

$y = P(x) = 20\sqrt{3x^2 + 25} - 80$

$a = 5$

$x$ range: $[0, 10]$

$y$ range: $[0, 300]$

**86.** *Pollution.* From Problem 78,

$y = P(x) = 10\sqrt{4x^2 + 36} - 50$

$a = 4$

$x$ range: $[0, 10]$

$y$ range: $[0, 200]$

**87.** *Learning.* From Problem 79,

$y = N'(t) = 6e^{-0.01t^2}$

$a = 0$

$t$ range: $[0, 10]$

$y$ range: $[0, 6]$

**88.** *Learning.* From Problem 80,

$y = N'(t) = 12e^{-0.005t^2}$

$a = 0$

$t$ range: $[0, 15]$

$y$ range: $[0, 12]$

# Section 2-2  TAYLOR SERIES

- Introduction
- Taylor Series
- Representation of Functions by Taylor Series

## ■ Introduction

If $f$ is a function with derivatives of all order at a point $a$, then we can construct the Taylor polynomial $p_n$ at $a$ for any integer $n$. Now we are interested in the relationship between the original function $f$ and the corresponding Taylor polynomial $p_n$ as $n$ assumes larger and larger values. If the Taylor polynomial is to be a useful tool for approximating functions, then the accuracy of the approximation should improve as we increase the size of $n$. Indeed, Figures 1, 2, and 3 and Tables 1 and 2 in the preceding section indicate that this is the case for the function $e^x$.

For a given value of $x$, we would like to know whether we can make $p_n(x)$ arbitrarily close to $f(x)$ by making $n$ sufficiently large. In other words, we want to know whether

$$\lim_{n \to \infty} p_n(x) = f(x) \tag{1}$$

It turns out that for most functions with derivatives of all order at a point, there is a set of values of $x$ for which equation (1) is valid. We will begin by considering a specific example to illustrate this.

Let

$$f(x) = \frac{1}{1-x} \quad \text{and} \quad a = 0$$

First, we find $p_n$:

**Step 1**

$$f(x) = (1-x)^{-1}$$

$$f'(x) = (-1)(1-x)^{-2}(-1) = (1-x)^{-2}$$

$$f''(x) = (-2)(1-x)^{-3}(-1) = 2(1-x)^{-3}$$

$$f^{(3)}(x) = (-3)(2)(1-x)^{-4}(-1) = 3!(1-x)^{-4}$$

$$\vdots$$

$$f^{(n)}(x) = n!(1-x)^{-n-1}$$

**Step 2**

$$f(0) = 1$$

$$f'(0) = 1$$

$$f''(0) = 2$$

$$f^{(3)}(0) = 3!$$

$$\vdots$$

$$f^{(n)}(0) = n!$$

**Step 3**

$$a_0 = f(0) = 1$$

$$a_1 = f'(0) = 1$$

$$a_2 = \frac{f''(0)}{2!} = \frac{2}{2!} = 1$$

$$a_3 = \frac{f^{(3)}(0)}{3!} = \frac{3!}{3!} = 1$$

$$\vdots$$

$$a_n = \frac{f^{(n)}(0)}{n!} = \frac{n!}{n!} = 1$$

**Step 4.**  *Write down the Taylor polynomial:*

$$p_n(x) = 1 + x + x^2 + x^3 + \cdots + x^n \tag{2}$$

$$= \sum_{k=0}^{n} x^k$$

Now we want to evaluate

$$\lim_{n \to \infty} p_n(x) = \lim_{n \to \infty} \sum_{k=0}^{n} x^k$$

As we saw in Chapter 6 of *Calculus* 11/e or Chapter 13 of *College Mathematics* 11/e, it is not possible to evaluate limits written in this form. The difficulty lies in the fact that the number of terms is increasing as $n$ increases. First, we must find a closed form for the summation that does not involve a sum of $n$ terms. [You may recognize that $p_n(x)$ is a finite geometric series with common ratio $x$. See Appendix B-2 of *Calculus* 11/e or *College Mathematics* 11/e].

If we multiply both sides of (2) by $x$ and subtract this new equation from the original, we obtain an equation that we can solve for $p_n(x)$. Notice how many terms drop out when the subtraction is performed.

$$
\begin{aligned}
p_n(x) &= 1 + x + x^2 + \cdots + x^{n-1} + x^n \\
xp_n(x) &= \quad\;\; x + x^2 + \cdots + x^{n-1} + x^n + x^{n+1} \\
\hline
p_n(x) - xp_n(x) &= 1 \qquad\qquad\qquad\qquad\qquad\quad - x^{n+1} \\
p_n(x)(1-x) &= 1 - x^{n+1} \\
p_n(x) &= \frac{1 - x^{n+1}}{1 - x} \qquad x \neq 1
\end{aligned}
$$

Since we had to exclude the value $x = 1$ to avoid division by 0, we must consider $x = 1$ as a special case:

$$
p_n(1) = \overbrace{1 + 1 + 1^2 + \cdots + 1^n}^{n+1\ \text{terms}}
$$
$$
= n + 1
$$

Thus,

$$
p_n(x) = \begin{cases} \dfrac{1 - x^{n+1}}{1 - x} & \text{if } x \neq 1 \\[2mm] n + 1 & \text{if } x = 1 \end{cases}
$$

If $x = 1$, it is clear that

$$
\lim_{n \to \infty} p_n(1) = \lim_{n \to \infty} (n + 1) = \infty
$$

That is, $\lim_{n \to \infty} p_n(1)$ does not exist. What happens to this limit for other values of $x$? It can be shown that

$$
\lim_{n \to \infty} x^{n+1} = \begin{cases} 0 & \text{if } -1 < x < 1 \\ 1 & \text{if } x = 1 \\ \text{Does not exist} & \text{if } x \leq -1 \text{ or } x > 1 \end{cases}
$$

Experimenting with a calculator for various values of $x$ and large values of $n$ should convince you that this statement is valid.

Thus,

$$
\lim_{n \to \infty} p_n(x) = \begin{cases} \dfrac{1}{1 - x} & \text{if } -1 < x < 1 \\ \text{Does not exist} & \text{if } x = 1 \\ \text{Does not exist} & \text{if } x \leq -1 \text{ or } x > 1 \end{cases}
$$

$\lim\limits_{n \to \infty} x^{n+1} = 0$

$\lim\limits_{n \to \infty} (n + 1)$ *does not exist*

$\lim\limits_{n \to \infty} x^{n+1}$ *does not exist*

## *Explore & Discuss* 1

(A) Construct a table with eight rows labeled with values of $n$ from 1 to 8, and seven columns labeled with the following values of $x$: $-1.5, -1, -0.5, 0, 0.5, 1$ and $1.5$. Complete the table by computing each entry $p_n(x)$.

(B) Explain how the table from part (A) provides supporting evidence for the statement that $\lim_{n \to \infty} p_n(x)$ equals $1/(1 - x)$ if $-1 < x < 1$, but does not exist if $x \leq -1$ or $x \geq 1$.

We can now conclude that

$$
\begin{aligned}
f(x) &= \frac{1}{1 - x} \\
&= \lim_{n \to \infty} p_n(x) \qquad -1 < x < 1 \\
&= \lim_{n \to \infty} \sum_{k=0}^{n} x^k
\end{aligned}
$$

and that the limit does not exist for any other values of $x$. This is often abbreviated by writing

$$\frac{1}{1-x} = 1 + x + x^2 + \cdots + x^n + \cdots \qquad -1 < x < 1 \qquad (3)$$

$$= \sum_{k=0}^{\infty} x^k$$

### INSIGHT

Equation (3) is often described in words as follows: The sum of an infinite geometric series, having first term 1 and common ratio $x$, is equal to $1/(1 - x)$, provided that the absolute value of the common ratio is less than 1.

The expression $\sum_{k=0}^{\infty} x^k$ is called the *Taylor series for f at* 0, and the expanded notation

$$1 + x + x^2 + \cdots + x^n + \cdots$$

is just another way to write this series. The Taylor series is said to *converge at x* if $\lim_{n \to \infty} p_n(x)$ exists and to *diverge at x* if this limit fails to exist. Thus, $\sum_{k=0}^{\infty} x^k$ converges for $-1 < x < 1$ and diverges for $x \le -1$ or $x \ge 1$. The set of values $\{x \mid -1 < x < 1\} = (-1, 1)$ where this Taylor series converges is called the *interval of convergence*. Since $f(x)$ is equal to the Taylor series for any $x$ in the interval of convergence, we say that $f$ is represented by its Taylor series throughout this interval. That is,

$$f(x) = \frac{1}{1-x}$$
$$= 1 + x + x^2 + \cdots + x^n + \cdots \qquad -1 < x < 1$$

There can be no relationship between $f$ and its Taylor series outside the interval of convergence since the series is not defined there (see Fig. 1).

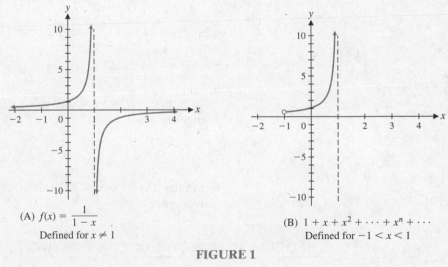

(A) $f(x) = \dfrac{1}{1-x}$

Defined for $x \ne 1$

(B) $1 + x + x^2 + \cdots + x^n + \cdots$

Defined for $-1 < x < 1$

**FIGURE 1**

### Explore & Discuss 2

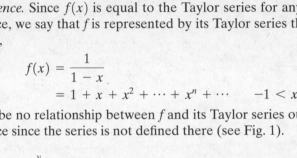

**FIGURE 2**

(A) The six functions $p_n(x) = 1 + x + \cdots + x^n$, $n = 1, 2, \ldots, 6$, are graphed in Figure 2. Which graph belongs to which function?

(B) Based solely on the graphs, would you conjecture that the interval of convergence of the Taylor series representation

$$\frac{1}{1-x} = 1 + x + x^2 + \cdots + x^n + \cdots$$

is $-1 < x < 1$? Explain.

## ■ Taylor Series

We now generalize the concepts introduced in the preceding discussion to arbitrary functions.

---

**DEFINITION** **Taylor Series**

If $f$ is a function with derivatives of all order at a point $a$, $p_n(x)$ is the $n$th-degree Taylor polynomial for $f$ at $a$, and $a_n = f^{(n)}(a)/n!$, $n = 0, 1, 2, \ldots$, then

$$\sum_{k=0}^{\infty} a_k(x - a)^k = a_0 + a_1(x - a) + a_2(x - a)^2 + \cdots + a_n(x - a)^n + \cdots$$

is the **Taylor series for $f$ at $a$.** The Taylor series **converges at $x$** if

$$\lim_{n \to \infty} p_n(x) = \lim_{n \to \infty} \sum_{k=0}^{n} a_k(x - a)^k$$

exists and **diverges at $x$** if this limit does not exist. The set of values of $x$ for which this limit exists is called the **interval of convergence.**

---

We must emphasize that we are not really adding up an infinite number of terms in a Taylor series, nor is a Taylor series an "infinite" polynomial. Rather, for $x$ in the interval of convergence,

$$\sum_{k=0}^{\infty} a_k(x - a)^k = \lim_{n \to \infty} \sum_{k=0}^{n} a_k(x - a)^k$$

Thus, when we write

$$\frac{1}{1 - x} = 1 + x + x^2 + \cdots + x^n + \cdots \qquad -1 < x < 1$$

we mean

$$\frac{1}{1 - x} = \lim_{n \to \infty} (1 + x + x^2 + \cdots + x^n) \qquad -1 < x < 1$$

In general, it is very difficult to find the interval of convergence by directly evaluating

$$\lim_{n \to \infty} \sum_{k=0}^{n} a_k(x - a)^k$$

For many Taylor series, the following theorem can be used to find the interval of convergence.

---

**THEOREM 1** **INTERVAL OF CONVERGENCE**

Let $f$ be a function with derivatives of all order at a point $a$, let $a_n = f^{(n)}(a)/n!$, $n = 0, 1, 2, \ldots$, and let

$$\sum_{k=0}^{\infty} a_k(x - a)^k = a_0 + a_1(x - a) + a_2(x - a)^2 + \cdots + a_n(x - a)^n + \cdots$$

be the Taylor series for $f$ at $a$. If $a_n \neq 0$ for $n \geq n_0$, then:

**Case 1.** If

$$\lim_{n \to \infty} \left| \frac{a_{n+1}}{a_n} \right| = L > 0 \qquad \text{and} \qquad R = \frac{1}{L}$$

then the series converges for $|x - a| < R$ and diverges for $|x - a| > R$.

---

**Case 2.**  If

$$\lim_{n\to\infty}\left|\frac{a_{n+1}}{a_n}\right| = 0$$

then the series converges for all values of $x$.

**Case 3.**  If

$$\lim_{n\to\infty}\left|\frac{a_{n+1}}{a_n}\right| = \infty$$

then the series converges only at $x = a$.

The various possibilities for the interval of convergence are illustrated in Figure 3. In case 1, the series may or may not converge at the end points $x = a - R$ and $x = a + R$. (Determination of the behavior of a series at the end points of the interval of convergence requires techniques that we will not discuss.)

**Case 1.**  $\displaystyle\lim_{n\to\infty}\left|\frac{a_{n+1}}{a_n}\right| = L > 0$

$R = 1/L$

**Case 2.**  $\displaystyle\lim_{n\to\infty}\left|\frac{a_{n+1}}{a_n}\right| = 0$

**Case 3.**  $\displaystyle\lim_{n\to\infty}\left|\frac{a_{n+1}}{a_n}\right| = \infty$

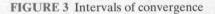

**FIGURE 3**  Intervals of convergence

Notice that the set of values where a series converges can be expressed in interval notation as $(a - R, a + R)$ in case 1 and as $(-\infty, \infty)$ in case 2. We were anticipating this result when we used the term *interval of convergence* to describe the set of points where a Taylor series converges.

Finally, the requirement that all the coefficients be nonzero from some point on ensures that the ratio $a_{n+1}/a_n$ is well defined.

REMARK
It is possible that

$$\lim_{n\to\infty}\left|\frac{a_{n+1}}{a_n}\right|$$

fails to exist so that none of the cases in Theorem 1 hold. In that event, the interval of convergence will still have one of the forms illustrated by Figure 3, but other techniques are required to determine which one.

INSIGHT

Assume that $a_n \neq 0$ for $n \geq n_0$ and consider case 1 of Theorem 1. The ratio of consecutive terms of the Taylor series is equal to

$$\frac{a_{n+1}(x - a)^{n+1}}{a_n(x - a)^n}$$

The absolute value of this ratio for large $n$, by the limit in case 1, is approximately $L|x - a|$. Therefore, the Taylor series is approximately an infinite geometric series with common ratio having absolute value $L|x - a|$. Such a series converges if $L|x - a| < 1$, or equivalently, if $|x - a| < 1/L = R$.

The next two examples illustrate the application of Theorem 1. The details of finding the Taylor series are omitted. In the next section we will discuss methods that greatly simplify the process of finding Taylor series. For now, we want to concentrate on understanding the significance of the interval of convergence.

**EXAMPLE 1**  **Finding the Interval of Convergence** The Taylor series at 0 for $f(x) = e^{2x}$ is

$$1 + 2x + \frac{2^2}{2!}x^2 + \cdots + \frac{2^n}{n!}x^n + \cdots$$

Find the interval of convergence.

**SOLUTION**
$$a_n = \frac{2^n}{n!} \qquad \text{Notice that } a_n > 0 \text{ for } n \geq 0.$$

$$a_{n+1} = \frac{2^{n+1}}{(n+1)!}$$

$$\frac{a_{n+1}}{a_n} = \frac{\dfrac{2^{n+1}}{(n+1)!}}{\dfrac{2^n}{n!}} \qquad \text{Form the ratio } a_{n-1}/a_n \text{ and simplify.}$$

$$= \frac{2^{n+1}}{(n+1)!} \cdot \frac{n!}{2^n} \qquad \frac{n!}{(n+1)!} = \frac{n!}{(n+1)n!} = \frac{1}{n+1}$$

$$= \frac{2}{n+1}$$

$$\lim_{n\to\infty}\left|\frac{a_{n+1}}{a_n}\right| = \lim_{n\to\infty}\left|\frac{2}{n+1}\right| \qquad \text{Absolute value signs can be dropped since } a_{n+1}/a_n > 0.$$

$$= \lim_{n\to\infty}\frac{2}{n+1} = 0*$$

Case 2 in Theorem 1 applies, and the series converges for all values of $x$. ■

**MATCHED PROBLEM 1** The Taylor series at 0 for $f(x) = e^{x/3}$ is

$$1 + \frac{1}{3}x + \frac{1}{3^2 2!}x^2 + \cdots + \frac{1}{3^n n!}x^n + \cdots$$

Find the interval of convergence.

**EXAMPLE 2**  **Finding the Interval of Convergence** The Taylor series at 0 for $f(x) = 1/(1+5x)^2$ is

$$1 - 2\cdot 5x + 3\cdot 5^2 x^2 - \cdots + (-1)^n(n+1)5^n x^n + \cdots$$

Find the interval of convergence.

---

*The limit expressions $\lim_{x\to\infty} g(x)$ and $\lim_{n\to\infty} g(n)$ are very similar. If $\lim_{x\to\infty} g(x) = L$ [that is, $g(x)$ can be made arbitrarily close to $L$ by making the *real number x* sufficiently large], then $\lim_{n\to\infty} g(n) = L$ [that is, $g(n)$ can be made arbitrarily close to $L$ by making the *integer n* sufficiently large]. In particular,

$$\lim_{n\to\infty}\frac{2}{n+1} = \lim_{x\to\infty}\frac{2}{x+1} = 0$$

Thus, the results of Section 5-3 of *Calculus* 11/e or Section 12-3 of *College Mathematics* 11/e on limits at infinity for rational functions can be helpful in evaluating limits of the form $\lim_{n\to\infty} g(n)$.

**SOLUTION**

$$a_n = (-1)^n(n + 1)5^n \qquad \textit{Notice that } a_n \neq 0 \textit{ for } n \geq 0.$$

$$a_{n+1} = (-1)^{n+1}(n + 2)5^{n+1}$$

$$\frac{a_{n+1}}{a_n} = \frac{(-1)^{n+1}(n + 2)5^{n+1}}{(-1)^n(n + 1)5^n}$$

$$= \frac{-5(n + 2)}{n + 1}$$

$$\lim_{n \to \infty} \left| \frac{a_{n+1}}{a_n} \right| = \lim_{n \to \infty} \left| \frac{-5(n + 2)}{n + 1} \right|$$

$$= \lim_{n \to \infty} \frac{5(n + 2)}{n + 1}$$

$$= \lim_{n \to \infty} \frac{5n + 10}{n + 1} \qquad \textit{Divide numerator and denominator by } n.$$

$$= \lim_{n \to \infty} \frac{5 + \dfrac{10}{n}}{1 + \dfrac{1}{n}} = 5$$

Applying case 1 in Theorem 1, we have $L = 5$, $R = \frac{1}{5}$, and the series converges for $-\frac{1}{5} < x < \frac{1}{5}$.

---

**MATCHED PROBLEM 2**

The Taylor series at 0 for $f(x) = 4/(2 + x)^2$ is

$$1 - \frac{2}{2}x + \frac{3}{2^2}x^2 - \cdots + \frac{(-1)^n(n + 1)}{2^n}x^n + \cdots$$

Find the interval of convergence.

---

## ■ Representation of Functions by Taylor Series

Referring to the function $f(x) = 1/(1 - x)$ discussed earlier in this section, our calculations showed that the Taylor series at 0 for $f$ is

$$\lim_{n \to \infty} p_n(x) = 1 + x + x^2 + \cdots + x^n + \cdots$$

with interval of convergence $(-1, 1)$. Furthermore, we showed that $f$ is represented by its Taylor series throughout the interval of convergence. That is,

$$f(x) = \frac{1}{1 - x} = \lim_{n \to \infty} p_n(x)$$

$$= 1 + x + x^2 + \cdots + x^n + \cdots \qquad -1 < x < 1$$

A similar statement can be made for most functions that have derivatives of all order at a point $a$, but not for all. For example, the absolute value function $g(x) = |x|$ has Taylor series (verify this)

$$1 + (x - 1) = x$$

at the point $a = 1$. Clearly, this very simple Taylor series converges for all $x$, but $g(x) = x$ only for $x \geq 0$. So $g$ is not represented by its Taylor series throughout the interval of convergence of the series. The problem with $g$ is that it is not differentiable at 0. Even if we require that a function $f$ has derivatives of all order at every point in the interval of convergence of its Taylor series, it is still possible that $f$ is not represented by its Taylor series. Since we will not consider such functions, we will make the following assumption.

AGREEMENT

> **Assumption on the Representation of Functions by Taylor Series**
>
> If $f$ is a function with derivatives of all order throughout the interval of convergence of its Taylor series at $a$, then **$f$ is represented by its Taylor series throughout the interval of convergence.** Thus, if $p_n(x)$ is the $n$th-degree Taylor polynomial at $a$ for $f$ and $a_n = f^{(n)}(a)/n!$, $n = 0, 1, 2, \ldots$, then
>
> $$f(x) = \lim_{n \to \infty} p_n(x) = \sum_{k=0}^{\infty} a_k(x - a)^k$$
>
> for any $x$ in the interval of convergence of the Taylor series at $a$ for $f$.

Consequently, the interval of convergence determines the values of $x$ for which the Taylor polynomials $p_n(x)$ can be used to approximate the values of the function $f$.

Referring to Examples 1 and 2, we can now write

$$e^{2x} = 1 + 2x + \frac{2^2}{2!}x^2 + \cdots + \frac{2^n}{n!}x^n + \cdots \qquad -\infty < x < \infty$$

and

$$\frac{1}{(1 + 5x)^2} = 1 - 2 \cdot 5x + 3 \cdot 5^2 x^2 - \cdots + (-1)^n(n + 1)5^n x^n + \cdots \qquad -\tfrac{1}{5} < x < \tfrac{1}{5}$$

---

**EXAMPLE 3** **Representation of a Function by Its Taylor Series** Let $f(x) = \ln x$.

(A) Find the $n$th-degree Taylor polynomial at $a = 1$ for $f$.

(B) Find the Taylor series at $a = 1$ for $f$.

(C) Determine the values of $x$ for which $f(x) = \lim\limits_{n \to \infty} p_n(x)$.

SOLUTION

(A) We use the four-step process from the preceding section to find the $n$th-degree Taylor polynomial:

**Step 1**

$$f(x) = \ln x$$

$$f'(x) = \frac{1}{x} = x^{-1}$$

$$f''(x) = (-1)x^{-2}$$

$$f^{(3)}(x) = (-1)(-2)x^{-3}$$
$$= (-1)^2 2! x^{-3}$$

$$f^{(4)}(x) = (-1)(-2)(-3)x^{-4}$$
$$= (-1)^3 3! x^{-4}$$
$$\vdots$$

$$f^{(n)}(x) = (-1)^{n-1}(n - 1)! x^{-n}$$

**Step 2**

$$f(1) = \ln 1 = 0$$

$$f'(1) = 1$$

$$f''(1) = -1$$

$$f^{(3)}(1) = (-1)^2 2!$$

$$f^{(4)}(1) = (-1)^3 3!$$
$$\vdots$$

$$f^{(n)}(1) = (-1)^{n-1}(n - 1)!$$

**Step 3**

$$a_0 = f(1) \quad = 0$$

$$a_1 = f'(1) \quad = 1$$

$$a_2 = \frac{f''(1)}{2!} = -\frac{1}{2}$$

$$a_3 = \frac{f^{(3)}(1)}{3!} = \frac{(-1)^2 2!}{3!} = \frac{1}{3}$$

$$a_4 = \frac{f^{(4)}(1)}{4!} = \frac{(-1)^3 3!}{4!} = \frac{-1}{4}$$
$$\vdots$$

$$a_n = \frac{f^{(n)}(1)}{n!} = \frac{(-1)^{n-1}(n - 1)!}{n!} = \frac{(-1)^{n-1}}{n}$$

**Step 4.**   The $n$th-degree Taylor polynomial at $a = 1$ for $f(x) = \ln x$ is

$$p_n(x) = (x - 1) - \frac{1}{2}(x - 1)^2 + \frac{1}{3}(x - 1)^3 - \cdots + \frac{(-1)^{n-1}}{n}(x - 1)^n$$

(B) Once the $n$th-degree Taylor polynomial for a function has been determined, finding the Taylor series is simply a matter of using the notation correctly. Using the Taylor polynomial from part (A), the Taylor series at $a = 1$ for $f(x) = \ln x$ can be written as

$$\lim_{n \to \infty} p_n(x) = (x - 1) - \frac{1}{2}(x - 1)^2 + \frac{1}{3}(x - 1)^3 - \cdots + \frac{(-1)^{n-1}}{n}(x - 1)^n + \cdots$$

or, using summation notation,* as

$$\lim_{n \to \infty} p_n(x) = \sum_{k=0}^{\infty} \frac{(-1)^{k-1}}{k}(x - 1)^k$$

(C) To determine the values for which $f(x) = \lim_{n \to \infty} p_n(x)$, we use Theorem 1 to find the interval of convergence of the Taylor series in part (B):

$$a_n = \frac{(-1)^{n-1}}{n}$$

$$a_{n+1} = \frac{(-1)^n}{n + 1}$$

$$\frac{a_{n+1}}{a_n} = \frac{\dfrac{(-1)^n}{n + 1}}{\dfrac{(-1)^{n-1}}{n}} = \frac{(-1)^n}{n + 1} \cdot \frac{n}{(-1)^{n-1}} = \frac{-n}{n + 1}$$

$$\lim_{n \to \infty} \left| \frac{a_{n+1}}{a_n} \right| = \lim_{n \to \infty} \left| \frac{-n}{n + 1} \right|$$

$$= \lim_{n \to \infty} \frac{n}{n + 1} \qquad \textit{Divide numerator and denominator by n.}$$

$$= \lim_{n \to \infty} \frac{1}{1 + \dfrac{1}{n}} = 1$$

From case 1 of Theorem 1, $L = 1$, $R = 1/L = 1$, and the series converges for $|x - 1| < 1$. Converting this to double inequalities, we have

$$|x - 1| < 1 \qquad \textit{|y| < c is equivalent to } -c < y < c.$$
$$-1 < x - 1 < 1 \qquad \textit{Add 1 to each side.}$$
$$0 < \quad x \quad < 2 \qquad \textit{Interval of convergence.}$$

Thus, $f(x) = \lim_{n \to \infty} p_n(x)$ for $0 < x < 2$ or, equivalently,

$$\ln x = (x - 1) - \frac{1}{2}(x - 1)^2 + \frac{1}{3}(x - 1)^3 - \cdots + \frac{(-1)^{n-1}}{n}(x - 1)^n + \cdots \qquad 0 < x < 2$$

■

---

*From this point on, we will use the expanded notation exclusively when writing Taylor series.

MATCHED PROBLEM 3 | Repeat Example 3 for $f(x) = 1/x$.

Table 1 displays the values of the Taylor polynomials for $f(x) = \ln x$ at $x = 1.5$ and at $x = 3$ for $n = 1, 2, \ldots, 10$ (see Example 3). Notice that $x = 1.5$ is in the interval of convergence of the Taylor series at 1 for $f(x)$ and that the values of $p_n(1.5)$ are approaching $\ln 1.5$ as $n$ increases. On the other hand, $x = 3$ is *not* in the interval of convergence and the values of $p_n(3)$ do not approach $\ln 3$. These calculations illustrate the importance of the interval of convergence.

The values of $x$ must be in the interval of convergence in order to use Taylor polynomials to approximate the values of a function.

**TABLE 1**

$f(x) = \ln x$

$$p_n(x) = (x - 1) - \frac{1}{2}(x - 1)^2 + \frac{1}{3}(x - 1)^3 - \cdots + \frac{(-1)^{n-1}}{n}(x - 1)^n$$

| $n$ | $p_n(1.5)$ | $p_n(3)$ |
|---|---|---|
| 1 | 0.5 | 2 |
| 2 | 0.375 | 0 |
| 3 | 0.416 667 | 2.666 667 |
| 4 | 0.401 042 | −1.333 333 |
| 5 | 0.407 292 | 5.066 667 |
| 6 | 0.404 688 | −5.6 |
| 7 | 0.405 804 | 12.685 714 |
| 8 | 0.405 315 | −19.314 286 |
| 9 | 0.405 532 | 37.574 603 |
| 10 | 0.405 435 | −64.825 397 |
| | $\ln 1.5 = 0.405\ 465$ | $\ln 3 = 1.098\ 612$ |

*Answers to Matched Problems*

**1.** $-\infty < x < \infty$

**2.** $-2 < x < 2$

**3.** (A) $1 - (x - 1) + (x - 1)^2 - \cdots + (-1)^n(x - 1)^n$

    (B) $1 - (x - 1) + (x - 1)^2 - \cdots + (-1)^n(x - 1)^n + \cdots$

    (C) $0 < x < 2$

# Exercise 2-2

**A** *In Problems 1–12, find the interval of convergence of the given Taylor series representation.*

**1.** $\dfrac{4}{1 - x} = 4 + 4x + 4x^2 + \cdots + 4x^n + \cdots$

**2.** $\dfrac{3}{1 - x} = 3 + 3x + 3x^2 + \cdots + 3x^n + \cdots$

**3.** $\dfrac{1}{1 + 7x} = 1 - 7x + 7^2x^2 - \cdots + (-1)^n 7^n x^n + \cdots$

**4.** $\dfrac{1}{1 + 2x} = 1 - 2x + 2^2x^2 - \cdots + (-1)^n 2^n x^n + \cdots$

**5.** $\dfrac{1}{6 - x} = \dfrac{1}{6} + \dfrac{1}{6^2}x + \dfrac{1}{6^3}x^2 + \cdots + \dfrac{1}{6^{n+1}}x^n + \cdots$

**6.** $\dfrac{2}{3 + x} = \dfrac{2}{3} - \dfrac{2}{3^2}x + \dfrac{2}{3^3}x^2 - \cdots + \dfrac{2(-1)^n}{3^{n+1}}x^n + \cdots$

**7.** $\ln(1 + x) = x - \dfrac{1}{2}x^2 + \dfrac{1}{3}x^3 - \cdots + \dfrac{(-1)^{n-1}}{n}x^n + \cdots$

**8.** $\ln(1 - x) = -x - \dfrac{1}{2}x^2 - \dfrac{1}{3}x^3 - \cdots - \dfrac{1}{n}x^n - \cdots$

**9.** $\dfrac{1}{(1 + x)^2} = 1 - 2x + 3x^2 - \cdots$
    $+ (-1)^n(n + 1)x^n + \cdots$

**10.** $\dfrac{x}{(1 - x)^2} = x + 2x^2 + 3x^3 + \cdots + nx^n + \cdots$

**11.** $e^x = 1 + x + \dfrac{1}{2!}x^2 + \cdots + \dfrac{1}{n!}x^n + \cdots$

**12.** $e^{-x} = 1 - x + \dfrac{1}{2!}x^2 - \cdots + \dfrac{(-1)^n}{n!}x^n + \cdots$

**13.** The Taylor series for $e^x$ at 0 is
$$e^x = 1 + x + \dfrac{1}{2!}x^2 + \dfrac{1}{3!}x^3 + \cdots + \dfrac{1}{n!}x^n + \cdots$$

(A) Find the smallest value of $n$ such that

$$|p_n(2) - e^2| < 0.1$$

(B) Does a value of $n$ exist such that

$$|p_n(100) - e^{100}| < 0.1? \text{ Explain.}$$

14. Suppose the function $f$ has the following Taylor series representation at 0:

$$f(x) = 1 + x + 2!x^2 + 3!x^3 + \cdots + n!x^n + \cdots$$

(A) Find the interval of convergence.

(B) Compute $p_n(0.2)$ for $n = 1, 2, 3, \ldots, 10$, where $p_n(x)$ is the $n$th-degree Taylor polynomial for $f$ at 0. Explain why your computations are consistent with your answer to part (A).

**B**  *In Problems 15–22, find the interval of convergence of the given Taylor series representation.*

15. $\ln(4 + x) = (x + 3) - \dfrac{1}{2}(x + 3)^2$

$$+ \dfrac{1}{3}(x + 3)^3 - \cdots + \dfrac{(-1)^{n+1}}{n}(x + 3)^n + \cdots$$

16. $\ln(9 - 2x) = -2(x - 4) - \dfrac{2^2}{2}(x - 4)^2$

$$- \dfrac{2^3}{3}(x - 4)^3 - \cdots - \dfrac{2^n}{n}(x - 4)^n - \cdots$$

17. $e^{2x} = 1 + 2x + \dfrac{2^2}{2!}x^2 + \cdots + \dfrac{2^n}{n!}x^n + \cdots$

18. $e^{-x/3} = 1 - \dfrac{1}{3}x + \dfrac{1}{2!3^2}x^2 - \cdots + \dfrac{(-1)^n}{n!3^n}x^n + \cdots$

19. $\dfrac{1}{6 - x} = \dfrac{1}{4} + \dfrac{1}{4^2}(x - 2) + \dfrac{1}{4^3}(x - 2)^2 + \cdots$

$$+ \dfrac{1}{4^{n+1}}(x - 2)^n + \cdots$$

20. $\dfrac{3}{2 + 3x} = -3 - 3^2(x + 1) - 3^3(x + 1)^2 - \cdots$

$$- 3^{n+1}(x + 1)^n - \cdots$$

21. $\dfrac{1}{6 - x} = -\dfrac{1}{2} + \dfrac{1}{2^2}(x - 8) - \dfrac{1}{2^3}(x - 8)^2$

$$+ \cdots + \dfrac{(-1)^n}{2^{n+1}}(x - 8)^n + \cdots$$

22. $\dfrac{3}{2 + 3x} = \dfrac{3}{5} - \dfrac{3^2}{5^2}(x - 1) + \dfrac{3^3}{5^3}(x - 1)^2 - \cdots$

$$+ \dfrac{(-1)^n 3^{n+1}}{5^{n+1}}(x - 1)^n + \cdots$$

23. (A) Graph the $n$th-degree Taylor polynomials at 0 for $f(x) = e^x, n = 1, 2, 3, 4, 5$. Use the viewing window Xmin $= -3$, Xmax $= 3$, Ymin $= -5$, and Ymax $= 15$. Based solely on the graphs, what would you conjecture the interval of convergence to be? Explain.

(B) What is the actual interval of convergence?

24. (A) Graph the $n$th-degree Taylor polynomials at 0 for $f(x) = \ln(1 + x), n = 1, 2, 3, 4, 5$. Use the viewing window Xmin $= -3$, Xmax $= 3$, Ymin $= -15$, and Ymax $= 15$. Based solely on the graphs, what would you conjecture the interval of convergence to be? Explain.

(B) What is the actual interval of convergence?

**C**  *In Problems 25–30, find the $n$th-degree Taylor polynomial at 0 for $f$, find the Taylor series at 0 for $f$, and determine the values of $x$ for which $f(x) = \lim\limits_{n \to \infty} p_n(x)$.*

25. $f(x) = e^{4x}$  26. $f(x) = e^{-x/2}$

27. $f(x) = \ln(1 + 2x)$  28. $f(x) = \ln(1 - 4x)$

29. $f(x) = \dfrac{2}{2 - x}$  30. $f(x) = \dfrac{3}{3 + x}$

*In Problems 31–34, find the $n$th-degree Taylor polynomial at the indicated value of $a$ for $f$, find the Taylor series at $a$ for $f$, and determine the values of $x$ for which $f(x) = \lim\limits_{n \to \infty} p_n(x)$.*

31. $f(x) = \ln(2 - x); a = 1$

32. $f(x) = \ln(2 + x); a = -1$

33. $f(x) = \dfrac{1}{1 - x}; a = 2$

34. $f(x) = \dfrac{1}{4 - x}; a = 3$

35. (A) Find the interval of convergence of the Taylor series representation

$$\dfrac{1}{4 - x} = \dfrac{1}{4} + \dfrac{1}{4^2}x + \dfrac{1}{4^3}x^2 + \cdots + \dfrac{1}{4^{n+1}}x^n + \cdots$$

(B) Explain why Theorem 1 is not directly applicable to the Taylor series representation

$$\dfrac{1}{4 - x^2} = \dfrac{1}{4} + \dfrac{1}{4^2}x^2 + \dfrac{1}{4^3}x^4 + \cdots + \dfrac{1}{4^{n+1}}x^{2n} + \cdots$$

(C) Use part (A) to find the interval of convergence of the Taylor series of part (B).

36. (A) Explain why Theorem 1 is not directly applicable to the Taylor series representation

$$e^{x^2} = 1 + x^2 + \dfrac{1}{2!}x^4 + \dfrac{1}{3!}x^6 + \cdots + \dfrac{1}{n!}x^{2n} + \cdots$$

(B) Use the Taylor series for $e^x$ at 0 to find the interval of convergence of the Taylor series of part (A).

37. (A) Explain why the interval of convergence of the Taylor series

$$\dfrac{1}{1 - x} + \dfrac{1}{1 - x^2} = 2 + x + 2x^2 + x^3$$

$$+ 2x^4 + x^5 + \cdots$$

cannot be determined by applying Theorem 1.

(B) If two Taylor series have the same interval of convergence, then so does their term-by-term sum. Use this fact to find the interval of convergence of the Taylor series of part (A).

38. (A) Explain why the interval of convergence of the Taylor series

$$e^x - e^{x^2} = x + \left(\dfrac{1}{2!} - 1\right)x^2 + \dfrac{1}{3!}x^3$$

$$+ \left(\dfrac{1}{4!} - \dfrac{1}{2!}\right)x^4 + \dfrac{1}{5!}x^5 + \cdots$$

cannot be determined by applying Theorem 1.

(B) If two Taylor series have the same interval of convergence, then so does their term-by-term difference. Use this fact to find the interval of convergence of the Taylor series of part (A).

## Section 2-3 OPERATIONS ON TAYLOR SERIES

- Basic Taylor Series
- Addition and Multiplication
- Differentiation and Integration
- Substitution

### ■ Basic Taylor Series

In the preceding sections, we used repeated differentiation and the formula $a_n = f^{(n)}(a)/n!$ to find Taylor polynomials and Taylor series for a variety of simple functions. In this section we will see how to use the series for these simple functions and some basic properties of Taylor series to find the Taylor series for more complicated functions. Table 1 lists the series we will use in this process for convenient reference.

Most of the discussion in this section involves Taylor series at 0. At the end of the section we will see that Taylor series at points other than 0 can be obtained from Taylor series at 0 by using a simple substitution process.

**TABLE 1** Basic Taylor Series

| Function $f(x)$ | Taylor Series at 0 | Interval of Convergence |
|---|---|---|
| $\dfrac{1}{1-x}$ | $= 1 + x + x^2 + \cdots + x^n + \cdots$ | $-1 < x < 1$ |
| $\dfrac{1}{1+x}$ | $= 1 - x + x^2 - \cdots + (-1)^n x^n + \cdots$ | $-1 < x < 1$ |
| $\ln(1-x)$ | $= -x - \dfrac{1}{2}x^2 - \dfrac{1}{3}x^3 - \cdots - \dfrac{1}{n}x^n - \cdots$ | $-1 < x < 1$ |
| $\ln(1+x)$ | $= x - \dfrac{1}{2}x^2 + \dfrac{1}{3}x^3 - \cdots + \dfrac{(-1)^{n-1}}{n}x^n + \cdots$ | $-1 < x < 1$ |
| $e^x$ | $= 1 + x + \dfrac{1}{2!}x^2 + \cdots + \dfrac{1}{n!}x^n + \cdots$ | $-\infty < x < \infty$ |
| $e^{-x}$ | $= 1 - x + \dfrac{1}{2!}x^2 - \cdots + \dfrac{(-1)^n}{n!}x^n + \cdots$ | $-\infty < x < \infty$ |

### ■ Addition and Multiplication

**PROPERTY 1** | **Addition**

Two Taylor series at 0 can be added term by term:

If
$$f(x) = a_0 + a_1 x + a_2 x^2 + \cdots + a_n x^n + \cdots$$
and
$$g(x) = b_0 + b_1 x + b_2 x^2 + \cdots + b_n x^n + \cdots$$
then
$$f(x) + g(x) = (a_0 + b_0) + (a_1 + b_1)x + (a_2 + b_2)x^2 + \cdots$$
$$+ (a_n + b_n)x^n + \cdots$$

This operation is valid in the intersection of the intervals of convergence of the series for $f$ and $g$.

**EXAMPLE 1** **Addition of Taylor Series** Find the Taylor series at 0 for $f(x) = e^x + \ln(1 + x)$, and find the interval of convergence.

**SOLUTION** From Table 1,

$$e^x = 1 + x + \frac{1}{2!}x^2 + \frac{1}{3!}x^3 + \cdots + \frac{1}{n!}x^n + \cdots \qquad -\infty < x < \infty$$

and

$$\ln(1 + x) = x - \frac{1}{2}x^2 + \frac{1}{3}x^3 - \cdots + \frac{(-1)^{n-1}}{n}x^n + \cdots \qquad -1 < x < 1$$

Adding the coefficients of corresponding powers of $x$, we have

$$f(x) = e^x + \ln(1 + x)$$

$$= 1 + 2x + \left(\frac{1}{2!} - \frac{1}{2}\right)x^2 + \left(\frac{1}{3!} + \frac{1}{3}\right)x^3 + \cdots$$

$$+ \left[\frac{1}{n!} + \frac{(-1)^{n-1}}{n}\right]x^n + \cdots$$

The series for $e^x$ converges for all $x$; however, the series for $\ln(1 + x)$ converges only for $-1 < x < 1$. Thus, the combined series converges for $-1 < x < 1$, the intersection of these two intervals of convergence. ▬

**MATCHED PROBLEM 1** Find the Taylor series at 0 for $f(x) = e^x + 1/(1 - x)$, and find the interval of convergence.

---

**PROPERTY 2**   **Multiplication**

A Taylor series at 0 can be multiplied term by term by an expression of the form $cx^r$, where $c$ is a nonzero constant and $r$ is a non-negative integer:

If

$$f(x) = a_0 + a_1x + a_2x^2 + \cdots + a_nx^n + \cdots$$

then

$$cx^r f(x) = ca_0x^r + ca_1x^{r+1} + ca_2x^{r+2} + \cdots + ca_nx^{r+n} + \cdots$$

The Taylor series for $cx^r f(x)$ has the same interval of convergence as the Taylor series for $f$.

---

**EXAMPLE 2**   **Multiplication of Taylor Series** Find the Taylor series at 0 for $f(x) = 2x^2e^{-x}$, and find the interval of convergence.

**SOLUTION** From Table 1,

$$e^{-x} = 1 - x + \frac{1}{2!}x^2 - \cdots + \frac{(-1)^n}{n!}x^n + \cdots \qquad -\infty < x < \infty$$

Multiplying each term of this series by $2x^2$, the series for $f(x)$ is

$$f(x) = 2x^2e^{-x}$$

$$= 2x^2\left[1 - x + \frac{1}{2!}x^2 - \cdots + \frac{(-1)^n}{n!}x^n + \cdots\right]$$

$$= 2x^2 - 2x^3 + \frac{2}{2!}x^4 - \cdots + \frac{2(-1)^n}{n!}x^{n+2} + \cdots$$

Since the series for $e^{-x}$ converges for all $x$, the series for $f(x)$ also converges for $-\infty < x < \infty$. ▬

MATCHED PROBLEM 2   Find the Taylor series at 0 for $f(x) = 3x^3 \ln(1 - x)$, and find the interval of convergence.

## ■ Differentiation and Integration

PROPERTY 3   **Differentiation**

A Taylor series at 0 can be differentiated term by term:

If

$$f(x) = a_0 + a_1x + a_2x^2 + a_3x^3 + \cdots + a_nx^n + \cdots$$

then

$$f'(x) = a_1 + 2a_2x + 3a_3x^2 + \cdots + na_nx^{n-1} + \cdots$$

The Taylor series for $f'$ has the same interval of convergence as the Taylor series for $f$.

---

### INSIGHT

Assume that $a_n \neq 0$ for $n \geq n_0$. Then the ratio of consecutive coefficients in the Taylor series for $f'$ is

$$\frac{(n + 1)a_{n+1}}{na_n}$$

and

$$\lim_{n\to\infty}\left|\frac{(n + 1)a_{n+1}}{na_n}\right| = \lim_{n\to\infty}\left|\frac{a_{n+1}}{a_n}\right| \quad \lim_{n\to\infty}\frac{n + 1}{n} = 1$$

Therefore, by Theorem 1 of the preceding section, the Taylor series for $f$ and $f'$ have the same interval of convergence.

---

EXAMPLE 3   **Differentiation of Taylor Series** Find the Taylor series at 0 for $f(x) = 1/(1 - x)^2$, and find the interval of convergence.

SOLUTION   We want to relate $f$ to the derivative of one of the functions in Table 1, so that we can apply property 3. Since

$$\frac{d}{dx}\left(\frac{1}{1 - x}\right) = \frac{d}{dx}(1 - x)^{-1}$$
$$= -(1 - x)^{-2}(-1)$$
$$= \frac{1}{(1 - x)^2}$$
$$= f(x)$$

we can apply property 3 to the series for $1/(1 - x)$. From Table 1,

$$\frac{1}{1 - x} = 1 + x + x^2 + x^3 + \cdots + x^n + \cdots \qquad -1 < x < 1$$

Differentiating both sides of this equation,

$$\frac{d}{dx}\left(\frac{1}{1 - x}\right) = \frac{d}{dx}(1 + x + x^2 + x^3 + \cdots + x^n + \cdots)$$
$$\frac{1}{(1 - x)^2} = \frac{d}{dx}1 + \frac{d}{dx}x + \frac{d}{dx}x^2 + \frac{d}{dx}x^3 + \cdots + \frac{d}{dx}x^n + \cdots$$
$$f(x) = 1 + 2x + 3x^2 + \cdots + nx^{n-1} + \cdots$$

Since the series for $1/(1 - x)$ converges for $-1 < x < 1$, the series for $f(x)$ also converges for $-1 < x < 1$. [Verify this statement by applying Theorem 1 in the preceding section to the series for $f(x)$.]

| MATCHED PROBLEM 3 | Find the Taylor series at 0 for $f(x) = 1/(1 + x)^2$, and find the interval of convergence.

*Explore & Discuss* **1**

The Taylor series representation for a function $f$ at 0 is

$$f(x) = x - \frac{1}{3!}x^3 + \frac{1}{5!}x^5 - \frac{1}{7!}x^7 + \cdots$$

$$+ \frac{(-1)^n}{(2n + 1)!}x^{2n+1} + \cdots \qquad -\infty < x < \infty$$

(A)  Find the Taylor series for $f'(x)$.

(B)  Find the Taylor series for $f''(x)$. How are $f$ and $f''$ related?

---

**PROPERTY 4  Integration**

A Taylor series at 0 can be integrated term by term:

If

$$f(x) = a_0 + a_1 x + a_2 x^2 + \cdots + a_n x^n + \cdots$$

then

$$\int f(x)\, dx = C + a_0 x + \frac{1}{2}a_1 x^2 + \frac{1}{3}a_2 x^3 + \cdots + \frac{1}{n + 1}a_n x^{n+1} + \cdots$$

where $C$ is the constant of integration. The Taylor series for $\int f(x)\, dx$ has the same interval of convergence as the Taylor series for $f$.

---

**EXAMPLE 4**

**Integration of Taylor Series** If $f$ is a function that satisfies $f'(x) = x^2 e^x$ and $f(0) = 2$, find the Taylor series at 0 for $f$. Find the interval of convergence.

**SOLUTION**  Using the series for $e^x$ from Table 1 and property 2, the series for $f'$ is

$$f'(x) = x^2 e^x$$

$$= x^2\left(1 + x + \frac{1}{2!}x^2 + \cdots + \frac{1}{n!}x^n + \cdots\right) \qquad -\infty < x < \infty$$

$$= x^2 + x^3 + \frac{1}{2!}x^4 + \cdots + \frac{1}{n!}x^{n+2} + \cdots$$

Integrating term by term produces a series for $f$:

$$f(x) = \int f'(x)\, dx$$

$$= \int \left(x^2 + x^3 + \frac{1}{2!}x^4 + \cdots + \frac{1}{n!}x^{n+2} + \cdots\right) dx$$

$$= \int x^2\, dx + \int x^3\, dx + \frac{1}{2!}\int x^4\, dx + \cdots + \frac{1}{n!}\int x^{n+2}\, dx + \cdots$$

$$= C + \frac{1}{3}x^3 + \frac{1}{4}x^4 + \frac{1}{(5)2!}x^5 + \cdots + \frac{1}{(n + 3)n!}x^{n+3} + \cdots$$

Now we use the condition $f(0) = 2$ to evaluate the constant of integration $C$:

$$2 = f(0)$$
$$= C + 0 + 0 + \cdots + 0 + \cdots$$
$$= C$$

Thus,

$$f(x) = 2 + \frac{1}{3}x^3 + \frac{1}{4}x^4 + \cdots + \frac{1}{(n+3)n!}x^{n+3} + \cdots$$

Since the series for $f'$ converges for $-\infty < x < \infty$, the series for $f$ also converges for $-\infty < x < \infty$. ▬

---

**MATCHED PROBLEM 4**   If $f$ is a function that satisfies $f'(x) = x \ln(1 + x)$ and $f(0) = 4$, find the Taylor series at 0 for $f$. Find the interval of convergence.

---

## ■ Substitution

**EXAMPLE 5**   **Using Substitution to Find Taylor Series**   Find the Taylor series at 0 for $f(x) = e^{-x^2}$, and find the interval of convergence.

**SOLUTION**   Suppose we try to solve this problem by finding the general form of the $n$th derivative of $f$:

$$f(x) = e^{-x^2}$$
$$f'(x) = -2xe^{-x^2}$$
$$f''(x) = -2e^{-x^2} + 4x^2e^{-x^2}$$
$$f^{(3)}(x) = 4xe^{-x^2} + 8xe^{-x^2} - 8x^3e^{-x^2}$$
$$= 12xe^{-x^2} - 8x^3e^{-x^2}$$
$$f^{(4)}(x) = 12e^{-x^2} - 24x^2e^{-x^2} - 24x^2e^{-x^2} + 16x^4e^{-x^2}$$
$$= 12e^{-x^2} - 48x^2e^{-x^2} + 16x^4e^{-x^2}$$

Since the higher-order derivatives are becoming very complicated and no general pattern is emerging, we will try another approach. How can we relate $f$ to one of the functions in Table 1? If we let $g(x) = e^{-x}$, then $f$ and $g$ are related by

$$f(x) = e^{-x^2} = g(x^2)$$

From Table 1,

$$g(x) = e^{-x}$$
$$= 1 - x + \frac{1}{2!}x^2 - \frac{1}{3!}x^3 + \cdots + \frac{(-1)^n}{n!}x^n + \cdots \quad -\infty < x < \infty$$

Substituting $x^2$ for $x$ in the series for $g$, we have

$$f(x) = g(x^2)$$
$$= 1 - x^2 + \frac{1}{2!}(x^2)^2 - \frac{1}{3!}(x^2)^3 + \cdots + \frac{(-1)^n}{n!}(x^2)^n + \cdots$$
$$= 1 - x^2 + \frac{1}{2!}x^4 - \frac{1}{3!}x^6 + \cdots + \frac{(-1)^n}{n!}x^{2n} + \cdots$$

Since the series for $g$ converges for all values of $x$, the series for $f$ must also converge for all values of $x$. ▬

*Explore & Discuss* **2**

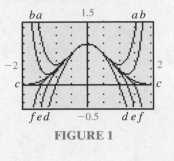

**FIGURE 1**

(A) The function $f(x) = e^{-x^2}$ and the Taylor polynomials

$$1 - x^2 + \frac{1}{2!}x^4 - \cdots + \frac{(-1)^n}{n!}x^{2n}$$

are graphed for $n = 1, 2, 3, 4, 5$ in Figure 1. Which curve belongs to which function?

(B) Does Figure 1 provide supporting evidence for the statement that the Taylor series for $f$ at 0 converges for all values of $x$? Explain.

| MATCHED PROBLEM 5 | Find the Taylor series at 0 for $f(x) = e^{x^3}$, and find the interval of convergence. |

Making a substitution in a known Taylor series to obtain a new series is a very useful technique. Since there are many different substitutions that can be used, it is difficult to make a general statement concerning the effect of a substitution on the interval of convergence. The following examples illustrate some of the possibilities that may occur.

**EXAMPLE 6**   **Using Substitution to Find Taylor Series**   Find the Taylor series at 0 for $f(x) = 1/(4 - x)$, and find the interval of convergence.

**SOLUTION**   If we factor a 4 out of the denominator of $f$, we can establish a relationship between $f(x)$ and $g(x) = 1/(1 - x)$. Thus,

$$f(x) = \frac{1}{4 - x}$$

$$= \left(\frac{1}{4}\right)\frac{1}{1 - (x/4)} \qquad g(x) = \frac{1}{1 - x}$$

$$= \frac{1}{4}g\left(\frac{x}{4}\right)$$

From Table 1,

$$g(x) = \frac{1}{1 - x}$$

$$= 1 + x + x^2 + \cdots + x^n + \cdots \qquad -1 < x < 1$$

Substituting $x/4$ for $x$ in this series and multiplying by $\frac{1}{4}$, we have

$$f(x) = \frac{1}{4}g\left(\frac{x}{4}\right)$$

$$= \frac{1}{4}\left[1 + \left(\frac{x}{4}\right) + \left(\frac{x}{4}\right)^2 + \cdots + \left(\frac{x}{4}\right)^n + \cdots\right]$$

$$= \frac{1}{4} + \frac{1}{4^2}x + \frac{1}{4^3}x^2 + \cdots + \frac{1}{4^{n+1}}x^n + \cdots$$

Since the original series for $g$ converges for $-1 < x < 1$ and we substituted $x/4$ for $x$ in that series, the series for $f$ converges for

$$-1 < \frac{x}{4} < 1 \quad \text{Multiply each member by 4.}$$

$$-4 < x < 4$$

**MATCHED PROBLEM 6**   Find the Taylor series at 0 for $f(x) = 1/(3 + x)$, and find the interval of convergence.

---

**EXAMPLE 7**   **Using Substitution to Find Taylor Series**   Find the Taylor series at 0 for $f(x) = \ln(1 + 4x^2)$, and find the interval of convergence.

**SOLUTION**   If we let $g(x) = \ln(1 + x)$, then

$$f(x) = \ln(1 + 4x^2)$$
$$= g(4x^2)$$

From Table 1, the series for $g$ is

$$g(x) = x - \frac{1}{2}x^2 + \frac{1}{3}x^3 - \cdots + \frac{(-1)^{n-1}}{n}x^n + \cdots \qquad -1 < x < 1$$

Substituting $4x^2$ for $x$ in this series, we have

$$f(x) = g(4x^2)$$
$$= 4x^2 - \frac{1}{2}(4x^2)^2 + \frac{1}{3}(4x^2)^3 - \cdots + \frac{(-1)^{n-1}}{n}(4x^2)^n + \cdots$$
$$= 4x^2 - \frac{4^2}{2}x^4 + \frac{4^3}{3}x^6 - \cdots + \frac{(-1)^{n-1}4^n}{n}x^{2n} + \cdots$$

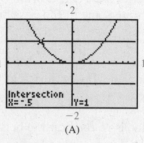

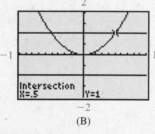

**FIGURE 2** Graphical solution of $-1 < 4x^2 < 1$

The series for $g$ converges for $-1 < x < 1$. Since we substituted $4x^2$ for $x$, the series for $f$ converges for

$$-1 < 4x^2 < 1$$

Inequalities of this type are easier to solve if we use absolute value notation:

| | |
|---|---|
| $-1 < 4x^2 < 1$ | Change to absolute value notation. |
| $\lvert 4x^2 \rvert < 1$ | Multiply by $\frac{1}{4}$. |
| $\lvert x^2 \rvert < \frac{1}{4}$ | |
| $\lvert x \rvert^2 < \frac{1}{4}$ | |
| $\lvert x \rvert < \frac{1}{2}$ | Convert to double inequalities. |
| $-\frac{1}{2} < x < \frac{1}{2}$ | See Figure 2 for a graphical solution. |

Thus, the series for $f$ converges for $-\frac{1}{2} < x < \frac{1}{2}$.

---

**MATCHED PROBLEM 7**   Find the Taylor series at 0 for $f(x) = 1/(1 + 9x^2)$, and find the interval of convergence.

---

Up to this point in this section we have restricted our attention to Taylor series at 0. How can we use the techniques discussed to find a Taylor series at a point $a \neq 0$? Properties 1–4 could be stated in terms of Taylor series at an arbitrary point $a$; however, there is an easier way to proceed. The method of substitution allows us to use Taylor series at 0 to find Taylor series at other points. The following example illustrates this technique.

---

**EXAMPLE 8**   **Using Substitution to Find Taylor Series**   Find the Taylor series at 1 for $f(x) = 1/(2 - x)$, and find the interval of convergence.

**SOLUTION**   In order to find a Taylor series for $f$ in powers of $x - 1$, we will use the substitution $t = x - 1$ to express $f$ as a function of $t$. If we find the Taylor series at 0 for this new function and then replace $t$ with $x - 1$, we will have obtained the Taylor series at 1 for $f$.

$$t = x - 1$$    Solve for x.

$$x = t + 1$$    Substitute for x in $f(x) = 1/(2 - x)$.

$$\frac{1}{2 - x} = \frac{1}{2 - (t + 1)}$$

$$= \frac{1}{1 - t}$$    Find the Taylor series at 0 for this function of t.

$$= 1 + t + t^2 + \cdots + t^n + \cdots$$    Substitute x − 1 for t.

$$-1 < t < 1$$

$$= 1 + (x - 1) + (x - 1)^2 + \cdots + (x - 1)^n + \cdots$$

Since $t = x - 1$ and the series in powers of $t$ converges for $-1 < t < 1$, the series in powers of $x - 1$ converges for

$$-1 < x - 1 < 1$$    Add 1 to each member.

$$0 < \quad x \quad < 2$$

▬

**MATCHED PROBLEM 8**   Find the Taylor series at 1 for $f(x) = 1/x$, and find the interval of convergence.

*Answers to Matched Problems*

**1.** $2 + 2x + \left(\frac{1}{2!} + 1\right)x^2 + \cdots + \left(\frac{1}{n!} + 1\right)x^n + \cdots, -1 < x < 1$

**2.** $-3x^4 - \frac{3}{2}x^5 - x^6 - \cdots - \frac{3}{n}x^{n+3} - \cdots, -1 < x < 1$

**3.** $1 - 2x + 3x^2 - \cdots + (-1)^{n+1}nx^{n-1} + \cdots, -1 < x < 1$

**4.** $4 + \frac{1}{3}x^3 - \frac{1}{8}x^4 + \frac{1}{15}x^5 - \cdots + \frac{(-1)^{n-1}}{n(n + 2)}x^{n+2} + \cdots, -1 < x < 1$

**5.** $1 + x^3 + \frac{1}{2!}x^6 + \cdots + \frac{1}{n!}x^{3n} + \cdots, -\infty < x < \infty$

**6.** $\frac{1}{3} - \frac{1}{3^2}x + \frac{1}{3^3}x^2 - \cdots + \frac{(-1)^n}{3^{n+1}}x^n + \cdots, -3 < x < 3$

**7.** $1 - 9x^2 + 9^2x^4 - \cdots + (-1)^n9^nx^{2n} + \cdots, -\frac{1}{3} < x < \frac{1}{3}$

**8.** $1 - (x - 1) + (x - 1)^2 - \cdots + (-1)^n(x - 1)^n + \cdots, 0 < x < 2$

# Exercise 2-3

*Solve all the problems in this exercise by performing operations on the Taylor series in Table 1. State the interval of convergence for each series you find.*

**A** *In Problems 1–12, find the Taylor series at 0.*

**1.** $f(x) = \frac{1}{1 - x} + \frac{1}{1 + x}$

**2.** $f(x) = e^x + e^{-x}$

**3.** $f(x) = \frac{1}{1 - x} + e^{-x}$

**4.** $f(x) = \frac{1}{1 + x} + \ln(1 + x)$

**5.** $f(x) = \frac{x^3}{1 + x}$

**6.** $f(x) = xe^x$

**7.** $f(x) = x^2 \ln(1 - x)$

**8.** $f(x) = \frac{5x^4}{1 - x}$

**9.** $f(x) = e^{x^2}$

**10.** $f(x) = \ln(1 - x^3)$

**11.** $f(x) = \ln(1 + 3x)$

**12.** $f(x) = \frac{x}{1 + x^2}$

B   *In Problems 13–20, find the Taylor series at 0.*

13. $f(x) = \dfrac{1}{2 - x}$

14. $f(x) = \dfrac{1}{5 + x}$

15. $f(x) = \dfrac{1}{1 - 8x^3}$

16. $f(x) = \ln(1 + 16x^2)$
17. $f(x) = 10^x$
18. $f(x) = \log(1 + x)$
19. $f(x) = \log_2(1 - x)$
20. $f(x) = 3^x$

21. $f(x) = \dfrac{1}{4 + x^2}$

22. $f(x) = \dfrac{1}{9 - x^2}$

23. Substituting $\sqrt{x}$ for $x$ in the Taylor series at 0 for $e^x$ gives the formula

$$e^{\sqrt{x}} = 1 + x^{1/2} + \frac{1}{2!}x + \frac{1}{3!}x^{3/2} + \cdots + \frac{1}{n!}x^{n/2} + \cdots$$

(A) For which values of $x$ is the formula valid?
(B) Is the formula a Taylor series at 0 for $e^{\sqrt{x}}$? Explain.

24. Substituting $1/x$ for $x$ in the Taylor series at 0 for $1/(1 - x)$ gives the formula

$$\frac{1}{1 - \dfrac{1}{x}} = 1 + x^{-1} + x^{-2} + \cdots + x^{-n} + \cdots$$

(A) For which values of $x$ is the formula valid?
(B) Explain why the formula is not a Taylor series at 0 for

$$f(x) = \frac{1}{1 - \dfrac{1}{x}}$$

(C) Find a Taylor series at 0 that equals

$$\frac{1}{1 - \dfrac{1}{x}}$$

for all $x \neq 0$ in the interval $-1 < x < 1$.

25. Find the Taylor series at 0 for

(A) $f(x) = \dfrac{1}{1 - x^2}$

(B) $g(x) = \dfrac{2x}{(1 - x^2)^2}$

[*Hint:* Compare $f'(x)$ and $g(x)$.]

26. Find the Taylor series at 0 for

(A) $f(x) = \dfrac{x}{1 - x^2}$

(B) $g(x) = \dfrac{1 + x^2}{(1 - x^2)^2}$

[*Hint:* Compare $f'(x)$ and $g(x)$.]

27. If $f(x)$ satisfies $f'(x) = 1/(1 + x^2)$ and $f(0) = 0$, find the Taylor series at 0 for $f(x)$.

28. If $f(x)$ satisfies $f'(x) = \ln(1 + x^2)$ and $f(0) = 1$, find the Taylor series at 0 for $f(x)$.

29. If $f(x)$ satisfies $f'(x) = x^2 \ln(1 - x)$ and $f(0) = 5$, find the Taylor series at 0 for $f(x)$.

30. If $f(x)$ satisfies $f'(x) = xe^x$ and $f(0) = -3$, find the Taylor series at 0 for $f(x)$.

31. Most graphing calculators include the hyperbolic sine function, $f(x) = \sinh x$, as a built-in function. The Taylor series for the hyperbolic sine at 0 is

$$\sinh x = x + \frac{1}{3!}x^3 + \frac{1}{5!}x^5 + \cdots$$
$$+ \frac{1}{(2n + 1)!}x^{2n+1} + \cdots \qquad -\infty < x < \infty$$

(A) Graph $f$ and its Taylor polynomials $p_1(x)$, $p_3(x)$, and $p_5(x)$ in the viewing window $-5 \le x \le 5$, $-30 \le y \le 30$.

(B) Explain why $\dfrac{d^2}{dx^2}(\sinh x) = \sinh x$.

32. Most graphing calculators include the hyperbolic cosine function, $g(x) = \cosh x$, as a built-in function. The Taylor series for the hyperbolic cosine at 0 is

$$\cosh x = 1 + \frac{1}{2!}x^2 + \frac{1}{4!}x^4 + \frac{1}{6!}x^6 + \cdots$$
$$+ \frac{1}{(2n)!}x^{2n} + \cdots \qquad -\infty < x < \infty$$

(A) Graph $g$ and its Taylor polynomials $p_2(x)$, $p_4(x)$, and $p_6(x)$ in the viewing window $-5 \le x \le 5$, $0 \le y \le 60$.

(B) Explain why $\dfrac{d^2}{dx^2}(\cosh x) = \cosh x$.

*In Problems 33–38, use the substitution $t = x - a$ to find the Taylor series at the indicated value of $a$.*

33. $f(x) = \dfrac{1}{4 - x}$; at 3

34. $f(x) = \dfrac{1}{3 + x}$; at $-2$

35. $f(x) = \ln x$; at 1
36. $f(x) = \ln(3 - x)$; at 2

37. $f(x) = \dfrac{1}{4 - 3x}$; at 1

38. $f(x) = \dfrac{1}{5 - 2x}$; at 2

C

39. Use the Taylor series at 0 for $1/(1 - x)$ and repeated applications of property 3 to find the Taylor series at 0 for

$$f(x) = \frac{1}{(1 - x)^3}$$

40. Use the Taylor series at 0 for $1/(1 + x)$ and repeated applications of property 3 to find the Taylor series at 0 for

$$f(x) = \frac{1}{(1 + x)^4}$$

41. Suppose that the Taylor series for $f$ at 0 is

$$f(x) = x + \frac{1}{3}x^3 + \frac{1}{5}x^5 + \cdots$$
$$+ \frac{1}{2n + 1}x^{2n+1} + \cdots \qquad -1 < x < 1$$

(A) Find the Taylor series for $f'$.

(B) Explain why $f'(x) = \dfrac{1}{1 - x^2}$ for $-1 < x < 1$.

**42.** Suppose that the Taylor series for $g$ at 0 is

$$g(x) = x - \frac{1}{3}x^3 + \frac{1}{5}x^5 - \cdots$$
$$+ \frac{(-1)^n}{2n + 1}x^{2n+1} + \cdots \quad -1 < x < 1$$

(A) Find the Taylor series for $g'$.

(B) Explain why $g'(x) = \dfrac{1}{1 + x^2}$ for $-1 < x < 1$.

**43.** Find the Taylor series at 0 for $f(x) = \dfrac{1 + x}{1 - x}$.

$$\left[ Note: \frac{1 + x}{1 - x} = \frac{1}{1 - x} + \frac{x}{1 - x} \right]$$

**44.** Find the Taylor series at 0 for $f(x) = \dfrac{1 - 2x}{1 + x}$.

$$\left[ Note: \frac{1 - 2x}{1 + x} = \frac{1}{1 + x} - \frac{2x}{1 + x} \right]$$

**45.** Find the Taylor series at 0 for $f(x) = \dfrac{1}{2}\ln\left(\dfrac{1 + x}{1 - x}\right)$.

$$\left[ Note: \ln\left(\frac{1 + x}{1 - x}\right) = \ln(1 + x) - \ln(1 - x) \right]$$

**46.** Find the Taylor series at 0 for $f(x) = \ln(1 - 2x + x^2)$.

**47.** Find the Taylor series at 0 for $f(x) = \dfrac{e^x + e^{-x}}{2}$.

**48.** Find the Taylor series at 0 for $f(x) = \dfrac{e^x - e^{-x}}{2}$.

**49.** If $f(x)$ satisfies $f''(x) = \ln(1 + x)$, $f'(0) = 3$, and $f(0) = -2$, find the Taylor series at 0 for $f(x)$.

**50.** If $f(x)$ satisfies $f''(x) = \dfrac{1}{1 - x^2}$, $f'(0) = 4$, and $f(0) = 5$, find the Taylor series at 0 for $f(x)$.

**51.** Find the Taylor series at any $a > 0$ for $f(x) = \ln x$.

**52.** Find the Taylor series at any $a$ for $f(x) = e^x$.

**53.** If $a$ and $b$ are constants ($b \neq 0, a \neq b^{-1}$), find the Taylor series at $a$ for

$$f(x) = \frac{1}{1 - bx}$$

**54.** If $a$ and $b$ are constants ($a \neq b$), find the Taylor series at $a$ for

$$f(x) = \frac{1}{b - x}$$

# Section 2-4  APPROXIMATIONS USING TAYLOR SERIES

- The Remainder
- Taylor's Formula for the Remainder
- Taylor Series with Alternating Terms
- Approximating Definite Integrals

## The Remainder

Now that we can find Taylor series for a variety of functions, we return to our original goal: approximating the values of a function.

If $x$ is in the interval of convergence of the Taylor series for a function $f$ and $p_n(x)$ is the $n$th-degree Taylor polynomial, then

$$f(x) = \lim_{n \to \infty} p_n(x)$$

and $p_n(x)$ can be used to approximate $f(x)$. We want to consider two questions:

**1.** If we select a particular value of $n$, how accurate is the approximation $f(x) \approx p_n(x)$?

**2.** If we want the approximation $f(x) \approx p_n(x)$ to have a specified accuracy, how do we select the proper value of $n$?

It turns out that both of these questions can be answered by examining the difference between $f(x)$ and $p_n(x)$. This difference is called the *remainder* and is defined in the following box.

| DEFINITION | The Remainder of a Taylor Series |
|---|---|

If $p_n$ is the $n$th-degree Taylor polynomial for $f$, then the **remainder** is

$$R_n(x) = f(x) - p_n(x)$$

The **error** in the approximation $f(x) \approx p_n(x)$ is

$$|f(x) - p_n(x)| = |R_n(x)|$$

If the Taylor series at 0 for $f$ is

$$f(x) = \overbrace{a_0 + a_1x + a_2x^2 + \cdots + a_nx^n}^{p_n(x)} + \overbrace{a_{n+1}x^{n-1} + \cdots}^{R_n(x)}$$

then

$$p_n(x) = a_0 + a_1x + a_2x^2 + \cdots + a_nx^n$$

and

$$R_n(x) = a_{n+1}x^{n+1} + a_{n+2}x^{n+2} + \cdots$$

Similar statements can be made for Taylor series at $a$.

It follows from our basic assumption (see page 71) for the functions we consider that

$$\lim_{n \to \infty} R_n(x) = \lim_{n \to \infty}[f(x) - p_n(x)] = f(x) - f(x) = 0$$

if and only if $x$ is in the interval of convergence of $f$.*

In general, it is difficult to find the exact value of $R_n(x)$. In fact, since $f(x) = p_n(x) + R_n(x)$, this is equivalent to finding the exact value of $f(x)$. Instead, we will discuss two methods for *estimating* the value of $R_n(x)$. The first method works in all cases, but can be difficult to apply, while the second method is easy to apply, but does not work in all cases.

*Explore & Discuss* **1**

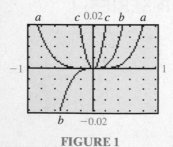

**FIGURE 1**

Let $p_1(x)$, $p_2(x)$, and $p_3(x)$ be Taylor polynomials for $f(x) = e^x$ at 0. The corresponding remainder functions $R_1(x)$, $R_2(x)$, and $R_3(x)$ are graphed in Figure 1.

(A) Which curve belongs to which function?

(B) Use graphical approximation techniques to estimate those values of $x$ for which the error in each of the approximations $f(x) \approx p_n(x)$ is less than 0.01.

## ■ Taylor's Formula for the Remainder

The first method for estimating the remainder of a Taylor series is based on *Taylor's formula for the remainder*.

| FORMULA | Taylor's Formula for the Remainder |
|---|---|

If $f$ has derivatives of all order at 0, then

$$R_n(x) = \frac{f^{(n+1)}(t)x^{n+1}}{(n+1)!}$$

for some number $t$ between 0 and $x$.

---

* In more advanced texts, the statement $\lim_{n \to \infty} R_n(x) = 0$ for $x$ in the interval of convergence is actually proved for functions such as $e^x$, making our basic assumption unnecessary.

A similar formula can be stated for the remainder of a Taylor series at an arbitrary point $a$. (We will not use this formula in this text.)

In most applications of the remainder formula, it is not possible to find the value of $t$. However, if we can find a number $M$ satisfying

$$|f^{(n+1)}(t)| < M \qquad \text{for all } t \text{ between 0 and } x$$

then we can estimate $R_n(x)$ as follows:

$$|R_n(x)| = \left| \frac{f^{(n+1)}(t)x^{n+1}}{(n+1)!} \right| < \frac{M|x|^{n+1}}{(n+1)!}$$

This technique is illustrated in the next example.

---

**EXAMPLE 1**

**Taylor's Formula for the Remainder**  Use the second-degree Taylor polynomial at 0 for $f(x) = e^x$ to approximate $e^{0.1}$. Use Taylor's formula for the remainder to estimate the error in this approximation.

**SOLUTION**  Since $f^{(n)}(x) = e^x$ for any $n$, we can write

$$f(x) = e^x = p_2(x) + R_2(x)$$

$$= \underbrace{1 + x + \frac{1}{2!}x^2}_{p_2(x)} + \underbrace{\frac{e^t x^3}{3!}}_{R_2(x)} \quad f^{(3)}(t) = e^t$$

for some number $t$ between 0 and $x$. Thus,

$$f(0.1) = e^{0.1}$$
$$= p_2(0.1) + R_2(0.1)$$
$$\approx p_2(0.1)$$
$$= 1 + 0.1 + \tfrac{1}{2}(0.1)^2$$
$$= 1.105$$

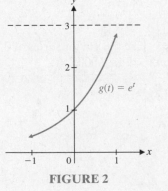

**FIGURE 2**

To estimate the error in this approximation, we must estimate

$$|R_2(0.1)| = \left| \frac{e^t (0.1)^3}{3!} \right| = \frac{e^t}{6,000}$$

where $0 \le t \le 0.1$. In order to estimate $|R_2(0.1)|$, we must estimate $g(t) = e^t$ for $0 \le t \le 0.1$. Since $e^t$ is always increasing, $e^t \le e^{0.1}$ for $0 \le t \le 0.1$. However, $e^{0.1}$ is the number we are trying to approximate. We do not want to use this number in our estimate of the error. (This situation occurs frequently in approximation problems involving the exponential function.) Instead, we will use the following rough estimate for $e^t$ (see Fig. 2):

$$\text{If} \quad t \le 1, \quad \text{then} \quad e^t \le 3. \tag{1}$$

Since estimate (1) holds for $t \le 1$, it certainly holds for $0 \le t \le 0.1$. Thus,

$$|R_2(0.1)| = \frac{e^t}{6,000} \le \frac{3}{6,000} = 0.0005$$

and we can conclude that the approximate value 1.105 is within $\pm 0.0005$ of the exact value of $e^{0.1}$.

---

**MATCHED PROBLEM 1**

Use the second-degree Taylor polynomial at 0 for $f(x) = e^x$ to approximate $e^{0.2}$. Use Taylor's formula for the remainder to estimate the error in the approximation.

> ### INSIGHT
>
> If $f$ is a polynomial of degree $n$ and $k \geq n$, then $f^{(k+1)}(t) = 0$ for all $t$. So by Taylor's formula for the remainder, $R_k(x) = 0$. Therefore, in the special case in which $f$ is a polynomial of degree $n$, each Taylor polynomial $p_k$, for $k \geq n$, is equal to $f$.

## ■ Taylor Series with Alternating Terms

Taylor's formula for the remainder can be difficult to apply for large values of $n$. For most functions, the formula for the $n$th derivative becomes very complicated as $n$ increases. For a certain type of problem, there is another method that does not require estimation of the $n$th derivative. If the series of numbers that is formed by evaluating a Taylor series at a given number $x_0$ *alternates in sign and decreases in absolute value*, then the remainder can be estimated by simply examining the numbers in the series. Series of numbers whose terms alternate in sign are called **alternating series.** The estimate for the remainder in an alternating series is given in Theorem 1.

### THEOREM 1 ERROR ESTIMATION FOR ALTERNATING SERIES

If $x_0$ is a number in the interval of convergence for

$$f(x) = a_0 + a_1x + a_2x^2 + \cdots + a_kx^k + \cdots$$

and the terms in the series

$$f(x_0) = a_0 + a_1x_0 + a_2x_0^2 + \cdots + a_kx_0^k + \cdots$$

are alternating in sign and decreasing in absolute value, then the error in the approximation

$$f(x_0) \approx a_0 + a_1x_0 + a_2x_0^2 + \cdots + a_nx_0^n$$

is strictly less than the absolute value of the next term. That is,

$$|R_n(x_0)| < |a_{n+1}x_0^{n+1}|$$

### EXAMPLE 2 Estimating the Remainder for Alternating Series

Use the Taylor series at 0 for $f(x) = e^{-x}$ to approximate $e^{-0.3}$ with an error of no more than 0.0005.

**SOLUTION** From Table 1 in Section 2-3, the Taylor series at 0 for $f(x) = e^{-x}$ is

$$e^{-x} = 1 - x + \frac{1}{2!}x^2 - \frac{1}{3!}x^3 + \cdots + \frac{(-1)^k}{k!}x^k + \cdots \qquad -\infty < x < \infty$$

If we substitute $x = 0.3$ in this series, we obtain

$$f(0.3) = e^{-0.3}$$
$$= 1 - 0.3 + \tfrac{1}{2}(0.3)^2 - \tfrac{1}{6}(0.3)^3 + \tfrac{1}{24}(0.3)^4 - \cdots$$
$$= 1 - 0.3 + 0.045 - 0.0045 + 0.0003375 - \cdots$$

Since the terms in this series are alternating in sign and decreasing in absolute value, Theorem 1 applies. If we use the first four terms in this series to approximate $e^{-0.3}$, then the error in this approximation is less than the absolute value of the fifth term. That is,

$$|R_3(0.3)| < 0.0003375 < 0.0005$$

Thus,

$$e^{-0.3} \approx 1 - 0.3 + 0.045 - 0.0045 = 0.7405$$

and the error in this approximation is less than the specified accuracy of 0.0005.

| MATCHED PROBLEM 2 | Use the Taylor series at 0 for $f(x) = e^{-x}$ to approximate $e^{-0.1}$ with an error of no more than 0.0005. |

As Example 2 illustrates, Theorem 1 is much easier to use than Taylor's formula for the remainder. Notice that we did not have to find an estimate for $f^{(n+1)}(t)$. However, it is very important to understand that Theorem 1 can be applied only if the terms in the series *alternate in sign after they have been evaluated at $x_0$*. For example, if we try to use the series for $e^{-x}$ to approximate $e^{0.3}$ by substituting $x_0 = -0.3$, we have

$$e^{0.3} = 1 - (-0.3) + \tfrac{1}{2}(-0.3)^2 - \tfrac{1}{6}(-0.3)^3 + \cdots$$
$$= 1 + 0.3 + \tfrac{1}{2}(0.3)^2 + \tfrac{1}{6}(0.3)^3 + \cdots$$

Since these numbers do not alternate in sign, Theorem 1 does not apply. Taylor's formula for the remainder would have to be used to estimate the error in approximations obtained from this series.

## ■ Approximating Definite Integrals

In order to find the exact value of a definite intergral $\int_a^b f(x)\,dx$, we must first find an antiderivative of the function $f$. But suppose we cannot find an anti-derivative of $f$ (it may not even exist in a convenient form). In Section 6-5 of *Calculus* 11/e or Section 13-5 of *College Mathematics* 11/e we saw that Riemann sums can be used to approximate definite integrals. Taylor series techniques provide an alternative method for approximating definite integrals that is often more efficient, and, in the case of alternating series, automatically determines the accuracy of the approximation.

| EXAMPLE 3 | **Using Taylor Series to Approximate Definite Integrals**   Approximate $\int_0^1 e^{-x^2}\,dx$ with a maximum error of 0.005. |

**SOLUTION**   If $F(x)$ is an antiderivative of $e^{-x^2}$, then

$$\int_0^1 e^{-x^2}\,dx = F(1) - F(0)$$

It is not possible to express $F(x)$ as a finite combination of simple functions; however, it is possible to find a Taylor series for $F$. This series can be used to approximate the values of $F$ and, consequently, of the definite integral.

**Step 1.**  *Find a Taylor series for the integrand:* From Table 1 in Section 2-3,

$$e^x = 1 + x + \frac{1}{2!}x^2 + \cdots + \frac{1}{n!}x^n + \cdots \quad -\infty < x < \infty$$

Thus,

$$e^{-x^2} = 1 + (-x^2) + \frac{1}{2!}(-x^2)^2 + \cdots + \frac{1}{n!}(-x^2)^n + \cdots$$
$$= 1 - x^2 + \frac{1}{2!}x^4 - \cdots + \frac{(-1)^n}{n!}x^{2n} + \cdots \quad -\infty < x < \infty$$

**Step 2.**  *Find the Taylor series for the antiderivative:* Integrating term by term, we have

$$F(x) = \int e^{-x^2}\,dx$$
$$= \int \left[ 1 - x^2 + \frac{1}{2!}x^4 - \cdots + \frac{(-1)^n}{n!}x^{2n} + \cdots \right] dx$$
$$= C + x - \frac{1}{3}x^3 + \frac{1}{10}x^5 - \cdots + \frac{(-1)^n}{n!(2n+1)}x^{2n+1} + \cdots \quad -\infty < x < \infty$$

**Step 3.** *Approximate the definite integral:* If we choose $C = 0$, then $F(0) = 0$ and

$$\int_0^1 e^{-x^2}\, dx = F(1) - F(0)$$

$$= \left[ x - \frac{1}{3}x^3 + \frac{1}{10}x^5 - \cdots + \frac{(-1)^n}{n!(2n+1)}x^{2n+1} + \cdots \right]\Big|_0^1$$

$$= \left[ 1 - \frac{1}{3} + \frac{1}{10} - \cdots + \frac{(-1)^n}{n!(2n+1)} + \cdots \right] - 0$$

$$= 1 - \frac{1}{3} + \frac{1}{10} - \frac{1}{42} + \frac{1}{216} - \cdots$$

$$= 1 - 0.333\,333 + 0.1 - 0.023\,809 + 0.004\,630 - \cdots$$

Since the Taylor series for $F(x)$ converges for all values of $x$, we can use this last series of numbers to approximate $F(1)$. Notice that the terms in this series are alternating in sign and decreasing in absolute value. Applying Theorem 1, we conclude that the error introduced by approximating $F(1)$ with the first four terms of this series will be no more than $0.004\,630$, the absolute value of the fifth term. Since this is less than the specified error of $0.005$, we have

$$\int_0^1 e^{-x^2}\, dx = F(1) - F(0)$$

$$\approx 1 - 0.333\,333 + 0.1 - 0.023\,809$$

$$\approx 0.743 \quad \textit{Rounded to three decimal places.}$$

| MATCHED PROBLEM 3 | Approximate $\int_0^{0.5} e^{-x^2}$ with a maximum error of $0.005$.

*Explore & Discuss* **2**   Suppose you wish to use a Taylor series for

$$f(x) = \frac{1}{1 - x^2} \qquad \text{to approximate} \qquad \int_2^3 \frac{dx}{1 - x^2}$$

(A) Explain why you would not use the Taylor series for $f$ at 0.

(B) If you use the Taylor series for $f$ at $a$, which value of $a$ would you expect to yield the best approximation to the integral? Explain.

EXAMPLE 4   **Income Distribution** Approximate the index of income concentration for the Lorenz curve given by

$$f(x) = \frac{11x^4}{10 + x^2}$$

with an error of no more than $0.005$.

SOLUTION   Referring to Section 7-1 of *Calculus* 11/e or Section 14-1 of *College Mathematics* 11/e, the index of income concentration for a Lorenz curve is twice the area between the graph of the Lorenz curve and the graph of the line $y = x$ (see Fig. 3). Thus, we must evaluate the integral

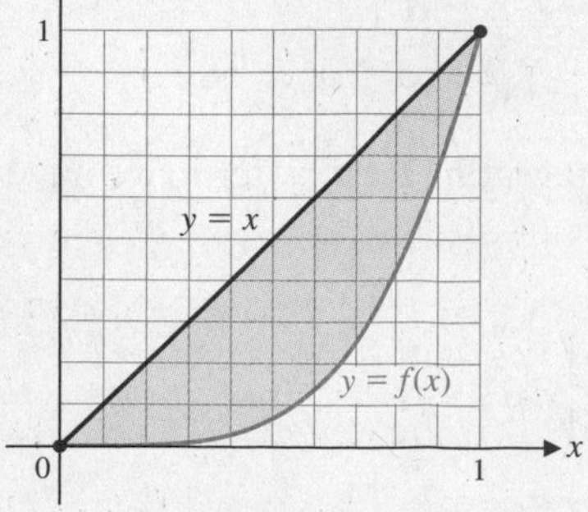

$$2\int_0^1 [x - f(x)]\, dx = 2\int_0^1 x\, dx - 2\int_0^1 f(x)\, dx$$

$$= \int_0^1 2x\, dx - \int_0^1 \frac{22x^4}{10 + x^2}\, dx \tag{2}$$

**FIGURE 3**

The first integral in (2) is easy to evaluate:

$$\int_0^1 2x \, dx = x^2 \Big|_0^1 = 1 - 0 = 1$$

Since the second integral in (2) cannot be evaluated by any of the techniques we have discussed, we will use a Taylor series to approximate this integral.

**Step 1.**   *Find a Taylor series for the integrand:*

$$\frac{22x^4}{10 + x^2} = \frac{22x^4}{10}\left[\frac{1}{1 + (x^2/10)}\right] \quad \text{Substitute } x^2/10 \text{ for } x \text{ in the series for } 1/(1 + x).$$

$$= 2.2x^4\left[1 - \frac{x^2}{10} + \left(\frac{x^2}{10}\right)^2 - \cdots + (-1)^n\left(\frac{x^2}{10}\right)^n + \cdots\right] \quad -1 < \frac{x^2}{10} < 1$$

$$= 2.2x^4 - \frac{2.2}{10}x^6 + \frac{2.2}{10^2}x^8 - \cdots + \frac{2.2(-1)^n}{10^n}x^{2n+4} + \cdots$$

To find the interval of convergence, we solve $-1 < x^2/10 < 1$ for $x$:

$$-1 < \frac{x^2}{10} < 1 \qquad \text{Change to absolute value notation.}$$

$$\left|\frac{x^2}{10}\right| < 1 \qquad \text{Multiply by 10.}$$

$$|x^2| < 10 \qquad \text{Take the square root of both sides.}$$

$$|x| < \sqrt{10} \qquad \text{Convert to double inequalities.}$$

$$-\sqrt{10} < x < \sqrt{10} \qquad \text{Interval of convergence.}$$

**Step 2.**   *Find the Taylor series for the antiderivative:* Using property 4,

$$\int \frac{22x^4}{10 + x^2} \, dx = \int\left[2.2x^4 - \frac{2.2}{10}x^6 + \frac{2.2}{10^2}x^8 - \cdots + \frac{2.2(-1)^n}{10^n}x^{2n+4} + \cdots\right] dx$$

$$= C + \frac{2.2}{5}x^5 - \frac{2.2}{7 \cdot 10}x^7 + \frac{2.2}{9 \cdot 10^2}x^9 - \cdots$$

$$+ \frac{2.2(-1)^n}{(2n + 5)10^n}x^{2n+5} + \cdots$$

This series also converges for $-\sqrt{10} < x < \sqrt{10}$.

**Step 3.**   *Approximate the definite integral:* Choosing $C = 0$ in the antiderivative, we have

$$\int_0^1 \frac{22x^4}{10 + x^2} \, dx = \left[\frac{2.2}{5}x^5 - \frac{2.2}{7 \cdot 10}x^7 + \frac{2.2}{9 \cdot 10^2}x^9 - \cdots\right.$$

$$\left.+ \frac{2.2(-1)^n}{(2n + 5)10^n}x^{2n+5} + \cdots\right]\Bigg|_0^1$$

$$= \left[\frac{2.2}{5} - \frac{2.2}{7 \cdot 10} + \frac{2.2}{9 \cdot 10^2} - \cdots + \frac{2.2(-1)^n}{(2n + 5)10^n} + \cdots\right] - 0$$

$$= 0.44 - 0.031\,429 + 0.002\,444 - \cdots$$

Since the limits of integration, 0 and 1, are within the interval of convergence of the Taylor series for the antiderivative, we can use this series to approximate the value of the definite integral. Notice that the numbers in this series are alternating in sign and decreasing in absolute value. Since the absolute value of the third term is less than 0.005, Theorem 1 implies that we can use the first two terms of this series to approximate the definite integral to the specified accuracy. Thus,

$$\int_0^1 \frac{22x^4}{10 + x^2} \, dx \approx 0.44 - 0.031\,429$$

$$\approx 0.409 \qquad \text{Rounded to three decimal places.}$$

Returning to (2), we have

$$2\int_0^1 [x - f(x)]\,dx = 2\int_0^1 x\,dx - 2\int_0^1 f(x)\,dx$$

$$\approx 1 - 0.409$$

$$= 0.591 \quad \text{Index of income concentration.}$$

—

MATCHED PROBLEM 4    Repeat Example 4 for $f(x) = \dfrac{11x^5}{10 + x^2}$.

*Answers to Matched Problems*

**1.** $e^{0.2} \approx 1 + 0.2 + \frac{1}{2}(0.2)^2 = 1.22; \; |R_2(0.2)| < 0.004$

**2.** $e^{-0.1} \approx 1 - 0.1 + 0.005 = 0.905$ within $\pm 0.000\,167$

**3.** $\displaystyle\int_0^{0.5} e^{-x^2}\,dx \approx 0.5 - \frac{1}{3}(0.5)^3 \approx 0.458$ within $\pm 0.003\,125$

**4.** $\displaystyle\int_0^1 \frac{22x^5}{10 + x^2}\,dx \approx \frac{2.2}{6} - \frac{2.2}{80} \approx 0.339$ within $\pm 0.0022$;

index of income concentration $\approx 0.661$

## Exercise 2-4

*In Problems 1–30, use Theorem 1 to perform the indicated error estimations.*

**A**   *Evaluating the Taylor series at 0 for $f(x) = e^{-x}$ at $x = 0.6$ produces the following series:*

$$e^{-0.6} = 1 - 0.6 + 0.18 - 0.036 + 0.0054$$
$$- 0.000\,648 + \cdots$$

*In Problems 1–4, use the indicated number of terms in this series to approximate $e^{-0.6}$, and then estimate the error in this approximation.*

**1.** Two terms      **2.** Three terms

**3.** Four terms      **4.** Five terms

*Evaluating the Taylor series at 0 for $f(x) = \ln(1 + x)$ at $x = 0.3$ produces the following series:*

$$\ln 1.3 = 0.3 - 0.045 + 0.009 - 0.002\,025 + 0.000\,486$$
$$- 0.000\,121\,5 + \cdots$$

*In Problems 5–8, use the indicated number of terms in this series to approximate $\ln 1.3$, and then estimate the error in this approximation.*

**5.** Two terms      **6.** Three terms

**7.** Four terms      **8.** Five terms

*In Problems 9–12, use the third-degree Taylor polynomial at 0 for $f(x) = e^{-x}$ to approximate each expression, and then estimate the error in the approximation.*

**9.** $e^{-0.2}$      **10.** $e^{-0.5}$

**11.** $e^{-0.03}$      **12.** $e^{-0.06}$

*In Problems 13–16, use the third-degree Taylor polynomial at 0 for $f(x) = \ln(1 + x)$ to approximate each expression, and then estimate the error in the approximation.*

**13.** $\ln 1.6$      **14.** $\ln 1.8$

**15.** $\ln 1.06$      **16.** $\ln 1.08$

**B**   *In Problems 17–24, use a Taylor polynomial at 0 to approximate each expression with an error of no more than 0.000 005. Select the polynomial of lowest degree that can be used to obtain this accuracy and state the degree of this polynomial.*

**17.** $e^{-0.1}$      **18.** $e^{-0.2}$

**19.** $e^{-0.01}$      **20.** $e^{-0.02}$

**21.** $\ln 1.2$      **22.** $\ln 1.1$

**23.** $\ln 1.02$      **24.** $\ln 1.01$

*In Problems 25–30, use a Taylor series at 0 to approximate each integral with an error of no more than 0.0005.*

**25.** $\displaystyle\int_0^{0.2} \frac{1}{1 + x^2}\,dx$      **26.** $\displaystyle\int_0^{0.5} \frac{x}{1 + x^4}\,dx$

**27.** $\displaystyle\int_0^{0.6} \ln(1 + x^2)\,dx$      **28.** $\displaystyle\int_0^{0.7} x\ln(1 + x^4)\,dx$

**29.** $\displaystyle\int_0^{0.4} x^2 e^{-x^2}\,dx$      **30.** $\displaystyle\int_0^{0.8} x^4 e^{-x^2}\,dx$

*In Problems 31–34, assume that $f(x)$ is a function such that $|f^{(n)}(x)| \leq 1$ for all n and all x [the trigonometric functions $f(x) = \sin x$ and $f(x) = \cos x$, for example, satisfy that property] and let $p_n(x)$ be the nth-degree Taylor polynomial for f at 0.*

**31.** Use Taylor's formula for the remainder to find the smallest value of n such that the error in the approximation of $f(2)$ by $p_n(2)$ is guaranteed to be less than 0.001.

**32.** Use Taylor's formula for the remainder to find the smallest value of n such that the error in the approximation of $f(10)$ by $p_n(10)$ is guaranteed to be less than 0.001.

**33.** Use Taylor's formula for the remainder to determine the values of x such that the error in the approximation of $f(x)$ by $p_5(x)$ is guaranteed to be less than 0.05.

**34.** Use Taylor's formula for the remainder to determine the values of x such that the error in the approximation of $f(x)$ by $p_{10}(x)$ is guaranteed to be less than 0.05.

**35.** Let $f(x) = xe^x$ and consider the third-degree Taylor polynomial $p_3(x)$ for f at 0. Use graphical approximation techniques to determine those values of x for which the error in the approximation $f(x) \approx p_3(x)$ is less than 0.005.

**36.** Let $f(x) = 1/(4 - x^2)$ and consider the fourth-degree Taylor polynomial $p_4(x)$ for f at 0. Use graphical approximation techniques to determine those values of x for which the error in the approximation $f(x) \approx p_4(x)$ is less than 0.0001.

C  *In Problems 37–40, use the second-degree Taylor polynomial at 0 for $f(x) = e^x$ to approximate the given number. Use Taylor's formula for the remainder to estimate the error in each approximation.*

**37.** $e^{-0.3}$    **38.** $e^{-0.4}$    **39.** $e^{0.05}$    **40.** $e^{0.01}$

**41.** The Taylor series at 0 for $f(x) = \sqrt{16 + x}$ converges for $-16 < x < 16$. Use the second-degree Taylor polynomial at 0 to approximate $\sqrt{17}$. Use Taylor's formula for the remainder to estimate the error in this approximation.

**42.** The Taylor series at 0 for $f(x) = \sqrt{(4 + x)^3}$ converges for $-4 < x < 4$. Use the third-degree Taylor polynomial at 0 to approximate $\sqrt{125}$. Use Taylor's formula for the remainder to estimate the error in this approximation.

**43.** To estimate

$$\int_{-2}^{0} \frac{1}{1 - x} \, dx$$

a student takes the first five terms of the Taylor series for $f(x) = 1/(1 - x)$ at 0 and integrates term by term. He obtains the estimate 5.067. A second student estimates the integral by noting that $F(x) = -\ln(1 - x)$ is an antiderivative for $f(x)$. She evaluates $F(x)$ between $-2$ and $0$ to obtain the estimate 1.099. Is either computation correct? How do you account for the large discrepancy between the two estimates?

**44.** To estimate

$$\int_{0}^{1.5} \frac{1}{1 + x^2} \, dx$$

a student takes the first five nonzero terms of the Taylor series for $f(x) = 1/(1 + x^2)$ at 0 and integrates term by term. He obtains the estimate 3.724. A second student doubts the estimate. She claims that since $1/(1 + x^2) \leq 1$ for $0 \leq x \leq 1.5$, the value of the integral must be less than 1.5. Is either student correct? How do you account for the large discrepancy between their estimates?

*There are different ways to approximate a function f by polynomials. If, for example, $f(a), f'(a)$, and $f''(a)$ are known, then we can construct the second-degree Taylor polynomial $p_2(x)$ at a for $f(x)$; $p_2(x)$ and $f(x)$ will have the same value at a and the same first and second derivatives at a. If, on the other hand, $f(x_1), f(x_2)$, and $f(x_3)$ are known, then we can compute the quadratic regression polynomial $q_2(x)$ for the points $(x_1, f(x_1)), (x_2, f(x_2)), (x_3, f(x_3))$; $q_2(x)$ and $f(x)$ will have the same values at $x_1, x_2, x_3$. Problems 45 and 46 concern these contrasting methods of approximation by polynomials.*

**45.** (A) Find the second-degree Taylor polynomial $p_2(x)$ at 0 for $f(x) = e^x$, and use a graphing utility to compute the quadratic regression polynomial $q_2(x)$ for the points $(-0.1, e^{-0.1})$, $(0, e^0)$, and $(0.1, e^{0.1})$.

(B) Use graphical approximation techniques to find the maximum error for $-0.1 \leq x \leq 0.1$ in approximating $f(x) = e^x$ by $p_2(x)$ and by $q_2(x)$.

(C) Which polynomial, $p_2(x)$ or $q_2(x)$, gives the better approximation to

$$\int_{-0.1}^{0.1} e^x dx?$$

**46.** (A) Find the fourth-degree Taylor polynomial $p_4(x)$ at 0 for $f(x) = \ln(1 + x)$, and use a graphing utility to compute the quartic regression polynomial $q_4(x)$ for the points $(0, \ln 1)$,

$$\left(-\tfrac{1}{2}, \ln(1 - \tfrac{1}{2})\right), \left(-\tfrac{3}{4}, \ln(1 - \tfrac{3}{4})\right),$$
$$\left(-\tfrac{7}{8}, \ln(1 - \tfrac{7}{8})\right), \left(-\tfrac{15}{16}, \ln(1 - \tfrac{15}{16})\right).$$

(B) Use graphical approximation techniques to find the maximum error for $-\tfrac{15}{16} \leq x \leq 0$ in approximating $f(x) = \ln(1 + x)$ by $p_4(x)$ and by $q_4(x)$.

(C) Which polynomial, $p_4(x)$ or $q_4(x)$, gives the better approximation to

$$\int_{-15/16}^{0} \ln(1 + x) \, dx?$$

## Applications

*In Problems 47–58, use Theorem 1 to perform the indicated error estimations.*

**47.** *Income distribution.* The income distribution for a certain country is represented by the Lorenz curve with the equation

$$f(x) = \frac{5x^6}{4 + x^2}$$

Approximate the index of income concentration to within ±0.005.

**48.** *Income distribution.* Repeat Problem 47 for

$$f(x) = \frac{10x^4}{9 + x^2}$$

**49.** *Marketing.* A soft drink manufacturer is ready to introduce a new diet soda by a national sales campaign. After test marketing the soda in a carefully selected city, the market research department estimates that sales (in millions of dollars) will increase at the monthly rate of

$$S'(t) = 10 - 10e^{-0.01t^2} \qquad 0 \le t \le 12$$

*t* months after the national campaign is started. Use the fourth-degree Taylor polynomial at 0 for $S'(t)$ to approximate the total sales during the first 4 months of the campaign, and estimate the error in this approximation.

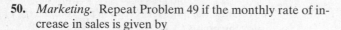

**50.** *Marketing.* Repeat Problem 49 if the monthly rate of increase in sales is given by

$$S'(t) = 10 - 10e^{-0.005t^2} \qquad 0 \le t \le 12$$

**51.** *Useful life.* A computer store rents time on desktop publishing systems. The total accumulated costs $C(t)$ and revenues $R(t)$ (in thousands of dollars) from a particular system satisfy

$$C'(t) = 4 \quad \text{and} \quad R'(t) = \frac{80}{16 + t^2}$$

where *t* is the time in years that the system has been available for rental. Find the useful life of the system, and approximate the total profit during the useful life to within ±0.005.

**52.** *Average price.* Given the demand equation

$$p = D(x) = 10 - 20\ln\left(1 + \frac{x^2}{2{,}500}\right) \qquad 0 \le x \le 40$$

approximate the average price (in dollars) over the demand interval [0, 20] to within ±0.005.

**53.** *Temperature.* The temperature (in degrees Celsius) in an artificial habitat is made to change according to the equation

$$C(t) = 20 + 800\ln\left(1 + \frac{t^2}{100}\right) \qquad 0 \le t \le 2$$

Use a Taylor series at 0 to approximate the average temperature over the time interval [0, 2] to within ±0.005.

**54.** *Temperature.* Repeat Problem 53 for

$$C(t) = 10 + 200\ln\left(1 + \frac{t^2}{50}\right) \qquad 0 \le t \le 2$$

**55.** *Medicine.* The rate of healing for a skin wound (in square centimeters per day) is given by

$$A'(t) = \frac{-75}{t^2 + 25}$$

The initial wound has an area of 12 square centimeters. Use the second-degree Taylor polynomial at 0 for $A'(t)$ to approximate the area of the wound after 2 days, and estimate the error in this approximation.

**56.** *Medicine.* Repeat Problem 55 for

$$A'(t) = \frac{-60}{t^2 + 20}$$

**57.** *Learning.* In a particular business college, it was found that an average student enrolled in an advanced typing class progresses at a rate of $N'(t) = 6e^{-0.01t^2}$ words per minute per week, *t* weeks after enrolling in a 15-week course. At the beginning of the course an average student could type 40 words per minute. Use the second-degree Taylor polynomial at 0 for $N'(t)$ to approximate the improvement in typing after 5 weeks in the course, and estimate the error in this approximation.

**58.** *Learning.* In the same business college, it was found that an average student in a beginning shorthand class progressed at a rate of $N'(t) = 12e^{-0.005t^2}$ words per minute per week, *t* weeks after enrolling in a 15-week course. At the beginning of the course none of the students could take any dictation by shorthand. Use the second-degree Taylor polynomial at 0 for $N'(t)$ to approximate the improvement after 5 weeks in the course, and estimate the error in this approximation.

# CHAPTER 2 REVIEW

## Important Terms, Symbols, and Concepts

### 2-1 Taylor Polynomials

If $f$ is a function that has $n$ derivatives at 0, then the **$n$th-degree Taylor polynomial for $f$ at 0** is

$$p_n(x) = f(0) + f'(0)x + \frac{f''(0)}{2!}x^2 + \cdots + \frac{f^{(n)}(0)}{n!}x^n$$

If $f$ has $n$ derivatives at $a$, then the **$n$th-degree Taylor polynomial for $f$ at $a$** is

$$p_n(x) = f(a) + f'(a)(x - a) + \frac{f''(a)}{2!}(x - a)^2 + \cdots$$

$$+ \frac{f^{(n)}(a)}{n!}(x - a)^n$$

### 2-2 Taylor Series

If $f$ is a function that has derivatives of all order at a point $a$ and $p_n(x)$ is the $n$th-degree Taylor polynomial for $f$ at $a$, then the **Taylor series for $f$ at $a$** is

$$f(a) + f'(a)(x - a) + \frac{f''(a)}{2!}(x - a)^2 + \cdots$$

$$+ \frac{f^{(n)}(a)}{n!}(x - a)^n + \cdots$$

The Taylor series **converges at $x$** if $\lim_{n \to \infty} p_n(x)$ exists, and **diverges at $x$** if the limit does not exist. The set of values of $x$ for which this limit exists is called the **interval of convergence.**

#### • *Theorem 1 Interval of Convergence*

Let $f$ be a function with derivatives of all order at a point $a$, let $a_n = f^{(n)}(a)/n!$ for $n = 0, 1, 2, \ldots$, and let $a_0 + a_1(x - a) + a_2(x - a)^2 + \cdots + a_n(x - a)^n + \cdots$ be the Taylor series for $f$ at $a$. If $a_n \neq 0$ for $n \geq n_0$, then:

**Case 1.** If $\lim_{n \to \infty} \left| \frac{a_{n+1}}{a_n} \right| = L > 0$ and $R = \frac{1}{L}$ then the series converges for $|x - a| < R$ and diverges for $|x - a| > R$.

**Case 2.** If $\lim_{n \to \infty} \left| \frac{a_{n+1}}{a_n} \right| = 0$, then the series converges for all values of $x$.

**Case 3.** If $\lim_{n \to \infty} \left| \frac{a_{n+1}}{a_n} \right| = \infty$, then the series converges only at $x = a$.

We assume that functions are represented by their Taylor series throughout the interval of convergence (there exist functions that do not satisfy this condition, but they are not considered in this supplement).

### 2-3 Operations on Taylor Series

**Property 1.** Two Taylor series can be added term by term. This operation is valid in the intersection of the intervals of convergence of the series for $f$ and $g$.

**Property 2.** A Taylor series for $f$ at 0 can be multiplied term by term by an expression of the form $cx^r$, where $c$ is a nonzero constant and $r$ is a non-negative integer. The resulting series has the same interval of convergence as the Taylor series for $f$.

**Property 3.** A Taylor series for $f$ at 0 can be differentiated term by term to obtain a Taylor series for $f'$. Both series have the same interval of convergence.

**Property 4.** A Taylor series at 0 can be integrated term by term to obtain a Taylor series for $\int f(x)\, dx$. Both series have the same interval of convergence.

Making a substitution in a known Taylor series to obtain a new series is a useful technique. A series for $e^{x^2}$, for example, can be obtained by substituting $x^2$ for $x$ in the Taylor series for $e^x$ at 0.

### 2-4 Approximations Using Taylor Series

If $p_n(x)$ is the $n$th-degree Taylor polynomial for $f$, then the **remainder** is $R_n(x) = f(x) - p_n(x)$. The **error** in the approximation $f(x) \approx p_n(x)$ is $|f(x) - p_n(x)| = |R_n(x)|$.

• *Taylor's Formula for the Remainder*

If $f$ has derivatives of all order at 0, then

$$R_n(x) = \frac{f^{(n+1)}(t)\,x^{n+1}}{(n+1)!} \text{ for some number } t \text{ between 0 and } x.$$

### THEOREM 1    ERROR ESTIMATION FOR ALTERNATING SERIES

If $x_0$ is a number in the interval of convergence for

$$f(x) = a_0 + a_1 x + a_2 x^2 + \cdots + a_k x^k + \cdots$$

and the terms in the series

$$f(x_0) = a_0 + a_1 x_0 + a_2 x_0^2 + \cdots + a_k x_0^k + \cdots$$

are alternating in sign and decreasing in absolute value, then

$$|R_n(x_0)| < |a_{n+1} x_0^{n+1}|$$

Taylor series techniques provide a method for approximating definite integrals which, in the case of alternating series, automatically determines the accuracy of the approximation.

## REVIEW EXERCISE

*Work through all the problems in this chapter review and check your answers in the back of this supplement. Answers to all review problems are there along with section numbers in italics to indicate where each type of problem is discussed. Where weaknesses show up, review appropriate sections in the text.*

*Unless directed otherwise, use Theorem 1 of Section 2-4 in all problems involving error estimation.*

A

1. Find $f^{(4)}(x)$ for $f(x) = \ln(x + 5)$.
2. Use the third-degree Taylor polynomial at 0 for $f(x) = \sqrt[3]{1 + x}$ and $x = 0.01$ to approximate $\sqrt[3]{1.01}$.
3. Use the third-degree Taylor polynomial at $a = 3$ for $f(x) = \sqrt{1 + x}$ and $x = 2.9$ to approximate $\sqrt{3.9}$.
4. Use the second-degree Taylor polynomial at 0 for $f(x) = \sqrt{9 + x^2}$ and $x = 0.1$ to approximate $\sqrt{9.01}$.

*Use Theorem 1 of Section 2-2 to find the interval of convergence of each Taylor series representation given in Problems 5–8.*

5. $\dfrac{1}{1 - 4x} = 1 + 4x + 4^2 x^2 + \cdots + 4^n x^n + \cdots$

6. $\dfrac{5}{x - 1} = 1 - \dfrac{1}{5}(x - 6) + \dfrac{1}{5^2}(x - 6)^2 - \cdots$
   $\qquad\qquad + \dfrac{(-1)^n}{5^n}(x - 6)^n + \cdots$

7. $\dfrac{2x}{(1 - x)^3} = 1 \cdot 2x + 2 \cdot 3x^2 + 3 \cdot 4x^3 + \cdots$
   $\qquad\qquad + n(n + 1)x^n + \cdots$

8. $e^{10x} = 1 + 10x + \dfrac{10^2}{2!}x^2 + \cdots + \dfrac{10^n}{n!}x^n + \cdots$

9. Find the $n$th derivative of $f(x) = e^{-9x}$.

B   *In Problems 10 and 11, use the formula $a_n = f^{(n)}(a)/n!$ to find the Taylor series at the indicated value of $a$. Use Theorem 1 of Section 2-2 to find the interval of convergence.*

10. $f(x) = \dfrac{1}{7 - x}$; at 0    11. $f(x) = \ln x$; at 2

*In Problems 12–16, use Table 1 of Section 2-3 and the properties of Taylor series to find the Taylor series of each function at the indicated value of $a$. Find the interval of convergence.*

12. $f(x) = \dfrac{1}{10 + x}$; at 0    13. $f(x) = \dfrac{x^2}{4 - x^2}$; at 0

14. $f(x) = x^2 e^{3x}$; at 0    15. $f(x) = x \ln(e + x)$; at 0

16. $f(x) = \dfrac{1}{4 - x}$; at 2

17. (A) Explain why Theorem 1 of Section 2-2 is not directly applicable to the Taylor series representation

$$\ln(1 - 5x^2) = -5x^2 - \frac{5^2}{2}x^4 - \frac{5^3}{3}x^6 - \cdots - \frac{5^n}{n}x^{2n} - \cdots$$

    (B) Use another method to find the interval of convergence of the Taylor series in part (A).

18. Substituting $\sqrt{x}$ for $x$ in the Taylor series at 0 for $1/(1 + x)$ gives the formula

$$\frac{1}{1 + \sqrt{x}} = 1 - x^{1/2} + x - x^{3/2} + \cdots + (-1)^n x^{n/2} + \cdots$$

    (A) For which values of $x$ is the formula valid?

    (B) Is the formula a Taylor series at 0 for $1/(1 + \sqrt{x})$? Explain.

*In Problems 19 and 20, find the Taylor series at 0 for $f(x)$ and use the relationship $g(x) = f'(x)$ to find the Taylor series at 0 for $g(x)$. Find the interval of convergence for both series.*

**19.** $f(x) = \dfrac{1}{2 - x}$; $g(x) = \dfrac{1}{(2 - x)^2}$

**20.** $f(x) = \dfrac{x^2}{1 + x^2}$; $g(x) = \dfrac{2x}{(1 + x^2)^2}$

*In Problems 21 and 22, find the Taylor series at 0, and find the interval of convergence.*

**21.** $f(x) = \displaystyle\int_0^x \dfrac{t^2}{9 + t^2}\, dt$  **22.** $f(x) = \displaystyle\int_0^x \dfrac{t^4}{16 - t^2}\, dt$

**23.** (A) Compute the Taylor series for
$$f(x) = x^3 - 3x^2 + 4 \text{ at } a = 1 \text{ and at } a = -1.$$
  (B) Explain how the two series are related.

**24.** (A) Explain why $f(x) = |x|$ does not have any Taylor polynomials at 0.
  (B) If $a \neq 0$, find the Taylor polynomials for $f(x) = |x|$ at $a$.

*In Problems 25 and 26, use the second-degree Taylor polynomial at 0 for $f(x) = e^x$ to approximate the indicated quantity. Use Taylor's formula for the remainder to estimate the error in the approximation.*

**25.** $e^{0.6}$  **26.** $e^{0.06}$

*In Problems 27 and 28, use a Taylor polynomial at 0 for $f(x) = \ln(1 + x)$ to estimate the indicated quantity to within $\pm 0.0005$. Give the degree of the Taylor polynomial of lowest degree that will provide this accuracy.*

**27.** $\ln 1.3$  **28.** $\ln 1.03$

**29.** If $f(x)$ satisfies $f'(x) = x\ln(1 - x)$ and $f(0) = 5$, find the Taylor series at 0 for $f(x)$.

**30.** If $f(x)$ satisfies $f''(x) = xe^{-x}$, $f'(0) = -4$, and $f(0) = 3$, find the Taylor series at 0 for $f(x)$.

C  *In Problems 31 and 32, approximate the integral to within $\pm 0.0005$.*

**31.** $\displaystyle\int_0^1 \dfrac{1}{16 + x^2}\, dx$  **32.** $\displaystyle\int_0^1 x^2 e^{-0.1x^2}\, dx$

*In Problems 33 and 34, assume that $f(x)$ is a function such that $|f^{(n)}(x)| \leq 10$ for all $n$ and all $x$, and let $p_n(x)$ be the nth-degree Taylor polynomial for $f$ at 0.*

**33.** Use Taylor's formula for the remainder to find the smallest value of $n$ such that the error in the approximation of $f(0.5)$ by $p_n(0.5)$ is guaranteed to be less than $10^{-6}$.

**34.** Use Taylor's formula for the remainder to determine the values of $x$ such that the error in the approximation of $f(x)$ by $p_8(x)$ is guaranteed to be less than $10^{-6}$.

**35.** Let $f(x) = e^{x^2}$ and consider the fourth-degree Taylor polynomial $p_4(x)$ for $f$ at 0. Use graphical approximation techniques to determine those values of $x$ for which the error in the approximation $f(x) \approx p_4(x)$ is less than 0.01.

**36.** Let $f(x) = \ln(1 - x^2)$ and consider the sixth-degree Taylor polynomial $p_6(x)$ for $f$ at 0. Use graphical approximation techniques to determine those values of $x$ for which the error in the approximation $f(x) \approx p_6(x)$ is less than 0.001.

## APPLICATIONS

**37.** *Average price.* Given the demand equation
$$p = D(x) = \tfrac{1}{10}\sqrt{2{,}500 - x^2}$$
use the second-degree Taylor polynomial at 0 to approximate the average price (in dollars) over the demand interval $[0, 15]$.

**38.** *Production.* The rate of production of an oil well (in thousands of dollars per year) is given by
$$R(t) = 6 + 3e^{-0.01t^2}$$
Use the second-degree Taylor polynomial at 0 to approximate the total production during the first 10 years of operation of the well.

**39.** *Income distribution.* The income distribution for a certain country is represented by the Lorenz curve with the equation
$$f(x) = \dfrac{9x^3}{8 + x^2}$$
Approximate the index of income concentration to within $\pm 0.005$.

**40.** *Marketing.* A cereal manufacturer is ready to introduce a new high-fiber cereal by a national sales campaign. After test-marketing the cereal in a carefully selected city, the market research department estimates that sales (in millions of dollars) will increase at the monthly rate of
$$S'(t) = 20 - 20e^{-0.001t^2} \qquad 0 \leq t \leq 12$$
$t$ months after the national campaign is started. Use the fourth-degree Taylor polynomial at 0 for $S'(t)$ to approximate the total sales during the first 8 months of the campaign, and estimate the error in this approximation.

**41.** *Medicine.* The rate of healing for a skin wound (in square centimeters per day) is given by
$$A'(t) = \dfrac{-100}{t^2 + 40}$$
The initial wound has an area of 15 square centimeters. Use the second-degree Taylor polynomial at 0 for $A'(t)$ to approximate the area of the wound after 2 days, and estimate the error in this approximation.

**42.** *Medicine.* A large injection of insulin is administered to a patient. The level of insulin in the blood-stream $t$ minutes after the injection is given approximately by

$$L(t) = \frac{5,000t^2}{10,000 + t^4}$$

Express the average insulin level over the time interval $[0, 5]$ as a definite integral, and use a Taylor series at 0 to approximate this integral to within $\pm 0.005$.

**43.** *Politics.* In a newly incorporated city, the number of voters (in thousands) $t$ years after incorporation is given by

$$N(t) = 10 + 2t - 5e^{-0.01t^2} \qquad 0 \le t \le 5$$

Express the average number of voters over the time interval $[0, 5]$ as a definite integral, and use a Taylor series at 0 to approximate this integral to within $\pm 0.05$.

# Probability and Calculus

CHAPTER **3**

3-1 Improper Integrals

3-2 Continuous Random Variables

3-3 Expected Value, Standard Deviation, and Median

3-4 Special Probability Distributions

Chapter 3 Review

Review Exercise

## INTRODUCTION

In Section 7-2 of *Calculus* 11/e or Section 14-2 of *College Mathematics* 11/e we briefly touched on applications of calculus to probability. In this chapter, we discuss this topic in much more detail. In the first section we introduce improper integrals, a new integral form that is fundamental to the work that follows. After discussing basic concepts in the next two sections, we consider several commonly used probability distributions, culminating with the very important normal distribution.

## Section 3-1   IMPROPER INTEGRALS

- Improper Integrals
- Application: Capital Value

### ■ Improper Integrals

We are now going to consider an integral form that has wide application in probability studies as well as other areas. Earlier, when we introduced the idea of a definite integral,

$$\int_a^b f(x)\, dx \tag{1}$$

we required $f$ to be continuous over a closed interval $[a, b]$. Now we are going to extend the meaning of (1) so that the interval $[a, b]$ may become infinite in length.

*Explore & Discuss* **1**

For each of the following functions, find the area under the graph from $x = 1$ to $x = b, b > 1$ (see Fig. 1). Discuss the behavior of this area as $b$ assumes larger and larger values.

(A) $f(x) = \dfrac{1}{x^{5/4}} = x^{-5/4}$   (B) $g(x) = \dfrac{1}{x^{4/5}} = x^{-4/5}$

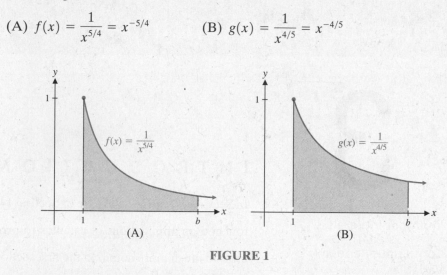

(A)          (B)

**FIGURE 1**

Let us investigate a particular example that will motivate several general definitions. What would be a reasonable interpretation for the following expression?

$$\int_1^\infty \frac{dx}{x^2}$$

Sketching a graph of $f(x) = 1/x^2, x \geq 1$ (see Fig. 2), we note that for any fixed $b > 1$, $\int_1^b f(x)\, dx$ is the area between the graph of $y = 1/x^2$, the $x$ axis, $x = 1$, and $x = b$.

Let us see what happens when we let $b \to \infty$; that is, when we compute the following limit:

$$\lim_{b \to \infty} \int_1^b \frac{dx}{x^2} = \lim_{b \to \infty}\left[ (-x^{-1}) \Big|_1^b \right]$$

$$= \lim_{b \to \infty}\left( -\frac{1}{b} + 1 \right) = 1$$

Did you expect this result? No matter how large $b$ is taken, the area under the graph from $x = 1$ to $x = b$ never exceeds 1, and in the limit it *is* 1 (see Fig. 3). This suggests that we write

$$\int_1^\infty \frac{dx}{x^2} = \lim_{b \to \infty} \int_1^b \frac{dx}{x^2} = 1$$

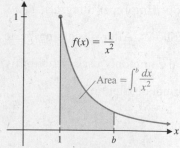

**FIGURE 2**

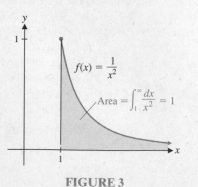

**FIGURE 3**

This integral is an example of an *improper integral.* In general, the forms

$$\int_{-\infty}^{b} f(x)\,dx \qquad \int_{a}^{\infty} f(x)\,dx \qquad \int_{-\infty}^{\infty} f(x)\,dx$$

where $f$ is continuous over the indicated interval, are called **improper integrals.** (These integrals are "improper" because the interval of integration is unbounded, as indicated by the use of $\infty$ for one or both limits of integration. There are other types of improper integrals involving certain types of points of discontinuity within the interval of integration, but these will not be considered here.) Each type of improper integral above is formally defined in the following box:

---

**DEFINITION**   **Improper Integrals**

If $f$ is continuous over the indicated interval and the limit exists, then

1. $\displaystyle\int_{a}^{\infty} f(x)\,dx = \lim_{b\to\infty} \int_{a}^{b} f(x)\,dx$

2. $\displaystyle\int_{-\infty}^{b} f(x)\,dx = \lim_{a\to-\infty} \int_{a}^{b} f(x)\,dx$

3. $\displaystyle\int_{-\infty}^{\infty} f(x)\,dx = \int_{-\infty}^{c} f(x)\,dx + \int_{c}^{\infty} f(x)\,dx$

where $c$ is any point on $(-\infty, \infty)$, provided *both* improper integrals on the right exist.

---

If the indicated limit exists, then the improper integral is said to exist or to **converge;** if the limit does not exist, then the improper integral is said not to exist or to **diverge** (and no value is assigned to it).

**EXAMPLE 1**   **Evaluating an Improper Integral**   Evaluate the following, if it converges:

$$\int_{2}^{\infty} \frac{dx}{x}$$

**SOLUTION**   $\displaystyle\int_{2}^{\infty} \frac{dx}{x} = \lim_{b\to\infty} \int_{2}^{b} \frac{dx}{x}$

$$= \lim_{b\to\infty} \left[ (\ln x)\Big|_{2}^{b} \right]$$

$$= \lim_{b\to\infty} (\ln b - \ln 2)$$

Since $\ln b \to \infty$ as $b \to \infty$, the limit does not exist. Hence, the improper integral diverges.

**MATCHED PROBLEM 1**   Evaluate the following, if it converges:

$$\int_{3}^{\infty} \frac{dx}{(x-1)^2}.$$

**EXAMPLE 2**   **Evaluating an Improper Integral**   Evaluate the following, if it converges:

$$\int_{-\infty}^{2} e^x\,dx$$

**SOLUTION** $\displaystyle\int_{-\infty}^{2} e^x \, dx = \lim_{a \to -\infty} \int_{a}^{2} e^x \, dx$

$$= \lim_{a \to -\infty} \left( e^x \Big|_{a}^{2} \right)$$

$$= \lim_{a \to -\infty} (e^2 - e^a) = e^2 - 0 = e^2 \quad \text{The integral converges.}$$

**MATCHED PROBLEM 2** Evaluate the following, if it converges:

$$\int_{-\infty}^{-1} x^{-2} \, dx.$$

**EXAMPLE 3** **Evaluating an Improper Integral** Evaluate the following, if it converges:

$$\int_{-\infty}^{\infty} \frac{2x}{(1 + x^2)^2} \, dx$$

**SOLUTION** $\displaystyle\int_{-\infty}^{\infty} \frac{2x}{(1 + x^2)^2} \, dx = \int_{-\infty}^{0} (1 + x^2)^{-2} 2x \, dx + \int_{0}^{\infty} (1 + x^2)^{-2} 2x \, dx$

$$= \lim_{a \to -\infty} \int_{a}^{0} (1 + x^2)^{-2} 2x \, dx + \lim_{b \to \infty} \int_{0}^{b} (1 + x^2)^{-2} 2x \, dx$$

$$= \lim_{a \to -\infty} \left[ \frac{(1 + x^2)^{-1}}{-1} \Big|_{a}^{0} \right] + \lim_{b \to \infty} \left[ \frac{(1 + x^2)^{-1}}{-1} \Big|_{0}^{b} \right]$$

$$= \lim_{a \to -\infty} \left[ -1 + \frac{1}{1 + a^2} \right] + \lim_{b \to \infty} \left[ -\frac{1}{1 + b^2} + 1 \right]$$

$$= -1 + 1 = 0 \quad \text{The integral converges.}$$

**MATCHED PROBLEM 3** Evaluate the following, if it converges:

$$\int_{-\infty}^{\infty} \frac{dx}{e^x}.$$

**INSIGHT**

It is impossible, in general, solely by inspecting the graph of the function, to determine whether an improper integral converges or diverges. For example, the functions

$$f(x) = \frac{1}{x} \quad \text{and} \quad g(x) = \frac{1}{x^{1.001}}$$

have graphs that are nearly indistinguishable on $[1, \infty)$. But by evaluating the appropriate limits, it can be shown that $\int_{0}^{\infty} f(x) \, dx$ diverges and $\int_{0}^{\infty} g(x) \, dx$ converges.

*Explore & Discuss* **2** Let

$$f(x) = \begin{cases} 2 - x & \text{if } 0 \le x \le 1 \\ 0 & \text{otherwise} \end{cases}$$

(A) Graph $y = f(x)$ and explain why $f$ is not continuous at $x = 0$ and at $x = 1$.

(B) Find $\displaystyle\int_{0}^{1} f(x) \, dx$.

(C) Discuss possible interpretations of the improper integral $\displaystyle\int_{-\infty}^{\infty} f(x) \, dx$.

In probability studies, we frequently encounter functions that are 0 over one or more intervals on the real axis. An improper integral involving such a function can be simplified by deleting the intervals where the function is 0 and considering the integral over the remaining intervals. The next example illustrates this process.

**EXAMPLE 4**   **Evaluating an Improper Integral**   Evaluate $\int_{-\infty}^{\infty} f(x)\,dx$ for

$$f(x) = \begin{cases} \dfrac{4}{(x+2)^2} & \text{if } x \geq 0 \\ 0 & \text{otherwise} \end{cases}$$

**SOLUTION**   The function $f$ is discontinuous at $x = 0$ (see Fig. 4). However, since $f(x) = 0$ for $x < 0$, we can proceed as follows:

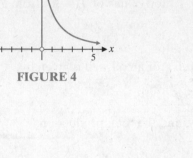

**FIGURE 4**

$$\int_{-\infty}^{\infty} f(x)\,dx = \int_{0}^{\infty} f(x)\,dx \qquad f(x) = 0 \text{ for } x < 0.$$

$$= \int_{0}^{\infty} \frac{4}{(x+2)^2}\,dx$$

$$= \lim_{b \to \infty} \int_{0}^{b} \frac{4}{(x+2)^2}\,dx$$

$$= \lim_{b \to \infty} \left[ -4(x+2)^{-1} \Big|_{0}^{b} \right]$$

$$= \lim_{b \to \infty} \left( \frac{-4}{b+2} + 2 \right)$$

$$= 2 \qquad \text{The integral converges.}$$

**MATCHED PROBLEM 4**   Evaluate $\int_{-\infty}^{\infty} f(x)\,dx$ for $f(x) = \begin{cases} 4x - x^2 & \text{if } 0 \leq x \leq 4 \\ 0 & \text{otherwise.} \end{cases}$

**EXAMPLE 5**   **Oil Production**   It is estimated that an oil well will produce oil at a rate of $R(t)$ million barrels per year $t$ years from now, as given by

$$R(t) = 8e^{-0.05t} - 8e^{-0.1t}$$

Estimate the total amount of oil that will be produced by this well.

**SOLUTION**   The total amount of oil produced in $T$ years of operation is $\int_{0}^{T} R(t)\,dt$ (see Fig. 5). At some point in the future, the annual production rate will become so low that it will

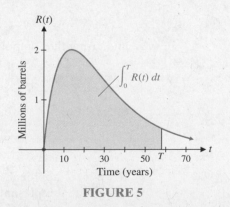

**FIGURE 5**

no longer be economically feasible to operate the well. However, since we do not know when this will occur, it is convenient to assume that the well is operated indefinitely so we can use an improper integral. Thus, the total amount of oil produced is approximately

$$\int_0^\infty R(t)\,dt = \lim_{T\to\infty}\int_0^T R(t)\,dt$$

$$= \lim_{T\to\infty}\int_0^T (8e^{-0.05t} - 8e^{-0.1t})\,dt$$

$$= \lim_{T\to\infty}\left[\left.(-160e^{-0.05t} + 80e^{-0.1t})\right|_0^T\right]$$

$$= \lim_{T\to\infty}(-160e^{-0.05T} + 80e^{-0.1T} + 80)$$

$$= 80 \quad\text{or}\quad 80{,}000{,}000 \text{ barrels}$$

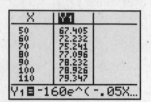

| X | Y1 |
|---|---|
| 50 | 67.405 |
| 60 | 72.232 |
| 70 | 75.241 |
| 80 | 77.096 |
| 90 | 78.232 |
| 100 | 78.926 |
| 110 | 79.347 |

Y₁⊟-160e^(-.05X...

**FIGURE 6** Total production over $[0, x]$

The total production over any finite time period $[0, T]$ is less than 80,000,000 barrels, and the larger $T$ is, the closer the total production will be to 80,000,000 barrels. Figure 6 displays the total production over $[0, T]$ for values of $T(= X)$ from 50 to 110 years. ▬

MATCHED PROBLEM 5  The annual production rate (in millions of barrels) for an oil well is given by

$$R(t) = 4e^{-0.1t} - e^{-0.2t}$$

Assuming that the well is operated indefinitely, find the total production.

## APPLICATION: CAPITAL VALUE

Recall that if money is invested at a rate $r$ compounded continuously for $T$ years, then the present value, $PV$, and the future value, $FV$, are related by the continuous compound interest formula (see Section 4-1 of *Calculus* 11/e or Section 11-1 of *College Mathematics* 11/e)

$$FV = PVe^{rT}$$

which also can be written as

$$PV = e^{-rT}FV$$

This relationship is valid for both single deposits and amounts generated by continuous income streams. Thus, if $f(t)$ is a continuous income stream, the future value is given by (see Section 7-2 of *Calculus* 11/e or Section 14-2 of *College Mathematics* 11/e)

$$FV = e^{rT}\int_0^T f(t)e^{-rt}\,dt \quad \text{Future value}$$

and, using the compound interest formula, the present value is given by

$$PV = e^{-rT}FV = e^{-rT}e^{rT}\int_0^T f(t)e^{-rt}\,dt$$

Since $e^{-rT}e^{rT} = 1$, we have

$$PV = \int_0^T f(t)e^{-rt}\,dt \quad \text{Present value}$$

If we let $T$ approach $\infty$ in this formula for present value, we obtain an improper integral that represents the *capital value* of the income stream.

**DEFINITION** **Capital Value of a Perpetual Income Stream**

A continuous income stream is called **perpetual** if it never stops producing income. The **capital value, CV,** of a perpetual income stream $f(t)$ at a rate $r$ compounded continuously is the present value over the time interval $[0, \infty)$. That is,

$$CV = \int_0^\infty f(t)e^{-rt}\, dt$$

Capital value provides a method for expressing the worth (in terms of today's dollars) of an investment that will produce income for an indefinite period of time.

**EXAMPLE 6** **Capital Value** A family has leased the oil rights of a property to a petroleum company in return for a perpetual annual payment of $1,200. Find the capital value of this lease at 5% compounded continuously.

**SOLUTION** The annual payments from the oil company produce a continuous income stream with rate of flow $f(t) = 1,200$ that continues indefinitely. (It is common practice to treat a sequence of equal periodic payments as a continuous income stream with a constant rate of flow, even if the income is received only at the end of each period.) Thus, the capital value is

$$CV = \int_0^\infty f(t)e^{-rt}\, dt$$

$$CV = \int_0^\infty 1,200e^{-0.05t}\, dt$$

$$= \lim_{T\to\infty} \int_0^T 1,200e^{-0.05t}\, dt$$

$$= \lim_{T\to\infty} \left[ -24,000e^{-0.05t} \Big|_0^T \right]$$

$$= \lim_{T\to\infty} (-24,000e^{-0.05T} + 24,000) = \$24,000 \quad \blacksquare$$

**MATCHED PROBLEM 6** Repeat Example 6 if the interest rate is 6% compounded continuously.

*Answers to Matched Problems* **1.** $\frac{1}{2}$ **2.** 1 **3.** Diverges **4.** $\frac{32}{3}$ **5.** 35,000,000 barrels **6.** $20,000

# Exercise 3-1

A *In Problems 1–16, find the value of each improper integral that converges.*

**1.** $\int_1^\infty \dfrac{dx}{x^3}$

**2.** $\int_1^\infty x\, dx$

**3.** $\int_1^\infty x^2\, dx$

**4.** $\int_1^\infty \dfrac{dx}{x^4}$

**5.** $\int_{-\infty}^0 e^{x/2}\, dx$

**6.** $\int_{-\infty}^{-1} \dfrac{dx}{x}$

**7.** $\int_4^\infty \dfrac{dx}{100\sqrt{x}}$

**8.** $\int_1^\infty e^{-2x}\, dx$

**9.** $\int_9^\infty \dfrac{dx}{x\sqrt{x}}$

**10.** $\int_8^\infty \dfrac{dx}{\sqrt[3]{x}}$

**11.** $\int_0^\infty \dfrac{dx}{(x+1)^{2/3}}$

**12.** $\int_0^\infty \dfrac{dx}{\sqrt{x+1}}$

**13.** $\int_1^\infty \dfrac{dx}{x^{0.99}}$

**14.** $\int_1^\infty \dfrac{dx}{x^{1.01}}$

**15.** $0.3\int_0^\infty e^{-0.3x}\, dx$

**16.** $0.01\int_0^\infty e^{-0.1x}\, dx$

*In Problems 17–22, graph $y = f(x)$ and find the value of $\int_{-\infty}^{\infty} f(x)\, dx$ if it converges.*

17. $f(x) = \begin{cases} 1 + x^2 & \text{if } 0 \le x \le 2 \\ 0 & \text{otherwise} \end{cases}$

18. $f(x) = \begin{cases} 2 + x & \text{if } 0 \le x \le 3 \\ 0 & \text{otherwise} \end{cases}$

19. $f(x) = \begin{cases} e^{-0.1x} & \text{if } x \ge 0 \\ 0 & \text{otherwise} \end{cases}$

20. $f(x) = \begin{cases} 2 - e^{-0.2x} & \text{if } x \ge 0 \\ 0 & \text{otherwise} \end{cases}$

21. $f(x) = \begin{cases} 4/(x + 2) & \text{if } x \ge 2 \\ 0 & \text{otherwise} \end{cases}$

22. $f(x) = \begin{cases} 4/(x + 1)^2 & \text{if } x \ge 1 \\ 0 & \text{otherwise} \end{cases}$

*In Problems 23–26, discuss the validity of each statement. If the statement is always true, explain why. If not, give a counterexample.*

23. If $f$ is a continuous increasing function on $[0, \infty)$ such that $f(x) > 0$ for all $x$, then $\int_0^{\infty} f(x)\, dx$ diverges.

24. If $f$ is a continuous decreasing function on $[0, \infty)$ such that $f(x) > 0$ for all $x$, then $\int_0^{\infty} f(x)\, dx$ converges.

25. If $f$ is a continuous function on $[1, \infty)$ such that $f(x) > 0$ for all $x$ and $f(x) \ge 10$ for $1 \le x \le 1{,}000$, then $\int_1^{\infty} f(x)\, dx$ diverges.

26. If $f$ is a continuous decreasing function on $[1, \infty)$ such that $0 < f(x) \le 0.001$ for all $x$, then $\int_1^{\infty} f(x)\, dx$ converges.

*In Problems 27–36, find $F(b)$, use a graphing utility to graph $F(b)$, and use the graph to estimate $\lim\limits_{b \to \infty} F(b)$ or, in Problems 31 and 32, $\lim\limits_{b \to -\infty} F(b)$.*

27. From Problem 1, $F(b) = \int_1^b \dfrac{dx}{x^3}$

28. From Problem 2, $F(b) = \int_1^b x\, dx$

29. From Problem 3, $F(b) = \int_1^b x^2\, dx$

30. From Problem 4, $F(b) = \int_1^b \dfrac{dx}{x^4}$

31. From Problem 5, $F(b) = \int_b^0 e^{x/2}\, dx$

32. From Problem 6, $F(b) = \int_b^{-1} \dfrac{dx}{x}$

33. From Problem 7, $F(b) = \int_4^b \dfrac{dx}{100\sqrt{x}}$

34. From Problem 8, $F(b) = \int_1^b e^{-2x}\, dx$

35. From Problem 9, $F(b) = \int_9^b \dfrac{dx}{x\sqrt{x}}$

36. From Problem 10, $F(b) = \int_8^b \dfrac{dx}{\sqrt[3]{x}}$

37. If $f$ is continuous on $[0, \infty)$ and $\int_0^{\infty} f(x)\, dx$ converges, does $\int_1^{\infty} f(x)\, dx$ converge? Explain.

38. If $f$ is continuous on $[0, \infty)$ and $\int_0^{\infty} f(x)\, dx$ diverges, does $\int_1^{\infty} f(x)\, dx$ diverge? Explain.

C  *In Problems 39–48, find the value of each improper integral that converges.*

39. $\displaystyle\int_0^{\infty} \dfrac{1}{k} e^{-x/k}\, dx,\ k > 0$

40. $\displaystyle\int_1^{\infty} \dfrac{1}{x^p}\, dx,\ p > 1$

41. $\displaystyle\int_{-\infty}^{\infty} \dfrac{x}{1 + x^2}\, dx$

42. $\displaystyle\int_{-\infty}^{\infty} xe^{-x^2}\, dx$

43. $\displaystyle\int_0^{\infty} (e^{-x} - e^{-2x})\, dx$

44. $\displaystyle\int_0^{\infty} (e^{-x} + 2e^x)\, dx$

45. $\displaystyle\int_{-\infty}^0 \dfrac{1}{\sqrt{1 - x}}\, dx$

46. $\displaystyle\int_{-\infty}^0 \dfrac{1}{\sqrt[3]{(1 - x)^4}}\, dx$

47. $\displaystyle\int_1^{\infty} \dfrac{\ln x}{x}\, dx$

48. $\displaystyle\int_0^{\infty} \dfrac{e^{-x}}{1 + e^{-x}}\, dx$

## Applications

49. *Capital value.* The perpetual annual rent for a property is \$6,000. Find the capital value at 5% compounded continuously.

50. *Capital value.* The perpetual annual rent for a property is \$10,000. Find the capital value at 4% compounded continuously.

51. *Capital value.* A trust fund produces a perpetual stream of income with rate of flow

$$f(t) = 1{,}500e^{0.04t}$$

Find the capital value at 7% compounded continuously.

52. *Capital value.* A trust fund produces a perpetual stream of income with rate of flow

$$f(t) = 1{,}000e^{0.02t}$$

Find the capital value at 6% compounded continuously.

53. *Capital value.* Refer to Problem 49. Discuss the effect on the capital value if the interest rate is increased to 6%, and then if decreased to 4%. Interpret these results.

54. *Capital value.* Refer to Problem 50. Discuss the effect on the capital value if the interest rate is increased to 5%, and then if decreased to 2%. Interpret these results.

**55.** *Production.* The rate of production of a natural gas well (in billions of cubic feet per year) is given by (see figure below)

$$R(t) = 3e^{-0.2t} - 3e^{-0.4t}$$

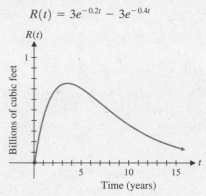

(A) Assuming that the well is operated indefinitely, find the total production.

(B) When will the output from the well reach 50% of the total production? Round answer to two decimal places.

**56.** *Production.* The rate of production of an oil well (in millions of barrels per year) is given by

$$R(t) = \frac{1{,}000t}{(50 + t^2)^2}$$

(A) Assuming that the well is operated indefinitely, find the total production.

(B) When will the output from the well reach 50% of the total production? Round answer to two decimal places.

**57.** *Pollution.* It has been estimated that the rate of seepage of toxic chemicals from a waste dump is $R(t)$ gallons per year $t$ years from now, where

$$R(t) = \frac{500}{(1 + t)^2}$$

Assuming that this seepage continues indefinitely, find the total amount of toxic chemicals that seep from the dump.

**58.** *Drug assimilation.* When a person takes a drug, the body does not assimilate all of the drug. One way to determine the amount of the drug assimilated is to measure the rate at which the drug is eliminated from the body. If the rate of elimination of the drug (in milliliters per minute) is given by

$$R(t) = 3e^{-0.1t} - 3e^{-0.3t}$$

where $t$ is the time in minutes since the drug was administered, how much of the drug is eliminated from the body?

**59.** *Immigration.* In order to control rising population, a country is planning to institute a new policy for immigration. Analysts estimate that the rate of immigration into the country $t$ years after the policy is instituted (in millions of immigrants per year) will be given by

$$R(t) = \frac{400}{(5 + t)^3}$$

Find the total number of immigrants that will enter the country under this policy, assuming that the policy is applied indefinitely.

## Section 3-2  CONTINUOUS RANDOM VARIABLES

- Continuous Random Variables
- Probability Density Functions
- Cumulative Distribution Functions

### ■ Continuous Random Variables

Probability theory is concerned with determining the long-run frequency of the outcomes of an experiment when the outcome each time the experiment is performed is uncertain. For example, when a fair coin is tossed, it is uncertain whether it will turn up heads or tails. However, probability theory enables us to predict that the coin will turn up heads approximately 50% of the time.

Assigning a value to each outcome of an experiment defines a function called a *random variable.* For example, consider the experiment of tossing three coins and the function $X$ whose values are the number of heads. The possible outcomes of the experiment and the corresponding values of $X$ are shown in Table 1. Since the set of possible outcomes for this experiment is a finite set, $X$ is called a **discrete random variable.**

Things are more complicated if an experiment has an infinite number of outcomes. For example, consider the experiment of turning on light bulbs and letting them remain lit until they burn out. One bulb may be defective and not light at all. Another may still be lit 50 or 100 years from now. Let $X$ be the function whose values

| TABLE 1 Number of Heads in the Toss of Three Coins | |
|---|---|
| Sample Space $S$ | Number of Heads $X(e_j)$ |
| $e_1$: TTT | 0 |
| $e_2$: TTH | 1 |
| $e_3$: THT | 1 |
| $e_4$: HTT | 1 |
| $e_5$: THH | 2 |
| $e_6$: HTH | 2 |
| $e_7$: HHT | 2 |
| $e_8$: HHH | 3 |

(in hours) are the life expectancies of the light bulbs. Theoretically, there is no reason to exclude any of the values in the interval $[0, \infty)$ as a possibility for the life expectancy of the bulb. Thus, we assume that the range of $X$ is $[0, \infty)$, and we call $X$ a *continuous random variable*.

---

**DEFINITION** **Continuous Random Variable**

A **continuous random variable** $X$ is a function that assigns a numeric value to each outcome of an experiment. The set of possible values of $X$ is an interval of real numbers. This interval may be open or closed, and it may be bounded or unbounded.

---

The term *continuous* is not used in the same sense here as it was used in Section 3-2 of *Calculus* 11/e or Section 10-2 of *College Mathematics* 11/e. In the present case, it refers to the fact that the values of the random variable form a continuous set of real numbers, such as $[0, \infty)$, rather than a discrete set, such as $\{0, 1, 2, 3\}$ or $\{2, 4, 8, \dots\}$.

*Explore & Discuss **1***

For each of the following experiments, determine whether the indicated random variable is discrete or continuous.

(A) Two dice are rolled, and $X$ is the sum of the dots on the upturned faces of the dice.

(B) A dart is thrown at a circular dart board, and $X$ is the distance from the dart to the center of the board.

(C) The drive-up window of a bank is open 10 hours a day, and $X$ is the number of customers served in one day.

(D) The drive-up window of a bank is open 10 hours a day, and $X$ is the time (in minutes) each customer must wait for service.

## ■ Probability Density Functions

If $X$ is a discrete random variable, then the probability that $X$ lies in a given interval can be computed by addition. For example, consider the experiment of tossing three coins, and let $X$ represent the number of heads. The *probability distribution* for $X$ is shown in Table 2 and illustrated in the histogram in Figure 1. We denote the probability that $X$ is between 0 and 2, inclusively, by $P(0 \le X \le 2)$ and compute it by addition:

$$P(0 \le X \le 2) = P(X = 0) + P(X = 1) + P(X = 2)$$
$$= \tfrac{1}{8} + \tfrac{3}{8} + \tfrac{3}{8} = \tfrac{7}{8}$$

**TABLE 2** Probability Distribution

| Number of Heads $x$ | 0 | 1 | 2 | 3 |
|---|---|---|---|---|
| Probability $p(x)$ | $\frac{1}{8}$ | $\frac{3}{8}$ | $\frac{3}{8}$ | $\frac{1}{8}$ |

If $X$ is a continuous random variable, the same approach will not work. Since $X$ can now assume any real number in the interval $[0, 2]$, it is impossible to write $P(0 \le X \le 2)$ as a finite or even an infinite sum. (What would be the second term in such a sum? That is, what is the "next" real number after 0? Think about this.) Instead, we introduce a new type of function, called a *probability density function,* and use integrals involving this function to compute the probability that a continuous random variable $X$ lies in a given interval. For example, we might use a probability density function to find the probability that the actual amount of soda in a 12-ounce can is between 11.9 and 12.1 ounces, or that the speed of a car involved in an accident was between 60 and 65 miles per hour.

For convenience in stating definitions and formulas, we will assume that the value of a continuous random variable can be any real number; that is, the range is $(-\infty, \infty)$.

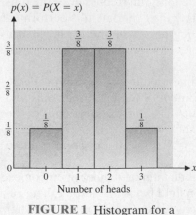

$p(x) = P(X = x)$

**FIGURE 1** Histogram for a probability distribution

**DEFINITION**   **Probability Density Function**

The function $f(x)$ is a **probability density function** for a continuous random variable $X$ if

**1.** $f(x) \geq 0$ for all $x \in (-\infty, \infty)$

**2.** $\displaystyle\int_{-\infty}^{\infty} f(x)\, dx = 1$

**3.** The probability that $X$ lies in the interval $[c, d]$ is given by

$$P(c \leq X \leq d) = \int_{c}^{d} f(x)\, dx$$

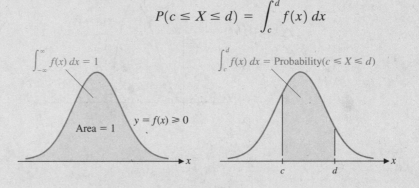

Range of $X = (-\infty, \infty) =$ Domain of $f$

---

**EXAMPLE 1**   **Using a Probability Density Function**  Let

$$f(x) = \begin{cases} 12x^2 - 12x^3 & \text{if } 0 \leq x \leq 1 \\ 0 & \text{otherwise} \end{cases}$$

Graph $f$ and verify that $f$ satisfies the first two conditions for a probability density function. Then find each of the following probabilities and illustrate graphically:

(A) $P(.4 \leq X \leq .7)$      (B) $P(X \leq .5)$

(C) $P(X \geq .6)$      (D) $P(X = .3)$

**SOLUTION**   The graph of $f$ (Fig. 2) shows that $f(x) \geq 0$ for all $x$. Also,

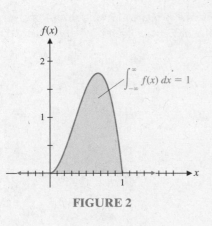

**FIGURE 2**

$$\int_{-\infty}^{\infty} f(x)\, dx = \int_{0}^{1} (12x^2 - 12x^3)\, dx = (4x^3 - 3x^4)\Big|_{0}^{1} = 1$$

(A) $P(.4 \leq X \leq .7) = \displaystyle\int_{.4}^{.7} f(x)\, dx$

$$= \int_{.4}^{.7} (12x^2 - 12x^3)\, dx = (4x^3 - 3x^4)\Big|_{.4}^{.7} = .4725$$

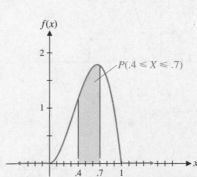

(B) $P(X \le .5) = \int_0^{.5} f(x)\,dx$     *Note that f(x) = 0 for x < 0.*

$$= \int_0^{.5} (12x^2 - 12x^3)\,dx = (4x^3 - 3x^4)\Big|_0^{.5} = .3125$$

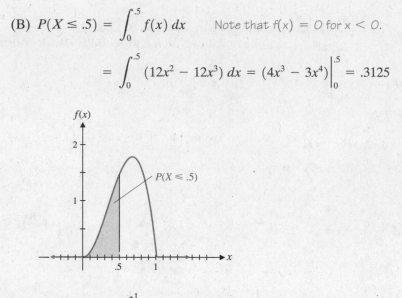

(C) $P(X \ge .6) = \int_{.6}^{1} f(x)\,dx$     *Note that f(x) = 0 for x > 1.*

$$= \int_{.6}^{1} (12x^2 - 12x^3)\,dx = (4x^3 - 3x^4)\Big|_{.6}^{1} = .5248$$

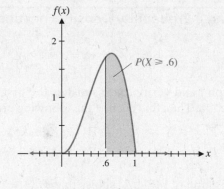

(D) $P(X = .3) = \int_{.3}^{.3} f(x)\,dx = 0$     *Definite integral property 1 (Section 6-4 of Calculus 11/e or Section 13-4 of College Mathematics 11/e)*

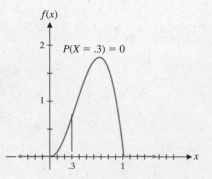

**MATCHED PROBLEM 1**   Let

$$f(x) = \begin{cases} 6x - 6x^2 & \text{if } 0 \le x \le 1 \\ 0 & \text{otherwise} \end{cases}$$

Graph $f$ and verify that $f$ satisfies the first two conditions for a probability density function. Then find each of the following probabilities and illustrate graphically:

(A) $P(.3 \le X \le .8)$    (B) $P(X \le .7)$

(C) $P(X \ge .4)$    (D) $P(X = .5)$

---

### INSIGHT

There is a fundamental difference between discrete and continuous random variables. In the discrete case, there is a *probability distribution* $P(x)$ that gives the probability of each possible value of the random variable. The probability, for example, that a roll of a fair die will give a 5, is equal to $\frac{1}{6}$. In the continuous case, the probability that an outcome is exactly 5 (not 5.01, not 4.999) is equal to 0. For in the continuous case it is the integral of the *probability density function* $f(x)$ that gives the probability that the outcome lies in a certain interval:

$$P(X = 5) = P(5 \le X \le 5) = \int_5^5 f(x)\, dx = 0$$

If $X$ is a continuous random variable and $c$ is any real number, then the probability that the outcome is exactly $c$ is

$$P(X = c) = P(c \le X \le c) = \int_c^c f(x)\, dx = 0$$

The fact that $P(X = c) = 0$ also implies that excluding either end point from an interval does not change the probability that the random variable lies in that interval; that is,

$$P(a < X < b) = P(a < X \le b) = P(a \le X < b) = P(a \le X \le b)$$

$$= \int_a^b f(x)\, dx$$

---

**EXAMPLE 2**    **Using a Probability Density Function**  Use the probability density function in Example 1 to compute $P(.1 < X \le .2)$ and $P(X > .9)$.

**SOLUTION**   $\displaystyle P(.1 < X \le .2) = \int_{.1}^{.2} f(x)\, dx \qquad f(x) = \begin{cases} 12x^2 - 12x^3 & \text{if } 0 \le x \le 1 \\ 0 & \text{otherwise} \end{cases}$

$$= \int_{.1}^{.2} (12x^2 - 12x^3)\, dx$$

$$= (4x^3 - 3x^4)\Big|_{.1}^{.2}$$

$$= .0272 - .0037 = .0235$$

$$P(X > .9) = \int_{.9}^{\infty} f(x)\, dx$$

$$= \int_{.9}^{1} (12x^2 - 12x^3)\, dx$$

$$= (4x^3 - 3x^4)\Big|_{.9}^{1}$$

$$= 1 - .9477 = .0523$$

| MATCHED PROBLEM 2 | Use the probability density function in Matched Problem 1 to compute the probabilities $P(.2 \leq X < .4)$ and $P(X < .8)$. |

---

EXAMPLE 3

**Shelf Life** The shelf life (in months) of a certain drug is a continuous random variable with probability density function (see Fig. 3)

$$f(x) = \begin{cases} 50/(x + 50)^2 & \text{if } x \geq 0 \\ 0 & \text{otherwise} \end{cases}$$

Find the probability that the drug has a shelf life of

(A) Between 10 and 20 months

(B) At most 30 months

(C) More than 25 months

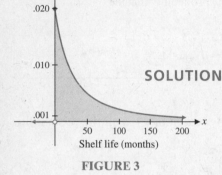

**FIGURE 3**

**SOLUTION** (A) $P(10 \leq X \leq 20) = \int_{10}^{20} f(x)\, dx = \int_{10}^{20} \dfrac{50}{(x + 50)^2}\, dx = \dfrac{-50}{x + 50}\Big|_{10}^{20}$

$= \left(-\tfrac{50}{70}\right) - \left(-\tfrac{50}{60}\right) = \tfrac{5}{42}$

(B) $P(X \leq 30) = \int_{-\infty}^{30} f(x)\, dx = \int_{0}^{30} \dfrac{50}{(x + 50)^2}\, dx = \dfrac{-50}{x + 50}\Big|_{0}^{30}$

$= \left(-\tfrac{50}{80}\right) - (-1) = \tfrac{3}{8}$

(C) $P(X > 25) = \int_{25}^{\infty} f(x)\, dx = \int_{25}^{\infty} \dfrac{50}{(x + 50)^2}\, dx$

This improper integral can be evaluated directly using the techniques discussed in Section 3-1. However, there is another method that does not involve evaluation of any improper integrals. Since $f$ is a probability density function, we can write

$$1 = \int_{-\infty}^{\infty} f(x)\, dx$$

$$= \int_{0}^{\infty} \dfrac{50}{(x + 50)^2}\, dx \qquad \int_{a}^{b} f(x)\, dx = \int_{a}^{c} f(x)\, dx + \int_{c}^{b} f(x)\, dx$$

$$= \int_{0}^{25} \dfrac{50}{(x + 50)^2}\, dx + \int_{25}^{\infty} \dfrac{50}{(x + 50)^2}\, dx$$

Solving this last equation for $\int_{25}^{\infty} [50/(x + 50)^2]\, dx$, we have

$$\int_{25}^{\infty} \dfrac{50}{(x + 50)^2}\, dx = 1 - \int_{0}^{25} \dfrac{50}{(x + 50)^2}\, dx$$

$$= 1 - \dfrac{-50}{x + 50}\Big|_{0}^{25}$$

$$= 1 - \left(\tfrac{-50}{75}\right) - 1 = \tfrac{2}{3}$$

Thus, $P(X > 25) = \tfrac{2}{3}$.

| MATCHED PROBLEM 3 | In Example 3, find the probability that the drug has a shelf life of |

(A) Between 50 and 100 months

(B) At most 20 months

(C) More than 10 months

## Cumulative Distribution Functions

*Explore & Discuss* **2**    Let

$$f(x) = \begin{cases} \frac{1}{2} & \text{if } 0 \le x \le 2 \\ 0 & \text{otherwise} \end{cases} \quad \text{and} \quad F(x) = \begin{cases} 0 & \text{if } x < 0 \\ \frac{1}{2}x & \text{if } 0 \le x \le 2 \\ 1 & \text{if } x > 2 \end{cases}$$

(A) Graph each function.

(B) Is $f$ a probability density function? Is $F$?

(C) Is there any relationship between $f$ and $F$?

(D) Evaluate $\int_{.5}^{1.5} f(x)\, dx$.

(E) Can you use $F$ to evaluate the integral in part (D)?

Each time we compute the probability for a continuous random variable, we must find the antiderivative of the probability density function. This antiderivative is used so often that it is convenient to give it a name.

---

**DEFINITION**   **Cumulative Distribution Function**

If $f$ is a probability density function, then the associated **cumulative distribution function $F$** is defined by

$$F(x) = P(X \le x) = \int_{-\infty}^{x} f(t)\, dt$$

Furthermore,

$$P(c \le X \le d) = F(d) - F(c)$$

Figure 4 gives a geometric interpretation of these ideas.

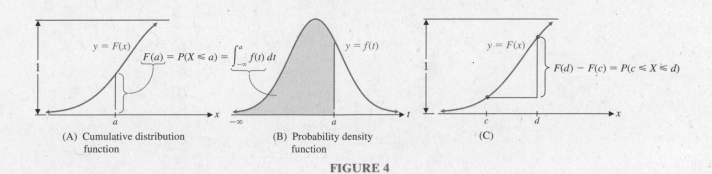

(A) Cumulative distribution function

(B) Probability density function

(C)

**FIGURE 4**

Notice that $F(x) = \int_{-\infty}^{x} f(t)\, dt$ is a function of $x$, the upper limit of integration, not $t$, the variable in the integrand. We state some important properties of cumulative distribution functions in the next box. These properties follow directly from the fact that $F(x)$ can be interpreted geometrically as the area under the graph of $y = f(t)$ from $-\infty$ to $x$ (see Fig. 4B on preceding page).

---

**PROPERTIES**   **Cumulative Distribution Functions**

If $f$ is a probability density function and

$$F(x) = \int_{-\infty}^{x} f(t)\, dt$$

is the associated cumulative distribution function, then

1. $F'(x) = f(x)$ wherever $f$ is continuous
2. $0 \le F(x) \le 1$,   $-\infty < x < \infty$
3. $F(x)$ is nondecreasing on $(-\infty, \infty)$*

---
*A function $F(x)$ is nondecreasing on $(a, b)$ if $F(x_1) \le F(x_2)$ for $a < x_1 < x_2 < b$.

---

**EXAMPLE 4**   **Using a Cumulative Distribution Function**  Find the cumulative distribution function for the probability density function in Example 1, and use it to compute $P(.1 \le X \le .9)$.

**SOLUTION**   If $x < 0$, then

$$F(x) = \int_{-\infty}^{x} f(t)\, dt \qquad f(x) = \begin{cases} 12x^2 - 12x^3 & \text{if } 0 \le x \le 1 \\ 0 & \text{otherwise} \end{cases}$$

$$= \int_{-\infty}^{x} 0\, dt = 0$$

If $0 \le x \le 1$, then

$$F(x) = \int_{-\infty}^{x} f(t)\, dt = \int_{-\infty}^{0} f(t)\, dt + \int_{0}^{x} f(t)\, dt$$

$$= 0 + \int_{0}^{x} (12t^2 - 12t^3)\, dt = (4t^3 - 3t^4)\Big|_{0}^{x}$$

$$= 4x^3 - 3x^4$$

If $x > 1$, then

$$F(x) = \int_{-\infty}^{x} f(t)\, dt = \int_{-\infty}^{0} f(t)\, dt + \int_{0}^{1} f(t)\, dt + \int_{1}^{x} f(t)\, dt$$

$$= 0 + 1 + 0 = 1$$

Thus,

$$F(x) = \begin{cases} 0 & \text{if } x < 0 \\ 4x^3 - 3x^4 & \text{if } 0 \le x \le 1 \\ 1 & \text{if } x > 1 \end{cases}$$

And

$$P(.1 \le X \le .9) = F(.9) - F(.1) = .9477 - .0037 = .944$$

See Figure 5.

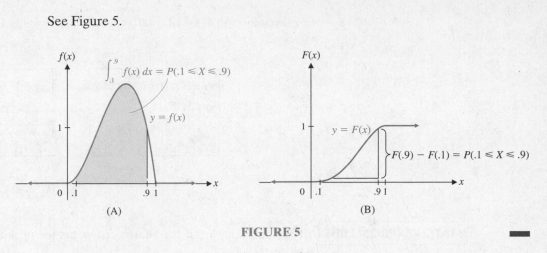

**FIGURE 5**

---

**MATCHED PROBLEM 4**  Find the cumulative distribution function for the probability density function in Matched Problem 1, and use it to compute $P(.3 \le X \le .7)$.

---

**EXAMPLE 5**  **Shelf Life**  Returning to the discussion of the shelf life of a drug in Example 3, suppose a pharmacist wants to be 95% certain that the drug is still good when it is sold. How long is it safe to leave the drug on the shelf?

**SOLUTION**  Let $x$ be the number of months the drug has been on the shelf when it is sold. The probability that the shelf life of the drug is less than the number of months it has been sitting on the shelf is $P(0 \le X \le x)$. The pharmacist wants this probability to be .05. Thus, we must solve the equation $P(0 \le X \le x) = .05$ for $x$. First, we will find the cumulative distribution function $F$. For $x < 0$, we see that $F(x) = 0$. For $x \ge 0$,

$$F(x) = \int_0^x \frac{50}{(50 + t)^2}\, dt = \frac{-50}{50 + t}\bigg|_0^x = \frac{-50}{50 + x} - (-1) = 1 - \frac{50}{50 + x}$$

$$= \frac{x}{50 + x}$$

Thus,

$$F(x) = \begin{cases} 0 & \text{if } x < 0 \\ x/(50 + x) & \text{if } x \ge 0 \end{cases}$$

Now, to solve the equation $P(0 \le X \le x) = .05$, we solve

$$F(x) - F(0) = .05 \qquad\qquad F(0) = 0$$

$$\frac{x}{50 + x} = .05$$

$$x = 2.5 + .05x$$

$$.95x = 2.5$$

$$x \approx 2.6$$

If the drug is sold during the first 2.6 months it is on the shelf, then the probability that it is still good is .95.

Figure 6 shows an alternative method for solving Example 5 on a graphing utility.

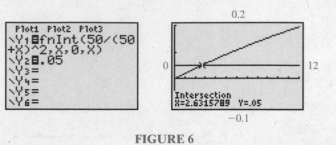

**FIGURE 6**

MATCHED PROBLEM 5    Repeat Example 5 if the pharmacist wants the probability that the drug is still good to be .99.

*Answers to Matched Problems*    **1.**

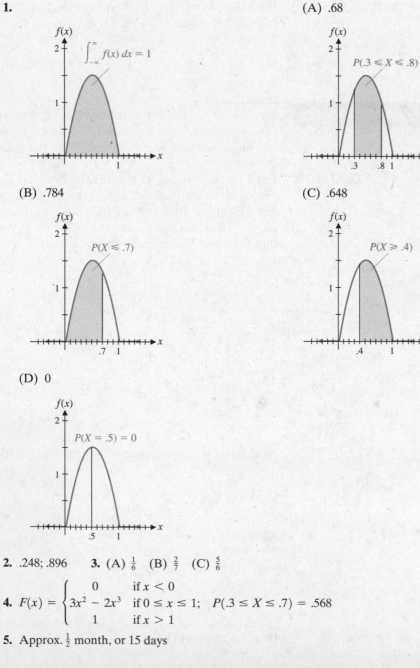

**2.** .248; .896    **3.** (A) $\frac{1}{6}$    (B) $\frac{2}{7}$    (C) $\frac{5}{6}$

**4.** $F(x) = \begin{cases} 0 & \text{if } x < 0 \\ 3x^2 - 2x^3 & \text{if } 0 \le x \le 1; \quad P(.3 \le X \le .7) = .568 \\ 1 & \text{if } x > 1 \end{cases}$

**5.** Approx. $\frac{1}{2}$ month, or 15 days

## Exercise 3-2

**A** *In Problems 1 and 2, graph f, and show that f satisfies the first two conditions for a probability density function.*

**1.**  $f(x) = \begin{cases} \frac{1}{8}x & \text{if } 0 \le x \le 4 \\ 0 & \text{otherwise} \end{cases}$

**2.**  $f(x) = \begin{cases} \frac{1}{9}x^2 & \text{if } 0 \le x \le 3 \\ 0 & \text{otherwise} \end{cases}$

*In Problems 3–6, is f a probability density function? Explain.*

**3.**  $f(x) = \begin{cases} 2 + 2x & \text{if } -1 \le x \le 0 \\ 0 & \text{otherwise} \end{cases}$

**4.**  $f(x) = \begin{cases} e^{-x} & \text{if } 0 \le x \le 4 \\ 0 & \text{otherwise} \end{cases}$

**5.**  $f(x) = \begin{cases} 0.2 & \text{if } 1 \le x \le 5 \\ 0 & \text{otherwise} \end{cases}$

**6.**  $f(x) = \begin{cases} 3x^2 - \dfrac{1}{2} & \text{if } -1 \le x \le 1 \\ 0 & \text{otherwise} \end{cases}$

**7.**  Use the function in Problem 1 to find the indicated probabilities. Illustrate each probability with a graph.

(A)  $P(1 \le X \le 3)$

(B)  $P(X \le 2)$

(C)  $P(X > 3)$

**8.**  Use the function in Problem 2 to find the indicated probabilities. Illustrate each probability with a graph.

(A)  $P(1 \le X \le 2)$

(B)  $P(X \ge 1)$

(C)  $P(X < 2)$

**9.**  Use the function in Problem 1 to find the indicated probabilities.

(A)  $P(X = 1)$

(B)  $P(X > 5)$

(C)  $P(X < 5)$

**10.**  Use the function in Problem 2 to find the indicated probabilities.

(A)  $P(X = 2)$

(B)  $P(X > 4)$

(C)  $P(X < 4)$

**11.**  Find and graph the cumulative distribution function associated with the function in Problem 1.

**12.**  Find and graph the cumulative distribution function associated with the function in Problem 2.

**13.**  Use the cumulative distribution function from Problem 11 to find the indicated probabilities.

(A)  $P(2 \le X \le 4)$

(B)  $P(0 < X < 2)$

**14.**  Use the cumulative distribution function from Problem 12 to find the indicated probabilities.

(A)  $P(0 \le X \le 1)$

(B)  $P(2 < X < 3)$

**15.**  Use the cumulative distribution function from Problem 11 to find the value of $x$ that satisfies each equation.

(A)  $P(0 \le X \le x) = \frac{1}{4}$

(B)  $P(0 \le X \le x) = \frac{1}{9}$

**16.**  Use the cumulative distribution function from Problem 12 to find the value of $x$ that satisfies each equation.

(A)  $P(0 \le X \le x) = \frac{1}{8}$

(B)  $P(0 \le X \le x) = \frac{1}{64}$

**B** *In Problems 17 and 18, graph f, and show that f satisfies the first two conditions for a probability density function.*

**17.**  $f(x) = \begin{cases} 2/(1 + x)^3 & \text{if } x \ge 0 \\ 0 & \text{otherwise} \end{cases}$

**18.**  $f(x) = \begin{cases} 2/(2 + x)^2 & \text{if } x \ge 0 \\ 0 & \text{otherwise} \end{cases}$

**19.**  Use the function in Problem 17 to find the indicated probabilities.

(A)  $P(1 \le X \le 4)$

(B)  $P(X > 3)$

(C)  $P(X \le 2)$

**20.**  Use the function in Problem 18 to find the indicated probabilities.

(A)  $P(2 \le X \le 8)$

(B)  $P(X \ge 3)$

(C)  $P(X < 1)$

**21.**  Find and graph the cumulative distribution function associated with the function in Problem 17.

**22.**  Find and graph the cumulative distribution function associated with the function in Problem 18.

**23.**  Use the cumulative distribution function from Problem 21 to find the value of $x$ that satisfies each equation.

(A)  $P(0 \le X \le x) = \frac{3}{4}$

(B)  $P(X \ge x) = \frac{1}{16}$

**24.**  Use the cumulative distribution function from Problem 22 to find the value of $x$ that satisfies each equation.

(A)  $P(0 \le X \le x) = \frac{3}{4}$     (B)  $P(X > x) = \frac{1}{3}$

*In Problems 25–28, find the associated cumulative distribution function. Graph both functions (on separate sets of axes).*

**25.**  $f(x) = \begin{cases} \frac{3}{2}x - \frac{3}{4}x^2 & \text{if } 0 \le x \le 2 \\ 0 & \text{otherwise} \end{cases}$

**26.**  $f(x) = \begin{cases} \frac{3}{4} - \frac{3}{4}x^2 & \text{if } -1 \le x \le 1 \\ 0 & \text{otherwise} \end{cases}$

**27.**  $f(x) = \begin{cases} \frac{1}{2} + \frac{1}{2}x^3 & \text{if } -1 \le x \le 1 \\ 0 & \text{otherwise} \end{cases}$

**28.**  $f(x) = \begin{cases} \frac{3}{4} - \frac{3}{8}\sqrt{x} & \text{if } 0 \le x \le 4 \\ 0 & \text{otherwise} \end{cases}$

*In Problems 29–32, use a graphing utility to approximate (to two decimal places) the value of x that satisfies the given equation for the indicated cumulative distribution function F(x).*

**29.** $P(0 \le X \le x) = .2$ for $F(x)$ from Problem 25

**30.** $P(-1 \le X \le x) = .4$ for $F(x)$ from Problem 26

**31.** $P(-1 \le X \le x) = .6$ for $F(x)$ from Problem 27

**32.** $P(0 \le X \le x) = .7$ for $F(x)$ from Problem 28

*For each function f in Problems 33–36, find a constant k so that kf is a probability density function, or explain why no such constant exists.*

**33.** $f(x) = \begin{cases} e^{-x/3} & \text{if } x \ge 0 \\ 0 & \text{otherwise} \end{cases}$

**34.** $f(x) = \begin{cases} e^{-x/3} & \text{if } x \le 0 \\ 0 & \text{otherwise} \end{cases}$

**35.** $f(x) = \begin{cases} 1/x^5 & \text{if } x \ge 1 \\ 0 & \text{otherwise} \end{cases}$

**36.** $f(x) = \begin{cases} x^3 & \text{if } 0 \le x \le 4 \\ 0 & \text{otherwise} \end{cases}$

**37.** Can the function

$$F(x) = \begin{cases} 2 - 2x & \text{if } 0 \le x \le 1 \\ 0 & \text{otherwise} \end{cases}$$

be a cumulative distribution function? Explain.

**38.** Can the function

$$F(x) = \begin{cases} 0.5 - 0.5x^3 & \text{if } -1 \le x \le 1 \\ 0 & \text{otherwise} \end{cases}$$

be a cumulative distribution function? Explain.

C *In Problems 39 and 40, use a graphing utility to graph f and compute the integrals $\int_{-N}^{N} f(x)\,dx$ for N = 1, 2, and 3. Discuss the evidence for concluding that f is a probability density function.*

**39.** $f(x) = \dfrac{1}{\sqrt{2\pi}}\, e^{-0.5x^2}$   **40.** $f(x) = \dfrac{2}{\sqrt{2\pi}}\, e^{-2(x-1.5)^2}$

*In Problems 41–44, F(x) is the cumulative distribution function for a continuous random variable X. Find the probability density function f(x) associated with each F(x).*

**41.** $F(x) = \begin{cases} 0 & \text{if } x < 0 \\ x^2 & \text{if } 0 \le x \le 1 \\ 1 & \text{if } x > 1 \end{cases}$

**42.** $F(x) = \begin{cases} 0 & \text{if } x < 1 \\ \frac{1}{2}x - \frac{1}{2} & \text{if } 1 \le x \le 3 \\ 1 & \text{if } x > 3 \end{cases}$

**43.** $F(x) = \begin{cases} 0 & \text{if } x < 0 \\ 6x^2 - 8x^3 + 3x^4 & \text{if } 0 \le x \le 1 \\ 1 & \text{if } x > 1 \end{cases}$

**44.** $F(x) = \begin{cases} 1 - (1/x^3) & \text{if } x \ge 1 \\ 0 & \text{otherwise} \end{cases}$

*In Problems 45 and 46, find the associated cumulative distribution function.*

**45.** $f(x) = \begin{cases} x & \text{if } 0 \le x \le 1 \\ 2 - x & \text{if } 1 < x \le 2 \\ 0 & \text{otherwise} \end{cases}$

**46.** $f(x) = \begin{cases} \frac{1}{4} & \text{if } 0 \le x \le 1 \\ \frac{1}{2} & \text{if } 1 < x \le 2 \\ \frac{1}{4} & \text{if } 2 < x \le 3 \\ 0 & \text{otherwise} \end{cases}$

## Applications

**47.** *Electricity consumption.* The daily demand for electricity (in millions of kilowatt-hours) in a large city is a continuous random variable with probability density function

$$f(x) = \begin{cases} .2 - .02x & \text{if } 0 \le x \le 10 \\ 0 & \text{otherwise} \end{cases}$$

(A) Evaluate $\int_2^6 f(x)\,dx$ and interpret the results.

(B) What is the probability that the daily demand for electricity is less than 8 million kilowatt-hours?

(C) What is the probability that 5 million kilowatt-hours will not be sufficient to meet the daily demand?

**48.** *Gasoline consumption.* The daily demand for gasoline (in millions of gallons) in a large city is a continuous random variable with probability density function

$$f(x) = \begin{cases} .4 - .08x & \text{if } 0 \le x \le 5 \\ 0 & \text{otherwise} \end{cases}$$

(A) Evaluate $\int_1^4 f(x)\,dx$ and interpret the results.

(B) What is the probability that the daily demand is less than 2 million gallons?

(C) What is the probability that 3 million gallons will not be sufficient to meet the daily demand?

**49.** *Time-sharing.* In a computer time-sharing network, the time it takes (in seconds) to respond to a user's request is a continuous random variable with probability density function given by

$$f(x) = \begin{cases} \frac{1}{10}e^{-x/10} & \text{if } x \ge 0 \\ 0 & \text{otherwise} \end{cases}$$

(A) Evaluate $\int_5^{10} f(x)\,dx$ and interpret the results.

(B) What is the probability that the computer responds within 1 second?

(C) What is the probability that a user must wait more than 4 seconds for a response?

**50.** *Waiting time.* The time (in minutes) a customer must wait in line at a bank is a continuous random variable with probability density function given by

$$f(x) = \begin{cases} \frac{1}{2}e^{-x/2} & \text{if } x \geq 0 \\ 0 & \text{otherwise} \end{cases}$$

(A) Evaluate $\int_2^6 f(x)\,dx$ and interpret the results.

(B) What is the probability that a customer waits less than 3 minutes?

(C) What is the probability that a customer waits more than 5 minutes?

**51.** *Demand.* The weekly demand for hamburger (in thousands of pounds) for a chain of supermarkets is a continuous random variable with probability density function given by

$$f(x) = \begin{cases} .003x\sqrt{100 - x^2} & \text{if } 0 \leq x \leq 10 \\ 0 & \text{otherwise} \end{cases}$$

(A) What is the probability that more than 4,000 pounds of hamburger are demanded?

(B) The manager of the meat department orders 8,000 pounds of hamburger. What is the probability that the demand will not exceed this amount?

(C) The manager wants the probability that the demand does not exceed the amount ordered to be .9. How much hamburger meat should be ordered?

**52.** *Demand.* The demand for a weekly sports magazine (in thousands of copies) in a certain city is a continuous random variable with probability density function given by

$$f(x) = \begin{cases} \frac{3}{125}x\sqrt{25 - x^2} & \text{if } 0 \leq x \leq 5 \\ 0 & \text{otherwise} \end{cases}$$

(A) The magazine's distributor in this city orders 3,000 copies of the magazine. What is the probability that the demand exceeds this number?

(B) What is the probability that the demand does not exceed 4,000 copies?

(C) If the distributor wants to be 95% certain that the demand does not exceed the number ordered, how many copies should be ordered?

**53.** *Life expectancy.* The life expectancy (in minutes) of a certain microscopic organism is a continuous random variable with probability density function given by

$$f(x) = \begin{cases} \frac{1}{5,000}(10x^3 - x^4) & \text{if } 0 \leq x \leq 10 \\ 0 & \text{otherwise} \end{cases}$$

(A) What is the probability that an organism lives for at least 7 minutes?

(B) What is the probability that an organism lives for at most 5 minutes?

**54.** *Life expectancy.* The life expectancy (in months) of plants of a certain species is a continuous random variable with probability density function given by

$$f(x) = \begin{cases} \frac{1}{36}(6x - x^2) & \text{if } 0 \leq x \leq 6 \\ 0 & \text{otherwise} \end{cases}$$

(A) What is the probability that one of these plants survives for at least 4 months?

(B) What is the probability that one of these plants survives for at most 5 months?

**55.** *Shelf life.* The shelf life (in days) of a perishable drug is a continuous random variable with probability density function given by

$$f(x) = \begin{cases} 800x/(400 + x^2)^2 & \text{if } x \geq 0 \\ 0 & \text{otherwise} \end{cases}$$

(A) What is the probability that the drug has a shelf life of at most 20 days?

(B) What is the probability that the shelf life exceeds 15 days?

(C) If the user wants the probability that the drug is still good to be .8, when is the last time it should be used?

**56.** *Shelf life.* Repeat Problem 55 if

$$f(x) = \begin{cases} 200x/(100 + x^2)^2 & \text{if } x \geq 0 \\ 0 & \text{otherwise} \end{cases}$$

**57.** *Learning.* The number of words per minute a beginner can type after 1 week of practice is a continuous random variable with probability density function given by

$$f(x) = \begin{cases} \frac{1}{20}e^{-x/20} & \text{if } x \geq 0 \\ 0 & \text{otherwise} \end{cases}$$

(A) What is the probability that a beginner can type at least 30 words per minute after 1 week of practice?

(B) What is the probability that a beginner can type at least 80 words per minute after 1 week of practice?

**58.** *Learning.* The number of hours it takes a chimpanzee to learn a new task is a continuous random variable with probability density function given by

$$f(x) = \begin{cases} \frac{4}{9}x^2 - \frac{4}{27}x^3 & \text{if } 0 \leq x \leq 3 \\ 0 & \text{otherwise} \end{cases}$$

(A) What is the probability that the chimpanzee learns the task in the first hour?

(B) What is the probability that the chimpanzee does not learn the task in the first 2 hours?

## Section 3-3   EXPECTED VALUE, STANDARD DEVIATION, AND MEDIAN

- Expected Value (or Mean) and Standard Deviation
- Alternative Formula for Variance
- Median

### ■ Expected Value (or Mean) and Standard Deviation

| TABLE 1 | | | | |
|---------|---|---|---|---|
| $x$ | 0 | 1 | 2 | 3 |
| $p(x)$ | $\frac{1}{8}$ | $\frac{3}{8}$ | $\frac{3}{8}$ | $\frac{1}{8}$ |

Statisticians use a measure of central tendency called the *mean* and a measure of dispersion called the *standard deviation* to help describe sets of data. These same terms are used to describe both discrete and continuous random variables. For example, if $X$ is the discrete random variable representing the number of heads when three coins are tossed (Table 1), then the mean and standard deviation are computed as follows:

$$\text{Mean} = 0 \cdot \tfrac{1}{8} + 1 \cdot \tfrac{3}{8} + 2 \cdot \tfrac{3}{8} + 3 \cdot \tfrac{1}{8} = \tfrac{3}{2}$$

$$\text{Variance} = (0 - \tfrac{3}{2})^2 \cdot \tfrac{1}{8} + (1 - \tfrac{3}{2})^2 \cdot \tfrac{3}{8} + (2 - \tfrac{3}{2})^2 \cdot \tfrac{3}{8} + (3 - \tfrac{3}{2})^2 \cdot \tfrac{1}{8}$$

$$= \tfrac{3}{4}$$

$$\text{Standard deviation} = \sqrt{\text{Variance}} = \sqrt{\tfrac{3}{4}} \approx .8660$$

Notice that these quantities were computed using finite sums and the probabilities in Table 1.

The extension of these ideas to continuous random variables makes use of integration and the probability density function.

**DEFINITION**

**Expected Value (or Mean) and Standard Deviation for a Continuous Random Variable**

Let $f(x)$ be the probability density function for a continuous random variable $X$. The **expected value, or mean, of $X$** is

$$\mu = E(X) = \int_{-\infty}^{\infty} x f(x)\, dx$$

The **variance** is

$$V(X) = \int_{-\infty}^{\infty} (x - \mu)^2 f(x)\, dx$$

and the **standard deviation** is

$$\sigma = \sqrt{V(X)}$$

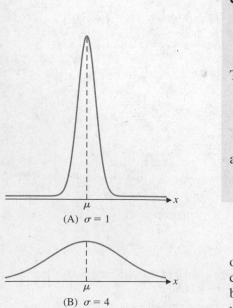

(A) $\sigma = 1$

(B) $\sigma = 4$

**FIGURE 1**

Just as in the discrete case, the mean of a continuous random variable is a measure of central tendency for the variable, and the standard deviation is a measure of the dispersion of the variable about the mean. This is illustrated in Figure 1. The probability density function in Figure 1A has a standard deviation of 1. Most of the area under the curve is near the mean. In Figure 1B, the standard deviation is four times as large, and the area under the graph is much more spread out.

**EXAMPLE 1**  **Computing Expected Value (Mean), Variance, and Standard Deviation**

Find the expected value (mean), variance, and standard deviation for

$$f(x) = \begin{cases} 12x^2 - 12x^3 & \text{if } 0 \le x \le 1 \\ 0 & \text{otherwise} \end{cases}$$

**SOLUTION**
$$\mu = E(X) = \int_{-\infty}^{\infty} xf(x)\, dx = \int_0^1 x(12x^2 - 12x^3)\, dx$$

$$= \int_0^1 (12x^3 - 12x^4)\, dx$$

$$= \left. (3x^4 - \tfrac{12}{5}x^5) \right|_0^1 = \tfrac{3}{5} \quad \text{Expected value (mean)}$$

$$V(X) = \int_{-\infty}^{\infty} (x - \mu)^2 f(x)\, dx = \int_0^1 (x - \tfrac{3}{5})^2 (12x^2 - 12x^3)\, dx$$

$$= \int_0^1 (x^2 - \tfrac{6}{5}x + \tfrac{9}{25})(12x^2 - 12x^3)\, dx$$

$$= \int_0^1 (\tfrac{108}{25}x^2 - \tfrac{468}{25}x^3 + \tfrac{132}{5}x^4 - 12x^5)\, dx$$

$$= \left. (\tfrac{36}{25}x^3 - \tfrac{117}{25}x^4 + \tfrac{132}{25}x^5 - 2x^6) \right|_0^1 = \tfrac{1}{25} \quad \text{Variance}$$

$$\sigma = \sqrt{V(X)} = \sqrt{\tfrac{1}{25}} = \tfrac{1}{5} \quad \text{Standard deviation}$$

**MATCHED PROBLEM 1**  Find the expected value (mean), variance, and standard deviation for

$$f(x) = \begin{cases} 6x - 6x^2 & \text{if } 0 \le x \le 1 \\ 0 & \text{otherwise} \end{cases}$$

*Explore & Discuss 1*  Consider the probability density function

$$f(x) = \begin{cases} \tfrac{1}{2}x & \text{if } 0 \le x \le 2 \\ 0 & \text{otherwise} \end{cases}$$

(A) Find the mean $\mu$.

(B) The vertical line $x = \mu$ divides the area under the graph of $y = f(x)$ into two regions. Do these regions have equal areas? If not, find a number $m$ with the property that the vertical line $x = m$ divides the area under the graph of $y = f(x)$ into two regions with equal areas.

The graph of the probability density function considered in Example 1 is shown in Figure 2A, with the areas to the left and right of the mean indicated (computations omitted). As you discovered in the preceding Explore–Discuss 1, the vertical line through the mean does not divide the region under the graph of $f$ into two regions with equal areas. The value of $x$ that accomplishes this is the *median*, which we will discuss later in this section. However, the mean does have a geometric interpretation. If the shaded region in Figure 2A were drawn on a piece of wood of uniform thickness and the wood cut around the outside of the region, then the resulting object

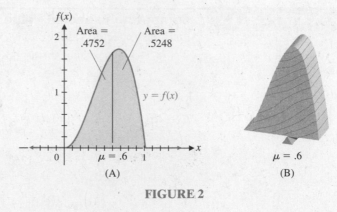

**FIGURE 2**

would balance on a wedge placed at the mean $\mu = .6$ (see Fig. 2B). For this reason, the mean $\mu$ is often referred to as a *measure of central tendency*.

---

EXAMPLE 2

**Life Expectancy** The life expectancy (in hours) for a particular brand of light bulbs is a continuous random variable with probability density function

$$f(x) = \begin{cases} \frac{1}{100} - \frac{1}{20,000}x & \text{if } 0 \le x \le 200 \\ 0 & \text{otherwise} \end{cases}$$

(A) What is the average life expectancy of one of these light bulbs?

(B) What is the probability that a bulb will last longer than this average?

SOLUTION

(A) Since the value of this random variable is the number of hours a bulb lasts, the average life expectancy is just the expected value of the random variable. Thus,

$$E(X) = \int_{-\infty}^{\infty} xf(x)\, dx = \int_{0}^{200} x\left(\frac{1}{100} - \frac{1}{20,000}x\right) dx$$

$$= \int_{0}^{200} \left(\frac{1}{100}x - \frac{1}{20,000}x^2\right) dx = \left(\frac{1}{200}x^2 - \frac{1}{60,000}x^3\right)\Big|_{0}^{200}$$

$$= \frac{200}{3} \approx 66.67 \text{ hours}$$

(B) The probability that a bulb lasts longer than $\frac{200}{3}$ hours is

$$P\left(X > \tfrac{200}{3}\right) = \int_{200/3}^{\infty} f(x)\, dx = \int_{200/3}^{200} \left(\frac{1}{100} - \frac{1}{20,000}x\right) dx$$

$$= \left(\frac{1}{100}x - \frac{1}{40,000}x^2\right)\Big|_{200/3}^{200}$$

$$= 1 - \frac{5}{9} = \frac{4}{9} \approx 0.44$$

---

MATCHED PROBLEM 2

Repeat Example 2 if the probability density function is

$$f(x) = \begin{cases} \frac{1}{200} - \frac{1}{90,000}x & \text{if } 0 \le x \le 300 \\ 0 & \text{otherwise} \end{cases}$$

## ■ Alternative Formula for Variance

The term $(x - \mu)^2$ in the formula for $V(X)$ introduces some complicated algebraic manipulations in the evaluation of the integral. We can use the properties of the definite integral to simplify this formula. Thus,

$$V(X) = \int_{-\infty}^{\infty} (x - \mu)^2 f(x)\, dx$$

$$= \int_{-\infty}^{\infty} (x^2 - 2x\mu + \mu^2) f(x)\, dx \qquad \text{Algebra}$$

$$= \int_{-\infty}^{\infty} [x^2 f(x) - 2x\mu f(x) + \mu^2 f(x)]\, dx \qquad \text{Algebra}$$

$$= \int_{-\infty}^{\infty} x^2 f(x)\, dx - \int_{-\infty}^{\infty} 2x\mu f(x)\, dx + \int_{-\infty}^{\infty} \mu^2 f(x)\, dx \qquad \text{Definite integral property}$$

$$= \int_{-\infty}^{\infty} x^2 f(x)\, dx - 2\mu \int_{-\infty}^{\infty} x f(x)\, dx + \mu^2 \int_{-\infty}^{\infty} f(x)\, dx \qquad \text{Definite integral property}$$

$$= \int_{-\infty}^{\infty} x^2 f(x)\, dx - 2\mu(\mu) + \mu^2(1) \qquad \int_{-\infty}^{\infty} x f(x)\, dx = \mu, \quad \int_{-\infty}^{\infty} f(x)\, dx = 1$$

$$= \int_{-\infty}^{\infty} x^2 f(x)\, dx - \mu^2$$

In general, it will be easier to evaluate $\int_{-\infty}^{\infty} x^2 f(x)\, dx$ than to evaluate $\int_{-\infty}^{\infty} (x - \mu)^2 f(x)\, dx$.

### THEOREM 1   ALTERNATIVE FORMULA FOR VARIANCE

$$V(X) = \int_{-\infty}^{\infty} x^2 f(x)\, dx - \mu^2$$

**EXAMPLE 3**   **Computing Variance**  Use the alternative formula for variance (Theorem 1) to compute the variance in Example 1.

**SOLUTION**   From Example 1, we have $\mu = \int_{-\infty}^{\infty} x f(x)\, dx = \frac{3}{5}$. Thus,

$$\int_{-\infty}^{\infty} x^2 f(x)\, dx = \int_{0}^{1} x^2 (12x^2 - 12x^3)\, dx \qquad f(x) = \begin{cases} 12x^2 - 12x^3 & \text{if } 0 \le x \le 1 \\ 0 & \text{otherwise} \end{cases}$$

$$= \int_{0}^{1} (12x^4 - 12x^5)\, dx$$

$$= \left( \tfrac{12}{5} x^5 - \tfrac{12}{6} x^6 \right) \Big|_0^1 = \tfrac{2}{5} = .4$$

$$V(X) = \int_{-\infty}^{\infty} x^2 f(x)\, dx - \mu^2 = \tfrac{2}{5} - \left( \tfrac{3}{5} \right)^2 = \tfrac{1}{25} = .04$$

**MATCHED PROBLEM 3**   Use the alternative formula for variance (Theorem 1) to compute the variance in Matched Problem 1.

---

EXAMPLE 4 **Computing Expected Value (Mean), Variance, and Standard Deviation**
Find the expected value (mean), variance, and standard deviation for

$$f(x) = \begin{cases} 3/x^4 & \text{if } x \geq 1 \\ 0 & \text{otherwise} \end{cases}$$

SOLUTION

$$\mu = \int_{-\infty}^{\infty} xf(x)\, dx = \int_{1}^{\infty} x\frac{3}{x^4}\, dx = \lim_{R \to \infty} \int_{1}^{R} \frac{3}{x^3}\, dx$$

$$= \lim_{R \to \infty}\left[-\frac{3}{2}\left(\frac{1}{x^2}\right)\right]\Big|_{1}^{R} = \lim_{R \to \infty}\left[-\frac{3}{2}\left(\frac{1}{R^2}\right) + \frac{3}{2}\right] = \frac{3}{2} = 1.5$$

$$\int_{-\infty}^{\infty} x^2 f(x)\, dx = \int_{1}^{\infty} x^2\frac{3}{x^4}\, dx = \lim_{R \to \infty} \int_{1}^{R} \frac{3}{x^2}\, dx = \lim_{R \to \infty}\left(-\frac{3}{x}\right)\Big|_{1}^{R}$$

$$= \lim_{R \to \infty}\left(-\frac{3}{R} + 3\right) = 3$$

$$V(X) = \int_{-\infty}^{\infty} x^2 f(x)\, dx - \mu^2 = 3 - \left(\frac{3}{2}\right)^2 = \frac{3}{4} = .75$$

$$\sigma = \sqrt{V(X)} = \sqrt{\frac{3}{4}} = \frac{\sqrt{3}}{2} \approx .8660$$

MATCHED PROBLEM 4  Find the expected value (mean), variance, and standard deviation for

$$f(x) = \begin{cases} 24/x^4 & \text{if } x \geq 2 \\ 0 & \text{otherwise} \end{cases}$$

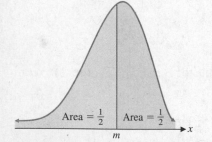

FIGURE 3 Large Calculator

## ■ Median

Another measurement often used to describe the properties of a random variable is the median. A **median** is a value of the random variable that divides the area under the graph of the probability density function into two equal parts (see Fig. 3). If $m$ is a median, then $m$ must satisfy

$$P(X \leq m) = \tfrac{1}{2}$$

Generally, this equation is solved by first finding the cumulative distribution function.

---

EXAMPLE 5 **Finding the Median**  Find the median of the continuous random variable with probability density function

$$f(x) = \begin{cases} 3/x^4 & \text{if } x \geq 1 \\ 0 & \text{otherwise} \end{cases}$$

SOLUTION  **Step 1.**  *Find the cumulative distribution function:* For $x < 1$, we have $F(x) = 0$. If $x \geq 1$, then

$$F(x) = \int_{-\infty}^{x} f(t)\, dt = \int_{1}^{x} \frac{3}{t^4}\, dt = -\frac{1}{t^3}\Big|_{1}^{x} = -\frac{1}{x^3} + 1 = 1 - \frac{1}{x^3}$$

**Step 2.** *Solve the equation* $P(X \leq m) = \frac{1}{2}$ *for m:*

$$F(m) = P(X \leq m)$$

$$1 - \frac{1}{m^3} = \frac{1}{2}$$

$$\frac{1}{2} = \frac{1}{m^3}$$

$$m^3 = 2$$

$$m = \sqrt[3]{2}$$

Thus, the median is $\sqrt[3]{2} \approx 1.26$.

Figure 4 shows an alternative method for solving Example 5 on a graphing utility.

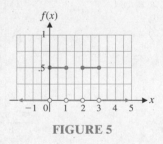

**FIGURE 4**

MATCHED PROBLEM 5   Find the median of the continuous random variable with probability density function

$$f(x) = \begin{cases} 24/x^4 & \text{if } x \geq 2 \\ 0 & \text{otherwise} \end{cases}$$

*Explore & Discuss* **2**   Consider the probability density function $f$ defined by the graph in Figure 5. Does a median exist? Is it unique? Explain.

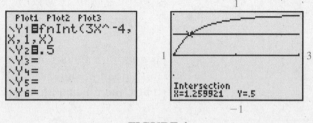

**FIGURE 5**

**EXAMPLE 6**   **Life Expectancy**   In Example 2, find the median life expectancy of a lightbulb.

**SOLUTION**   **Step 1.**   *Find the cumulative distribution function:* If $x < 0$, we have $F(x) = 0$. If $0 \leq x \leq 200$, then

$$F(x) = \int_{-\infty}^{x} f(t)\, dt \qquad f(x) = \begin{cases} \frac{1}{100} - \frac{1}{20{,}000}x & \text{if } 0 \leq x \leq 200 \\ 0 & \text{otherwise} \end{cases}$$

$$= \int_{0}^{x} \left( \frac{1}{100} - \frac{1}{20{,}000} t \right) dt$$

$$= \left( \frac{1}{100} t - \frac{1}{40{,}000} t^2 \right) \Big|_{0}^{x}$$

$$= \frac{1}{100} x - \frac{1}{40{,}000} x^2$$

If $x > 200$, then

$$F(x) = \int_{-\infty}^{x} f(t)\, dt = \int_{-\infty}^{0} f(t)\, dt + \int_{0}^{200} f(t)\, dt + \int_{200}^{x} f(t)\, dt$$

$$= 0 + 1 + 0 = 1$$

Thus,

$$F(x) = \begin{cases} 0 & \text{if } x < 0 \\ \frac{1}{100}x - \frac{1}{40,000}x^2 & \text{if } 0 \le x \le 200 \\ 1 & \text{if } x > 200 \end{cases}$$

**Step 2.** *Solve the equation* $P(X \le m) = \frac{1}{2}$ *for* $m$:

$$F(m) = P(X \le m) = \frac{1}{2}$$

$$\frac{1}{100}m - \frac{1}{40,000}m^2 = \frac{1}{2}$$

$$m^2 - 400m + 20,000 = 0 \quad \text{\footnotesize The solution must occur for } 0 \le m \le 200.$$

This quadratic equation has two solutions, $200 + 100\sqrt{2}$ and $200 - 100\sqrt{2}$. Since $m$ must lie in the interval $[0, 200]$, the second solution is the correct answer.

Thus, the median life expectancy is $200 - 100\sqrt{2} \approx 58.58$ hours. ◾

**MATCHED PROBLEM 6**

In Matched Problem 2, find the median life expectancy of a lightbulb.

Figure 6 shows the mean and the median for two of the probability density functions discussed in this section (see Examples 2, 4, 5, and 6). As we noted earlier, the mean and the median generally are not equal.

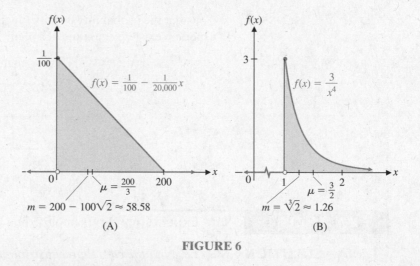

**FIGURE 6**

| INSIGHT |
| --- |

Although the median is smaller than the mean for the probability distributions of Figure 6, for other distributions the median is larger than the mean. To obtain examples of such distributions, simply reflect the graphs of Figure 6 in the $y$ axis.

*Answers to Matched Problems*

1. $\mu = \frac{1}{2}$; $V(X) = \frac{1}{20}$; $\sigma \approx .2236$

2. (A) 125 hr   (B) $\frac{133}{288}$

3. $\frac{1}{20}$

4. $\mu = 3$; $V(X) = 3$; $\sigma = \sqrt{3} \approx 1.732$

5. $m = \sqrt[3]{16} = 2\sqrt[3]{2} \approx 2.52$

6. $m = 450 - 150\sqrt{5} \approx 114.59$ hr

# Exercise 3-3

A **In Problems 1–6, find the mean, variance, and standard deviation.**

**1.** $f(x) = \begin{cases} 2 & \text{if } 0 \le x \le 0.5 \\ 0 & \text{otherwise} \end{cases}$

**2.** $f(x) = \begin{cases} \frac{1}{4} & \text{if } -1 \le x \le 3 \\ 0 & \text{otherwise} \end{cases}$

**3.** $f(x) = \begin{cases} \frac{1}{8}x & \text{if } 0 \le x \le 4 \\ 0 & \text{otherwise} \end{cases}$

**4.** $f(x) = \begin{cases} 18x & \text{if } 0 \le x \le \frac{1}{3} \\ 0 & \text{otherwise} \end{cases}$

**5.** $f(x) = \begin{cases} 4 - 2x & \text{if } 1 \le x \le 2 \\ 0 & \text{otherwise} \end{cases}$

**6.** $f(x) = \begin{cases} 2 - \frac{1}{2}x & \text{if } 2 \le x \le 4 \\ 0 & \text{otherwise} \end{cases}$

**In Problems 7–12, find the median.**

**7.** $f(x) = \begin{cases} 0.2 & \text{if } -2 \le x \le 3 \\ 0 & \text{otherwise} \end{cases}$

**8.** $f(x) = \begin{cases} 10 & \text{if } 2.5 \le x \le 2.6 \\ 0 & \text{otherwise} \end{cases}$

**9.** $f(x) = \begin{cases} \frac{1}{6}x & \text{if } 2 \le x \le 4 \\ 0 & \text{otherwise} \end{cases}$

**10.** $f(x) = \begin{cases} \frac{1}{4}x & \text{if } 1 \le x \le 3 \\ 0 & \text{otherwise} \end{cases}$

**11.** $f(x) = \begin{cases} \frac{1}{2} - \frac{1}{8}x & \text{if } 0 \le x \le 4 \\ 0 & \text{otherwise} \end{cases}$

**12.** $f(x) = \begin{cases} 2 - 2x & \text{if } 0 \le x \le 1 \\ 0 & \text{otherwise} \end{cases}$

B **Problems 13 and 14 refer to random variables $X_1$ and $X_2$ with probability density functions $f_1$ and $f_2$, respectively, where**

$$f_1(x) = \begin{cases} 0.5 & \text{if } 0 \le x \le 2 \\ 0 & \text{otherwise} \end{cases}$$

$$f_2(x) = \begin{cases} 0.25 & \text{if } -2 \le x \le 2 \\ 0 & \text{otherwise} \end{cases}$$

**13.** Explain how you can predict from the graphs of $f_1$ and $f_2$ which random variable, $X_1$ or $X_2$, has the greater mean. Check your prediction by computing the mean of each.

**14.** Explain how you can predict from the graphs of $f_1$ and $f_2$ which random variable, $X_1$ or $X_2$, has the greater variance. Check your prediction by computing the variance of each.

**Problems 15 and 16 refer to random variables $X_1$ and $X_2$ with probability density functions $g_1$ and $g_2$, respectively, where**

$$g_1(x) = \begin{cases} 0.5 & \text{if } 0 \le x < 1 \\ 0.25 & \text{if } 1 \le x \le 3 \\ 0 & \text{otherwise} \end{cases}$$

$$g_2(x) = \begin{cases} 0.5 & \text{if } 0 \le x < 1 \\ 0.125 & \text{if } 1 \le x \le 5 \\ 0 & \text{otherwise} \end{cases}$$

**15.** Explain how you can predict from the graphs of $g_1$ and $g_2$ which random variable, $X_1$ or $X_2$, has the greater median. Check your prediction by computing the median of each.

**16.** Explain how you can predict from the graphs of $g_1$ and $g_2$ which random variable, $X_1$ or $X_2$, has the greater mean. Check your prediction by computing the mean of each.

**In Problems 17–20, find the mean, variance, and standard deviation.**

**17.** $f(x) = \begin{cases} 4/x^5 & \text{if } x \ge 1 \\ 0 & \text{otherwise} \end{cases}$

**18.** $f(x) = \begin{cases} 5/x^6 & \text{if } x \ge 1 \\ 0 & \text{otherwise} \end{cases}$

**19.** $f(x) = \begin{cases} 64/x^5 & \text{if } x \ge 2 \\ 0 & \text{otherwise} \end{cases}$

**20.** $f(x) = \begin{cases} 81/x^4 & \text{if } x \ge 3 \\ 0 & \text{otherwise} \end{cases}$

**In Problems 21 and 22, use a graphing utility to approximate the mean, variance, and standard deviation by replacing integrals over $(-\infty, \infty)$ by integrals over $[-10, 10]$.**

**21.** $f(x) = \dfrac{2}{\sqrt{2\pi}} e^{-2(x-1)^2}$

**22.** $f(x) = \dfrac{1}{1.4\sqrt{2\pi}} e^{-(x+2.1)^2/3.92}$

**In Problems 23–30, find the median.**

**23.** $f(x) = \begin{cases} 1/x & \text{if } 1 \le x \le e \\ 0 & \text{otherwise} \end{cases}$

**24.** $f(x) = \begin{cases} 1/(2x) & \text{if } 1 \le x \le e^2 \\ 0 & \text{otherwise} \end{cases}$

**25.** $f(x) = \begin{cases} 4/(2 + x)^2 & \text{if } 0 \le x \le 2 \\ 0 & \text{otherwise} \end{cases}$

**26.** $f(x) = \begin{cases} 2/(1 + x)^2 & \text{if } 0 \le x \le 1 \\ 0 & \text{otherwise} \end{cases}$

**27.** $f(x) = \begin{cases} 1/(1 + x)^2 & \text{if } x \ge 0 \\ 0 & \text{otherwise} \end{cases}$

**28.** $f(x) = \begin{cases} 3/(3 + x)^2 & \text{if } x \ge 0 \\ 0 & \text{otherwise} \end{cases}$

**29.** $f(x) = \begin{cases} 2e^{-2x} & \text{if } x \ge 0 \\ 0 & \text{otherwise} \end{cases}$

**30.** $f(x) = \begin{cases} e^{-x} & \text{if } x \ge 0 \\ 0 & \text{otherwise} \end{cases}$

*In Problems 31 and 32, use the graph of the cumulative distribution function F to find the median. (Assume each portion of the graph of F is a straight line segment.) Is the median unique? Explain.*

**31.**

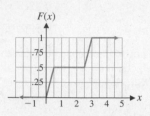

**32.**

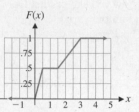

*In Problems 33 and 34, use the graph of the probability density function f to find the cumulative distribution function F. (Assume each portion of the graph of f is a straight line segment.) Find the median. Is the median unique? Explain.*

**33.**

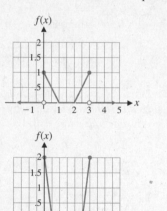

**34.**

f(x)

C *In Problems 35 and 36, $f(x)$ is a continuous probability density function with mean $\mu$ and standard deviation $\sigma$; a and b are constants. Evaluate each integral, expressing the result in terms of a, b, $\mu$, and $\sigma$.*

**35.** $\displaystyle\int_{-\infty}^{\infty} (ax + b)f(x)\, dx$

**36.** $\displaystyle\int_{-\infty}^{\infty} (x - a)^2 f(x)\, dx$

*The **quartile points** for a probability density function are the values $x_1$, $x_2$, $x_3$ that divide the area under the graph of the function into four equal parts. Find the quartile points for the probability density functions in Problems 37–40.*

**37.** $f(x) = \begin{cases} \frac{1}{2}x & \text{if } 0 \leq x \leq 2 \\ 0 & \text{otherwise} \end{cases}$

**38.** $f(x) = \begin{cases} 3x^2 & \text{if } 0 \leq x \leq 1 \\ 0 & \text{otherwise} \end{cases}$

**39.** $f(x) = \begin{cases} 3/(3 + x)^2 & \text{if } x \geq 0 \\ 0 & \text{otherwise} \end{cases}$

**40.** $f(x) = \begin{cases} 1/(1 + x)^2 & \text{if } x \geq 0 \\ 0 & \text{otherwise} \end{cases}$

*In Problems 41–44, use a graphing utility to approximate the median of the indicated probability density function f (to two decimal places).*

**41.** $f(x) = \begin{cases} 4x - 4x^3 & \text{if } 0 \leq x \leq 1 \\ 0 & \text{otherwise} \end{cases}$

**42.** $f(x) = \begin{cases} 3x - 3x^5 & \text{if } 0 \leq x \leq 1 \\ 0 & \text{otherwise} \end{cases}$

**43.** $f(x) = \begin{cases} \dfrac{1}{2x^2} + \dfrac{3}{2x^4} & \text{if } x \geq 1 \\ 0 & \text{otherwise} \end{cases}$

**44.** $f(x) = \begin{cases} \dfrac{2}{3x^3} + \dfrac{2}{x^4} & \text{if } x \geq 1 \\ 0 & \text{otherwise} \end{cases}$

## Applications

**45.** *Profit.* A building contractor's profit (in thousands of dollars) on each unit in a subdivision is a continuous random variable with probability density function given by

$$f(x) = \begin{cases} 0.08(20 - x) & \text{if } 15 \leq x \leq 20 \\ 0 & \text{otherwise} \end{cases}$$

(A) Find the contractor's expected profit.

(B) Find the median profit.

**46.** *Electricity consumption.* The daily consumption of electricity (in millions of kilowatt-hours) in a large city is a continuous random variable with probability density function

$$f(x) = \begin{cases} .2 - .02x & \text{if } 0 \leq x \leq 10 \\ 0 & \text{otherwise} \end{cases}$$

(A) Find the expected daily consumption of electricity.

(B) Find the median daily consumption of electricity.

**47.** *Waiting time.* The time (in minutes) a customer must wait in line at a bank is a continuous random variable with probability density function given by

$$f(x) = \begin{cases} \frac{1}{3}e^{-x/3} & \text{if } x \geq 0 \\ 0 & \text{otherwise} \end{cases}$$

Find the median waiting time.

**48.** *Product life.* The life expectancy (in years) of an automobile battery is a continuous random variable with probability density function given by

$$f(x) = \begin{cases} \frac{1}{2}e^{-x/2} & \text{if } x \geq 0 \\ 0 & \text{otherwise} \end{cases}$$

Find the median life expectancy.

**49.** *Water consumption.* The daily consumption of water (in millions of gallons) in a small city is a continuous random variable with probability density function given by

$$f(x) = \begin{cases} 1/(1 + x^2)^{3/2} & \text{if } x \geq 0 \\ 0 & \text{otherwise} \end{cases}$$

Find the expected daily consumption.

**50.** *Gasoline consumption.* The daily consumption of gasoline (in millions of gallons) in a large city is a continuous random variable with probability density function

$$f(x) = \begin{cases} 4/(4 + x^2)^{3/2} & \text{if } x \geq 0 \\ 0 & \text{otherwise} \end{cases}$$

Find the expected daily consumption of gasoline.

**51.** *Life expectancy.* The life expectancy of a certain microscopic organism (in minutes) is a continuous random variable with probability density function given by

$$f(x) = \begin{cases} \frac{1}{5,000}(10x^3 - x^4) & \text{if } 0 \leq x \leq 10 \\ 0 & \text{otherwise} \end{cases}$$

Find the mean life expectancy of one of these organisms.

**52.** *Life expectancy.* The life expectancy (in months) of plants of a certain species is a continuous random variable with probability density function given by

$$f(x) = \begin{cases} \frac{1}{36}(6x - x^2) & \text{if } 0 \leq x \leq 6 \\ 0 & \text{otherwise} \end{cases}$$

Find the mean life expectancy of one of these plants.

**53.** *Shelf life.* The shelf life (in days) of a perishable drug is a continuous random variable with probability density function given by

$$f(x) = \begin{cases} 800x/(400 + x^2)^2 & \text{if } x \geq 0 \\ 0 & \text{otherwise} \end{cases}$$

Find the median shelf life.

**54.** *Shelf life.* Repeat Problem 53 if

$$f(x) = \begin{cases} 200x/(100 + x^2)^2 & \text{if } x \geq 0 \\ 0 & \text{otherwise} \end{cases}$$

**55.** *Learning.* The number of hours it takes a chimpanzee to learn a new task is a continuous random variable with probability density function given by

$$f(x) = \begin{cases} \frac{4}{9}x^2 - \frac{4}{27}x^3 & \text{if } 0 \leq x \leq 3 \\ 0 & \text{otherwise} \end{cases}$$

What is the expected number of hours it will take a chimpanzee to learn the task?

**56.** *Voter turnout.* The number of registered voters (in thousands) who vote in an off-year election in a small town is a continuous random variable with probability density function given by

$$f(x) = \begin{cases} 1 - \frac{1}{8}x & \text{if } 4 \leq x \leq 8 \\ 0 & \text{otherwise} \end{cases}$$

(A) Find the expected voter turnout.

(B) Find the median voter turnout.

## Section 3-4   SPECIAL PROBABILITY DISTRIBUTIONS

- Uniform Distributions
- Exponential Distributions
- Normal Distributions

Now that we have developed some general properties of probability density functions, we can examine some specific functions that are used extensively in applications.

### Uniform Distributions

*Explore & Discuss 1*

Consider a random variable $X$ with probability density function

$$f(x) = \begin{cases} \frac{1}{10} & \text{if } 0 \leq x \leq 10 \\ 0 & \text{otherwise} \end{cases}$$

Graph $f$ and use the graph to find the indicated probabilities.

(A) Find $P(0 \leq X \leq 1)$, and $P(2 \leq X \leq 3)$, $P(6 \leq X \leq 7)$, $P(9 \leq X \leq 10)$.

(B) If $a \geq 0, n \geq 0$, and $a + n \leq 10$, find $P(a \leq X \leq a + n)$.

(C) If $I$ is an interval of length $n$ that lies in the interval $[0, 10]$, what is the probability that the random variable $X$ lies in $I$? Does this probability depend on the location of the interval $I$? Explain.

Simple probability density functions like the example given in the preceding Explore–Discuss 1 turn out to be quite useful. The next example will lead to some general observations about functions of this form.

EXAMPLE 1 **Waiting Time** A bus arrives every 30 minutes at a particular bus stop. If an individual arrives at the bus stop at a random time (that is, with no knowledge of the bus schedule), then the random variable $X$, representing the time this individual must spend waiting for the next bus, is said to be *uniformly distributed* on the interval $[0, 30]$. This means that the probability that $X$ lies in a small interval of fixed length is independent of the location of that interval within $[0, 30]$. Thus, the probability of waiting between 0 and 5 minutes is the same as the probability of waiting between 5 and 10 minutes or the probability of waiting between 18 and 23 minutes. It can be shown that the probability density function for this uniformly distributed random variable is

$$f(x) = \begin{cases} \frac{1}{30} & \text{if } 0 \le x \le 30 \\ 0 & \text{otherwise} \end{cases}$$

Thus,

$$P(0 \le X \le 5) = \int_0^5 \frac{1}{30}\, dx = \frac{x}{30}\Big|_0^5 = \frac{5}{30} - \frac{0}{30} = \frac{1}{6}$$

$$P(5 \le X \le 10) = \int_5^{10} \frac{1}{30}\, dx = \frac{x}{30}\Big|_5^{10} = \frac{10}{30} - \frac{5}{30} = \frac{1}{6}$$

$$P(18 \le X \le 23) = \int_{18}^{23} \frac{1}{30}\, dx = \frac{x}{30}\Big|_{18}^{23} = \frac{23}{30} - \frac{18}{30} = \frac{1}{6}$$

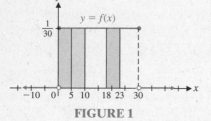

**FIGURE 1**

Each of these probabilities is represented as an area under the graph of $f$ in Figure 1.

MATCHED PROBLEM 1 Use the probability density function given in Example 1 to find the probability that the individual waits:

(A) Between 0 and 10 minutes

(B) Between 10 and 20 minutes

(C) Between 17 and 27 minutes

In general, if the outcomes of an experiment lie in an interval $[a, b]$ and if the probability of the outcome lying in a small interval of fixed length is independent of the location of this small interval within $[a, b]$, then we say that the continuous random variable for this experiment is **uniformly distributed** on the interval $[a, b]$. The **uniform probability density function** is

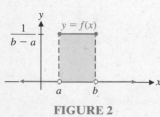

**FIGURE 2**

$$f(x) = \begin{cases} \dfrac{1}{b - a} & \text{if } a \le x \le b \\ 0 & \text{otherwise} \end{cases}$$

See Figure 2. Since $f(x) \ge 0$ and

$$\int_{-\infty}^{\infty} f(x)\, dx = \int_a^b \frac{1}{b - a}\, dx = \frac{x}{b - a}\Big|_a^b = \frac{b}{b - a} - \frac{a}{b - a} = 1$$

$f$ satisfies the necessary conditions for a probability density function.

If $F$ is the associated cumulative distribution function, then for $x < a$, $F(x) = 0$. For $a \le x \le b$, we have

$$F(x) = \int_{-\infty}^{x} f(t)\, dt = \int_{a}^{x} \frac{1}{b-a}\, dt = \frac{t}{b-a}\Big|_{a}^{x}$$

$$= \frac{x}{b-a} - \frac{a}{b-a} = \frac{x-a}{b-a}$$

For $x > b$, $F(x) = 1$.

Using the techniques discussed in the preceding section, it can be shown that (see Problems 49–52 in Exercise 3-4)

$$\mu = m = \frac{1}{2}(a+b) \quad \text{and} \quad \sigma = \frac{1}{\sqrt{12}}(b-a)$$

These properties are summarized in the next box.

**SUMMARY    UNIFORM PROBABILITY DENSITY FUNCTION**

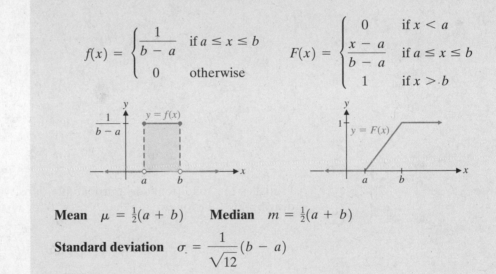

$$f(x) = \begin{cases} \dfrac{1}{b-a} & \text{if } a \le x \le b \\ 0 & \text{otherwise} \end{cases} \qquad F(x) = \begin{cases} 0 & \text{if } x < a \\ \dfrac{x-a}{b-a} & \text{if } a \le x \le b \\ 1 & \text{if } x > b \end{cases}$$

**Mean** $\mu = \frac{1}{2}(a+b)$   **Median** $m = \frac{1}{2}(a+b)$

**Standard deviation** $\sigma = \dfrac{1}{\sqrt{12}}(b-a)$

**EXAMPLE 2**

**Electrical Current** Standard electrical current is uniformly distributed between 110 and 120 volts. What is the probability that the current is between 113 and 118 volts?

**SOLUTION** Since we are told that the current is uniformly distributed on the interval $[110, 120]$, we choose the uniform probability density function

$$f(x) = \begin{cases} \frac{1}{10} & \text{if } 110 \le x \le 120 \\ 0 & \text{otherwise} \end{cases}$$

Then

$$P(113 \le X \le 118) = \int_{113}^{118} \frac{1}{10}\, dx = \frac{x}{10}\Big|_{113}^{118} = \frac{118}{10} - \frac{113}{10} = \frac{1}{2}$$

**MATCHED PROBLEM 2**   In Example 2, what is the probability that the current is at least 116 volts?

## ■ Exponential Distributions

*Explore & Discuss 1*    Determine a relationship between the positive constants $a$ and $b$ that will make the following function a probability density function.

$$f(x) = \begin{cases} ae^{-bx} & \text{if } x \geq 0 \\ 0 & \text{otherwise} \end{cases}$$

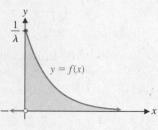

A continuous random variable has an **exponential distribution** and is referred to as an **exponential random variable** if its probability density function is the **exponential probability density function,**

$$f(x) = \begin{cases} (1/\lambda)e^{-x/\lambda} & \text{if } x \geq 0 \\ 0 & \text{otherwise} \end{cases}$$

where $\lambda$ is a positive constant. Exponential random variables are used in a variety of applications, including studies of the length of telephone conversations, the time customers spend waiting in line at a bank, and the life expectancy of machine parts.

Since $f(x) \geq 0$ and

$$\int_{-\infty}^{\infty} f(x)\, dx = \int_{0}^{\infty} \frac{1}{\lambda} e^{-x/\lambda}\, dx$$

$$= \lim_{R \to \infty} \int_{0}^{R} \frac{1}{\lambda} e^{-x/\lambda}\, dx$$

$$= \lim_{R \to \infty} \left. (-e^{-x/\lambda}) \right|_{0}^{R}$$

$$= \lim_{R \to \infty} (-e^{-R/\lambda} + 1) = 1$$

$f$ satisfies the conditions for a probability density function. If $F$ is the cumulative distribution function, we see that $F(x) = 0$ for $x < 0$. For $x \geq 0$, we have

$$F(x) = \int_{-\infty}^{x} f(t)\, dt = \int_{0}^{x} \frac{1}{\lambda} e^{-t/\lambda}\, dt$$

$$= \left. -e^{-t/\lambda} \right|_{0}^{x} = 1 - e^{-x/\lambda}$$

To find the median, we solve

$$F(m) = P(X \leq m) = \frac{1}{2}$$

$$1 - e^{-m/\lambda} = \frac{1}{2}$$

$$\frac{1}{2} = e^{-m/\lambda}$$

$$\ln \frac{1}{2} = -\frac{m}{\lambda}$$

$$m = -\lambda \ln \frac{1}{2} = \lambda \ln 2 \qquad \text{Note:} \quad \ln \frac{1}{2} = -\ln 2$$

Integration by parts can be used to show that $\mu = \lambda$ and $\sigma = \lambda$. The calculations are not included here. These results are summarized in the next box.

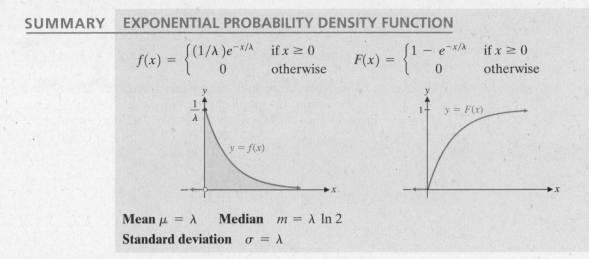

SUMMARY  **EXPONENTIAL PROBABILITY DENSITY FUNCTION**

$$f(x) = \begin{cases} (1/\lambda)e^{-x/\lambda} & \text{if } x \geq 0 \\ 0 & \text{otherwise} \end{cases} \qquad F(x) = \begin{cases} 1 - e^{-x/\lambda} & \text{if } x \geq 0 \\ 0 & \text{otherwise} \end{cases}$$

**Mean** $\mu = \lambda$     **Median** $m = \lambda \ln 2$

**Standard deviation** $\sigma = \lambda$

---

**EXAMPLE 3**

**Arrival Rates** The length of time between calls received by the switchboard in a large legal firm is an exponential random variable. The average length of time between calls is 20 seconds. If a call has just been received, what is the probability that no calls are received in the next 30 seconds?

**SOLUTION**

Let $X$ be the random variable that represents the length of time between calls received (in seconds). Since the average length of time is 20 seconds, we have $\mu = \lambda = 20$. Thus, the probability density function for $X$ is

$$f(x) = \begin{cases} \frac{1}{20}e^{-x/20} & \text{if } x \geq 0 \\ 0 & \text{otherwise} \end{cases}$$

and

$$P(X \geq 30) = \int_{30}^{\infty} \frac{1}{20}e^{-x/20}\, dx \qquad \int_{0}^{30} f(x)\, dx + \int_{30}^{\infty} f(x)\, dx = 1$$

$$= 1 - \int_{0}^{30} \frac{1}{20}e^{-x/20}\, dx$$

$$= 1 - (-e^{-x/20})\Big|_{0}^{30} = e^{-1.5} \approx .223$$

**MATCHED PROBLEM 3**

In Example 3, if a call has just been received, what is the probability that no calls are received in the next 10 seconds?

## ■ Normal Distributions

We now consider the most important of all the probability density functions, the *normal probability density function*. This function is at the heart of a great deal of statistical theory, and it is also a useful tool in its own right for solving problems.

A continuous random variable $X$ has a **normal distribution** and is referred to as a **normal random variable** if its probability density function is the **normal probability density function**

$$f(x) = \frac{1}{\pi\sqrt{2\pi}} e^{-(x-\mu)^2/(2\sigma^2)} \tag{1}$$

where $\mu$ is any constant and $\sigma$ is any positive constant. It can be shown, but not easily, that

$$\int_{-\infty}^{\infty} f(x)\, dx = 1$$

and

$$E(X) = \int_{-\infty}^{\infty} xf(x)\,dx = \mu \qquad V(X) = \int_{-\infty}^{\infty} (x - \mu)^2 f(x)\,dx = \sigma^2$$

Thus, $\mu$ is the mean of the normal probability density function and $\sigma$ is the standard deviation. The graph of $f(x)$ is always a bell-shaped curve called a **normal curve.** Figure 3 illustrates three normal curves for different values of $\mu$ and $\sigma$.

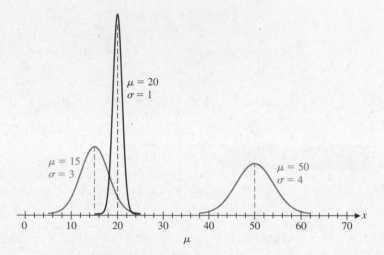

**FIGURE 3** Normal probability distributions

INSIGHT

Equation (1) for the normal probability density function is fairly complicated and involves both $\pi \approx 3.1416$ and $e \approx 2.7183$. Given values of $\mu$ and $\sigma$, however, the function is completely specified, and we could plot points or use a graphing calculator to produce its graph. Substituting $x + h$ for $x$ in equation (1) produces an equation of the same form but with a different value of $\mu$. Therefore, in the terminology of Section 2-2 of *Calculus* 11/e or *College Mathematics* 11/e, any horizontal translation of a normal curve is also a normal curve.

The standard deviation measures the dispersion of the normal probability density function about the mean. A small standard deviation indicates a tight clustering about the mean and thus a tall, narrow curve; a large standard deviation indicates a large deviation from the mean and thus a broad, flat curve. Notice that each of the normal curves in Figure 3 is symmetric about a vertical line through the mean. This is true for any normal curve. Thus, the line $x = \mu$ divides the region under a normal curve into two regions with equal areas. Since the total area under a normal curve is always 1, the area of each of these regions is .5. This implies that the median of a normal random variable is always equal to the mean (see Fig. 4).

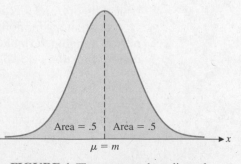

**FIGURE 4** The mean and median of a
normal random variable

The properties of the normal probability density function are summarized in the next box for ease of reference.

**SUMMARY** | **NORMAL PROBABILITY DENSITY FUNCTION**

$$f(x) = \frac{1}{\sigma\sqrt{2\pi}} e^{-(x-\mu)^2/(2\sigma^2)} \qquad \sigma > 0$$

**Mean** $\mu$
**Median** $\mu$
**Standard deviation** $\sigma$

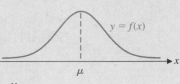

$y = f(x)$

The graph of $f(x)$ is symmetric with respect to the line $x = \mu$.

The cumulative distribution function for a normal random variable is given formally by

$$F(x) = \frac{1}{\sigma\sqrt{2\pi}} \int_{-\infty}^{x} e^{-(t-\mu)^2/(2\sigma^2)} \, dt$$

It is not possible to express $F(x)$ as a finite combination of the functions with which we are familiar. Furthermore, we cannot use antidifferentiation to evaluate probabilities such as

$$P(c \le X \le d) = \frac{1}{\sigma\sqrt{2\pi}} \int_{c}^{d} e^{-(x-\mu)^2/(2\sigma^2)} \, dx$$

```
fnInt((1/√(2π))e
^(-X²/2),X,0,1.5
)
        .4331927987
```

**FIGURE 5**

However, numerical integration can be used to approximate probabilities of this type, as illustrated in Figure 5, where we use a graphing utility to approximate $P(0 \le X \le 1.5)$ for the normal probability density function with $\mu = 0$ and $\sigma = 1$.

It is also possible to use a table to approximate probabilities involving the normal probability density function. Remarkably, the area under a normal curve between a mean $\mu$ and a given number of standard deviations to the right (or left) of $\mu$ is the same, regardless of the shape of the normal curve. For example, the area under the normal curve with $\mu = 3, \sigma = 5$ from $\mu = 3$ to $\mu + 1.5\sigma = 10.5$ is equal to the area under the normal curve with $\mu = 15, \sigma = 2$ from $\mu = 15$ to $\mu = 1.5\sigma = 18$ (see Fig. 6 below, noting that the shaded regions have the same areas, or equivalently, the same numbers of pixels). Therefore, such areas for any normal curve can be easily determined from the areas for the **standard normal curve,** that is, the normal curve with mean 0 and standard deviation 1. In fact, if $z$ represents the number of standard deviations that a measurement $x$ is from a mean $\mu$, then the area under a normal curve from $\mu$ to $\mu + z\sigma$ equals the area under the standard normal curve from 0 to $z$ (see Fig. 7). Table III in Appendix C lists those areas for the standard normal curve.

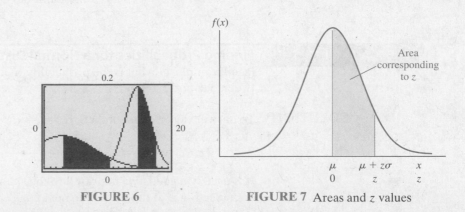

**FIGURE 6**        **FIGURE 7** Areas and $z$ values

In the discussion that follows we will concentrate on using Table III in Appendix C to approximate areas under the normal curve. If desired, numerical integration on a graphing utility can be used to obtain the same results.

<br>

**EXAMPLE 4**   **Finding Probabilities for a Normal Distribution**   A manufacturing process produces lightbulbs with life expectancies that are normally distributed with a mean of 500 hours and a standard deviation of 100 hours. What percentage of the lightbulbs can be expected to last between 500 and 670 hours?

**SOLUTION**   To answer this question, we first determine how many standard deviations 670 is from 500, the mean. This is easily done by dividing the distance between 500 and 670 by 100, the standard deviation. Thus,

$$z = \frac{670 - 500}{100} = \frac{170}{100} = 1.70$$

That is, 670 is 1.7 standard deviations from 500, the mean. Referring to Table III in Appendix C, we see that .4554 corresponds to $z = 1.70$. Also, because the total area under a normal curve is 1, we conclude that 45.54% of the lightbulbs produced will last between 500 and 670 hours (see Fig. 8).

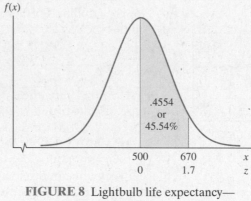

**FIGURE 8**  Lightbulb life expectancy—
positive $z$

**MATCHED PROBLEM 4**   In Example 4, what percentage of the lightbulbs can be expected to last between 500 and 750 hours?

<br>

In general, to find how many standard deviations a measurement $x$ is from a mean $\mu$, first determine the distance between $x$ and $\mu$, and then divide by $\sigma$:

$$z = \frac{\text{Distance between } x \text{ and } \mu}{\text{Standard deviation}} = \frac{x - \mu}{\mu}$$

<br>

**EXAMPLE 5**   **Finding Probabilities for a Normal Distribution**   From all lightbulbs produced (see Example 4), what is the probability of a lightbulb chosen at random lasting between 380 and 500 hours?

**SOLUTION**   To answer this, we first find $z$:

$$z = \frac{x - \mu}{\sigma} = \frac{380 - 500}{100} = -1.20$$

It is usually a good idea to draw a rough sketch of a normal curve and insert relevant data (see Fig. 9 at the top of the next page).

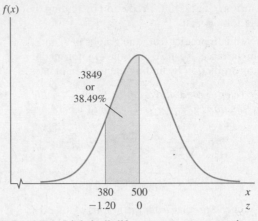

FIGURE 9 Lightbulb life expectancy—negative $z$

Table III in Appendix C does not include negative values for $z$, but because normal curves are symmetrical with respect to a vertical line through the mean, we simply use the absolute value (positive value) of $z$ for the table. Thus, the area corresponding to $z = -1.20$ is the same as the area corresponding to $z = 1.20$, which is .3849, so the probability of a lightbulb chosen at random lasting between 380 and 500 hours is .3849.

**MATCHED PROBLEM 5**  In Example 5, what is the probability of a lightbulb chosen at random lasting between 400 and 500 hours?

*Answers to Matched Problems*  **1.** (A) $\frac{1}{3}$ (B) $\frac{1}{3}$ (C) $\frac{1}{3}$  **2.** $\frac{2}{5}$  **3.** $e^{-0.5} \approx .607$  **4.** 49.38%  **5.** .3413

# Exercise 3-4

*A  In Problems 1–4, find the probability density function $f$ and the associated cumulative distribution function $F$ for the continuous random variable $X$ if*

  **1.** $X$ is uniformly distributed on $[0, 2]$.

  **2.** $X$ is uniformly distributed on $[3, 6]$.

  **3.** $X$ is an exponential random variable with $\lambda = \frac{1}{2}$.

  **4.** $X$ is an exponential random variable with $\lambda = \frac{1}{4}$.

*In Problems 5–8, find the mean, median, and standard deviation of the continuous random variable $X$ if*

  **5.** $X$ is uniformly distributed on $[1, 5]$.

  **6.** $X$ is uniformly distributed on $[2, 8]$.

  **7.** $X$ is an exponential random variable with $\lambda = 5$.

  **8.** $X$ is an exponential random variable with $\lambda = 3$.

*In Problems 9–16, use Table III in Appendix C to find the area under the standard normal curve from 0 to the indicated value of $z$.*

  **9.** 1.5          **10.** 0.8

  **11.** −0.72       **12.** 3.18

  **13.** 2.01        **14.** −2.39

  **15.** −1.93       **16.** 1.67

*In Problems 17–24, given a normal distribution with mean 100 and standard deviation 10, use Table III in Appendix C to find the area under this normal curve from the mean to the indicated measurement.*

  **17.** 120         **18.** 135

  **19.** 95          **20.** 113

  **21.** 106         **22.** 87

  **23.** 78          **24.** 99

*B  Problems 25 and 26 refer to random variables $X_1$ and $X_2$ with probability density functions $f_1$ and $f_2$, respectively, where*

$$f_1(x) = \frac{1}{\sqrt{2\pi}} e^{-x^2/2}$$

$$f_2(x) = \begin{cases} 0.4 & \text{if } -1.25 \le x \le 1.25 \\ 0 & \text{otherwise} \end{cases}$$

  **25.** Explain how you can predict from the graphs of $f_1$ and $f_2$ which random variable, $X_1$ or $X_2$, has the greater standard deviation. Check your prediction by finding the standard deviation of each.

26. Explain how you can predict from the graphs of $f_1$ and $f_2$ which random variable, $X_1$ or $X_2$, has the greater median. Check your prediction by finding the median of each.

*Problems 27 and 28 refer to random variables $X_1$ and $X_2$ with probability density functions $g_1$ and $g_2$, respectively, where*

$$g_1(x) = \frac{1}{\sqrt{2\pi}} e^{-(x-1)^2/2}$$

$$g_2(x) = \begin{cases} e^{-x} & \text{if } x \geq 0 \\ 0 & \text{otherwise} \end{cases}$$

27. Explain how you can predict from the graphs of $g_1$ and $g_2$ which random variable, $X_1$ or $X_2$, has the greater median. Check your prediction by finding the median of each.

28. Explain how you can predict from the graphs of $g_1$ and $g_2$ which random variable, $X_1$ or $X_2$, has the greater mean. Check your prediction by finding the mean of each.

*In Problems 29–32, X is a continuous random variable with mean $\mu$. Find $\mu$, and then find $P(X \leq \mu)$ if*

29. $X$ is uniformly distributed on $[0, 4]$.
30. $X$ is uniformly distributed on $[0, 10]$.
31. $X$ is an exponential random variable with $\lambda = 1$.
32. $X$ is an exponential random variable with $\lambda = 2$.

*In Problems 33–36, X is a continuous random variable with mean $\mu$ and standard deviation $\sigma$. Find $\mu$ and $\sigma$, and then find $P(\mu - \sigma \leq X \leq \mu + \sigma)$ if*

33. $X$ is uniformly distributed on $[-5, 5]$.
34. $X$ is uniformly distributed on $[-2, 2]$.
35. $X$ is an exponential random variable with $m = 6 \ln 2$.
36. $X$ is an exponential random variable with $m = 4 \ln 2$.

*Given a normal random variable X with mean 70 and standard deviation 8, find the indicated probabilities in Problems 37–44.*

37. $P(60 \leq X \leq 80)$
38. $P(50 \leq X \leq 90)$
39. $P(62 \leq X \leq 74)$
40. $P(66 \leq X \leq 78)$
41. $P(X \geq 88)$
42. $P(X \geq 90)$
43. $P(X \leq 60)$
44. $P(X \leq 56)$

*Problems 45–48 refer to the normal random variable X with mean $\mu$ and standard deviation $\sigma$. Use a graphing utility with a built-in numerical integration routine to compute the indicated probabilities.*

45. Approximate $P(\mu - \sigma \leq X \leq \mu + \sigma)$ to four decimal places for $\mu = 0$ and

  (A) $\sigma = .5$  (B) $\sigma = 1$  (C) $\sigma = 2$

  Write a brief verbal explanation of the result indicated by these calculations.

46. Approximate $P(\mu - 2\sigma \leq X \leq \mu + 2\sigma)$ to four decimal places for $\mu = 0$ and

  (A) $\sigma = .5$  (B) $\sigma = 1$  (C) $\sigma = 2$

Write a brief verbal explanation of the result indicated by these calculations.

47. Approximate $P(\mu - 3\sigma \leq X \leq \mu + 3\sigma)$ to four decimal places for $\mu = 0$ and

  (A) $\sigma = .5$  (B) $\sigma = 1$  (C) $\sigma = 2$

  Write a brief verbal explanation of the result indicated by these calculations.

48. In Problems 45–47, if $\mu$ is changed from 0 to 5, will any of the probabilities change? Explain.

C *Problems 49–52 refer to the uniformly distributed random variable X with probability density function*

$$f(x) = \begin{cases} \dfrac{1}{b - a} & \text{if } a \leq x \leq b \\ 0 & \text{otherwise} \end{cases}$$

49. Show that $\mu = (a + b)/2$.
50. Show that $m = (a + b)/2$.
51. Show that $\int_{-\infty}^{\infty} x^2 f(x) \, dx = (b^2 + ab + a^2)/3$.
52. Show that $V(X) = (b - a)^2/12$.

53. Use numerical integration on a graphing utility to approximate the mean, median, and standard deviation of the probability density function

$$f(x) = \begin{cases} 5e^{-5x} & \text{if } x \geq 0 \\ 0 & \text{otherwise} \end{cases}$$

  Explain why it is reasonable to replace integrals over the interval $(-\infty, \infty)$ by integrals over the interval $[0, 10]$. Verify that your answers agree with the results expected for exponential probability density functions.

54. Use numerical integration on a graphing utility to approximate the mean, median, and standard deviation of the probability density function

$$f(x) = \begin{cases} 0.36e^{-0.36x} & \text{if } x \geq 0 \\ 0 & \text{otherwise} \end{cases}$$

  Explain why it is reasonable to replace integrals over the interval $(-\infty, \infty)$ by integrals over the interval $[0, 100]$. Verify that your answers agree with the results expected for exponential probability density functions.

*Problems 55–58 refer to a random variable X with probability density function*

$$f(x) = \begin{cases} p/x^{p+1} & \text{if } x \geq 1 \\ 0 & \text{otherwise} \end{cases}$$

*where p is a positive constant. This random variable is often referred to as a **Pareto*** random variable.*

55. Verify that $f$ satisfies the necessary conditions for a probability density function and find the mean, if it exists. What restrictions must you place on $p$ to ensure that the mean exists?

56. Find the standard deviation, if it exists. What restrictions must you place on $p$ to ensure that the standard deviation exists?

---

*Named after Vilfredo Pareto (1848–1923), an Italian sociologist and economist. Although best known for his theories on political behavior, Pareto also developed new applications of mathematics to economics.

**57.** Find the median.

**58.** The mean and the median for a uniformly distributed random variable are equal, as are the mean and standard deviation for an exponentially distributed random variable. Use a graph to illustrate the relationship between the mean, median, and standard deviation for a Pareto random variable. Approximate abscissas of intersection points to two decimal places.

## Applications

**59.** *Waiting time.* The time (in minutes) applicants must wait to receive a driver's examination is uniformly distributed on the interval $[0, 40]$. What is the probability that an applicant must wait more than 25 minutes?

**60.** *Waiting time.* The time (in minutes) passengers must wait for a commuter plane in a large airport is uniformly distributed on the interval $[0, 60]$. What is the probability that a passenger waits less than 20 minutes?

**61.** *Communications.* The length of time for telephone conversations (in minutes) is exponentially distributed. The average (mean) length of a conversation is 3 minutes. What is the probability that a conversation lasts less than 2 minutes?

**62.** *Waiting time.* The waiting time (in minutes) for customers at a drive-in bank is an exponential random variable. The average (mean) time a customer waits is 4 minutes. What is the probability that a customer waits more than 5 minutes?

**63.** *Service time.* The time between failures of a photo zcopier is an exponential random variable. Half the copiers require service during the first 2 years of operation. What is the probability that a copier requires service during the first year of operation?

**64.** *Component failure.* The life expectancy (in years) of a component in a computer is an exponential random variable. Half the components fail in the first 3 years. The company that manufactures the component offers a 1-year warranty. What is the probability that a component will fail during the warranty period?

**65.** *Sales.* The annual sales for salespeople at a business machine company are normally distributed. The average (mean) annual sales are $200,000 with a standard deviation of $20,000. What percentage of the salespeople would be expected to make annual sales of $240,000 or more?

**66.** *Guarantees.* The life expectancy of a car battery is normally distributed. The average (mean) lifetime is 170 weeks with a standard deviation of 10 weeks. If the company guarantees the battery for 3 years, what percentage of the batteries sold would be expected to be returned before the end of the warranty period?

**67.** *Quality control.* A manufacturing process produces a precision part whose length is a normal random variable with mean 100 millimeters and standard deviation 2 millimeters. All parts deviating by more than 5 millimeters from the mean must be rejected. On average, what percentage of the parts must be rejected?

**68.** *Quality control.* An automated manufacturing process produces a component whose width is a normal random variable with mean 7.55 centimeters and standard devia-

tion 0.02 centimeter. All components deviating by more than 0.05 centimeter from the mean must be rejected. On average, what percentage of the parts must be rejected?

**69.** *Survival time.* The time of death (in years) after patients have contracted a certain disease is exponentially distributed. The probability that a patient dies within 1 year is .3.

(A) What is the expected time of death?

(B) What is the probability that a patient survives longer than the expected time of death?

**70.** *Survival time.* Repeat Problem 69 if the probability that a patient dies within 1 year is .5.

**71.** *Medicine.* The healing time of a certain type of incision is normally distributed with an average healing time of 240 hours and a standard deviation of 20 hours. What percentage of the people having this incision would heal in 8 days or less?

**72.** *Agriculture.* The height of a hay crop is normally distributed with an average height of 38 inches and a standard deviation of 1.5 inches. What percentage of the crop will be 40 inches or more?

**73.** *Learning.* The time (in minutes) it takes an adult to memorize a sequence of random digits is an exponential random variable. The average time is 2 minutes. What is the probability that it takes an adult over 5 minutes to memorize the digits?

**74.** *Psychology.* The time (in seconds) it takes rats to find their way through a maze is exponentially distributed. The average time is 30 seconds. What is the probability that it takes a rat over 1 minute to find a path through the maze?

**75.** *Grading on a curve.* An instructor grades on a curve by assuming the grades on a test are normally distributed. If the average grade is 70 and the standard deviation is 8, find the test scores for each grade interval if the instructor wishes to assign grades as follows: 10% A's, 20% B's, 40% C's, 20% D's, and 10% F's. Round answers to one decimal place.

**76.** *Psychology.* A test devised to measure aggressive–passive personalities was standardized on a large group of people. The scores were normally distributed with a mean of 50 and a standard deviation of 10. If we want to designate the highest 10% as aggressive, the next 20% as moderately aggressive, the middle 40% as average, the next 20% as moderately passive, and the lowest 10% as passive, what range of scores will be covered by these five designations? Round answers to one decimal place.

# CHAPTER 3 REVIEW

## Important Terms, Symbols, and Concepts

### 3-1 Improper Integrals

If $f$ is continuous over the indicated interval and the limit exists, then the improper integral on the left in each of the following equations is defined by

1. $\displaystyle\int_a^{\infty} f(x)\,dx = \lim_{b\to\infty}\int_a^b f(x)\,dx$

2. $\displaystyle\int_{-\infty}^b f(x)\,dx = \lim_{a\to-\infty}\int_a^b f(x)\,dx$

3. $\displaystyle\int_{-\infty}^{\infty} f(x)\,dx = \int_{-\infty}^c f(x)\,dx + \int_c^{\infty} f(x)\,dx$

where $c$ is any point on $(-\infty, \infty)$, provided both integrals on the right exist.

The improper integral **converges** if the limit exists; otherwise it **diverges** (and no value is assigned to it).

A continuous income stream is **perpetual** if it never stops producing income. The **capital value, $CV$,** of a perpetual income stream $f(t)$, at a rate $r$ compounded continuously, is defined by

$$CV = \int_0^{\infty} f(t)e^{-rt}\,dt$$

Capital value provides a method for expressing the worth (in terms of today's dollars) of an investment that will produce income indefinitely.

### 3-2 Continuous Random Variables

A **continuous random variable** $X$ is a function that assigns a numeric value to each outcome of an experiment. The set of possible values of $X$ is an interval of real numbers. This interval may be open or closed, bounded or unbounded.

The function $f(x)$ is a **probability density function** for a continuous random variable $X$ if:

1. $f(x) \geq 0$ for all $x$ in $(-\infty, \infty)$

2. $\displaystyle\int_{-\infty}^{\infty} f(x)\,d = 1$

3. The probability that $X$ lies in the interval $[c, d]$ is:

$$P(c \leq X \leq d) = \int_c^d f(x)\,dx$$

If $f$ is a probability density function, then the associated **cumulative distribution function $F$** is:

$$F(x) = P(X \leq x) = \int_{-\infty}^x f(t)\,dt$$

The cumulative distribution function $F$ has the properties

1. $F'(x) = f(x)$ wherever $f$ is continuous
2. $0 \leq F(x) \leq 1$ for $x$ in $(-\infty, \infty)$
3. $F(x)$ is nondecreasing on $(-\infty, \infty)$

### 3-3 Expected Value, Standard Deviation, and Median

Let $f(x)$ be the probability density function for a continuous random variable $X$. The **expected value, or mean, of $X$** is

$$\mu = E(X) = \int_{-\infty}^{\infty} xf(x)\,dx$$

The **variance** is:

$$V(X) = \int_{-\infty}^{\infty} (x - \mu)^2 f(x)\,dx = \int_{-\infty}^{\infty} x^2 f(x)\,dx - \mu^2$$

and the **standard deviation** is $\sigma = \sqrt{V(X)}$

The **median** is a value of the random variable that divides the area under the graph of the probability density function into two equal parts, each of area $\frac{1}{2}$.

## 3-4  Special Probability Distributions

A continuous random variable has a **uniform distribution** on $[a, b]$, and a **uniform probability density function,** if

$$f(x) = \begin{cases} \dfrac{1}{b-a} & \text{if } a \le x \le b \\ 0 & \text{otherwise} \end{cases}$$

A continuous random variable has an **exponential distribution,** and an **exponential probability density function,** if

$$f(x) = \begin{cases} (1/\lambda)e^{-x/\lambda} & \text{if } x \ge 0 \\ 0 & \text{otherwise} \end{cases}$$

A continuous random variable has a **normal distribution,** and a **normal probability density function,** if

$$f(x) = \frac{1}{\sigma\sqrt{2\pi}}\, e^{-(x-\mu)^2/(2\sigma^2)}$$

where $\mu$ (the mean) is any constant and $\sigma$ (the standard deviation) is any positive constant. The graph of a normal probability density function is called a **normal curve.** The **standard normal curve** has mean 0 and standard deviation 1. The area under any normal curve from $\mu$ to $x = \mu + z\sigma$ is equal to the area under the standard normal curve from 0 to $z$, where $z = \dfrac{x - \mu}{\sigma}$.

## REVIEW EXERCISE

Work through all the problems in this chapter review and check your answers in the back of this supplement. Answers to all review problems are there along with section numbers in italics to indicate where each type of problem is discussed. Where weaknesses show up, review appropriate sections in the text.

A  In Problems 1–3, evaluate each improper integral, if it converges.

1.  $\displaystyle\int_0^\infty e^{-2x}\,dx$

2.  $\displaystyle\int_0^\infty \frac{1}{x+1}\,dx$

3.  $\displaystyle\int_1^\infty \frac{16}{x^3}\,dx$

Problems 4–7 refer to the continuous random variable X with probability density function

$$f(x) = \begin{cases} 1 - \tfrac{1}{2}x & \text{if } 0 \le x \le 2 \\ 0 & \text{otherwise} \end{cases}$$

4.  Find $P(0 \le X \le 1)$ and illustrate with a graph.

5.  Find the mean, variance, and standard deviation.

6.  Find and graph the associated cumulative distribution function.

7.  Find the median.

8.  If $X$ is a normal random variable with a mean of 100 and a standard deviation of 10, find $P(100 \le X \le 118)$.

In Problems 9 and 10, find the probability density function f and the associated cumulative distribution function F for the continuous random variable X if

9.  $X$ is uniformly distributed on $[5, 15]$.

10.  $X$ is an exponential random variable with $\lambda = \tfrac{1}{5}$.

In Problems 11 and 12, is f a probability density function? Explain.

11.  $f(x) = e^{-x^2}$

12.  $f(x) = \begin{cases} 0.75(x^2 - 4x + 3) & \text{if } 0 \le x \le 4 \\ 0 & \text{otherwise} \end{cases}$

B  Problems 13–16 refer to the continuous random variable X with probability density function

$$f(x) = \begin{cases} \tfrac{5}{2}x^{-7/2} & \text{if } x \ge 1 \\ 0 & \text{otherwise} \end{cases}$$

13.  Find $P(1 \le X \le 4)$ and illustrate with a graph.

14.  Find the mean, variance, and standard deviation.

15.  Find and graph the associated cumulative distribution function.

16.  Find the median.

Problems 17–20 refer to an exponentially distributed random variable X.

17.  If $P(4 \le X) = e^{-2}$, find the probability density function.

18.  Find $P(0 \le X \le 2)$.

19.  Find the associated cumulative distribution function.

20.  Find the mean, standard deviation, and median.

21.  Given a normal distribution with mean 50 and standard deviation 6, find the area under the normal curve:

(A)  Between 41 and 62

(B)  From 59 on

**22.** Given a normal random variable $X$ with mean 82 and standard deviation 8, find:

(A) $P(84 \le X \le 94)$

(B) $P(X \ge 60)$

*In Problems 23 and 24, evaluate each improper integral, if it converges.*

**23.** $\displaystyle\int_{-\infty}^{0} e^x \, dx$ **24.** $\displaystyle\int_{0}^{\infty} \frac{1}{(x+3)^2} \, dx$

**25.** If $f$ is continuous on $[-1, \infty)$ and $\int_{-1}^{\infty} f(x) \, dx$ converges, does $\int_{1}^{\infty} f(x) \, dx$ converge? Explain.

*In Problems 26 and 27, find a constant $k$ so that $kf$ is a probability density function, or explain why no such $k$ exists.*

**26.** $f(x) = \begin{cases} e^{-10x} & \text{if } x \ge 0 \\ 0 & \text{otherwise} \end{cases}$

**27.** $f(x) = \begin{cases} e^{10x} & \text{if } x \ge 0 \\ 0 & \text{otherwise} \end{cases}$

**28.** If $X$ is an exponentially distributed random variable with median $m = 3 \ln 2$ and mean $\mu$, find $P(X \le \mu)$.

**29.** Find the mean and the median of the continuous random variable with probability density function

$$f(x) = \begin{cases} 50/(x+5)^3 & \text{if } x \ge 0 \\ 0 & \text{otherwise} \end{cases}$$

**30.** Use a graphing utility to approximate (to two decimal places) the median of the continuous random variable with probability density function

$$f(x) = \begin{cases} \dfrac{.8}{x^2} + \dfrac{.8}{x^5} & \text{if } x \ge 1 \\ 0 & \text{otherwise} \end{cases}$$

**31.** Find $\displaystyle\int_{-\infty}^{\infty} \frac{e^x}{(1 + e^x)^2} \, dx$, if it converges.

**32.** If $f$ is a continuous probability density function with mean $\mu$ and standard deviation $\sigma$ and $a$, $b$, and $c$ are constants, evaluate the following integral. Express the result in terms of $\mu, \sigma, a, b,$ and $c$.

$$\int_{-\infty}^{\infty} (ax^2 + bx + c)f(x) \, dx$$

*Problems 33–36 refer to random variables $X_1$ and $X_2$ with probability density functions $f_1$ and $f_2$, respectively, where*

$$f_1(x) = \begin{cases} 0.25xe^{-x/2} & \text{if } x \ge 0 \\ 0 & \text{otherwise} \end{cases}$$

$$f_2(x) = \begin{cases} 0.0625x^2e^{-x/2} & \text{if } x \ge 0 \\ 0 & \text{otherwise} \end{cases}$$

**33.** Explain how you can predict from the graphs of $f_1$ and $f_2$ which random variable, $X_1$ or $X_2$, has the greater mean.

**34.** Explain how you can predict from the graphs of $f_1$ and $f_2$ which random variable, $X_1$ or $X_2$, has the greater variance.

**35.** Check your prediction in Problem 33 by using numerical integration to approximate the means of $X_1$ and $X_2$.

**36.** Check your prediction in Problem 34 by using numerical integration to approximate the variances of $X_1$ and $X_2$.

## APPLICATIONS

**37.** *Production.* The rate of production of an oil well (in millions of barrels per year) is given by (see the figure below)

$$R(t) = 12e^{-0.3t} - 12e^{-0.6t}$$

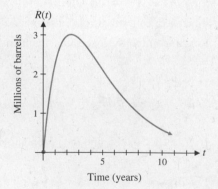

Time (years)

(A) Assuming that the well is operated indefinitely, find the total production.

(B) When will the output from the well reach 50% of the total production? Round answer to two decimal places.

**38.** *Demand.* The manager of a movie theater has determined that the weekly demand for popcorn (in pounds) is a continuous random variable with probability density function

$$f(x) = \begin{cases} .02(1 - .01x) & \text{if } 0 \le x \le 100 \\ 0 & \text{otherwise} \end{cases}$$

(A) Evaluate $\int_{40}^{100} f(x) \, dx$ and interpret.

(B) If the manager has 50 pounds of popcorn on hand at the beginning of the week, what is the probability that this will be enough to meet the weekly demand?

(C) If the manager wants the probability that the supply on hand exceeds the weekly demand to be .96, how much popcorn must be on hand at the beginning of the week?

**39.** *Capital value.* The perpetual annual rent for a property is $2,400. Find the capital value at 6% compounded continuously.

**40.** *Credit applications.* The percentage of applications (expressed as a decimal between 0 and 1) for a national credit card processed on the same day the applications are received is a continuous random variable with probability density function

$$f(x) = \begin{cases} 6x(1 - x) & \text{if } 0 \le x \le 1 \\ 0 & \text{otherwise} \end{cases}$$

(A) What is the probability that at least 20% of the applications received are processed the same day they arrive?

(B) What is the expected percentage of applications processed the same day they arrive?

(C) What is the median percentage of applications processed the same day they arrive?

**41.** *Computer failure.* A computer manufacturer has determined that the time between failures for its computer is an exponentially distributed random variable with a mean failure time of 4,000 hours. Suppose a particular computer has just been repaired.

(A) What is the probability that the computer operates for the next 4,000 hours without failure?

(B) What is the probability that the computer fails in the next 1,000 hours?

**42.** *Radial tire failure.* The life expectancy (in miles) of a certain brand of radial tires is a normal random variable with a mean of 35,000 miles and a standard deviation of 5,000 miles. What is the probability that a tire fails during the first 25,000 miles of use?

**43.** *Personnel screening.* The scores on a screening test for new technicians are normally distributed with mean 100 and standard deviation 10. Find the approximate percentage of applicants taking the test who score

(A) Between 92 and 108

(B) 115 or higher

**44.** *Medicine.* The shelf life (in months) of a certain drug is a continuous random variable with probability density function

$$f(x) = \begin{cases} 10/(x + 10)^2 & \text{if } x \ge 0 \\ 0 & \text{otherwise} \end{cases}$$

(A) Evaluate $\int_2^8 f(x)\, dx$ and interpret.

(B) What is the probability that the drug is still usable after 5 months?

(C) What is the median shelf life?

**45.** *Life expectancy.* The life expectancy (in months) after dogs have contracted a certain disease is an exponentially distributed random variable. The probability of surviving more than 1 month is $e^{-2}$. After contracting the disease,

(A) What is the probability of the dog surviving more than 2 months?

(B) What is the mean life expectancy?

**46.** *Drug assimilation.* The rate at which the body eliminates a drug (in milliliters per hour) is given by

$$R(t) = 15e^{-0.2t} - 15e^{-0.3t}$$

where $t$ is the number of hours since the drug was administered.

(A) What is the total amount of the drug that is eliminated by the body?

(B) How long will it take for 50% of the total amount to be eliminated? Round answer to two decimal places.

**47.** *Testing.* The IQ scores for 6-year-old children in a certain area are normally distributed with a mean of 108 and a standard deviation of 12. What percentage of the children can be expected to have IQ scores of 135 or more?

**48.** *Politics.* The rate of change of the voting population of a city with respect to time $t$ (in years) is estimated to be

$$N'(t) = \frac{100t}{(1 + t^2)^2}$$

where $N(t)$ is in thousands. If $N(0)$ is the current voting population, how much will this population increase during the next 3 years? If the population continues to grow at this rate indefinitely, what is the total increase in the voting population?

# APPENDIX C   Tables

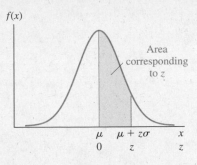

f(x)

Area corresponding to z

$\mu$   $\mu + z\sigma$   x
0      z             z

**TABLE III** Area Under the Standard Normal Curve

| | | | | (Table Entries Represent the Area Under the Standard Normal Curve from 0 to z, z ≥ 0) | | | | | |
|---|---|---|---|---|---|---|---|---|---|---|
| z | .00 | .01 | .02 | .03 | .04 | .05 | .06 | .07 | .08 | .09 |
| 0.0 | 0.0000 | 0.0040 | 0.0080 | 0.0120 | 0.0160 | 0.0199 | 0.0239 | 0.0279 | 0.0319 | 0.0359 |
| 0.1 | 0.0398 | 0.0438 | 0.0478 | 0.0517 | 0.0557 | 0.0596 | 0.0636 | 0.0675 | 0.0714 | 0.0753 |
| 0.2 | 0.0793 | 0.0832 | 0.0871 | 0.0910 | 0.0948 | 0.0987 | 0.1026 | 0.1064 | 0.1103 | 0.1141 |
| 0.3 | 0.1179 | 0.1217 | 0.1255 | 0.1293 | 0.1331 | 0.1368 | 0.1406 | 0.1443 | 0.1480 | 0.1517 |
| 0.4 | 0.1554 | 0.1591 | 0.1628 | 0.1664 | 0.1700 | 0.1736 | 0.1772 | 0.1808 | 0.1844 | 0.1879 |
| 0.5 | 0.1915 | 0.1950 | 0.1985 | 0.2019 | 0.2054 | 0.2088 | 0.2123 | 0.2157 | 0.2190 | 0.2224 |
| 0.6 | 0.2257 | 0.2291 | 0.2324 | 0.2357 | 0.2389 | 0.2422 | 0.2454 | 0.2486 | 0.2517 | 0.2549 |
| 0.7 | 0.2580 | 0.2611 | 0.2642 | 0.2673 | 0.2704 | 0.2734 | 0.2764 | 0.2794 | 0.2823 | 0.2852 |
| 0.8 | 0.2881 | 0.2910 | 0.2939 | 0.2967 | 0.2995 | 0.3023 | 0.3051 | 0.3078 | 0.3106 | 0.3133 |
| 0.9 | 0.3159 | 0.3186 | 0.3212 | 0.3238 | 0.3264 | 0.3289 | 0.3315 | 0.3340 | 0.3365 | 0.3389 |
| 1.0 | 0.3413 | 0.3438 | 0.3461 | 0.3485 | 0.3508 | 0.3531 | 0.3554 | 0.3577 | 0.3599 | 0.3621 |
| 1.1 | 0.3643 | 0.3665 | 0.3686 | 0.3708 | 0.3729 | 0.3749 | 0.3770 | 0.3790 | 0.3810 | 0.3830 |
| 1.2 | 0.3849 | 0.3869 | 0.3888 | 0.3907 | 0.3925 | 0.3944 | 0.3962 | 0.3980 | 0.3997 | 0.4015 |
| 1.3 | 0.4032 | 0.4049 | 0.4066 | 0.4082 | 0.4099 | 0.4115 | 0.4131 | 0.4147 | 0.4162 | 0.4177 |
| 1.4 | 0.4192 | 0.4207 | 0.4222 | 0.4236 | 0.4251 | 0.4265 | 0.4279 | 0.4292 | 0.4306 | 0.4319 |
| 1.5 | 0.4332 | 0.4345 | 0.4357 | 0.4370 | 0.4382 | 0.4394 | 0.4406 | 0.4418 | 0.4429 | 0.4441 |
| 1.6 | 0.4452 | 0.4463 | 0.4474 | 0.4484 | 0.4495 | 0.4505 | 0.4515 | 0.4525 | 0.4535 | 0.4545 |
| 1.7 | 0.4554 | 0.4564 | 0.4573 | 0.4582 | 0.4591 | 0.4599 | 0.4608 | 0.4616 | 0.4625 | 0.4633 |
| 1.8 | 0.4641 | 0.4649 | 0.4656 | 0.4664 | 0.4671 | 0.4678 | 0.4686 | 0.4693 | 0.4699 | 0.4706 |
| 1.9 | 0.4713 | 0.4719 | 0.4726 | 0.4732 | 0.4738 | 0.4744 | 0.4750 | 0.4756 | 0.4761 | 0.4767 |
| 2.0 | 0.4772 | 0.4778 | 0.4783 | 0.4788 | 0.4793 | 0.4798 | 0.4803 | 0.4808 | 0.4812 | 0.4817 |
| 2.1 | 0.4821 | 0.4826 | 0.4830 | 0.4834 | 0.4838 | 0.4842 | 0.4846 | 0.4850 | 0.4854 | 0.4857 |
| 2.2 | 0.4861 | 0.4864 | 0.4868 | 0.4871 | 0.4875 | 0.4878 | 0.4881 | 0.4884 | 0.4887 | 0.4890 |
| 2.3 | 0.4893 | 0.4896 | 0.4898 | 0.4901 | 0.4904 | 0.4906 | 0.4909 | 0.4911 | 0.4913 | 0.4916 |
| 2.4 | 0.4918 | 0.4920 | 0.4922 | 0.4925 | 0.4927 | 0.4929 | 0.4931 | 0.4932 | 0.4934 | 0.4936 |
| 2.5 | 0.4938 | 0.4940 | 0.4941 | 0.4943 | 0.4945 | 0.4946 | 0.4948 | 0.4949 | 0.4951 | 0.4952 |
| 2.6 | 0.4953 | 0.4955 | 0.4956 | 0.4957 | 0.4959 | 0.4960 | 0.4961 | 0.4962 | 0.4963 | 0.4964 |
| 2.7 | 0.4965 | 0.4966 | 0.4967 | 0.4968 | 0.4969 | 0.4970 | 0.4971 | 0.4972 | 0.4973 | 0.4974 |
| 2.8 | 0.4974 | 0.4975 | 0.4976 | 0.4977 | 0.4977 | 0.4978 | 0.4979 | 0.4979 | 0.4980 | 0.4981 |
| 2.9 | 0.4981 | 0.4982 | 0.4982 | 0.4983 | 0.4984 | 0.4984 | 0.4985 | 0.4985 | 0.4986 | 0.4986 |
| 3.0 | 0.4987 | 0.4987 | 0.4987 | 0.4988 | 0.4988 | 0.4989 | 0.4989 | 0.4989 | 0.4990 | 0.4990 |
| 3.1 | 0.4990 | 0.4991 | 0.4991 | 0.4991 | 0.4992 | 0.4992 | 0.4992 | 0.4992 | 0.4993 | 0.4993 |
| 3.2 | 0.4993 | 0.4993 | 0.4994 | 0.4994 | 0.4994 | 0.4994 | 0.4994 | 0.4995 | 0.4995 | 0.4995 |
| 3.3 | 0.4995 | 0.4995 | 0.4995 | 0.4996 | 0.4996 | 0.4996 | 0.4996 | 0.4996 | 0.4996 | 0.4997 |
| 3.4 | 0.4997 | 0.4997 | 0.4997 | 0.4997 | 0.4997 | 0.4997 | 0.4997 | 0.4997 | 0.4997 | 0.4998 |
| 3.5 | 0.4998 | 0.4998 | 0.4998 | 0.4998 | 0.4998 | 0.4998 | 0.4998 | 0.4998 | 0.4998 | 0.4998 |
| 3.6 | 0.4998 | 0.4998 | 0.4999 | 0.4999 | 0.4999 | 0.4999 | 0.4999 | 0.4999 | 0.4999 | 0.4999 |
| 3.7 | 0.4999 | 0.4999 | 0.4999 | 0.4999 | 0.4999 | 0.4999 | 0.4999 | 0.4999 | 0.4999 | 0.4999 |
| 3.8 | 0.4999 | 0.4999 | 0.4999 | 0.4999 | 0.4999 | 0.4999 | 0.4999 | 0.4999 | 0.4999 | 0.4999 |
| 3.9 | 0.5000 | 0.5000 | 0.5000 | 0.5000 | 0.5000 | 0.5000 | 0.5000 | 0.5000 | 0.5000 | 0.5000 |

## Section D-1   INTERPOLATING POLYNOMIALS AND DIVIDED DIFFERENCES

- Introduction
- The Interpolating Polynomial
- Divided Difference Tables
- Application

Given two points in the plane with distinct $x$ coordinates, we can use the point-slope form of the equation of a line to find a polynomial whose graph passes through these two points. If we are given a set of three, four, or more points with distinct $x$ coordinates, is there a polynomial whose graph will pass through all the given points? In this section, we will see that the answer to this question is yes, and we will discuss several methods for finding this polynomial, called the *interpolating polynomial*. The principal use of interpolating polynomials is to approximate $y$ coordinates for points not in the given set. For example, a retail sales firm may have obtained a table of prices at various demands by examining past sales records. Prices for demands not in the table can be approximated by an interpolating polynomial. Interpolating polynomials also have applications to computer graphics. If a computer user selects a set of points on a drawing, the interpolating polynomial can be used to produce a smooth curve that passes through the selected points.

### ■ Introduction

We usually write polynomials in standard form using either increasing or decreasing powers of $x$. Both of the following polynomials are written in standard form:

$$p(x) = 1 + x^2 - 2x^3 \qquad q(x) = -3x^5 + 2x^4 - 5x$$

In this section, we will find it convenient to write polynomials in a different form. The following activity will give you some experience with this new form.

*Explore & Discuss* **1**

Consider the points in Table 1.

(A) Let $p_1(x) = a_0 + a_1(x - 1)$. Determine $a_0$ and $a_1$ so that the graph of $y = p_1(x)$ passes through the first two points in Table 1.

(B) Let $p_2(x) = a_0 + a_1(x - 1) + a_2(x - 1)(x - 2)$. Determine $a_0, a_1,$ and $a_2$ so that the graph of $y = p_2(x)$ passes through the first three points in Table 1.

(C) Let
$$p_3(x) = a_0 + a_1(x - 1) + a_2(x - 1)(x - 2) \\ + a_3(x - 1)(x - 2)(x - 3).$$

Determine $a_0, a_1, a_2,$ and $a_3$ so that the graph of $y = p_3(x)$ passes through all four points in Table 1.

**TABLE 1**

| $x$ | 1 | 2 | 3 | 4 |
|---|---|---|---|---|
| $y$ | 4 | 7 | 4 | 1 |

The following example will illustrate basic concepts.

EXAMPLE 1   **Approximating Revenue**  A manufacturing company has defined the revenue function for one of its products by examining past records and listing the revenue (in thousands of dollars) for certain levels of production (in thousands of units). Use the revenue function defined by Table 2 to estimate the revenue if 3,000 units are produced and if 7,000 units are produced.

**135**

**SOLUTION** One way to approximate values of a function defined by a table is to use a **piecewise linear approximation.** To form the piecewise linear approximation for Table 2, we simply use the point-slope formula to find the equation of the line joining each successive pair of points in the table (see Fig. 1).

| TABLE 2 | Revenue $R$ Defined as a Function of Production $x$ by a Table | | |
|---|---|---|---|
| $x$ | 1 | 4 | 6 | 8 |
| $R(x)$ | 65 | 80 | 40 | 16 |

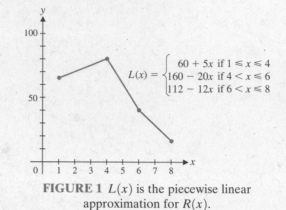

$$L(x) = \begin{cases} 60 + 5x & \text{if } 1 \leqslant x \leqslant 4 \\ 160 - 20x & \text{if } 4 < x \leqslant 6 \\ 112 - 12x & \text{if } 6 < x \leqslant 8 \end{cases}$$

**FIGURE 1** $L(x)$ is the piecewise linear approximation for $R(x)$.

This type of approximation is very useful in certain applications, but it has several disadvantages. First, the piecewise linear approximation usually has a sharp corner at each point in the table and thus is not differentiable at these points. Second, the piecewise linear approximation requires the use of a different formula between each successive pair of points in the table (see Fig. 1).

Instead of using the piecewise linear approximation, we will outline a method that will produce a polynomial whose values agree with $R(x)$ at each point in Table 2. This will provide us with a differentiable function given by a single formula that can be used to approximate $R(x)$ for any value of $x$ between 1 and 8.

Suppose $p(x)$ is a polynomial whose values agree with the values of $R(x)$ at the four $x$ values given in Table 2. Instead of expressing $p(x)$ in terms of powers of $x$, the standard method for writing polynomial forms, we use the first three $x$ values in the table to write

$$p(x) = a_0 + a_1(x - 1) + a_2(x - 1)(x - 4) + a_3(x - 1)(x - 4)(x - 6)$$

As we will see, writing $p(x)$ in this special form will greatly simplify our work.

Since $p(x)$ is to agree with $R(x)$ at each $x$ value in Table 2, we can write the following equations involving the coefficients $a_0$, $a_1$, $a_2$, and $a_3$:

$$65 = R(1) = p(1) = a_0 \tag{1}$$

$$80 = R(4) = p(4) = a_0 + 3a_1 \tag{2}$$

$$40 = R(6) = p(6) = a_0 + 5a_1 + 10a_2 \tag{3}$$

$$16 = R(8) = p(8) = a_0 + 7a_1 + 28a_2 + 56a_3 \tag{4}$$

From equation (1), we see that $a_0 = 65$. Solving equation (2) for $a_1$ and substituting for $a_0$, we have

$$a_1 = \frac{1}{3}(80 - a_0) = \frac{1}{3}(80 - 65) = 5$$

Proceeding the same way with equations (3) and (4), we have

$$a_2 = \frac{1}{10}(40 - a_0 - 5a_1) = \frac{1}{10}(40 - 65 - 25) = -5$$

$$a_3 = \frac{1}{56}(16 - a_0 - 7a_1 - 28a_2) = \frac{1}{56}(16 - 65 - 35 + 140) = 1$$

Thus,

$$p(x) = 65 + 5(x - 1) - 5(x - 1)(x - 4) + (x - 1)(x - 4)(x - 6)$$

**TABLE 2**

| $x$ | 1 | 4 | 6 | 8 |
|---|---|---|---|---|
| $R(x)$ | 65 | 80 | 40 | 16 |

The polynomial $p(x)$ agrees with $R(x)$ at each $x$ value in Table 2 (verify this) and can be used to approximate $R(x)$ for values of $x$ between 1 and 8 (see Fig. 2).

If 3,000 units are produced, then the revenue can be approximated by evaluating $p(3)$:

$$R(3) \approx p(3) = 65 + 5(2) - 5(2)(-1) + (2)(-1)(-3)$$
$$= 91 \text{ or } \$91,000$$

If 7,000 units are produced, then

$$R(7) \approx p(7) = 65 + 5(6) - 5(6)(3) + (6)(3)(1)$$
$$= 23 \text{ or } \$23,000$$

**FIGURE 2**  $p(x) = 65 + 5(x - 1) - 5(x - 1)(x - 4) + (x - 1)(x - 4)(x - 6)$

---

**MATCHED PROBLEM 1**  Refer to Example 1. Approximate the revenue if 2,000 units are produced and if 5,000 units are produced.

---

Since the revenue function in Example 1 was defined by a table, we have no information about this function for any value of $x$ other than those listed in the table. Thus, we cannot say anything about the accuracy of the approximations obtained by using $p(x)$. As we mentioned earlier, the piecewise linear approximation might provide a better approximation in some cases. The primary advantage of using $p(x)$ is that we have a differentiable function that is defined by a single equation and agrees with the revenue function at every value of $x$ in the table.

## ■ The Interpolating Polynomial

The procedure we used to find a polynomial approximation for the revenue function in Example 1 can be applied to any function that is defined by a table. The polynomial that is obtained in this way is referred to as the *interpolating polynomial*. The basic concepts are summarized in the next box.

---

**DEFINITION**  **The Interpolating Polynomial**

If $f(x)$ is the function defined by the following table of $n + 1$ points,

| $x$ | $x_0$ | $x_1$ | ... | $x_n$ |
|---|---|---|---|---|
| $f(x)$ | $y_0$ | $y_1$ | ... | $y_n$ |

then the **interpolating polynomial** for $f(x)$ is the polynomial $p(x)$ of degree less than or equal to $n$ that satisfies

$$p(x_0) = y_0 = f(x_0)$$

$$p(x_1) = y_1 = f(x_1)$$

$$\vdots \qquad \vdots$$

$$p(x_n) = y_n = f(x_n)$$

**Newton's form** for the interpolating polynomial is

$$p(x) = a_0 + a_1(x - x_0) + a_2(x - x_0)(x - x_1) + \cdots$$
$$+ a_n(x - x_0)(x - x_1) \cdots (x - x_{n-1})$$

Notice that if we graph the points in the defining table and the interpolating polynomial $p(x)$ on the same set of axes, then the graph of $p(x)$ will pass through every point given in the table (see Fig. 2). Is it possible to find a polynomial that is different from $p(x)$ and also has a graph that passes through all the points in the table? In more advanced texts, it is shown that

The interpolating polynomial is the only polynomial of degree less than or equal to $n$ whose graph will pass through every point in the table.

Any other polynomial whose graph goes through all these points must be of degree greater than $n$. The steps we used in finding the interpolating polynomial are summarized in the following box.

**PROCEDURE**

**Steps for Finding the Interpolating Polynomial**

**Step 1.** Write Newton's form for $p(x)$.

**Step 2.** Use the conditions $p(x_i) = y_i, i = 0, 1, \ldots, n$ to write $n + 1$ linear equations for the coefficients. Do not change the order of these equations. This system of equations is called a **lower triangular system.**

**Step 3.** Starting with the first and proceeding down the list, solve each equation for the coefficient with the largest subscript and substitute all the previously determined coefficients. This method of solving for the coefficients is called **forward substitution.**

**EXAMPLE 2**

**Finding the Interpolating Polynomial** Find the interpolating polynomial for the function defined by the following table:

| $x$ | 0 | 1 | 2 | 3 |
|-----|---|---|----|----|
| $f(x)$ | 5 | 4 | -3 | -4 |

**SOLUTION**    **Step 1.**    Newton's form for $p(x)$ is

$$p(x) = a_0 + a_1 x + a_2 x(x - 1) + a_3 x(x - 1)(x - 2)$$

**Step 2.**    The lower triangular system is

$$a_0 \qquad\qquad\qquad\qquad = \quad 5 \quad p(0) = f(0) = \ 5$$

$$a_0 + a_1 \qquad\qquad\qquad = \quad 4 \quad p(1) = f(1) = \ 4$$

$$a_0 + 2a_1 + 2a_2 \qquad\quad = -3 \quad p(2) = f(2) = -3$$

$$a_0 + 3a_1 + 6a_2 + 6a_3 = -4 \quad p(3) = f(3) = -4$$

**Step 3.** Solving this system by forward substitution, we have

$$a_0 = 5$$

$$a_1 = 4 - a_0 = 4 - 5 = -1$$

$$a_2 = \frac{1}{2}(-3 - a_0 - 2a_1) = \frac{1}{2}(-3 - 5 + 2) = -3$$

$$a_3 = \frac{1}{6}(-4 - a_0 - 3a_1 - 6a_2) = \frac{1}{6}(-4 - 5 + 3 + 18) = 2$$

Thus, Newton's form for this interpolating polynomial is

$$p(x) = 5 - x - 3x(x - 1) + 2x(x - 1)(x - 2)$$

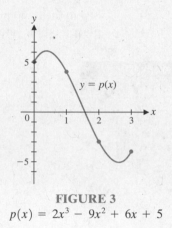

**FIGURE 3**
$$p(x) = 2x^3 - 9x^2 + 6x + 5$$

This form of $p(x)$ is suitable for evaluating the polynomial. For other operations, such as differentiation, integration, or graphing, it may be preferable to perform the indicated operations, collect like terms, and express $p(x)$ in the standard polynomial form

$$p(x) = 2x^3 - 9x^2 + 6x + 5$$

The graph of $p(x)$ is shown in Figure 3.

———

**MATCHED PROBLEM 2**  Find the interpolating polynomial for the function defined by the following table:

| $x$ | −1 | 0 | 1 | 2 |
|---|---|---|---|---|
| $f(x)$ | 5 | 3 | 3 | 11 |

*Explore & Discuss* **2**  Given the following polynomial and table,

$$p(x) = x^3 - 10x^2 + 29x - 17$$

| $x$ | 1 | 3 | 6 |
|---|---|---|---|
| $y$ | 3 | 7 | 13 |

(A) Show that the graph of $p(x)$ passes through each point in the table.

(B) Is $p(x)$ the interpolating polynomial for this table? If not, what is the interpolating polynomial for this table?

(C) Discuss the relationship between the number of points in a table and the degree of the interpolating polynomial for that table.

## ■ Divided Difference Tables

We now present a simple computational procedure for finding the coefficients $a_0, a_1, \ldots, a_n$ in Newton's form for an interpolating polynomial. To introduce this method, we return to Table 2 in Example 1, which we restate here.

| TABLE 2 | | | | |
|---|---|---|---|---|
| $x$ | 1 | 4 | 6 | 8 |
| $f(x)$ | 65 | 80 | 40 | 16 |

The coefficients in Newton's form for the interpolating polynomial for this table were $a_0 = 65$, $a_1 = 5$, $a_2 = -5$, and $a_3 = 1$. We will now construct a table, called a *divided difference table,* which will produce these coefficients with a minimum of computation. To begin, we place the $x$ and $y$ values in the first two columns of a new table. Then we compute the ratio of the change in $y$ to the change in $x$ for each successive pair of points in the table, and place the result on the line between the two points (see Table 3). These ratios are called the **first divided differences.**

To form the next column in the table, we repeat this process, using the change in the first divided differences in the numerator and the change in *two* successive values of $x$ in the denominator. These ratios are called the **second divided differences** and are placed on the line between the corresponding first divided differences (see Table 4).

**TABLE 3** First Divided Differences

| $x_k$ | $y_k$ | First Divided Difference |
|---|---|---|
| 1 | 65 | |
| | | $\dfrac{80-65}{4-1} = \dfrac{15}{3} = 5$ |
| 4 | 80 | |
| | | $\dfrac{40-80}{6-4} = \dfrac{-40}{2} = -20$ |
| 6 | 40 | |
| | | $\dfrac{16-40}{8-6} = \dfrac{-24}{2} = -12$ |
| 8 | 16 | |

**TABLE 4** Second Divided Differences

| $x_k$ | $y_k$ | First Divided Difference | Second Divided Difference |
|---|---|---|---|
| 1 | 65 | | |
| | | 5 | |
| 4 | 80 | | $\dfrac{-20-5}{6-1} = \dfrac{-25}{5} = -5$ |
| | | -20 | |
| 6 | 40 | | $\dfrac{-12-(-20)}{8-4} = \dfrac{8}{4} = 2$ |
| | | -12 | |
| 8 | 16 | | |

To form the next column of the table, we form the ratio of the change in the second divided differences to the change in *three* successive values of $x$. These ratios are called the **third divided differences** and are placed on the line between the corresponding second divided differences (see Table 5). Since our table has only two second divided differences, there is only one third divided difference and this process is now complete.

**TABLE 5** Third Divided Differences

| $x_k$ | $y_k$ | First Divided Difference | Second Divided Difference | Third Divided Difference |
|---|---|---|---|---|
| 1 | 65 | | | |
| | | 5 | | |
| 4 | 80 | | -5 | |
| | | -20 | | $\dfrac{2-(-5)}{8-1} = \dfrac{7}{7} = 1$ |
| 6 | 40 | | 2 | |
| | | -12 | | |
| 8 | 16 | | | |

We have presented each step in constructing the divided difference table here in a separate table to clearly illustrate this process. In applications of this technique, these steps are combined into a single table. With a little practice, you should be able to proceed quickly from the defining table for the function (Table 2) to the final form of the divided difference table (Table 6 on the next page).

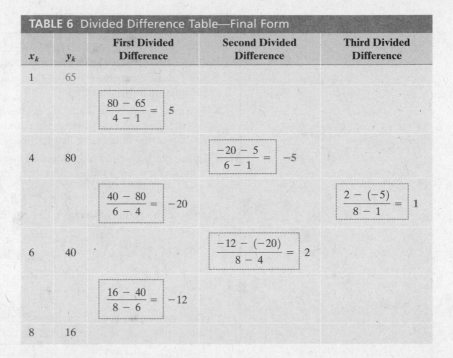

**TABLE 6**  Divided Difference Table—Final Form

| $x_k$ | $y_k$ | First Divided Difference | Second Divided Difference | Third Divided Difference |
|---|---|---|---|---|
| 1 | 65 | | | |
| | | $\dfrac{80 - 65}{4 - 1} = 5$ | | |
| 4 | 80 | | $\dfrac{-20 - 5}{6 - 1} = -5$ | |
| | | $\dfrac{40 - 80}{6 - 4} = -20$ | | $\dfrac{2 - (-5)}{8 - 1} = 1$ |
| 6 | 40 | | $\dfrac{-12 - (-20)}{8 - 4} = 2$ | |
| | | $\dfrac{16 - 40}{8 - 6} = -12$ | | |
| 8 | 16 | | | |

Now that we have computed the divided difference table, how do we use it? If we write the first number from each column of the divided difference table, beginning with the second column:

$$65 \qquad 5 \qquad -5 \qquad 1$$

we see that these numbers are the coefficients of the interpolating polynomial for Table 2 (see Example 1). Thus, Table 6 contains all the information we need to write the interpolating polynomial:

$$p(x) = 65 + 5(x - 1) - 5(x - 1)(x - 4) + (x - 1)(x - 4)(x - 6)$$

The divided difference table provides an alternate method for finding interpolating polynomials that generally requires fewer computations and can be implemented easily on a computer. The ideas introduced in the preceding discussion are summarized in the following box.

**PROCEDURE**

**Divided Difference Tables and Interpolating Polynomials**

Given the defining table for a function $f(x)$ with $n + 1$ points,

| $x$ | $x_0$ | $x_1$ | $\cdots$ | $x_n$ |
|---|---|---|---|---|
| $f(x)$ | $y_0$ | $y_1$ | $\cdots$ | $y_n$ |

where $x_0 < x_1 < \cdots < x_n$, then the **divided difference table** is computed as follows:

*Column 1*:   $x$ values from the defining table

*Column 2*:   $y$ values from the defining table

*Column 3*:   First divided differences computed using columns 1 and 2

*Column 4*:   Second divided differences computed using columns 1 and 3

$\vdots$

*Column $n + 2$*:   $n$th divided differences computed using columns 1 and $n + 1$

The coefficients in Newton's form for the interpolating polynomial,

$$p(x) = a_0 + a_1(x - x_0) + a_2(x - x_0)(x - x_1) + \cdots$$
$$+ a_n(x - x_0)(x - x_1) \cdots (x - x_{n-1})$$

are the first numbers in each column of the divided difference table, beginning with column 2.

## INSIGHT

1. The points in the defining table must be arranged with increasing $x$ values before computing the divided difference table. If the $x$ values are out of order, then the divided difference table will not contain the coefficients of Newton's form for the interpolating polynomial.

2. Since each column in the divided difference table uses all the values in the preceding column, it is necessary to compute all the numbers in every column, even though we are interested only in the first number in each column.

3. Other methods can be used to find interpolating polynomials. Referring to Table 1, we could write $p(x)$ in standard polynomial notation

$$p(x) = b_3 x^3 + b_2 x^2 + b_1 x + b_0$$

and use the points in the table to write the following system of linear equations:

$$
\begin{aligned}
p(1) = & \quad b_3 + \quad b_2 + \quad b_1 + b_0 = 65 \\
p(4) = & \quad 64b_3 + 16b_2 + 4b_1 + b_0 = 80 \\
p(6) = & \quad 216b_3 + 36b_2 + 6b_1 + b_0 = 40 \\
p(8) = & \quad 512b_3 + 64b_2 + 8b_1 + b_0 = 16
\end{aligned}
$$

The computations required to solve this system of equations are far more complicated than those involved in finding the divided difference table.

---

### EXAMPLE 3   Using a Divided Difference Table

**TABLE 7**

| $x$ | 0 | 1 | 2 | 3 | 4 |
|---|---|---|---|---|---|
| $f(x)$ | 35 | 25 | 19 | −7 | −29 |

(A) Find the divided difference table for the points in Table 7.

(B) Use the divided difference table to find the interpolating polynomial.

**SOLUTION**   (A) The divided difference table is as follows.

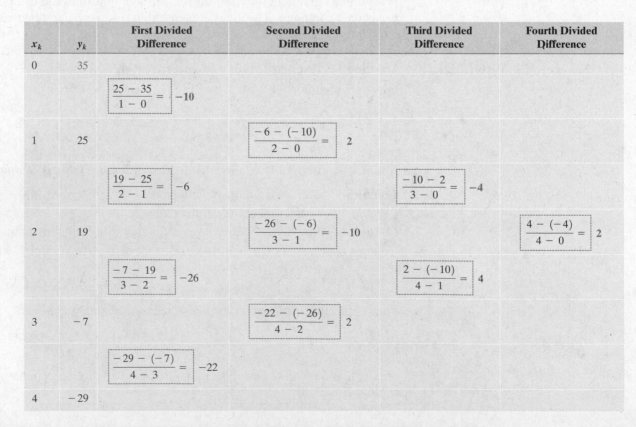

| $x_k$ | $y_k$ | First Divided Difference | Second Divided Difference | Third Divided Difference | Fourth Divided Difference |
|---|---|---|---|---|---|
| 0 | 35 | | | | |
| | | $\dfrac{25 - 35}{1 - 0} = -10$ | | | |
| 1 | 25 | | $\dfrac{-6 - (-10)}{2 - 0} = 2$ | | |
| | | $\dfrac{19 - 25}{2 - 1} = -6$ | | $\dfrac{-10 - 2}{3 - 0} = -4$ | |
| 2 | 19 | | $\dfrac{-26 - (-6)}{3 - 1} = -10$ | | $\dfrac{4 - (-4)}{4 - 0} = 2$ |
| | | $\dfrac{-7 - 19}{3 - 2} = -26$ | | $\dfrac{2 - (-10)}{4 - 1} = 4$ | |
| 3 | −7 | | $\dfrac{-22 - (-26)}{4 - 2} = 2$ | | |
| | | $\dfrac{-29 - (-7)}{4 - 3} = -22$ | | | |
| 4 | −29 | | | | |

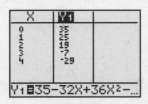

**FIGURE 4**

(B) Newton's form for the interpolating polynomial is

$$p(x) = a_0 + a_1 x + a_2 x(x-1) + a_3 x(x-1)(x-2)$$
$$+ a_4 x(x-1)(x-2)(x-3)$$

Substituting the values from the divided difference table for the coefficients in Newton's form, we have

$$p(x) = 35 - 10x + 2x(x-1) - 4x(x-1)(x-2)$$
$$+ 2x(x-1)(x-2)(x-3)$$
$$= 35 - 32x + 36x^2 - 16x^3 + 2x^4 \quad \text{Standard form}$$

Multiplication details for the standard form are omitted. Figure 4 verifies that the values of the interpolating polynomial agree with the values in Table 7. ■

MATCHED PROBLEM 3

**TABLE 8**

| $x$ | 0 | 1 | 2 | 3 | 4 |
|---|---|---|---|---|---|
| $f(x)$ | 5 | 1 | -1 | -7 | 1 |

(A) Find the divided difference table for the points in Table 8.

(B) Use the divided difference table to find the interpolating polynomial.

*Explore & Discuss* **3**

A graphing utility can be used to calculate a divided difference table. Figure 5A shows a program on a TI-83 that calculates divided difference tables and Figure 5B shows the input and output generated when we use this program to solve Example 3.

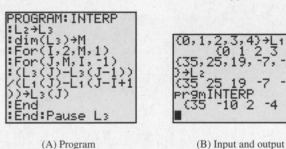

(A) Program  (B) Input and output

**FIGURE 5**

Enter this program into your TI-83 and use it to solve Matched Problem 3. If you have a different graphing utility, consult the manual for the correct format for each statement in the program.

## APPLICATION

EXAMPLE 4

**Inventory** A store orders 8,000 units of a new product. The inventory $I$ on hand $t$ weeks after the order arrived is given in the following table:

| Inventory | | | | | |
|---|---|---|---|---|---|
| $t$ | 0 | 2 | 4 | 6 | 8 |
| $I(t)$ | 8,000 | 5,952 | 3,744 | 1,568 | 0 |

Use the interpolating polynomial to approximate the inventory after 5 weeks and the average inventory during the first 5 weeks after the order arrived.

**SOLUTION**   The divided difference table is as follows:

| $t_k$ | $y_k$ | First Divided Difference | Second Divided Difference | Third Divided Difference | Fourth Divided Difference |
|---|---|---|---|---|---|
| 0 | 8,000 | | | | |
| | | $-1,024$ | | | |
| 2 | 5,952 | | $-20$ | | |
| | | $-1,104$ | | 4 | |
| 4 | 3,744 | | 4 | | 1 |
| | | $-1,088$ | | 12 | |
| 6 | 1,568 | | 76 | | |
| | | $-784$ | | | |
| 8 | 0 | | | | |

The interpolating polynomial is

$$p(t) = 8,000 - 1,024t - 20t(t - 2) + 4t(t - 2)(t - 4)$$
$$+ t(t - 2)(t - 4)(t - 6)$$

or, after simplifying,

$$p(t) = t^4 - 8t^3 - 1,000t + 8,000$$

The inventory after 5 weeks is given approximately by

$$p(5) = 5^4 - 8(5^3) - 1,000(5) + 8,000 \doteq 2,625 \text{ units}$$

The average inventory during the first 5 weeks is given approximately by

$$\frac{1}{5} \int_0^5 p(t)\, dt = \frac{1}{5} \int_0^5 (t^4 - 8t^3 - 1,000t + 8,000)\, dt$$

$$= \frac{1}{5} \left( \frac{1}{5} t^5 - 2t^4 - 500t^2 + 8,000t \right) \Big|_0^5$$

$$= \frac{1}{5} (625 - 1,250 - 12,500 + 40,000) - \frac{1}{5}(0)$$

$$= 5,375 \text{ units}$$

**MATCHED PROBLEM 4**   Refer to Example 4. Approximate the inventory after 7 weeks and the average inventory during the first 7 weeks.

*Answers to Matched Problems*   **1.** $p(2) = 88$ or $\$88,000$; $p(5) = 61$ or $\$61,000$

**2.** $p(x) = 5 - 2(x + 1) + (x + 1)x + (x + 1)x(x - 1)$

**3.** (A)

| $x_k$ | $y_k$ | First Divided Difference | Second Divided Difference | Third Divided Difference | Fourth Divided Difference |
|-------|-------|--------------------------|---------------------------|--------------------------|---------------------------|
| 0 | 5 | | | | |
| | | $-4$ | | | |
| 1 | 1 | | 1 | | |
| | | $-2$ | | $-1$ | |
| 2 | $-1$ | | $-2$ | | 1 |
| | | $-6$ | | 3 | |
| 3 | $-7$ | | 7 | | |
| | | 8 | | | |
| 4 | 1 | | | | |

(B) $p(x) = 5 - 4x + x(x - 1) - x(x - 1)(x - 2)$
$\qquad\qquad + x(x - 1)(x - 2)(x - 3)$
$\qquad = 5 - 13x + 15x^2 - 7x^3 + x^4$

**4.** 657 units; 4,294.2 units

# Exercise D-1

*A  In Problems 1–4,*
(A) *Write Newton's form for the interpolating polynomial.*
(B) *Write the associated lower triangular system for the coefficients.*
(C) *Use forward substitution to find the interpolating polynomial.*

**1.**

| $x$ | 1 | 3 | 4 |
|-----|---|---|----|
| $f(x)$ | 2 | 6 | 11 |

**2.**

| $x$ | $-1$ | 1 | 2 |
|-----|------|---|---|
| $f(x)$ | 1 | 3 | 7 |

**3.**

| $x$ | $-1$ | 0 | 2 | 4 |
|-----|------|---|---|----|
| $f(x)$ | 6 | 5 | 15 | $-39$ |

**4.**

| $x$ | $-1$ | 0 | 2 | 3 |
|-----|------|---|---|---|
| $f(x)$ | 5 | 1 | 5 | 1 |

*In Problems 5–10, find the divided difference table and then find the interpolating polynomial.*

**5.**

| $x$ | 1 | 2 | 3 |
|-----|---|---|----|
| $f(x)$ | 4 | 8 | 14 |

**6.**

| $x$ | 1 | 2 | 3 |
|-----|---|---|---|
| $f(x)$ | 1 | 3 | 7 |

**7.**

| $x$ | $-1$ | 0 | 1 | 2 |
|-----|------|---|---|---|
| $f(x)$ | $-3$ | 1 | 3 | 9 |

**8.**

| $x$ | $-1$ | 0 | 1 | 2 |
|-----|------|---|---|---|
| $f(x)$ | 5 | 6 | 3 | 2 |

**9.**

| $x$ | $-2$ | 1 | 2 | 4 |
|-----|------|----|----|----|
| $f(x)$ | 25 | 10 | 17 | 13 |

**10.**

| $x$ | $-1$ | 0 | 3 | 5 |
|-----|------|----|----|---|
| $f(x)$ | 17 | 10 | 25 | 5 |

*B*

**11.** Can a table with three points have a linear interpolating polynomial? A quadratic interpolating polynomial? A cubic interpolating polynomial? Explain.

**12.** Can a table with four points have a linear interpolating polynomial? A quadratic interpolating polynomial? A cubic interpolating polynomial? A quartic interpolating polynomial? Explain.

*In Problems 13–20, use the interpolating polynomial to approximate the value of the function defined by the table at the indicated values of x.*

**13.**

| $x$ | −4 | 0 | 4 | 8 |
|---|---|---|---|---|
| $f(x)$ | −64 | 32 | 0 | 224 |

(A) $f(2) \approx ?$  (B) $f(6) \approx ?$

**14.**

| $x$ | −5 | 0 | 5 | 10 |
|---|---|---|---|---|
| $f(x)$ | 250 | 50 | 100 | −350 |

(A) $f(-3) \approx ?$  (B) $f(8) \approx ?$

**15.**

| $x$ | −1 | 0 | 1 | 4 |
|---|---|---|---|---|
| $f(x)$ | 0 | 0 | 0 | 15 |

(A) $f(2) \approx ?$  (B) $f(3) \approx ?$

**16.**

| $x$ | −2 | 0 | 2 | 6 |
|---|---|---|---|---|
| $f(x)$ | 0 | 0 | 0 | −96 |

(A) $f(1) \approx ?$  (B) $f(4) \approx ?$

**17.**

| $x$ | −4 | −2 | 0 | 2 | 4 |
|---|---|---|---|---|---|
| $f(x)$ | 24 | 2 | 0 | −6 | 8 |

(A) $f(-3) \approx ?$  (B) $f(1) \approx ?$

**18.**

| $x$ | −6 | −2 | 0 | 2 | 6 |
|---|---|---|---|---|---|
| $f(x)$ | 19 | 3 | 10 | 3 | 19 |

(A) $f(1) \approx ?$  (B) $f(5) \approx ?$

**19.**

| $x$ | −3 | −2 | −1 | 1 | 2 | 3 |
|---|---|---|---|---|---|---|
| $f(x)$ | −24 | −6 | 0 | 0 | 6 | 24 |

(A) $f(-0.5) \approx ?$  (B) $f(2.5) \approx ?$

**20.**

| $x$ | −3 | −2 | −1 | 0 | 1 | 2 | 3 |
|---|---|---|---|---|---|---|---|
| $f(x)$ | 40 | 0 | 0 | 4 | 0 | 0 | 40 |

(A) $f(-2.5) \approx ?$  (B) $f(1.5) \approx ?$

*In Problems 21–30, find the interpolating polynomial. Graph the interpolating polynomial and the points in the given table on the same set of axes.*

**21.**

| $x$ | −2 | 0 | 2 |
|---|---|---|---|
| $f(x)$ | 2 | 0 | 2 |

**22.**

| $x$ | −2 | 0 | 2 |
|---|---|---|---|
| $f(x)$ | 2 | 0 | −2 |

**23.**

| $x$ | 0 | 1 | 2 |
|---|---|---|---|
| $f(x)$ | −4 | −2 | 0 |

**24.**

| $x$ | 0 | 1 | 2 |
|---|---|---|---|
| $f(x)$ | −4 | −3 | 0 |

**25.**

| $x$ | −1 | 0 | 2 | 3 |
|---|---|---|---|---|
| $f(x)$ | 0 | 2 | 0 | −4 |

**26.**

| $x$ | −3 | −1 | 0 | 1 |
|---|---|---|---|---|
| $f(x)$ | 0 | 4 | 3 | 0 |

**27.**

| $x$ | −2 | −1 | 0 | 1 | 2 |
|---|---|---|---|---|---|
| $f(x)$ | 1 | 5 | 3 | 1 | 5 |

**28.**

| $x$ | −2 | −1 | 0 | 1 | 2 |
|---|---|---|---|---|---|
| $f(x)$ | −8 | 0 | 2 | 4 | 12 |

**29.**

| $x$ | −2 | −1 | 0 | 1 | 2 |
|---|---|---|---|---|---|
| $f(x)$ | −3 | 0 | 5 | 0 | −3 |

**30.**

| $x$ | −1 | 0 | 1 | 2 | 3 |
|---|---|---|---|---|---|
| $f(x)$ | 6 | 2 | 0 | −6 | 2 |

*In problems 31–34, use the quartic regression routine on a graphing utility to fit a fourth degree polynomial to the tables in the indicated problems. Compare this polynomial with the interpolating polynomial.*

**31.** Problem 27      **32.** Problem 28

**33.** Problem 29      **34.** Problem 30

C

**35.** The following table was obtained from the function $f(x) = \sqrt{x}$:

| $x$ | 1 | 4 | 9 |
|---|---|---|---|
| $f(x)$ | 1 | 2 | 3 |

Find the interpolating polynomial for this table. Compare the values of the interpolating polynomial $p(x)$ and the original function $f(x) = \sqrt{x}$ by completing the table below. Use a calculator to evaluate $\sqrt{x}$ and round each value to one decimal place.

| $x$ | 1 | 2 | 3 | 4 | 5 | 6 | 7 | 8 | 9 |
|---|---|---|---|---|---|---|---|---|---|
| $p(x)$ | 1 | | | 2 | | | | | 3 |
| $\sqrt{x}$ | 1 | | | 2 | | | | | 3 |

**36.** The following table was obtained from the function $f(x) = 6/\sqrt{x}$:

| $x$ | 1 | 4 | 9 |
|---|---|---|---|
| $f(x)$ | 6 | 3 | 2 |

Find the interpolating polynomial for this table. Compare the values of the interpolating polynomial $p(x)$ and the original function $f(x) = 6/\sqrt{x}$ by completing the table below. Use a calculator to evaluate $6/\sqrt{x}$ and round each value to one decimal place.

| $x$ | 1 | 2 | 3 | 4 | 5 | 6 | 7 | 8 | 9 |
|---|---|---|---|---|---|---|---|---|---|
| $p(x)$ | 6 | | 3 | | | | | | 2 |
| $6/\sqrt{x}$ | 6 | | 3 | | | | | | 2 |

**37.** The following table was obtained from the function $f(x) = 10x/(1 + x^2)$:

| $x$ | $-2$ | $-1$ | 0 | 1 | 2 |
|---|---|---|---|---|---|
| $f(x)$ | $-4$ | $-5$ | 0 | 5 | 4 |

Find the interpolating polynomial $p(x)$ for this table. Graph $p(x)$ and $f(x)$ on the same set of axes.

**38.** The following table was obtained from the function $f(x) = (9 - x^2)/(1 + x^2)$:

| $x$ | $-2$ | $-1$ | 0 | 1 | 2 |
|---|---|---|---|---|---|
| $f(x)$ | 1 | 4 | 9 | 4 | 1 |

Find the interpolating polynomial $p(x)$ for this table. Graph $p(x)$ and $f(x)$ on the same set of axes.

**39.** Find the equation of the parabola whose graph passes through the points $(-x_1, y_1)$, $(0, y_2)$, and $(x_1, y_1)$, where $x_1 > 0$ and $y_1 \neq y_2$.

**40.** Find the equation of the parabola whose graph passes through the points $(0, 0)$, $(x_1, y_1)$, and $(2x_1, 0)$, where $x_1 > 0$ and $y_1 \neq 0$.

## Applications

**41.** *Cash reserves.* Suppose the cash reserves $C$ (in thousands of dollars) for a small business are given by the following table, where $t$ is the number of months after the first of the year.

| $t$ | 0 | 4 | 8 | 12 |
|---|---|---|---|---|
| $C(t)$ | 2 | 32 | 38 | 20 |

(A) Find the interpolating polynomial for this table.

(B) Use the interpolating polynomial to approximate (to the nearest thousand dollars) the cash reserves after 6 months.

(C) Use the interpolating polynomial to approximate (to the nearest hundred dollars) the average cash reserves for the first quarter.

**42.** *Inventory.* A hardware store orders 147 lawn mowers. The inventory $I$ of lawn mowers on hand $t$ months after the order arrived is given in the table.

| $t$ | 0 | 1 | 2 | 3 |
|---|---|---|---|---|
| $I(t)$ | 147 | 66 | 19 | 0 |

(A) Find the interpolating polynomial for this table.

(B) Use the interpolating polynomial to approximate (to the nearest integer) the average number of lawn mowers on hand for this three-month period.

**43.** *Income distribution.* The income distribution for the United States in 1999 is represented by the Lorenz curve $y = f(x)$, where $f(x)$ is given in the table.

| $x$ | 0 | 0.2 | 0.8 | 1 |
|---|---|---|---|---|
| $f(x)$ | 0 | 0.04 | 0.52 | 1 |

(A) Find the interpolating polynomial for this table.

(B) Use the interpolating polynomial to approximate (to four decimal places) the index of income concentration.

**44.** *Income distribution.* Refer to Problem 43. After making a series of adjustments for things like taxes, fringe benefits, and returns on home equity, the income distribution for the United States in 1999 is represented by the Lorenz curve $y = g(x)$, where $g(x)$ is given in the table.

| $x$ | 0 | 0.2 | 0.8 | 1 |
|---|---|---|---|---|
| $g(x)$ | 0 | 0.06 | 0.54 | 1 |

(A) Find the interpolating polynomial for this table.

(B) Use the interpolating polynomial to approximate (to four decimal places) the index of income concentration.

**45.** *Maximum revenue.* The revenue $R$ (in thousands of dollars) from the sale of $x$ thousand table lamps is given in the table.

| $x$ | 2 | 4 | 6 |
|---|---|---|---|
| $R(x)$ | 24.4 | 36 | 34.8 |

(A) Find the interpolating polynomial for this table.

(B) Use the interpolating polynomial to approximate (to the nearest thousand dollars) the revenue if 5,000 table lamps are produced.

(C) Use the interpolating polynomial to approximate (to the nearest integer) the production level that will maximize the revenue.

**46.** *Minimum average cost.* The cost $C$ (in thousands of dollars) of producing $x$ thousand microwave ovens is given in the table.

| $x$ | 1 | 3 | 5 |
|---|---|---|---|
| $C(x)$ | 215 | 535 | 1,055 |

(A) Find the interpolating polynomial for this table.

(B) Use the interpolating polynomial to approximate (to the nearest thousand dollars) the cost of producing 4,000 ovens.

(C) Use the interpolating polynomial to approximate (to the nearest integer) the production level that will minimize the average cost.

**47.** *Temperature.* The temperature $C$ (in degrees celsius) in an artificial habitat after $t$ hours is given in the table.

| $t$ | 0 | 1 | 2 | 3 | 4 |
|-----|-----|-----|-----|-----|-----|
| $C(t)$ | 14 | 13 | 16 | 17 | 10 |

(A) Find the interpolating polynomial for this table.

(B) Use the interpolating polynomial to approximate (to the nearest tenth of a degree) the average temperature over this 4-hour period.

**48.** *Drug concentration.* The concentration $C$ (in milligrams per cubic centimeter) of a particular drug in a patient's bloodstream $t$ hours after the drug is taken is given in the table.

| $t$ | 0 | 1 | 2 | 3 | 4 |
|-----|-----|-----|-----|-----|-----|
| $C(t)$ | 0 | 0.032 | 0.036 | 0.024 | 0.008 |

(A) Find the interpolating polynomial for this table.

(B) Use the interpolating polynomial to approximate (to two decimal places) the number of hours it will take for the drug concentration to reach its maximum level.

**49.** *Bacteria control.* A lake that is used for recreational swimming is treated periodically to control harmful bacteria growth. The concentration $C$ (in bacteria per cubic centimeter) $t$ days after a treatment is given in the table.

| $t$ | 0 | 2 | 4 | 6 |
|-----|-----|-----|-----|-----|
| $C(t)$ | 450 | 190 | 90 | 150 |

(A) Find the interpolating polynomial for this table.

(B) Use the interpolating polynomial to approximate (to two decimal places) the number of days it will take for the bacteria concentration to reach its minimum level.

**50.** *Medicine-respiration.* Physiologists use a machine called a pneumotachograph to produce a graph of the rate of

flow $R$ on air into the lungs (inspiration) and out (expiration). The figure gives the graph of the inspiration phase of the breathing cycle of an individual at rest.

(A) Use the values given by the graph at $t = 0, 1, 2$, and 3 to find the interpolating polynomial for $R$.

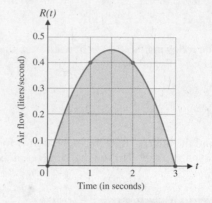

(B) Use the interpolating polynomial to approximate (to one decimal place) the total volume of air inhaled.

**51.** *Voter registration.* The number $N$ of registered voters in a precinct over a 30-year period is given in the table.

| $t$ | 0 | 10 | 20 | 30 |
|-----|-----|-----|-----|-----|
| $N(t)$ | 10,000 | 13,500 | 20,000 | 23,500 |

(A) Find the interpolating polynomial for this table.

(B) Use the interpolating polynomial to approximate (to the nearest thousand) the average number of voters over the first 20 years of this period.

**52.** *Voter registration.* The number $N$ of registered voters in a precinct over a 10-year period is given in the table.

| $t$ | 0 | 4 | 6 | 10 |
|-----|-----|-----|-----|-----|
| $N(t)$ | 15,000 | 18,800 | 22,200 | 26,000 |

(A) Find the interpolating polynomial for this table.

(B) Use the interpolating polynomial to approximate (to the nearest integer) the year $t$ in which the rate of increase in the number of voters is most rapid.

# ANSWERS

## CHAPTER 1

### Exercise 1-1

**11.** (C)    **13.** (A)    **15.**    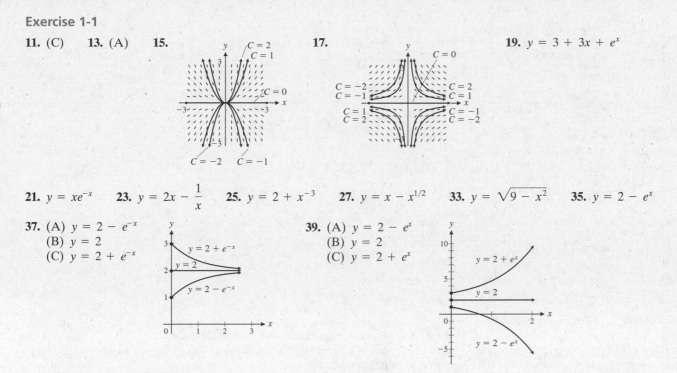    **17.**    **19.** $y = 3 + 3x + e^x$

**21.** $y = xe^{-x}$    **23.** $y = 2x - \dfrac{1}{x}$    **25.** $y = 2 + x^{-3}$    **27.** $y = x - x^{1/2}$    **33.** $y = \sqrt{9 - x^2}$    **35.** $y = 2 - e^x$

**37.** (A) $y = 2 - e^{-x}$
    (B) $y = 2$
    (C) $y = 2 + e^{-x}$

**39.** (A) $y = 2 - e^x$
    (B) $y = 2$
    (C) $y = 2 + e^x$

**41.** $dN/dt = 650$; $t =$ Time (in years), $N = N(t) =$ Number of employees at time $t$, $dN/dt = N' =$ Rate at which the number of employees is changing

**43.** $dT/dt = -k(T - 72)$; $t =$ Time, $T = T(t) =$ Temperature of pizza at time $t$, $k =$ Positive constant of proportionality, $dT/dt = T' =$ Rate at which the temperature of the pizza is changing ($dT/dt$ is negative because the temperature of the pizza is decreasing)

**45.** (A) $y = \dfrac{10}{1 + 9e^{-x}}$
    (B) $y = 10$
    (C) $y = \dfrac{10}{1 - 0.5e^{-x}}$

**47.** (A) $y = Cx^3 + 2$ for any $C$
    (B) No particular solution exists.
    (C) $y = 2 - x^3$

**49.** (A) 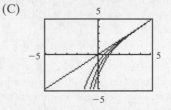    (C)

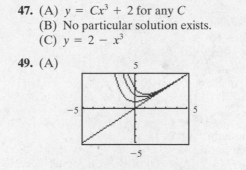

    (B) Each graph decreases to a local minimum and then increases, approaching the line $y = x$ as $x$ approaches $\infty$.

    (D) Each graph is increasing for all $x$ and approaches the line $y = x$ as $x$ approaches $\infty$.

**51.** (A) $\bar{p} = 5$
(B) $p(t) = 5 - 4e^{-0.1t}$
$p(t) = 5 + 5e^{-0.1t}$

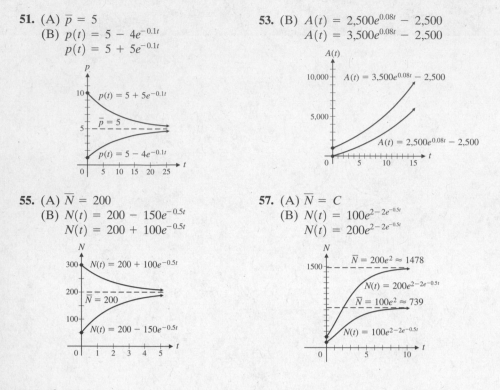

**53.** (B) $A(t) = 2{,}500e^{0.08t} - 2{,}500$
$A(t) = 3{,}500e^{0.08t} - 2{,}500$

**55.** (A) $\bar{N} = 200$
(B) $N(t) = 200 - 150e^{-0.5t}$
$N(t) = 200 + 100e^{-0.5t}$

**57.** (A) $\bar{N} = C$
(B) $N(t) = 100e^{2 - 2e^{-0.5t}}$
$N(t) = 200e^{2 - 2e^{-0.5t}}$

## Exercise 1-2

**1.** $dy/dt = 100{,}000$   **3.** $dy/dt = k(10{,}000 - y)$

**5.** The annual sales now are \$2 million, and sales are increasing at a rate proportional to the annual sales.

**7.** In a community of 5,000 people, a single person began the spread of a rumor that is spreading at a rate proportional to the product of the number of people who have heard the rumor and the number who have not.

**9.** $\dfrac{1}{y^2}\,y' = \dfrac{1}{x}$   **11.** $\dfrac{1}{y}\,y' = \dfrac{3 - x}{x}$

**13.** General solution: $y = x^3 + C$; particular solution: $y = x^3 - 1$
**15.** General solution: $y = 2 \ln x + C$; particular solution: $y = 2 \ln x + 2$
**17.** General solution: $y = Ce^x$; particular solution: $y = 10e^x$
**19.** General solution: $y = 25 - Ce^{-x}$; particular solution: $y = 25 - 20e^{-x}$
**21.** General solution: $y = Cx$; particular solution: $y = 5x$
**23.** General solution: $y = (3x + C)^{1/3}$; particular solution: $y = (3x + 24)^{1/3}$
**25.** General solution: $y = Ce^{e^x}$; particular solution: $y = 3e^{e^x}$
**27.** General solution: $y = \ln(e^x + C)$; particular solution: $y = \ln(e^x + 1)$
**29.** General solution: $y = Ce^{x^2/2} - 1$; particular solution: $y = 3e^{x^2/2} - 1$
**31.** General solution: $y = 2 - 1/(e^x + C)$; particular solution: $y = 2 - e^{-x}$
**33.** $y + \frac{1}{3}y^3 = x + \frac{1}{3}x^3 + C$   **35.** $\frac{1}{2}\ln|1 + y^2| = \ln|x| + (x^2/2) + C$ or $\ln(1 + y^2) = \ln(x^2) + x^2 + C$

**37.** $\ln(1 + e^y) = \frac{1}{2}x^2 + C$   **39.** $y = \sqrt{1 + (\ln x)^2}$   **41.** $y = \frac{1}{4}[x + \ln(x^2) + 3]^2$   **43.** $y = \sqrt{\ln(x^2 + e)}$
**45.** The differential equation cannot be written in the form $f(y)y' = g(x)$.
**47.** The integrand $e^{x^2}$ does not have an elementary antiderivative.
**49.** (A) $M = 3/(1 - e^{-k})$; $M = 4/(1 - e^{-1.5k})$   (B) 7.3   **51.** $5{,}000\,e^{0.275} \approx \$6{,}582.65$

**53.** (A) $\dfrac{7 \ln 0.5}{\ln 0.8} \approx 22$ days   (B) $100{,}000(1 - e^{0.5 \ln 0.5}) \approx 29{,}300$ people

**55.** $\dfrac{30 \ln 0.1}{\ln 0.5} \approx 100$ days

**57.** In 2010, total personal income is about \$15,452 billion and this amount is increasing at the rate of \$940.6 billion per year.

**59.** After 15 years, the total sales are about \$2.8 million and this amount is increasing at the rate of \$0.12 million per year. It will take 29 years for the sales to grow to \$4 million.

**61.** $k = 0.2$; $M = \$11$ million   **63.** $\dfrac{2 \ln(5/12)}{\ln(5/6)} \approx 9.6$ min   **65.** $25 + 300e^{4 \ln(2/3)} \approx 84.26°F$

**67.** (A) $100e^{5 \ln 1.4} \approx 538$ bacteria   (B) $\dfrac{\ln 10}{\ln 1.4} \approx 6.8$ hr

**69.** (A) $\dfrac{50{,}000}{1 + 499e^{2\ln(99/499)}} \approx 2{,}422$ people  (B) $\dfrac{10\ln(1/499)}{\ln(99/499)} \approx 38.4$ days

**71.** 10 P.M.  **73.** $I = As^k$  **75.** (A) $\dfrac{1{,}000}{1 + 199e^{7\ln(99/199)}} \approx 400$ people  (B) $\dfrac{\ln(3/3{,}383)}{\ln(99/199)} \approx 10$ days

## Exercise 1-3

**1.** Yes; $f(x) = -k,\, g(x) = 0$
**3.** No; the equation has a $y^2$ term, so it is not first-order linear.
**5.** $\int (x^3y' + 3x^2y)\,dx = x^3y$  **7.** $\int (e^{-3x}y' - 3e^{-3x}y)\,dx = e^{-3x}y$  **9.** $\int (x^4y' + 4x^3y)\,dx = x^4y$
**11.** $\int (e^{-0.5x}y' - 0.5e^{-0.5x}y)\,dx = e^{-0.5x}y$  **13.** $I(x) = e^{2x};\, y = 2 + Ce^{-2x};\, y = 2 - e^{-2x}$
**15.** $I(x) = e^x;\, y = -e^{-2x} + Ce^{-x};\, y = -e^{-2x} + 4e^{-x}$  **17.** $I(x) = e^{-x};\, y = 2xe^x + Ce^x;\, y = 2xe^x - 4e^x$
**19.** $I(x) = e^x;\, y = 3x^3e^{-x} + Ce^{-x};\, y = 3x^3e^{-x} + 2e^{-x}$  **21.** $I(x) = x;\, y = x + (C/x);\, y = x$
**23.** $I(x) = x^2;\, y = 2x^3 + (C/x^2);\, y = 2x^3 - (32/x^2)$  **25.** $I(x) = e^{x^2/2};\, y = 5 + Ce^{-x^2/2}$
**27.** $I(x) = e^{-2x};\, y = -2x - 1 + Ce^{2x}$  **29.** $I(x) = x;\, y = e^x - (e^x/x) + (C/x)$
**31.** $I(x) = x;\, y = \tfrac{1}{2}x\ln x - \tfrac{1}{4}x + (C/x)$  **33.** $I(x) = x^{3/2};\, y = 4x + Cx^{-3/2}$
**35.** (B) The solution is wrong. The term $1/x$ in front of the integral and the term $x$ inside the integral cannot be canceled.
  (C) $y = \tfrac{1}{4}x^3 + \tfrac{2}{3}x^2 + \tfrac{1}{2}x + (C/x)$
**37.** (B) The solution is wrong. A constant of integration is omitted.
  (C) $y = \tfrac{1}{2}e^{-x} + Ce^{-3x}$
**39.** $y = 1 + (C/x)$  **41.** $y = C(1 + x^2) - 1$  **43.** $y = Ce^{x^2} - 1$  **45.** $y = Ce^{kt}$  **47.** $y = (b/a) + Ce^{-ax}$
**49.** $A(t) = 62{,}500 - 42{,}500e^{0.048t}$; 8.035 years, \$24,105  **51.** \$15,124.60
**53.** $A(t) = 35{,}000e^{0.07t} - 30{,}000$; \$19,667.36  **55.** 5.23%  **57.** $p(t) = 20 + 10e^{-3t};\, \bar{p} = 20$
**59.** 622 lb; $\dfrac{622}{250} \approx 2.5$ lb/gal
**61.** $600 - 200e^{-0.5} \approx 479$ lb; $3 - e^{-0.5} \approx 2.4$ lb/gal  **63.** 6.2 hr
**65.** $120 + 40e^{-0.15} \approx 154$ lb; $-(\ln\tfrac{3}{4})/0.005 \approx 58$ days; weight will approach 120 lb if the diet is maintained for a long period of time
**67.** $17.5\left(\dfrac{125 - 130e^{-0.15}}{1 - e^{-0.15}}\right) \approx 1{,}647$ cal
**69.** Student $A$: $0.9 - 0.8e^{-4.8} \approx 0.8934$ or 89.34%; student $B$: $0.7 - 0.3e^{-4.8} \approx 0.6975$ or 69.75%

## Chapter 1 Review Exercise

**3.** (B) *(1-1)*  **4.** (A) *(1-1)*

**5.**

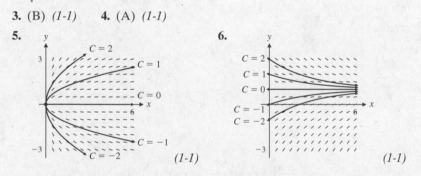

*(1-1)*  **6.** *(1-1)*

**7.** $dy/dt = -k(y - 5),\, k > 0$ *(1-2)*  **8.** $dy/dt = ky,\, k > 0$ *(1-2)*
**9.** A single person began the spread of a disease that is spreading at a rate proportional to the product of the number of people who have contracted the disease and the number who have not. *(1-2)*
**10.** The amount of radioactive material now is 100 g, and the amount is decreasing at a rate proportional to the amount present. *(1-2)*
**11.** First-order linear form: $y' + (3x^2 - e^x)y = (1 - x)e^x - 9$ *(1-2, 1-3)*
**12.** Neither form *(1-2, 1-3)*
**13.** Separation of variables form: $(y + 2)^{-1}y' = x - 3$; first-order linear form: $y' + (3 - x)y = 2x - 6$ *(1-2, 1-3)*
**14.** Separation of variables form: $y^2e^yy' = 4xe^x$ *(1-2, 1-3)*
**15.** $y = C/x^4$ *(1-2)*  **16.** $y = \tfrac{1}{6}x^2 + (C/x^4)$ *(1-3)*  **17.** $y = -1/(x^3 + C)$ *(1-2)*  **18.** $y = e^x + Ce^{2x}$ *(1-3)*
**19.** $y = \tfrac{1}{2}x^7 + Cx^5$ *(1-3)*  **20.** $y = C(2 + x) - 3$ *(1-2)*  **21.** $y = 10 - 10e^{-x}$ *(1-2, 1-3)*
**22.** $y = x - 1 + e^{-x}$ *(1-3)*  **23.** $y = e^2e^{-2e^{-x}}$ *(1-2)*  **24.** $y = (x^2 + 4)/(x + 4)$ *(1-3)*
**25.** $y = \sqrt{x^2 + 16} - 4$ *(1-2)*  **26.** $y = \tfrac{1}{3}x\ln x - \tfrac{1}{9}x + \tfrac{19}{9}(1/x^2)$ *(1-3)*  **27.** $y = \sqrt{1 + 2x^2}$ *(1-2)*
**28.** $y = (2x + 1)e^{-x^2}$ *(1-3)*  **29.** $y = -2 + Cx^4$ *(1-2, 1-3)*

**32.** $y = 2 + Cx^5$

(A) $y = 2 + Cx^5$ for any constant C
(B) No particular solution exists.
(C) $y = 2 - x^5$ *(1-3)*

**33.** $y = \sqrt{x^2 + 16}$ *(1-2)*

**34.** (A)

(B) The graphs are increasing and cross the $x$ axis only at $x = 0$.

(C)

(D) Each graph has a local maximum, a local minimum, and crosses the $x$ axis three times. *(1-1)*

**35.** (A) $M = 7/(1 - e^{-5k}), M = 4/(1 - e^{-2k})$  (B) 9.2 *(1-2)*

**36.** $500e^{-0.25 \ln 20} \approx \$236.44$ *(1-2)*   **37.** (A) 5 yr  (B) \$74,000 *(1-2)*

**38.** (A) $p = 50 + Ae^{-t}$   (B) $\bar{p} = 50$

(C)

*(1-1, 1-2)*

**39.** $A(t) = 100{,}000 - 40{,}000e^{0.05t}$; $(\ln 2.5)/0.05 \approx 18.326$ yr; \$91,630 *(1-3)*   **40.** 4.82% *(1-3)*

**41.** $y = 100 + te^{-t} - 100e^{-t}$ *(1-3)*   **42.** (A) 211.1 lb  (B) 10.3 hr *(1-3)*

**43.** (A) 337 birds  (B) 357 birds *(1-2)*

**44.** (A) $200 - 199e^{(5/2)\ln(190/199)} \approx 23$ people   (B) $\dfrac{2\ln(100/199)}{\ln(190/199)} \approx 30$ days *(1-2)*

# CHAPTER 2

**Exercise 2-1**

**1.** $-6/x^4$   **3.** $-e^{-x}$   **5.** $-486/(1 + 3x)^4$   **7.** $\dfrac{-15}{16}(1 + x)^{-7/2}$

**9.** $1 - x + \frac{1}{2}x^2 - \frac{1}{6}x^3 + \frac{1}{24}x^4$   **11.** $1 + 3x + 3x^2 + x^3$   **13.** $2x - 2x^2 + \frac{8}{3}x^3$   **15.** $1 + \frac{1}{3}x - \frac{1}{9}x^2 + \frac{5}{81}x^3$

**17.** (A) $p_3(x) = -1; -0.562 < x < 0.562$   (B) $p_4(x) = x^4 - 1$; for all $x$

**19.** $1 + (x - 1) + \frac{1}{2}(x - 1)^2 + \frac{1}{6}(x - 1)^3 + \frac{1}{24}(x - 1)^4$

**21.** $1 + 3(x - 1) + 3(x - 1)^2 + (x - 1)^3$   **23.** $3(x - \frac{1}{3}) - \frac{9}{2}(x - \frac{1}{3})^2 + 9(x - \frac{1}{3})^3$

**25.** $1 - 2x + 2x^2 - \frac{4}{3}x^3$; 0.604 166 67   **27.** $4 + \frac{1}{8}x - \frac{1}{512}x^2$; 4.123 046 9

**29.** $1 + \frac{1}{2}(x - 1) - \frac{1}{8}(x - 1)^2 + \frac{1}{16}(x - 1)^3 - \frac{5}{128}(x - 1)^4$; 1.095 437 5   **31.** $n!\,(4 - x)^{-(n+1)}$

**33.** $3^n e^{3x}$   **35.** $-(n - 1)!(6 - x)^{-n}$

**37.** $\dfrac{1}{4} + \dfrac{1}{4^2}x + \dfrac{1}{4^3}x^2 + \dfrac{1}{4^4}x^3 + \cdots + \dfrac{1}{4^{n+1}}x^n$   **39.** $1 + 3x + \dfrac{3^2}{2!}x^2 + \dfrac{3^3}{3!}x^3 + \cdots + \dfrac{3^n}{n!}x^n$

**41.** $-\dfrac{1}{6}x - \dfrac{1}{6^2 2}x^2 - \dfrac{1}{6^3 3}x^3 - \cdots - \dfrac{1}{6^n n}x^n$   **43.** $-1 - (x + 1) - (x + 1)^2 - (x + 1)^3 - \cdots - (x + 1)^n$

**45.** $(x - 1) - \dfrac{1}{2}(x - 1)^2 + \dfrac{1}{3}(x - 1)^3 - \dfrac{1}{4}(x - 1)^4 + \cdots + \dfrac{(-1)^{n+1}}{n}(x - 1)^n$

**47.** $\dfrac{1}{e^2} - \dfrac{1}{e^2}(x - 2) + \dfrac{1}{2!e^2}(x - 2)^2 - \dfrac{1}{3!e^2}(x - 2)^3 + \cdots + \dfrac{(-1)^n}{n!e^2}(x - 2)^n$

**49.** $3 \ln 3 + (1 + \ln 3)(x - 3) + \dfrac{1}{3 \cdot 2}(x - 3)^2 - \dfrac{1}{3^2 \, 2 \cdot 3}(x - 3)^3$

$\qquad + \dfrac{1}{3^3 \, 3 \cdot 4}(x - 3)^4 - \dfrac{1}{3^4 \, 4 \cdot 5}(x - 3)^5 + \cdots + \dfrac{(-1)^n}{3^{n-1}(n - 1) \cdot n}(x - 3)^n$

**51.** (A) $p_3(x) = 5x^2 + 1$  (B) 2  **53.** $n = 0, 3, 6$

**55.** $p_1(x) = x, \, p_2(x) = x - \frac{1}{2}x^2, \, p_3(x) = x - \frac{1}{2}x^2 + \frac{1}{3}x^3$

| $x$ | $p_1(x)$ | $p_2(x)$ | $p_3(x)$ | $f(x)$ |
|---|---|---|---|---|
| −0.2 | −0.2 | −0.22 | −0.222 667 | −0.223 144 |
| −0.1 | −0.1 | −0.105 | −0.105 333 | −0.105 361 |
| 0 | 0 | 0 | 0 | 0 |
| 0.1 | 0.1 | 0.095 | 0.095 333 | 0.095 31 |
| 0.2 | 0.2 | 0.18 | 0.182 667 | 0.182 322 |

| $x$ | $\lvert p_1(x) - f(x)\rvert$ | $\lvert p_2(x) - f(x)\rvert$ | $\lvert p_3(x) - f(x)\rvert$ |
|---|---|---|---|
| −0.2 | 0.023 144 | 0.003 144 | 0.000 477 |
| −0.1 | 0.005 361 | 0.000 361 | 0.000 028 |
| 0 | 0 | 0 | 0 |
| 0.1 | 0.004 69 | 0.000 31 | 0.000 023 |
| 0.2 | 0.017 678 | 0.002 322 | 0.000 345 |

**57.** 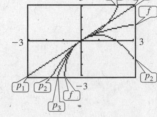  **59.** $-1.323 < x < 1.168$

**61.** $e^a \left[ 1 + (x - a) + \dfrac{1}{2!}(x - a)^2 + \dfrac{1}{3!}(x - a)^3 + \cdots + \dfrac{1}{n!}(x - a)^n \right] = e^a \displaystyle\sum_{k=0}^{n} \dfrac{1}{k!}(x - a)^k$

**63.** $\dfrac{1}{a} - \dfrac{1}{a^2}(x - a) + \dfrac{1}{a^3}(x - a)^2 - \dfrac{1}{a^4}(x - a)^3 + \cdots + \dfrac{(-1)^n}{a^{n+1}}(x - a)^n = \displaystyle\sum_{k=0}^{n} \dfrac{(-1)^k}{a^{k+1}}(x - a)^k$

**65.** $n \geq$ degree of $f$  **67.** Yes  **69.** $p_2(x) = 10 - 0.0005x^2$; \$9.85

**71.** $p_2(x) = 8 - \frac{3}{40}(x - 60) - \frac{1}{1024}(x - 60)^2$; \$7.12  **73.** $p_2(t) = 10 - 0.8t^2$; approx. \$18 million

**75.** $p_2(t) = -3 + \frac{3}{25}t^2$; 6.32 cm$^2$  **77.** $p_2(x) = 120 + 30(x - 5) + \frac{3}{4}(x - 5)^2$; 126.25 ppm

**79.** $p_2(t) = 6 - 0.06t^2$; 27.5 wpm

**81.**  **83.**  **85.**  **87.**
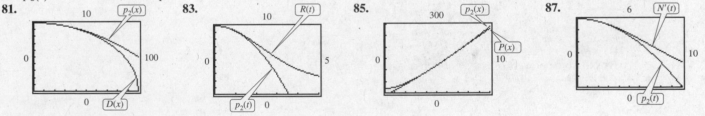

## Exercise 2-2

**1.** $\lvert x \rvert < 1$ or $-1 < x < 1$  **3.** $\lvert x \rvert < \frac{1}{7}$ or $-\frac{1}{7} < x < \frac{1}{7}$  **5.** $\lvert x \rvert < 6$ or $-6 < x < 6$  **7.** $\lvert x \rvert < 1$ or $-1 < x < 1$

**9.** $\lvert x \rvert < 1$ or $-1 < x < 1$  **11.** $-\infty < x < \infty$  **13.** (A) 6  (B) Yes  **15.** $\lvert x + 3 \rvert < 1$ or $-4 < x < -2$

**17.** $-\infty < x < \infty$  **19.** $\lvert x - 2 \rvert < 4$ or $-2 < x < 6$  **21.** $\lvert x - 8 \rvert < 2$ or $6 < x < 10$

**23.** (A)  (B) $-\infty < x < \infty$

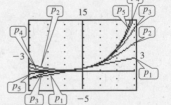

**25.** $p_n(x) = 1 + 4x + \dfrac{4^2}{2!}x^2 + \cdots + \dfrac{4^n}{n!}x^n$; $1 + 4x + \dfrac{4^2}{2!}x^2 + \cdots + \dfrac{4^n}{n!}x^n + \cdots$; $-\infty < x < \infty$

**27.** $p_n(x) = 2x - \dfrac{2^2}{2}x^2 + \dfrac{2^3}{3}x^3 - \cdots + \dfrac{(-1)^{n-1}2^n}{n}x^n$; $2x - \dfrac{2^2}{2}x^2 + \dfrac{2^3}{3}x^3 - \cdots + \dfrac{(-1)^{n-1}2^n}{n}x^n + \cdots$; $\lvert x \rvert < \frac{1}{2}$ or $-\frac{1}{2} < x < \frac{1}{2}$

**29.** $p_n(x) = 1 + \dfrac{1}{2}x + \dfrac{1}{2^2}x^2 + \cdots + \dfrac{1}{2^n}x^n$; $1 + \dfrac{1}{2}x + \dfrac{1}{2^2}x^2 + \cdots + \dfrac{1}{2^n}x^n + \cdots$; $\lvert x \rvert < 2$ or $-2 < x < 2$

**31.** $p_n(x) = -(x-1) - \frac{1}{2}(x-1)^2 - \frac{1}{3}(x-1)^3 - \cdots - \frac{1}{n}(x-1)^n;$

$\qquad -(x-1) - \frac{1}{2}(x-1)^2 - \frac{1}{3}(x-1)^3 - \cdots - \frac{1}{n}(x-1)^n - \cdots; |x-1| < 1 \text{ or } 0 < x < 2$

**33.** $p_n(x) = -1 + (x-2) - (x-2)^2 + \cdots + (-1)^{n-1}(x-2)^n;$

$\qquad -1 + (x-2) - (x-2)^2 + \cdots + (-1)^{n-1}(x-2)^n + \cdots; |x-2| < 1 \text{ or } 1 < x < 3$

**35.** (A) $|x| < 4$ or $-4 < x < 4$

(B) The coefficients of the odd powers of $x$ are 0.

(C) $|x| < 2$ or $-2 < x < 2$

**37.** (A) $\displaystyle\lim_{n\to\infty} \left| \frac{a_{n+1}}{a_n} \right|$ does not exist  (B) $|x| < 1$ or $-1 < x < 1$

## Exercise 2-3

**1.** $2 + 2x^2 + 2x^4 + \cdots + 2x^{2n} + \cdots;\ -1 < x < 1$

**3.** $2 + \frac{3}{2}x^2 + \frac{5}{6}x^3 + \cdots + \left[ 1 + \frac{(-1)^n}{n!} \right] x^n + \cdots;\ -1 < x < 1$

**5.** $x^3 - x^4 + x^5 - \cdots + (-1)^n x^{n+3} + \cdots;\ -1 < x < 1$

**7.** $-x^3 - \frac{1}{2}x^4 - \frac{1}{3}x^5 - \cdots - \frac{1}{n}x^{n+2} - \cdots;\ -1 < x < 1$

**9.** $1 + x^2 + \frac{1}{2}x^4 + \cdots + \frac{1}{n!}x^{2n} + \cdots;\ -\infty < x < \infty$

**11.** $3x - \frac{9}{2}x^2 + 9x^3 - \cdots + (-1)^{n-1}\frac{3^n}{n}x^n + \cdots;\ -\frac{1}{3} < x < \frac{1}{3}$

**13.** $\frac{1}{2} + \frac{1}{2^2}x + \frac{1}{2^3}x^2 + \cdots + \frac{1}{2^{n+1}}x^n + \cdots,\ -2 < x < 2$

**15.** $1 + 8x^3 + 8^2x^6 + \cdots + 8^n x^{3n} + \cdots,\ -\frac{1}{2} < x < \frac{1}{2}$

**17.** $10^x = 1 + (\ln 10)x + \frac{(\ln 10)^2}{2!}x^2 + \frac{(\ln 10)^3}{3!}x^3 + \cdots + \frac{(\ln 10)^n}{n!}x^n + \cdots;\ -\infty < x < \infty$

**19.** $-\frac{1}{\ln 2}x - \frac{1}{2\ln 2}x^2 - \frac{1}{3\ln 2}x^3 - \cdots - \frac{1}{n\ln 2}x^n - \cdots;\ -1 < x < 1$

**21.** $\frac{1}{4} - \frac{1}{4^2}x^2 + \frac{1}{4^3}x^4 - \cdots + \frac{(-1)^n}{4^{n+1}}x^{2n} + \cdots,\ -2 < x < 2$

**23.** (A) $x \ge 0$  (B) No; some powers of $x$ are not non-negative integers.

**25.** (A) $1 + x^2 + x^4 + x^6 + \cdots + x^{2n} + \cdots,\ -1 < x < 1$

(B) $2x + 4x^3 + 6x^5 + \cdots + 2nx^{2n-1} + \cdots,\ -1 < x < 1$

**27.** $x - \frac{1}{3}x^3 + \frac{1}{5}x^5 - \cdots + \frac{(-1)^n}{2n+1}x^{2n+1} + \cdots,\ -1 < x < 1$

**29.** $5 - \frac{1}{4}x^4 - \frac{1}{10}x^5 - \cdots - \frac{1}{n(n-3)}x^n - \cdots;\ -1 < x < 1$

**31.** (A)

(B) Differentiating the Taylor series for sinh $x$ twice yields the series for sinh $x$.

**33.** $1 + (x-3) + (x-3)^2 + \cdots + (x-3)^n + \cdots,\ 2 < x < 4$

**35.** $(x-1) - \frac{1}{2}(x-1)^2 + \frac{1}{3}(x-1)^3 - \cdots + \frac{(-1)^{n-1}}{n}(x-1)^n + \cdots,\ 0 < x < 2$

**37.** $1 + 3(x-1) + 3^2(x-1)^2 + \cdots + 3^n(x-1)^n + \cdots,\ \frac{2}{3} < x < \frac{4}{3}$

**39.** $1 + 3x + 6x^2 + \cdots + \frac{n(n-1)}{2}x^{n-2} + \cdots,\ -1 < x < 1$

**41.** (A) $1 + x^2 + x^4 + \cdots + x^{2n} + \cdots,\ -1 < x < 1$

(B) $f'$ and $\frac{1}{1-x^2}$ have identical Taylor series at 0.

**43.** $1 + 2x + 2x^2 + \cdots + 2x^n + \cdots, -1 < x < 1$

**45.** $x + \frac{1}{3}x^3 + \frac{1}{5}x^5 + \cdots + \frac{1}{2n+1}x^{2n+1} + \cdots, -1 < x < 1$

**47.** $1 + \frac{1}{2!}x^2 + \frac{1}{4!}x^4 + \cdots + \frac{1}{(2n)!}x^{2n} + \cdots, -\infty < x < \infty$

**49.** $-2 + 3x + \frac{1}{6}x^3 - \frac{1}{24}x^4 + \cdots + \frac{(-1)^{n+1}}{n(n-1)(n-2)}x^n + \cdots; -1 < x < 1$

**51.** $\ln a + \frac{1}{a}(x-a) - \frac{1}{2a^2}(x-a)^2 + \cdots + \frac{(-1)^{n-1}}{na^n}(x-a)^n + \cdots, 0 < x < 2a$

**53.** $\frac{1}{1-ab} + \frac{b}{(1-ab)^2}(x-a) + \frac{b^2}{(1-ab)^3}(x-a)^2 + \cdots + \frac{b^n}{(1-ab)^{n+1}}(x-a)^n + \cdots; |x-a| < \left|\frac{1-ab}{b}\right|$

## Exercise 2-4

**1.** $0.4; |R_1(0.6)| \leq 0.18$     **3.** $0.544; |R_3(0.6)| \leq 0.0054$     **5.** $0.255; |R_2(0.3)| \leq 0.009$

**7.** $0.261\,975; |R_4(0.3)| \leq 0.000\,486$     **9.** $0.818\,667; |R_3(0.2)| \leq 0.000\,067$

**11.** $0.970\,446; |R_3(0.03)| \leq 0.000\,000\,033$     **13.** $0.492; |R_3(0.6)| \leq 0.0324$

**15.** $0.058\,272; |R_3(0.06)| \leq 0.000\,003\,24$     **17.** $0.904\,833; n = 3$     **19.** $0.990\,050; n = 2$     **21.** $0.182\,320; n = 6$

**23.** $0.019\,800; n = 2$     **25.** $0.1973$     **27.** $0.0656$     **29.** $0.0193$     **31.** $n = 9$     **33.** $|x| < 1.817$

**35.** $-0.427 < x < 0.405$     **37.** $0.745; |R_2(0.3)| \leq 0.0135$     **39.** $1.051\,25; |R_2(0.05)| \leq 0.000\,063$

**41.** $4.123\,047; |R_2(1)| \leq 0.000\,061$

**43.** The second computation is correct. The use of the Taylor series for $f$ at 0 is invalid because the lower limit of integration lies outside the interval of convergence of the series.

**45.** (A) $p_2(x) = 1 + x + \frac{1}{2!}x^2$

    (B) $\text{Max}\,|f(x) - p_2(x)| = 0.000\,171$

        $\text{Max}\,|f(x) - q_2(x)| = 0.000\,065$

    (C) $q_2(x)$

```
QuadReg
y=ax²+bx+c
a=.5004168056
b=1.0016675
c=1
R²=1
```

**47.** $\int_0^1 \frac{10x^6}{4+x^2}\,dx \approx 0.302$; index of income concentration $\approx 0.698$

**49.** $S(4) \approx 2.031$ within $\pm 0.004$, or \$2,031,000 within $\pm$\$4,000     **51.** 2 yr; \$1,270     **53.** $30.539°C$

**55.** $A(2) \approx 6.32\,\text{cm}^2$ within $\pm 0.031$     **57.** $N(5) \approx 27.5$ wpm within $\pm 0.1875$

## Chapter 2 Review Exercise

**1.** $-6/(x+5)^4$ *(2-1)*     **2.** $1 + \frac{1}{3}x - \frac{1}{9}x^2 + \frac{5}{81}x^3; 1.003\,322$ *(2-1)*

**3.** $2 + \frac{1}{4}(x-3) - \frac{1}{64}(x-3)^2 + \frac{1}{512}(x-3)^3; 1.974\,842$ *(2-1)*

**4.** $3 + \frac{1}{6}x^2; 3.001\,667$ *(2-1)*     **5.** $-\frac{1}{4} < x < \frac{1}{4}$ *(2-2)*     **6.** $1 < x < 11$ *(2-2)*

**7.** $-1 < x < 1$ *(2-2)*     **8.** $-\infty < x < \infty$ *(2-2)*     **9.** $(-1)^n 9^n e^{-9x}$ *(2-1)*

**10.** $\frac{1}{7} + \frac{1}{7^2}x + \frac{1}{7^3}x^2 + \cdots + \frac{1}{7^{n+1}}x^n + \cdots, -7 < x < 7$ *(2-2)*

**11.** $\ln 2 + \frac{1}{2}(x-2) - \frac{1}{8}(x-2)^2 + \cdots + \frac{(-1)^{n-1}}{n2^n}(x-2)^n + \cdots, 0 < x < 4$ *(2-2)*

**12.** $\frac{1}{10} - \frac{1}{10^2}x + \frac{1}{10^3}x^2 - \cdots + \frac{(-1)^n}{10^{n+1}}x^n + \cdots, -10 < x < 10$ *(2-3)*

**13.** $\frac{1}{4}x^2 + \frac{1}{4^2}x^4 + \frac{1}{4^3}x^6 + \cdots + \frac{1}{4^{n+1}}x^{2n+2} + \cdots, -2 < x < 2$ *(2-3)*

**14.** $x^2 + 3x^3 + \frac{3^2}{2!}x^4 + \cdots + \frac{3^n}{n!}x^{n+2} + \cdots, -\infty < x < \infty$ *(2-3)*

**15.** $x + \frac{1}{e}x^2 - \frac{1}{2e^2}x^3 + \cdots + \frac{(-1)^{n-1}}{ne^n}x^{n+1} + \cdots, -e < x < e$ *(2-3)*

**16.** $\frac{1}{2} + \frac{1}{2^2}(x-2) + \frac{1}{2^3}(x-2)^2 + \cdots + \frac{1}{2^{n+1}}(x-2)^n + \cdots, 0 < x < 4$ *(2-3)*

**17.** (A) The coefficients of the odd powers of $x$ are 0.   (B) $-\sqrt{5}/5 < x < \sqrt{5}/5$ *(2-2, 2-3)*

**18.** (A) $0 \leq x < 1$   (B) No; some powers of $x$ are not non-negative integers. *(2-2, 2-3)*

**19.** $f(x) = \dfrac{1}{2} + \dfrac{1}{2^2}x + \dfrac{1}{2^3}x^2 + \dfrac{1}{2^4}x^3 + \cdots + \dfrac{1}{2^{n+1}}x^n + \cdots, -2 < x < 2$

$g(x) = \dfrac{1}{2^2} + \dfrac{1}{2^2}x + \dfrac{3}{2^4}x^2 + \cdots + \dfrac{n}{2^{n+1}}x^{n-1} + \cdots, -2 < x < 2$ *(2-3)*

**20.** $f(x) = x^2 - x^4 + x^6 - \cdots + (-1)^n x^{2n+2} + \cdots, -1 < x < 1;$

$g(x) = 2x - 4x^3 + 6x^5 - \cdots + (2n+2)(-1)^n x^{2n+1} + \cdots, -1 < x < 1$ *(2-3)*

**21.** $\dfrac{1}{3\cdot 9}x^3 - \dfrac{1}{5\cdot 9^2}x^5 + \dfrac{1}{7\cdot 9^3}x^7 - \cdots + \dfrac{(-1)^n}{(2n+3)9^{n+1}}x^{2n+3} + \cdots; -3 < x < 3$ *(2-3)*

**22.** $\dfrac{1}{5\cdot 16}x^5 + \dfrac{1}{7\cdot 16^2}x^7 + \dfrac{1}{9\cdot 16^3}x^9 + \cdots + \dfrac{1}{(2n+5)16^{n+1}}x^{2n+5} + \cdots; -4 < x < 4$ *(2-3)*

**23.** (A) $g(x) = 2 - 3(x-1) + (x-1)^3$; $h(x) = 9(x+1) - 6(x+1)^2 + (x+1)^3$
(B) The two series represent the same polynomial function for $-\infty < x < \infty$. *(2-2)*

**24.** (A) $f$ is not differentiable at 0.
(B) If $a > 0$, $p_n(x) = x$ for $n \geq 1$. If $a < 0$, $p_n(x) = -x$ for $n \geq 1$. *(2-2)*

**25.** $1.78; |R_2(0.6)| \leq 0.108$ *(2-4)*     **26.** $1.0618; |R_2(0.06)| \leq 0.000\,108$ *(2-4)*

**27.** $0.261\,975; n = 4$ *(2-4)*     **28.** $0.03; n = 1$ *(2-4)*

**29.** $5 - \dfrac{1}{3}x^3 - \dfrac{1}{8}x^4 - \cdots - \dfrac{1}{n(n-2)}x^n - \cdots; -1 < x < 1$ *(2-3)*

**30.** $3 - 4x + \dfrac{1}{6}x^3 - \dfrac{1}{12}x^4 + \cdots + \dfrac{(-1)^{n+1}}{(n-3)!(n-1)n}x^n + \cdots; -\infty < x < \infty$ *(2-3)*

**31.** $0.0612$ *(2-4)*     **32.** $0.314$ *(2-4)*     **33.** $n = 7$ *(2-4)*     **34.** $|x| < 0.6918$ *(2-4)*     **35.** $-0.616 < x < 0.616$ *(2-4)*

**36.** $-0.488 < x < 0.488$ *(2-4)*     **37.** $p_2(x) = 5 - \frac{1}{1,000}x^2$; $4.93$ *(2-4)*     **38.** $p_2(t) = 9 - 0.03t^2$; $80,000$ *(2-4)*

**39.** $\displaystyle\int_0^1 \dfrac{18x^3}{8+x^2}\,dx \approx 0.516$; index of income concentration $\approx 0.484$ *(2-4)*

**40.** $S(8) \approx 3.348$ within $\pm 0.001$, or $3,348,000$ within $\pm \$1,000$ *(2-4)*

**41.** $A(2) \approx 10.17$ cm$^2$ within $\pm 0.01$ *(2-4)*     **42.** $\dfrac{1}{5}\displaystyle\int_0^5 \dfrac{5,000t^2}{10,000+t^4}\,dt \approx 4.06$ *(2-4)*

**43.** $\dfrac{1}{5}\displaystyle\int_0^5 (10 + 2t - 5e^{-0.01t^2})\,dt \approx 10.4$ *(2-4)*

# CHAPTER 3

## Exercise 3-1

**1.** $\dfrac{1}{2}$   **3.** Diverges   **5.** 2   **7.** Diverges   **9.** $\dfrac{2}{3}$   **11.** Diverges   **13.** Diverges   **15.** 1

**17.** $\frac{14}{3}$   **19.** 10   **21.** Diverges

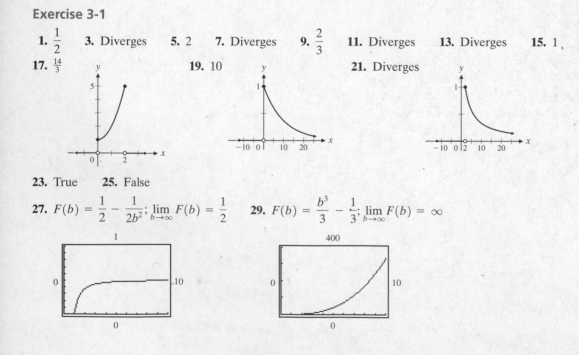

**23.** True   **25.** False

**27.** $F(b) = \dfrac{1}{2} - \dfrac{1}{2b^2}$; $\lim\limits_{b\to\infty} F(b) = \dfrac{1}{2}$   **29.** $F(b) = \dfrac{b^3}{3} - \dfrac{1}{3}$; $\lim\limits_{b\to\infty} F(b) = \infty$

**31.** $F(b) = 2 - 2e^{b/2}$; $\lim\limits_{b \to -\infty} F(b) = 2$     **33.** $F(b) = \dfrac{\sqrt{b}}{50} - \dfrac{1}{25}$; $\lim\limits_{b \to \infty} F(b) = \infty$

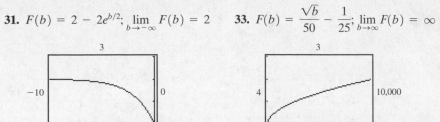

**35.** $F(b) = \dfrac{2}{3} - \dfrac{2}{\sqrt{b}}$; $\lim\limits_{b \to \infty} F(b) = \dfrac{2}{3}$     **37.** Yes     **39.** 1     **41.** Diverges     **43.** $\frac{1}{2}$     **45.** Diverges
**47.** Diverges     **49.** $120,000     **51.** $50,000
**53.** Increasing the interest rate to 6% decreases the capital value to $100,000, while decreasing the interest rate to 4% increases the capital value to $150,000. In general, increasing the interest rate decreases the amount of capital required to establish the income stream.

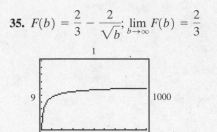

**55.** (A) 7.5 billion ft$^3$   (B) 6.14 yr     **57.** 500 gal     **59.** Approx. 8,000,000 immigrants

## Exercise 3-2

**1.** $f(x) \geq 0$ from graph
$$\int_0^4 f(x)\,dx = 1$$
**3.** Yes     **5.** No

**7.** (A) $\displaystyle\int_1^3 \tfrac{1}{8}x\,dx = \tfrac{1}{2}$     (B) $\displaystyle\int_0^2 \tfrac{1}{8}x\,dx = \tfrac{1}{4}$     (C) $\displaystyle\int_3^4 \tfrac{1}{8}x\,dx = \tfrac{7}{16}$

**9.** (A) $\displaystyle\int_1^1 f(x)\,dx = 0$   (B) $\displaystyle\int_5^\infty f(x)\,dx = 0$   (C) $\displaystyle\int_{-\infty}^5 f(x)\,dx = 1$

**11.** $F(x) = \begin{cases} 0 & \text{if } x < 0 \\ \frac{1}{16}x^2 & \text{if } 0 \leq x \leq 4 \\ 1 & \text{if } x > 4 \end{cases}$

**13.** (A) $F(4) - F(2) = \tfrac{3}{4}$   (B) $F(2) - F(0) = \tfrac{1}{4}$     **15.** (A) 2   (B) $\tfrac{4}{3}$

**17.** $f(x) \geq 0$ from graph
$$\int_0^\infty \frac{2}{(1+x)^3}\,dx = 1$$
**19.** (A) $\displaystyle\int_1^4 \frac{2}{(1+x)^3}\,dx = .21$
(B) $\displaystyle\int_3^\infty \frac{2}{(1+x)^3}\,dx = \frac{1}{16}$
(C) $\displaystyle\int_0^2 \frac{2}{(1+x)^3}\,dx = \frac{8}{9}$

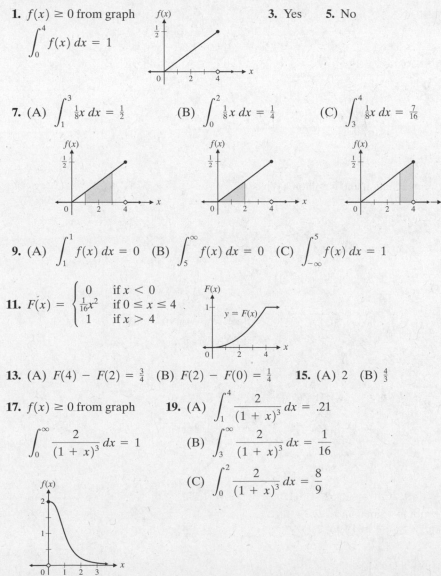

**21.** $F(x) = \begin{cases} 0 & \text{if } x < 0 \\ 1 - \dfrac{1}{(1+x)^2} & \text{if } x \geq 0 \end{cases}$

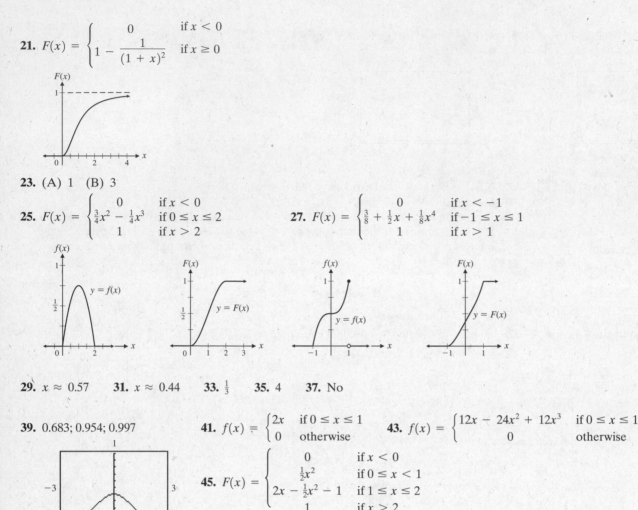

**23.** (A) 1  (B) 3

**25.** $F(x) = \begin{cases} 0 & \text{if } x < 0 \\ \frac{3}{4}x^2 - \frac{1}{4}x^3 & \text{if } 0 \leq x \leq 2 \\ 1 & \text{if } x > 2 \end{cases}$

**27.** $F(x) = \begin{cases} 0 & \text{if } x < -1 \\ \frac{3}{8} + \frac{1}{2}x + \frac{1}{8}x^4 & \text{if } -1 \leq x \leq 1 \\ 1 & \text{if } x > 1 \end{cases}$

**29.** $x \approx 0.57$   **31.** $x \approx 0.44$   **33.** $\frac{1}{3}$   **35.** 4   **37.** No

**39.** 0.683; 0.954; 0.997

**41.** $f(x) = \begin{cases} 2x & \text{if } 0 \leq x \leq 1 \\ 0 & \text{otherwise} \end{cases}$   **43.** $f(x) = \begin{cases} 12x - 24x^2 + 12x^3 & \text{if } 0 \leq x \leq 1 \\ 0 & \text{otherwise} \end{cases}$

**45.** $F(x) = \begin{cases} 0 & \text{if } x < 0 \\ \frac{1}{2}x^2 & \text{if } 0 \leq x < 1 \\ 2x - \frac{1}{2}x^2 - 1 & \text{if } 1 \leq x \leq 2 \\ 1 & \text{if } x > 2 \end{cases}$

**47.** (A) The probability that the daily demand is between 2 million and 6 million kWh is .48.

(B) $\displaystyle\int_0^8 (.2 - .02x)\, dx = .96$   (C) $\displaystyle\int_5^{10} (.2 - .02x)\, dx = .25$

**49.** (A) The probability that it takes the computer more than 5 and less than 10 sec to respond is .239.

(B) $\displaystyle\int_0^1 \frac{1}{10}e^{-x/10}\, dx = 1 - e^{-1/10} \approx .0952$   (C) $\displaystyle\int_4^\infty \frac{1}{10}e^{-x/10}\, dx = e^{-2/5} \approx .6703$

**51.** (A) $\displaystyle\int_4^{10} .003x\sqrt{100 - x^2}\, dx = \frac{(84)^{3/2}}{1,000} \approx .7699$   (B) $\displaystyle\int_0^8 .003x\sqrt{100 - x^2}\, dx = .784$   (C) $\sqrt{100 - (100)^{2/3}} \approx 8{,}858$ lb

**53.** (A) $\displaystyle\int_7^{10} \frac{1}{5,000}(10x^3 - x^4)\, dx = .47178$   (B) $\displaystyle\int_0^5 \frac{1}{5,000}(10x^3 - x^4)\, dx = \frac{3}{16} = .1875$

**55.** (A) $\displaystyle\int_0^{20} \frac{800x}{(400 + x^2)^2}\, dx = .5$   (B) $\displaystyle\int_{15}^\infty \frac{800x}{(400 + x^2)^2}\, dx = .64$   (C) 10 days

**57.** (A) $\displaystyle\int_{30}^\infty \frac{1}{20}e^{-x/20}\, dx = e^{-1.5} \approx .223$   (B) $\displaystyle\int_{80}^\infty \frac{1}{20}e^{-x/20}\, dx = e^{-4} \approx .018$

## Exercise 3-3

**1.** $\frac{1}{4}; \frac{1}{48}; .144$   **3.** $\frac{8}{3}; \frac{8}{9}; .943$   **5.** $\frac{4}{3}; \frac{1}{18}; .236$   **7.** $\frac{1}{2}$   **9.** $\sqrt{10} \approx 3.162$   **11.** $4 - 2\sqrt{2} \approx 1.172$

**13.** $X_1$ has mean 1; $X_2$ has mean 0.   **15.** $X_1$ and $X_2$ both have median 1.

**17.** $\frac{4}{3}; \frac{2}{9}; .471$   **19.** $\frac{8}{3}; \frac{8}{9}; .943$   **21.** 1.00; 0.25; 0.50   **23.** $e^{.5} \approx 1.649$

**25.** $\frac{2}{3}$   **27.** 1   **29.** $(\ln 2)/2 \approx .347$

**31.** $F(x) = \frac{1}{2}$ for any $x$ satisfying $0.5 \leq x \leq 2.5$. Median is not unique.

**33.** $F(x) = \begin{cases} 0 & \text{if } x < 0 \\ x - \frac{1}{2}x^2 & \text{if } 0 \le x < 1 \\ \frac{1}{2} & \text{if } 1 \le x \le 2; \\ \frac{1}{2}x^2 - 2x + \frac{5}{2} & \text{if } 2 < x \le 3 \\ 1 & \text{if } x > 3 \end{cases}$  $F(x) = \frac{1}{2}$ for any $x$ satisfying $1 \le x \le 2$. Median is not unique.

**35.** $a\mu + b$  **37.** $x_1 = 1; x_2 = \sqrt{2} \approx 1.414; x_3 = \sqrt{3} \approx 1.732$  **39.** $x_1 = 1; x_2 = 3; x_3 = 9$

**41.** .54  **43.** 1.47  **45.** (A) $\frac{50}{3} \approx \$16.667$, or $\$16,667$  (B) $20 - 5(\sqrt{2}/2) \approx \$16.464$, or $\$16,464$

**47.** $3 \ln 2 \approx 2.079$ min  **49.** 1 million gal  **51.** $\frac{20}{3} \approx 6.7$ min

**53.** 20 days  **55.** 1.8 hr

## Exercise 3-4

**1.** $f(x) = \begin{cases} \frac{1}{2} & \text{if } 0 \le x \le 2 \\ 0 & \text{otherwise} \end{cases}$  **3.** $f(x) = \begin{cases} 2e^{-2x} & \text{if } x \ge 0 \\ 0 & \text{otherwise} \end{cases}$  **5.** $\mu = 3; m = 3; \sigma = 2/\sqrt{3} \approx 1.155$

**7.** $\mu = 5; m = 5 \ln 2 \approx 3.466; \sigma = 5$

$F(x) = \begin{cases} 0 & \text{if } x < 0 \\ x/2 & \text{if } 0 \le x \le 2 \\ 1 & \text{if } x > 2 \end{cases}$  $F(x) = \begin{cases} 1 - e^{-2x} & \text{if } x \ge 0 \\ 0 & \text{otherwise} \end{cases}$

**9.** 0.4332  **11.** 0.2642  **13.** 0.4778

**15.** 0.4732  **17.** 0.4772  **19.** 0.1915

**21.** 0.2257  **23.** 0.4861

**25.** $X_1$ has standard deviation 1; $X_2$ has standard deviation .722.

**27.** $X_1$ has median 1; $X_2$ has median .693.  **29.** $\mu = 2; \frac{1}{2}$

**31.** $\mu = 1; 1 - e^{-1} \approx .632$  **33.** $\mu = 0; \sigma = 5/\sqrt{3} \approx 2.887; 1/\sqrt{3} \approx .577$

**35.** $\mu = 6; \sigma = 6; 1 - e^{-2} \approx .865$  **37.** .7888  **39.** .5328  **41.** .0122  **43.** .1056

**45.** (A) .6827
  (B) .6827
  (C) .6827  The probability of being within 1
    standard deviation of the mean is always .6827.

**47.** (A) .9973
  (B) .9973
  (C) .9973  The probability of being within 3 standard deviations
    of the mean is always .9973.

**53.** .2; .139; .2  **55.** $\mu = p/(p - 1)$, provided $p > 1$. The mean does not exist if $0 \le p \le 1$.

**57.** $m = \sqrt[p]{2}$  **59.** $\frac{3}{8} = .375$  **61.** $1 - e^{-2/3} \approx .487$  **63.** $1 - (1/\sqrt{2}) \approx .293$

**65.** 2.28%  **67.** 1.247%  **69.** (A) $-1/(\ln .7) \approx 2.8$ yr  (B) $e^{-1} \approx .368$  **71.** 0.82%  **73.** $e^{-2.5} \approx .082$

**75.** A's, 80.2 or greater; B's, 74.2–80.2; C's, 65.8–74.2; D's, 59.8–65.8; F's, 59.8 or lower

## Chapter 3 Review Exercise

**1.** $\frac{1}{2}$ *(3-1)*  **2.** Diverges *(3-1)*  **3.** 8 *(3-1)*  **5.** $\mu = \frac{2}{3} \approx 0.6667; V(X) = \frac{2}{9} \approx .2222;$
**4.** .75
$\sigma = \sqrt{2}/3 \approx .4714$ *(3-3)*

**6.** $F(x) = \begin{cases} 0 & \text{if } x < 0 \\ x - \frac{1}{4}x^2 & \text{if } 0 \le x \le 2 \\ 1 & \text{if } x > 2 \end{cases}$

**7.** $2 - \sqrt{2} \approx .5858$ *(3-3)*  **8.** .4641 *(3-4)*

**9.** $f(x) = \begin{cases} \frac{1}{10} & \text{if } 5 \le x \le 15 \\ 0 & \text{otherwise} \end{cases}$  **10.** $f(x) = \begin{cases} 5e^{-5x} & \text{if } x \ge 0 \\ 0 & \text{otherwise} \end{cases}$

$F(x) = \begin{cases} 0 & \text{if } x < 5 \\ \frac{1}{10}(x - 5) & \text{if } 5 \le x \le 15 \\ 1 & \text{if } x > 15 \end{cases}$ *(3-4)*  $F(x) = \begin{cases} 1 - e^{-5x} & \text{if } x \ge 0 \\ 0 & \text{otherwise} \end{cases}$ *(3-4)*

**11.** No; the area under the graph of $f$ is not 1. *(3-2)*  **12.** No; $f$ is negative for some values of $x$. *(3-2)*

**13.** $\frac{31}{32} \approx .9688$  **14.** $\mu = \frac{5}{3} \approx 1.6667; V(X) = \frac{20}{9} \approx 2.2222;$  **15.** $F(x) = \begin{cases} 1 - x^{-5/2} & \text{if } x \ge 1 \\ 0 & \text{otherwise} \end{cases}$
$\sigma = \frac{2}{3}\sqrt{5} \approx 1.4907$ *(3-3)*

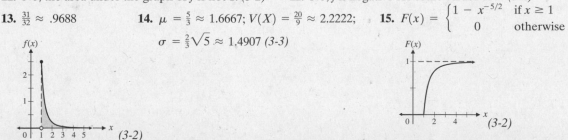

**16.** $2^{2/5} \approx 1.3195$ *(3-3)*     **17.** $f(x) = \begin{cases} \frac{1}{2}e^{-x/2} & \text{if } x \geq 0 \\ 0 & \text{otherwise} \end{cases}$ *(3-4)*     **18.** $1 - e^{-1} \approx .6321$ *(3-2)*

**19.** $F(x) = \begin{cases} 1 - e^{-x/2} & \text{if } x \geq 0 \\ 0 & \text{otherwise} \end{cases}$ *(3-4)*     **20.** $\mu = 2; \sigma = 2; m = 2\ln 2 \approx 1.3863$ *(3-4)*

**21.** (A) .9104   (B) .0668 *(3-4)*     **22.** (A) .3345   (B) .9970 *(3-4)*     **23.** 1 *(3-1)*     **24.** $\frac{1}{3}$ *(3-1)*

**25.** Yes *(3-1)*     **26.** $k = 10$ *(3-2)*     **27.** No such constant exists because $\int_0^\infty f(x)\,dx$ diverges. *(3-2)*

**28.** $P(X \leq 3) = 1 - e^{-1} \approx .6321$ *(3-4)*     **29.** $\mu = 5; m = 5\sqrt{2} - 5 \approx 2.0711$ *(3-3)*

**30.** 1.68 *(3-3)*     **31.** 1 *(3-1)*     **32.** $a\sigma^2 + a\mu^2 + b\mu + c$ *(3-3)*

**33.** In comparison with $f_1$, the graph of $f_2$ is shifted to the right, so $X_2$ has the greater mean. *(3-3)*

**34.** In comparison with $f_1$, the area under the graph of $f_2$ is more spread out, so $X_2$ has the greater variance. *(3-3)*

**35.** $X_2$ has mean 6; $X_1$ has mean 4. *(3-3)*     **36.** $X_2$ has variance 12; $X_1$ has variance 8. *(3-3)*

**37.** (A) 20,000,000 barrels   (B) 4.09 yr *(3-1)*

**38.** (A) The probability that the weekly demand is between 40 and 100 pounds is .36.

(B) $\frac{3}{4} = .75$   (C) 80 lb *(3-2)*     **39.** $40,000 *(3-1)*     **40.** (A) .896   (B) 50%   (C) 50% *(3-2, 3-3)*

**41.** (A) $e^{-1} \approx .3679$   (B) $1 - e^{-.25} \approx .2212$ *(3-4)*     **42.** .0228 *(3-4)*     **43.** (A) 60.46%   (B) 7.35% *(3-4)*

**44.** (A) The probability that the shelf life of the drug is between 2 and 8 months is $\frac{5}{18} \approx .2778$.   (B) $\frac{2}{3} \approx .6667$

(C) 10 months *(3-2, 3-3)*     **45.** (A) $e^{-4} \approx .0183$   (B) $\frac{1}{2}$ month *(3-2, 3-3)*

**46.** (A) 25 ml   (B) 6.93 hr *(3-1)*     **47.** 1.22% *(3-4)*     **48.** 45,000; 50,000 *(3-1)*

# APPENDIX D

## Exercise D-1

**1.** (A) $p(x) = a_0 + a_1(x - 1) + a_2(x - 1)(x - 3)$

(B) $a_0 = 2, a_1 = 2, a_2 = 1$

(C) $p(x) = 2 + 2(x - 1) + (x - 1)(x - 3)$

**3.** (A) $p(x) = a_0 + a_1(x + 1) + a_2(x + 1)x + a_3(x + 1)x(x - 2)$

(B) $a_0 = 6, a_1 = -1, a_2 = 2, a_3 = -2$

(C) $p(x) = 6 - (x + 1) + 2(x + 1)x - 2(x + 1)x(x - 2)$

**5.**

| $x_k$ | $y_k$ | 1st D.D. | 2nd D.D. |
|---|---|---|---|
| 1 | 4 | | |
| | | 4 | |
| 2 | 8 | | 1 |
| | | 6 | |
| 3 | 14 | | |

$p(x) = 4 + 4(x - 1) + (x - 1)(x - 2)$

**7.**

| $x_k$ | $y_k$ | 1st D.D. | 2nd D.D. | 3rd D.D. |
|---|---|---|---|---|
| −1 | −3 | | | |
| | | 4 | | |
| 0 | 1 | | −1 | |
| | | 2 | | 1 |
| 1 | 3 | | 2 | |
| | | 6 | | |
| 2 | 9 | | | |

$p(x) = -3 + 4(x + 1) - (x + 1)x + (x + 1)x(x - 1)$

**9.**

| $x_k$ | $y_k$ | 1st D.D. | 2nd D.D. | 3rd D.D. |
|---|---|---|---|---|
| −2 | 25 | | | |
| | | −5 | | |
| 1 | 10 | | 3 | |
| | | 7 | | −1 |
| 2 | 17 | | −3 | |
| | | −2 | | |
| 4 | 13 | | | |

$p(x) = 25 - 5(x + 2) + 3(x + 2)(x - 1) - (x + 2)(x - 1)(x - 2)$

**13.** $p(x) = -64 + 24(x + 4) - 4(x + 4)x + (x + 4)x(x - 4)$;   (A) 8   (B) 56

**15.** $p(x) = \frac{1}{4}(x + 1)x(x - 1)$;   (A) 1.5   (B) 6

**17.** $p(x) = 24 - 11(x + 4) + \frac{5}{2}(x + 4)(x + 2) - \frac{1}{2}(x + 4)(x + 2)x + \frac{1}{8}(x + 4)(x + 2)x(x - 2)$;

(A) $\frac{57}{8} = 7.125$   (B) $-\frac{23}{8} = -2.875$

**19.** $p(x) = -24 + 18(x + 3) - 6(x + 3)(x + 2) + (x + 3)(x + 2)(x + 1)$;   (A) $\frac{3}{8} = 0.375$   (B) $\frac{105}{8} = 13.125$

**21.** $p(x) = 2 - (x + 2) + \frac{1}{2}(x + 2)x = \frac{1}{2}x^2$    **23.** $p(x) = -4 + 2x$

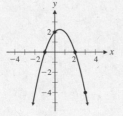

**25.** $p(x) = 2(x + 1) - (x + 1)x = 2 + x - x^2$

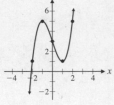

**27.** $p(x) = 1 + 4(x + 2) - 3(x + 2)(x + 1) + (x + 2)(x + 1)x = x^3 - 3x + 3$

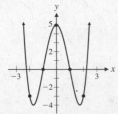

**29.** $p(x) = -3 + 3(x + 2) + (x + 2)(x + 1) - 2(x + 2)(x + 1)x + (x + 2)(x + 1)x(x - 1) = 5 - 6x^2 + x^4$

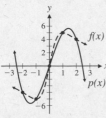

**31.** $p(x) = x^3 - 3x + 3$; the two polynomials are identical.
**33.** $p(x) = x^4 - 6x^2 + 5$; the two polynomials are identical.
**35.** $p(x) = 1 + \frac{1}{3}(x - 1) - \frac{1}{60}(x - 1)(x - 4)$

| $x$ | 1 | 2 | 3 | 4 | 5 | 6 | 7 | 8 | 9 |
|---|---|---|---|---|---|---|---|---|---|
| $p(x)$ | 1 | 1.4 | 1.7 | 2 | 2.3 | 2.5 | 2.7 | 2.9 | 3 |
| $\sqrt{x}$ | 1 | 1.4 | 1.7 | 2 | 2.2 | 2.4 | 2.6 | 2.8 | 3 |

**37.** $p(x) = -4 - (x + 2) + 3(x + 2)(x + 1) - (x + 2)(x + 1)x = 6x - x^3$    **39.** $y = y_2 + (y_1 - y_2)x^2/x_1^2$

**41.** (A)  $p(t) = 2 + 7.5t - 0.75t(t - 4) = 2 + 10.5t - 0.75t^2$
　　(B)  $C(6) \approx 38$ or $38,000$
　　(C)  approx. $15,500

**43.** (A) $p(x) = 0.2x + 0.75x(x - 0.2) + 1.25x(x - 0.2)(x - 0.8) = 0.25x - 0.5x^2 + 1.25x^3$

   (B) 0.1146

**45.** (A) $p(x) = 24.4 + 5.8(x - 2) - 1.6(x - 2)(x - 4) = 15.4x - 1.6x^2$

   (B) $R(5) \approx 37$ or \$37,000

   (C) approx. 4,813 lamps

**47.** (A) $p(t) = 14 - t + 2t(t - 1) - t(t - 1)(t - 2) = 14 - 5t + 5t^2 - t^3$

   (B) approx. 14.7°C

**49.** (A) $p(t) = 450 - 130t + 20t(t - 2) = 450 - 170t + 20t^2$

   (B) approx. 4.25 days

**51.** (A) $p(t) = 10,000 + 350t + 15t(t - 10) - t(t - 10)(t - 20)$

         $= 10,000 + 45t^2 - t^3$

   (B) approx. 14,000 registered voters

# 1 DIFFERENTIAL EQUATIONS

**1.** Substitute $y = Cx^2$, $y' = 2Cx$ into the given differential equation:
$$x(2Cx) = 2(Cx^2)$$
$$2Cx^2 = 2Cx^2$$
Thus, $y = Cx^2$ is the general solution.

**3.** Substitute $y = \dfrac{C}{x}$, $y' = -\dfrac{C}{x^2}$ into the given differential equation:
$$x\left(-\frac{C}{x^2}\right) = -\frac{C}{x}$$
$$-\frac{C}{x} = -\frac{C}{x}$$
Thus, $y = \dfrac{C}{x}$ is the general solution.

**7.** $y = ce^{x^2}$; $y' = 2xy$
$$y = ce^{x^2}$$
$$y' = c(2x)e^{x^2} = 2cxe^{x^2}$$
$$2xy = (2x)ce^{x^2} = 2cxe^{x^2}$$
Thus, $y = ce^{x^2}$ satisfies $y' = 2xy$.

**9.** Substituting $y = 4 + Ce^t$, $y' = Ce^t$ into the differential equation, we have
$$y' = y - 4$$
$$Ce^t = (4 + Ce^t) - 4$$
$$Ce^t = Ce^t$$

**11.** (C)    **13.** (A)

**15.**    **17.**

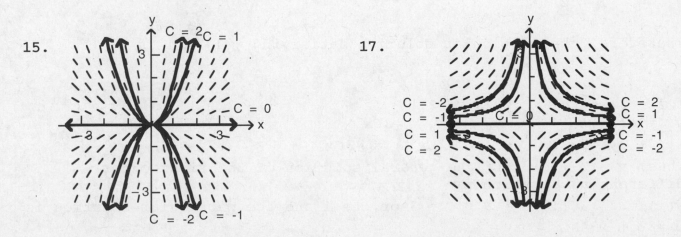

**19.** $y' = y - 3x$; $y(0) = 4$

Substituting $y = 3 + 3x + Ce^x$, $y' = 3 + Ce^x$ in the differential equation, we have

$$3 + Ce^x = 3 + 3x + Ce^x - 3x$$
$$3 + Ce^x = 3 + Ce^x$$

Thus, $y = 3 + 3x + Ce^x$ is the general solution.
Letting $x = 0$ and $y = 4$ in the general solution yields

$$4 = 3 + C$$
$$C = 1$$

Therefore, the particular solution satisfying $y(0) = 4$ is

$$y = 3 + 3x + e^x$$

**21.** $y' = -y + e^{-x}$; $y(0) = 0$

Substituting $y = xe^{-x} + Ce^{-x}$, $y' = e^{-x} - xe^{-x} - Ce^{-x}$ into the differential equation, we have

$$e^{-x} - xe^{-x} - Ce^{-x} = -(xe^{-x} + Ce^{-x}) + e^{-x}$$
$$e^{-x} - xe^{-x} - Ce^{-x} = -xe^{-x} - Ce^{-x} + e^{-x}$$

Thus, $y = xe^{-x} + Ce^{-x}$ is the general solution.
Letting $x = 0$ and $y = 0$ in the general solution yields

$$0 = 0 + C$$
$$C = 0$$

Therefore, the particular solution satisfying $y(0) = 0$ is

$$y = xe^{-x}$$

**23.** $xy' = 4x - y$; $y(1) = 1$

Substituting $y = 2x + \dfrac{C}{x}$, $y' = 2 - \dfrac{C}{x^2}$ into the differential equation, we have

$$x\left[2 - \frac{C}{x^2}\right] = 4x - \left[2x + \frac{C}{x}\right]$$
$$2x - \frac{C}{x} = 2x - \frac{C}{x}$$

Thus, $y = 2x + \dfrac{C}{x}$ is the general solution.

Letting $x = 1$ and $y = 1$ in the general solution yields

$$1 = 2 + C$$
$$C = -1$$

Therefore, the particular solution satisfying $y(1) = 1$ is

$$y = 2x - \frac{1}{x}$$

**25.** $y = 2 + Cx^{-3}$; $xy' + 3y = 6$; $y(1) = 3$

$y' = C(-3)x^{-4} = -3Cx^{-4}$, $xy' = -3Cx^{-3}$

$xy' + 3y = -3Cx^{-3} + 3(2 + Cx^{-3}) = -3Cx^{-3} + 6 + 3Cx^{-3} = 6$

Thus, $y = 2 + Cx^{-3}$ is the general solution of the first order differential equation $xy' + 3y = 6$.

$y(1) = 2 + C(1)^{-3} = 2 + C = 3$ or $C = 1$ and the particular solution is $y = 2 + x^{-3}$.

**27.** $y = x + Cx^{1/2}$; $2xy' - y = x$; $y(1) = 0$

$$y' = 1 + C\left(\frac{1}{2}\right)x^{-1/2} = 1 + \frac{C}{2}x^{-1/2}$$

$$2xy' - y = 2x\left(1 + \frac{C}{2}x^{-1/2}\right) - (x + Cx^{1/2})$$

$$= 2x + Cx^{1/2} - x - Cx^{1/2} = x$$

Thus, $y = x + Cx^{1/2}$ is the general solution of the first order differential equation $2xy' - y = x$.

$y(1) = 1 + C = 0$ or $C = -1$ and the particular solution is $y = x - x^{1/2}$.

**29.** Differentiating $y^3 + xy - x^3 = C$ implicitly, we have

$$D_x(y^3 + xy - x^3) = D_x C$$

$$D_x y^3 + D_x(xy) - D_x x^3 = 0$$

$$3y^2 y' + xy' + y - 3x^2 = 0$$

$$(3y^2 + x)y' = 3x^2 - y.$$

Thus, $y$ is a solution of the given differential equation.

**31.** Differentiating $xy + e^{y^2} - x^2 = C$ implicitly, we have

$$D_x(xy + e^{y^2} - x^2) = D_x C$$

$$D_x(xy) + D_x e^{y^2} - D_x x^2 = 0$$

$$xy' + y + e^{y^2} D_x y^2 - 2x = 0$$

$$xy' + y + 2yy' e^{y^2} - 2x = 0$$

$$(x + 2ye^{y^2})y' = 2x - y.$$

Thus, $y$ is a solution of the given differential equation.

**33.** Differentiating $y^2 + x^2 = C$ implicitly, we have

$$D_x(y^2 + x^2) = D_x C$$

$$D_x y^2 + D_x x^2 = 0$$

$$2yy' + 2x = 0$$

$$yy' = -x$$

Thus, $y$ is a solution of the given differential equation. Substituting $x = 0$ and $y = 3$ in $y^2 + x^2 = C$ yields

$$3^2 + 0^2 = C \quad \text{or} \quad C = 9.$$

Therefore, the particular solution satisfying $y(0) = 3$ is a solution of the equation

$$y^2 + x^2 = 9 \quad \text{or} \quad y^2 = 9 - x^2.$$

This equation has two continuous solutions

$$y_1(x) = \sqrt{9 - x^2} \quad \text{and} \quad y_2(x) = -\sqrt{9 - x^2}.$$

Clearly, $y_1(x) = \sqrt{9 - x^2}$ is the solution that satisfies the initial condition $y(0) = 3$.

**35.** Differentiating $\ln(2 - y) = x + C$ implicitly, we have
$$D_x\,[\ln(2 - y)] = D_x(x + C)$$
$$\frac{1}{2 - y}\,(-y') = 1$$
$$-y' = 2 - y$$
$$y' = y - 2.$$

Thus, $y$ is a solution of the given differential equation. Substituting
$x = 0$ and $y = 1$ in $\ln(2 - y) = x + C$ yields
$$\ln(2 - 1) = 0 + C \quad \text{or} \quad \ln 1 = C \quad \text{and} \quad C = 0.$$

Therefore, the particular solution satisfying $y(0) = 1$ is a solution of the equation
$$\ln(2 - y) = x.$$
We can solve this equation for $y$ by taking the exponential of both sides. We have
$$e^{\ln(2-y)} = e^x \quad \text{or} \quad 2 - y = e^x \quad \text{and} \quad y = 2 - e^x.$$

**37.** Given the general solution $y = 2 + Ce^{-x}$.

(A) Substituting $x = 0$ and $y = 1$ in the general solution yields
$$1 = 2 + Ce^0$$
$$C = -1.$$
Thus, the particular solution satisfying $y(0) = 1$ is $y_a = 2 - e^{-x}$.

(B) Substituting $x = 0$ and $y = 2$ in the general solution yields
$$2 = 2 + Ce^0$$
$$C = 0.$$
Thus, the particular solution satisfying $y(0) = 2$ is $y_b = 2$.

(C) Substituting $x = 0$ and $y = 3$ in the general solution yields
$$3 = 2 + Ce^0$$
$$C = 1.$$
Thus, the particular solution satisfying
$y(0) = 3$ is $y_c = 2 + e^{-x}$.

The graphs of the particular solutions for $x \geq 0$ are shown at the right.

**39.** Given the general solution $y = 2 + Ce^x$.

(A) Substituting $x = 0$ and $y = 1$ in the general solution yields
$$1 = 2 + Ce^0$$
$$C = -1$$
Thus, the particular solution satisfying $y(0) = 1$ is $y_a = 2 - e^x$.

(B) Substituting $x = 0$ and $y = 2$ in the general solution yields

$$2 = 2 + Ce^0$$
$$C = 0.$$

Thus, the particular solution satisfying $y(0) = 2$ is $y_b = 2$.

(C) Substituting $x = 0$ and $y = 3$ in the general solution yields

$$3 = 2 + Ce^0$$
$$C = 1.$$

Thus, the particular solution satisfying $y(0) = 3$ is $y_c = 2 + e^x$.

The graphs of these solutions for $x \geq 0$ are shown above.

**41.** $\dfrac{dN}{dt} = 650$

$t$ = Time (in years)
$N = N(t)$ = Number of employees at time $t$
$\dfrac{dN}{dt} = N'$ = Rate at which the number of employees is changing.

**43.** $\dfrac{dT}{dt} = -k(T - 72)$

$t$ = Time
$T = T(t)$ = Temperature of pizza at time $t$
$k$ = Positive constant of proportionality
$\dfrac{dT}{dt} = T'$ = Rate at which the temperature of the pizza is changing;
$\dfrac{dT}{dt}$ is negative because the temperature of the pizza is decreasing.

**45.** Given the general solution $y = \dfrac{10}{1 + Ce^{-x}}$.

(A) Substituting $x = 0$ and $y = 1$ in the general solution yields

$$1 = \frac{10}{1 + Ce^0} = \frac{10}{1 + C}.$$

Therefore, $1 + C = 10$ or $C = 9$.
Thus, the particular solution satisfying $y(0) = 1$ is $y_a = \dfrac{10}{1 + 9e^{-x}}$.

(B) Substituting $x = 0$ and $y = 10$ in the general solution yields

$$10 = \frac{10}{1 + Ce^0} = \frac{10}{1 + C}.$$

Therefore, $10 + 10C = 10$ and $C = 0$.
Thus, the particular solution satisfying $y(0) = 10$ is $y_b = 10$.

(C) Substituting $x = 0$ and $y = 20$ in the general solution yields
$$20 = \frac{10}{1 + Ce^0} = \frac{10}{1 + C}.$$
Therefore, $20 + 20C = 10$ and $C = -0.5$.
Thus, the particular solution satisfying $y(0) = 20$ is
$$y_c = \frac{10}{1 - 0.5e^{-x}}.$$

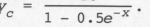

The graphs of these solutions for $x \geq 0$ are shown at the right.

47. Given the general solution $y = Cx^3 + 2$.
   (A) Substituting $x = 0$, $y = 2$ in the general solution yields
   $$2 = C \cdot 0 + 2$$
   This equation is satisfied for all values of $C$. Thus, $y = Cx^3 + 2$ satisfies the initial condition $y(0) = 2$ for any $C$.
   (B) Substituting $x = 0$, $y = 0$ in the general solution yields
   $$0 = C \cdot 0 + 2 \quad \text{or} \quad 0 = 2.$$
   This equation is <u>not</u> satisfied for any value of $C$. There is <u>no</u> particular solution of the differential equation which satisfies the initial condition $y(0) = 0$.

   (C) Substituting $x = 1$, $y = 1$ in the general solution yields
   $$1 = C \cdot 1 + 2 \quad \text{and} \quad C = -1$$
   Thus, $y = 2 - x^3$ is the particular solution of the differential equation which satisfies the initial condition $y(1) = 1$.

49. (A) The graphs of $y = x + e^{-x}$, $y = x + 2e^{-x}$, $y = x + 3e^{-x}$ and the graph of $y = x$ are shown at the right.

   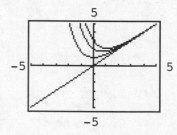

   (B) Each of the graphs $y = x + Ce^{-x}$, $C = 1, 2, 3$, lies above the line $y = x$, each decreases to a local minimum and then increases, approaching $y = x$ as $x$ approaches $\infty$.

   (C) The graphs of $y = x - e^{-x}$, $y = x - 2e^{-x}$, $y = x - 3e^{-x}$ and the graph of $y = x$ are shown at the right.

   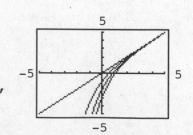

   (D) Each of the graphs $y = x - Ce^{-x}$, $C = 1, 2, 3$, is increasing and approaches $y = x$ as $x$ approaches $\infty$.

**51.** (A) Substitute $p = 5 - Ce^{-0.1t}$, $\dfrac{dp}{dt} = 0.1Ce^{-0.1t}$ into the differential equation:

$$0.1Ce^{-0.1t} = 0.5 - 0.1(5 - Ce^{-0.1t})$$
$$= 0.5 - 0.5 + 0.1Ce^{-0.1t}$$
$$= 0.1Ce^{-0.1t}$$

Thus, $p = 5 - Ce^{-0.1t}$ is the general solution and

$$\overline{p} = \lim_{t \to \infty} p(t) = \lim_{t \to \infty}(5 - Ce^{-0.1t}) = 5$$

(B) Setting $t = 0$ and $p = 1$ in the general solution yields:
$$1 = 5 - Ce^0 = 5 - C$$
and $\quad C = 4$

Thus, the particular solution satisfying the initial condition $p(0) = 1$ is:
$$p(t) = 5 - 4e^{-0.1t}$$

Setting $t = 0$ and $p = 10$ in the general solution yields:
$$10 = 5 - Ce^0 = 5 - C$$
and $\quad C = -5$

Thus, the particular solution satisfying the initial condition $p(0) = 10$ is:
$$p(t) = 5 + 5e^{-0.1t}$$

The graphs of these solutions are shown above.

(C) From part (A), the equilibrium price $\overline{p} = 5$.

If $p(0) < 5$, then the price increases and approaches 5 as a limit.
If $p(0) > 5$, then the price decreases and approaches 5 as a limit.
If $p(0) = 5$, then $C = 0$ and $p(t) = 5$ for all $t$.

**53.** (A) Substitute $A(t) = Ce^{0.08t} - 2,500$, $\dfrac{dA}{dt} = 0.08Ce^{0.08t}$ into the differential equation:

$$0.08Ce^{0.08t} = 0.08(Ce^{0.08t} - 2,500) + 200$$
$$= 0.08Ce^{0.08t} - 200 + 200$$
$$= 0.08Ce^{0.08t}$$

Thus, $A(t) = Ce^{0.08t} - 2,500$ is the general solution.

(B) Setting $t = 0$ and $A = 0$ in the general solution yields:
$$0 = Ce^0 - 2,500 = C - 2,500$$
and $\quad C = 2,500$

Thus, the particular solution satisfying $A(0) = 0$ is:

$$A_0(t) = 2{,}500e^{0.08t} - 2{,}500$$

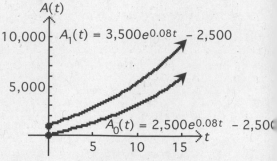

Setting $t = 0$ and $A = 1{,}000$ in the general solution yields:

$$1000 = Ce^0 - 2{,}500 = C - 2{,}500$$
and $\quad C = 3{,}500$

Thus, the particular solution satisfying $A(0) = 1{,}000$ is:

$$A_1(t) = 3{,}500e^{0.08t} - 2{,}500$$

The graphs of these solutions are shown above.

(C) Each of the particular solutions, $A_0(t)$, $A_1(t)$, in part (B) approaches infinity as $t$ approaches infinity. Also, the difference between $A_0(t)$ and $A_1(t)$:

$$A_1(t) - A_0(t) = 1{,}000e^{0.08t}$$

grows exponentially. Thus, the larger $A(0)$ is, the greater the amount in the account.

**55.** (A) Substitute $N(t) = 200 - Ce^{-0.5t}$, $\dfrac{dN}{dt} = 0.5Ce^{-0.5t}$ into the differential equation:

$$\begin{aligned}
0.5Ce^{-0.5t} &= 100 - 0.5(200 - Ce^{-0.5t}) \\
&= 100 - 100 + 0.5Ce^{-0.5t} \\
&= 0.5Ce^{-0.5t}
\end{aligned}$$

Thus, $N(t) = 200 - Ce^{-0.5t}$ is the general solution.
Now, $\overline{N} = \lim\limits_{t \to \infty}(200 - Ce^{-0.5t}) = 200 - C \lim\limits_{t \to \infty} e^{-0.5t} = 200$
The equilibrium size of the population is: $\overline{N} = 200$.

(B) Setting $t = 0$ and $N = 50$ in the general solution yields
$$50 = 200 - Ce^0 = 200 - C$$
and $\quad C = 150$

Thus, the particular solution satisfying $N(0) = 50$ is:
$$N(t) = 200 - 150e^{-0.5t}$$

Setting $t = 0$ and $N = 300$ in the general solution yields:
$$300 = 200 - Ce^0 = 200 - C$$
and $\quad C = -100$

Thus, the particular solution satisfying $N(0) = 300$ is:
$$N(t) = 200 + 100e^{-0.5t}$$

The graphs of these solutions are shown at the right.

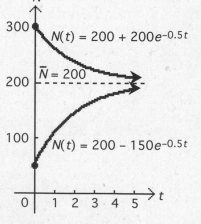

(C) If $N(0) < 200$, then number $N(t)$ of bacteria increases and approaches 200 as a limit; if $N(0) > 200$, then the number of bacteria decreases and approaches 200 as a limit. If $N(0) = 200$, then the number of bacteria remains constant for all $t$.

**57.** (A) Substitute $N(t) = Ce^{-2e^{-0.5t}}$, $\dfrac{dN}{dt} = Ce^{-2e^{-0.5t}}(-2e^{-0.5t})(-0.5) =$

$Ce^{-0.5t}e^{-2e^{-0.5t}}$ into the differential equation:

$$Ce^{-0.5t}e^{-2e^{-0.5t}} = (Ce^{-2e^{-0.5t}})e^{-0.5t}$$
$$= Ce^{-0.5t}e^{-2e^{-0.5t}}$$

Thus, $N(t) = Ce^{-2e^{-0.5t}}$ is the general solution.

Now, $\overline{N} = \lim\limits_{t \to \infty} Ce^{-2e^{-0.5t}} = C \lim\limits_{t \to \infty} e^{-2e^{-0.5t}}$

$$= C \lim\limits_{t \to \infty} e^{-2/e^{0.5t}} = Ce^0 = C$$

(B) Setting $t = 0$ and $N = 100$ in the general solution yields:
$$100 = Ce^{-2e^0} = Ce^{-2}$$
and $\quad C = 100e^2$

Thus, the particular solution satisfying $N(0) = 100$ is:
$$N(t) = 100e^{2-2e^{-0.5t}}$$

Setting $t = 0$ and $N = 200$ in the general solution yields:
$$200 = Ce^{-2e^0} = Ce^{-2}$$
and $\quad C = 200e^2$

Thus, the particular solution satisfying $N(0) = 200$ is:
$$N(t) = 200e^{2-2e^{-0.5t}}$$

The graphs of these solutions are shown at the right.

(C) As $t$ approaches infinity, the number $N(t)$ of individuals who have heard the rumor approaches
$$\overline{N} = N(0)e^2$$

## EXERCISE 1-2

**1.** $\dfrac{dy}{dt} = 100{,}000$

**3.** $\dfrac{dy}{dt} = k(10{,}000 - y)$, $k > 0$ constant

**5.** The annual sales of a company are $2 million initially, and are increasing at a rate (10%) proportional to the annual sales.

**7.** In a community of 5,000 people, one person started a rumor that is spreading at a rate proportional to the product of the number of people who have heard the rumor and the number who have not.

**9.** $xy' = y^2$

$\dfrac{1}{y^2}y' = \dfrac{1}{x}$ $\quad\left(\text{multiply by } \dfrac{1}{xy^2}\right)$

**11.** $xy' + xy = 3y$

$\qquad xy' = 3y - xy$

$\qquad xy' = y(3 - x)$

$\qquad \dfrac{1}{y}y' = \dfrac{3 - x}{x}$ $\quad\left(\text{multiply by } \dfrac{1}{xy}\right)$

**13.** $y' = 3x^2$

$\quad \int dy = \int 3x^2 \, dx$

$\qquad y = x^3 + C$ $\quad$ *General solution*

Applying the initial condition $y(0) = -1$, we have

$\qquad -1 = 0 + C$ which implies $C = -1$

Thus, $y = x^3 - 1$ $\quad$ *Particular solution*

**15.** $y' = \dfrac{2}{x}$

$\quad \int dy = \int \dfrac{2}{x} \, dx$

$\qquad y = 2 \ln|x| + C$ $\quad$ *General solution*

Applying the initial condition $y(1) = 2$, we have

$\qquad 2 = 2 \ln 1 + C = 2 \cdot 0 + C = C$

so $C = 2$

Thus,

$\qquad y = 2 \ln|x| + 2$ $\quad$ *Particular solution*

**17.** $y' = y;\ y(0) = 10.$

$\dfrac{y'}{y} = 1.$ Separate the variables

$\int \dfrac{dy}{y} = \int dx$

$\ln|y| = x + C$

$\qquad y = e^{x+C}$

$\qquad\ = e^C e^x$ (put $K = e^C$)

$\qquad\ = Ke^x.$ *General solution* Applying the initial condition $y(0) = 10$, we have

$\qquad 10 = Ke^0$ or $K = 10.$

Thus,

$\qquad y = 10e^x.$ *Particular solution*

**19.** $y' = 25 - y;\ y(0) = 5.$

$\dfrac{y'}{25 - y} = 1.$ Separate the variables

$\int \dfrac{dy}{25 - y} = \int dx$

$-\ln|25 - y| = x + C$

$\ \ln|25 - y| = -(x + C)$

$\qquad 25 - y = e^{-x-C}$

$\qquad 25 - y = e^{-C}e^{-x}$

$\qquad 25 - y = Ke^{-x}$ (put $K = e^{-C}$)

$\qquad\ \ \ y = 25 - Ke^{-x}.$

$\qquad$ *General solution*

Applying the initial condition $y(0) = 5$, we have

$\qquad 5 = 25 - Ke^0$ or $K = 20.$

Thus,

$\qquad y = 25 - 20e^{-x}.$

$\qquad$ *Particular solution*

**21.** $y' = \dfrac{y}{x}$; $y(1) = 5$, $x > 0$.

$\dfrac{y'}{y} = \dfrac{1}{x}$  Separate the variables

$\displaystyle\int \dfrac{dy}{y} = \int \dfrac{dx}{x}$

$\ln|y| = \ln x + C$

(<u>Note</u>: $x > 0$, so $|x| = x$.)

$\quad y = e^{\ln x + C} = e^C e^{\ln x}$

$\quad\quad = Kx.$

(<u>Note</u>: $e^{\ln x} = x$, put $K = e^C$.)

Thus, $\quad y = Kx$. *General solution*

Applying the initial condition
$y(1) = 5$, we have
$$5 = K(1) \quad\text{or}\quad K = 5.$$
Therefore,
$$y = 5x. \text{ *Particular solution*}$$

**23.** $y' = \dfrac{1}{y^2}$; $y(1) = 3$.

$y^2 y' = 1$

$\displaystyle\int y^2 \, dy = \int dx$

$\dfrac{y^3}{3} = x + C$

$y^3 = 3x + 3C$

$\quad = 3x + A.$ (put $A = 3C$)

Thus, $y = (3x + A)^{1/3}$. *General solution*

Applying the initial condition
$y(1) = 3$, we have
$$3 = (3 + A)^{1/3}$$
$$27 = 3 + A \quad\text{and}\quad A = 24.$$
Therefore,
$$y = (3x + 24)^{1/3}. \text{ *Particular solution*}$$

**25.** $y' = ye^x$; $y(0) = 3e$

$\dfrac{y'}{y} = e^x$

$\displaystyle\int \dfrac{dy}{y} = \int e^x dx$

$\ln|y| = e^x + C$

$\quad y = e^{e^x + C}$

$\quad\quad = e^C e^{e^x}$

$\quad\quad = Ke^{e^x}$. *General solution*

Applying the initial
condition $y(0) = 3e$, we have

$3e = Ke^{e^0}$

$\quad = Ke$ and $K = 3$.

Thus,

$\quad y = 3e^{e^x}$.  *Particular solution*

**27.** $y' = \dfrac{e^x}{e^y}$; $y(0) = \ln 2$.

$e^y y' = e^x$

$\displaystyle\int e^y dy = \int e^x dx$

$e^y = e^x + C$

$\quad y = \ln(e^x + C)$ *General solution*

Applying the initial condition
$y(0) = \ln 2$, we have

$\ln 2 = \ln(e^0 + C)$

$\quad = \ln(1 + C)$.

Therefore,

$1 + C = 2$ and $C = 1$.

Thus,

$\quad y = \ln(e^x + 1)$. *Particular solution*

**29.** $y' = xy + x; \quad y(0) = 2$
$y' = x(y + 1)$

$$\int \frac{dy}{y + 1} = \int x \, dx$$

$$\ln|y + 1| = \frac{x^2}{2} + C$$

$$y + 1 = e^{x^2/2 + C}$$

$$y + 1 = e^C e^{x^2/2}$$

$$y + 1 = K e^{x^2/2}$$

$$y = K e^{x^2/2} - 1. \ \textit{General solution}$$

Applying the initial condition
$y(0) = 2$, we have

$$2 = Ke^0 - 1 = K - 1 \text{ and } K = 3.$$

Thus,

$$y = 3e^{x^2/2} - 1$$
$$\textit{Particular solution}$$

**31.** $y' = (2 - y)^2 e^x; \quad y(0) = 1.$

$$\frac{y'}{(2 - y)^2} = e^x$$

$$\int \frac{dy}{(2 - y)^2} = \int e^x \, dx$$

$$\frac{1}{2 - y} = e^x + C$$

$$2 - y = \frac{1}{e^x + C}$$

$$y = 2 - \frac{1}{e^x + C} \cdot \ \textit{General solution}$$

Applying the initial condition
$y(0) = 1$, we have

$$1 = 2 - \frac{1}{e^0 + C}$$

$$\frac{1}{1 + C} = 1$$
$$1 + C = 1 \quad \text{and} \quad C = 0.$$

Thus,

$$y = 2 - \frac{1}{e^x} \quad \text{or} \quad y = 2 - e^{-x}$$

$$\textit{Particular solution}$$

**33.** $y' = \dfrac{1 + x^2}{1 + y^2}$

$(1 + y^2)y' = 1 + x^2 \quad$ Separate the variables
$\int (1 + y^2) \, dy = \int (1 + x^2) \, dx$

$$y + \frac{y^3}{3} = x + \frac{x^3}{3} + C. \qquad \textit{General solution}$$

**35.** $xyy' = (1 + x^2)(1 + y^2)$

$$\frac{yy'}{1 + y^2} = \frac{1 + x^2}{x} \qquad \text{Separate the variables}$$

$$\int \frac{y \, dy}{1 + y^2} = \int \left( \frac{1}{x} + x \right) dx$$

$$\frac{1}{2} \int \frac{2y \, dy}{1 + y^2} = \ln|x| + \frac{x^2}{2} + C$$

$$\frac{1}{2} \ln(1 + y^2) = \ln|x| + \frac{x^2}{2} + C, \quad \textit{General solution}$$

or $\ln(1 + y^2) = \ln(x^2) + x^2 + C.$

**37.** $x^2 e^y y' = x^3 + x^3 e^y$

$x^2 e^y y' = x^3(1 + e^y)$

$\dfrac{e^y y'}{1 + e^y} = x \qquad$ Separate the variables

$\displaystyle\int \dfrac{e^y y'}{1 + e^y} = \int x \, dx \qquad$ (Note: Put $u = 1 + e^y$, $du = e^y dy$.)

$\ln(1 + e^y) = \dfrac{x^2}{2} + C. \quad$ *General solution*

**39.** $xyy' = \ln x;\; y(1) = 1$

$yy' = \dfrac{\ln x}{x}$

$\displaystyle\int y \, dy = \frac{1}{x}\int \frac{\ln x}{x \cdot x} \, dx \qquad$ (Note: Put $u = \ln x$,

$\dfrac{y^2}{2} = \dfrac{[\ln x]^2}{2} + C$

$y^2 = [\ln x]^2 + 2C$ or $y^2 = [\ln x]^2 + A. \quad$ *General solution.*

Solving this equation for $y$, we have

$y = \pm\sqrt{[\ln x]^2 + A}$

Applying the initial condition $y(1) = 1$, we choose the function

$\quad y = \sqrt{[\ln x]^2 + A} \quad$ and get

$\quad 1 = \sqrt{[\ln x]^2 + A}$

$\quad 1 = \sqrt{0 + A} \quad$ and $\quad A = 1.$

Thus, $y = \sqrt{[\ln x]^2 + 1}. \quad$ *Particular solution*

**41.** $xy' = x\sqrt{y} + 2\sqrt{y};\; y(1) = 4$

$xy' = \sqrt{y}\,(x + 2$

$\dfrac{y'}{\sqrt{y}} = \dfrac{x + 2}{x} \quad$ Separate the variables

$\displaystyle\int \dfrac{dy}{\sqrt{y}} = \int\left(1 + \dfrac{2}{x}\right) dx$

$2y^{1/2} = x + 2\ln|x| + C$

$y^{1/2} = \dfrac{1}{2}(x + 2\ln|x| + C)$

$y = \dfrac{1}{4}(x + 2\ln|x| + C)^2$ or

$y = \dfrac{1}{4}[x + \ln(x^2) + C]^2.$

*General solution*

Applying the initial condition $y(1) = 4$, we have

$4 = \dfrac{1}{4}[1 + \ln(1^2) + C]^2$

$16 = (1 + C)^2$

$1 + C = 4$

$C = 3.$

Thus,

$y = \dfrac{1}{4}[x + \ln(x^2) + 3]^2.$

*Particular solution*

**43.** $yy' = xe^{-y^2}$; $y(0) = 1$

$ye^{y^2}y' = x$

$\int ye^{y^2}\,dy = \int x\,dx$    (<u>Note</u>: Put $u = y^2$, $dy = 2y\,dy$.)

$\dfrac{1}{2}e^{y^2} = \dfrac{x^2}{2} + C$

$e^{y^2} = x^2 + 2C$

$e^{y^2} = x^2 + A$

$y^2 = \ln(x^2 + A)$. *General*
                           *solution*

Solving this equation for $y$, we have $y = \pm\sqrt{\ln(x^2 + A)}$.

We choose the function $y = \sqrt{\ln(x^2 + A)}$ and apply the initial condition
$y(0) = 1$ to obtain

$1 = \sqrt{\ln(0 + A)}$

$\ln A = 1$

$A = e$

Thus, $y = \sqrt{\ln(x^2 + e)}$.    *Particular solution*

**45.** $y' + xy = 2$

This differential equation cannot be written in the form
$f(y)y' = g(x)$.

**47.** $3yy' = e^{x^2+y} = e^{x^2}e^y$

$3ye^{-y}y' = e^{x^2}$; $f(y) = 3ye^{-y}$, $g(x) = e^{x^2}$

$\int 3ye^{-y}\,dy = \int ex^2\,dx$

The integrand $g(x) = e^{x^2}$ does not have an elementary antiderivative.

**49.** $y = M(1 - e^{-kt})$, $M > 0$, $k > 0$ constants.

(A) At $t = 1$, $y = 3$:

$3 = M(1 - e^{-k})$   so   $M = \dfrac{3}{1 - e^{-k}}$

At $t = 1.5$, $y = 4$

$4 = M(1 - e^{1.5k})$ so $M = \dfrac{4}{1 - e^{-1.5k}}$

(B) Using a graphing utility to graph the two equations in part (A),
we find that $M \approx 7.3$ (and $k \approx 0.53$).

**51.** The model for this problem is $\dfrac{dA}{dt} = 0.055A$; $A(0) = 5000$.

$$\frac{dA}{A} = 0.055dt$$

$$\int \frac{dA}{A} = \int 0.055dt$$

$$\ln|A| = 0.055t + C$$

$$A = e^{0.055t+C}$$

$$\phantom{A} = A^C e^{0.055t}$$

$$\phantom{A} = Ke^{0.055t}. \qquad \textit{General solution}$$

Applying the initial condition $A(0) = 5000$, we have

$$5000 = Ke^0, \quad K = 5000.$$

Thus, $A(t) = 5000e^{0.055t}$ is the particular solution.

When $t = 5$ years,

$$A(5) = 5000e^{0.055(5)} = 5000e^{0.275} \approx \$6,582.65.$$

**53.** (A) Let $s(t)$ denote the number of people who have heard about the new product. Then the model for this problem is

$$\frac{ds}{dt} = k(100,000 - s); \quad s(0) = 0, \ s(7) = 20,000, \ k > 0.$$

$$\frac{ds}{(100,000 - s)} = k \, dt$$

$$\int \frac{ds}{100,000 - s} = \int k \, dt$$

$$-\ln|100,000 - s| = kt + C \qquad (0 < s < 100,000)$$

$$\ln|100,000 - s| = -kt - C$$

$$100,000 - s = e^{-kt-C}$$

$$100,000 - s = e^{-C}e^{-kt}$$

$$\phantom{100,000 - s} = Ae^{-kt}.$$

Thus,

$$s(t) = 100,000 - Ae^{-kt}. \quad (\textit{General solution})$$

We use the conditions $s(0) = 0$, $s(7) = 20,000$ to determine the constants $A$ and $k$.

$$s(0) = 0 = 100,000 - Ae^0, \quad A = 100,000.$$

Thus, $s(t) = 100,000 - 100,000e^{-kt}$.

$$s(7) = 20,000 = 100,000 - 100,000e^{-7k}$$

$$100,000e^{-7k} = 80,000$$

$$e^{-7k} = 0.8$$

$$-7k = \ln(0.8)$$

$$k = \frac{-\ln(0.8)}{7}.$$

Therefore, the particular solution is

$$s(t) = 100,000 - 100,000e^{[\ln(0.8)/7]t}.$$

Finally, we want to find time $t$ such that $s(t) = 50,000$:

$$50,000 = 100,000 - 100,000e^{[\ln(0.8)/7]t}$$

$$e^{[\ln(0.8)/7]t} = 0.5$$

$$\frac{\ln 0.8}{7}t = \ln(0.5).$$

Thus, $t = \dfrac{7\ln(0.5)}{\ln(0.8)} \approx 22$ days.

(B) From part (A), we know that

$$s(t) = 100,000(1 - e^{-kt})$$

Since we want 50,000 people to be aware of the product after 14 days, we have

$$50,000 = 100,000(1 - e^{-14k})$$

$$1 - e^{-14k} = \frac{1}{2}$$

$$e^{-14k} = \frac{1}{2}$$

$$-14k = \ln(0.5)$$

$$k = -\frac{1}{14}\ln(0.5)$$

Thus, $s(t) = 100,000[1 - e^{(t/14)\ln(0.5)}]$

Now, $s(7) = 100,000[1 - e^{(7/14)\ln(0.5)}]$
$$= 100,000[1 - e^{0.5\,\ln(0.5)}] \approx 29,300$$

Therefore, approximately 29,300 people must become aware of the product during the first 7 days in order to ensure that 50,000 will be aware of the product after 14 days.

55. Let $s(t)$ denote the percentage of the deodorizer that is present at time $t$. Then the model for this problem is

$$\frac{ds}{dt} = ks; \quad s(0) = 1, \; s(30) = 0.5, \; k < 0.$$

Separating the variables, we have

$$\frac{ds}{s} = k\,dt$$

$$\int\frac{ds}{s} = \int k\,dt$$

$$\ln|s| = kt + C$$

$$s = e^{kt+C}$$

$$s = Ae^{kt}. \quad \textit{General solution}$$

We use the conditions $s(0) = 1$, $s(30) = 0.5$, to evaluate the constants $A$ and $k$.

$$s(0) = 1 = Ae^0, \; A = 1.$$

Thus, $s(t) = e^{kt}$

$\qquad s(30) = 0.5 = e^{30k}$

$\qquad\quad 30k = \ln(0.5)$

$\qquad\qquad k = \dfrac{\ln(0.5)}{30}.$

Therefore, $s(t) = e^{[\ln(0.5)/30]t}$ *Particular solution*

Now, we want to find time $t$ such that $s(t) = 0.1$:

$\qquad\qquad 0.1 = e^{[\ln(0.5)/30]t}$

$\dfrac{\ln(0.5)}{30}t = \ln(0.1)$

$\qquad\qquad t = \dfrac{30\ln(0.1)}{\ln(0.5)} \approx 100$ days.

**57.** Let $y$ be the total personal income and take 1995 as the base year (0), then $y(0) = 6{,}201$ billion, $y(5) = 8{,}407$ billion and $\dfrac{dy}{dt} = ky$.

From the last equation we have

$\qquad \dfrac{dy}{y} = k\,dt$ or $\ln y = kt + C$ or

$\qquad y = e^{kt+C} = e^C e^{kt} = Ae^{kt}$

$\qquad y(0) = A = 6{,}201$ and hence $y(t) = 6{,}201e^{kt}$

$\qquad y(5) = 8{,}407 = 6{,}201e^{k(5)}$ or $e^{5k} = \dfrac{8{,}407}{6{,}201}$ or $k = \dfrac{1}{5}\ln\!\left(\dfrac{8{,}407}{6{,}201}\right).$

Thus, $y(t) = 6{,}201e^{\frac{1}{5}\ln\left(\frac{8,407}{6,201}\right)t}$

Now, for $t = 15$, we have

$\qquad y(15) = 6{,}201e^{\frac{1}{5}\ln\left(\frac{8,407}{6,201}\right)(15)} \approx 15{,}452$ billion

Note that $\dfrac{dy}{dt} = 6{,}201\left[\dfrac{1}{5}\ln\!\left(\dfrac{8{,}407}{6{,}201}\right)\right]e^{\frac{1}{5}t\,\ln\left(\frac{8,407}{6,201}\right)}$ and

$\dfrac{dy}{dt}\bigg|_{15} \approx 940.6$ billion per year increase.

**59.** Let $y$ be the annual sales of the company, then $y(0) = 0$, $y(4) = 1$ million and

$\qquad \dfrac{dy}{dt} = k(5 - y).$

From this equation we have

$\qquad \dfrac{dy}{5 - y} = k\,dt$ or $\ln(5 - y)^{-1} = kt + C$ or

$\qquad 5 - y = e^{-(kt+C)}$ or $y = 5 - Ae^{-kt}.$

$\qquad y(0) = 5 - A = 0$, so $A = 5$ and hence

$\qquad\quad y = 5(1 - e^{-kt}).$

$\qquad y(4) = 5(1 - e^{-4k}) = 1$ or $1 - e^{-4k} = 0.2$ or

$\qquad e^{-4k} = 0.8$ and $k = -\dfrac{1}{4}\ln(0.8) = -0.25\ln(0.8).$

Thus,
$$y(t) = 5(1 - e^{0.25t \ln(0.8)})$$

For $t = 15$, $y(15) = 5(1 - e^{0.25(15)\ln(0.8)}) = 2.8$ million.
$$\frac{dy}{dt} = -5(0.25)\ln(0.8)e^{0.25t \ln(0.8)} \quad \text{and}$$

$$\left.\frac{dy}{dt}\right|_{15} \approx 0.12 \text{ million per year increase.}$$

In order for the sales to grow to \$4 million, we have to solve the following equation for $t$:
$$4 = 5(1 - e^{0.25t \ln(0.8)})$$
From this equation we have:
$$5e^{0.25t \ln(0.8)} = 1$$
or
$$0.25t \ln(0.8) = \ln(0.2)$$
or
$$t = \frac{\ln(0.2)}{0.25 \ln(0.8)} \approx 29 \text{ years.}$$

**61.** $S(t) = M(1 - e^{-kt})$, $M > 0$, $k > 0$ constant
We have:
$$S(1) = M(1 - e^{-k}) = 2$$
and $\quad S(3) = M(1 - e^{-3k}) = 5$

From the first equation, $M = \dfrac{2}{1 - e^{-k}}$. Substituting this in the second equation yields:
$$\frac{2}{1 - e^{-k}}(1 - e^{-3k}) = 5$$
$$2 - 2e^{-3k} = 5 - 5e^{-k}$$
and $\quad 2e^{-3k} - 5e^{-k} + 3 = 0 \qquad (1)$

Now let $t = e^{-k}$ ($k = -\ln t$). Then equation (1) becomes
$$2t^3 - 5t + 3 = 0$$
which factors into
$$(t - 1)(2t^2 + 2t - 3) = 0$$
The solutions are $t = 1$, $t \approx 0.82$ and $t \approx -1.82$.

Since $\ln(-1.82)$ is undefined and $\ln(1)$ implies $k = 0$, it follows that $k = -\ln(0.82) \approx 0.2$.

Now, $M = \dfrac{2}{(1 - e^{-0.2})} \approx 11$.

Thus, $k \approx 0.2$ and $M \approx \$11$ million.

**63.** Let $T(t)$ denote the temperature of the bar at time $t$. Then the model for this problem is

$$\frac{dT}{dt} = k(T - 800), \ k < 0; \ T(0) = 80, \ T(2) = 200.$$

Separating the variables, we have

$$\frac{dT}{T - 800} = k \ dt$$

$$\int \frac{dT}{T - 800} = \int k \ dt$$

$$\ln|T - 800| = kt + C$$

$$T - 800 = e^{kt+C}$$

$$T = 800 + Ae^{kt} \quad \textit{General solution}$$

Using the conditions $T(0) = 80$ and $T(2) = 200$ to evaluate the constants $A$ and $k$, we have

$$T(0) = 80 = 800 + Ae^0, \ A = -720.$$

Thus, $T(t) = 800 - 720e^{kt}$.

$$T(2) = 200 = 800 - 720e^{2k}$$

$$720e^{2k} = 600$$

$$e^{2k} = \frac{5}{6}$$

$$2k = \ln\left(\frac{5}{6}\right)$$

$$k = \frac{1}{2}\ln\left(\frac{5}{6}\right)$$

Therefore, $T(t) = 800 - 720e^{1/2 \ \ln(5/6)t}$.

Finally, we want to find $t$ such that $T(t) = 500$. Thus,

$$500 = 800 - 720e^{1/2 \ \ln(5/6)t}$$

$$720e^{1/2 \ \ln(5/6)t} = 300$$

$$e^{1/2 \ \ln(5/6)t} = \frac{5}{12}$$

$$\frac{1}{2}\ln\left(\frac{5}{6}\right)t = \ln\left(\frac{5}{12}\right)$$

$$t = \frac{2 \ln\left(\frac{5}{12}\right)}{\ln\left(\frac{5}{6}\right)} \approx 9.6 \ \text{minutes}.$$

**65.** Let $T(t)$ denote the temperature of the pie at time $t$. Then the model for this problem is:

$$\frac{dT}{dt} = k(T - 25), \quad k < 0; \quad T(0) = 325, \quad T(1) = 225.$$

Separating the variables, we have $\dfrac{dT}{T - 25} = k \, dt$

$$\int \frac{dT}{T - 25} = \int k \, dt$$
$$\ln|T - 25| = kt + C$$
$$T - 25 = e^{kt+C}$$
$$T = 25 + Ae^{kt}$$

Using the initial conditions $T(0) = 325$ and $T(1) = 225$ to evaluate the constants $A$ and $k$, we have:

$$T(0) = 325 = 25 + Ae^0, \quad A = 300$$

Thus, $T(t) = 25 + 300e^{kt}$.

$$T(1) = 225 = 25 + 300e^k$$
$$300e^k = 200$$
$$e^k = \frac{2}{3}$$
$$k = \ln\left(\frac{2}{3}\right)$$

Therefore, $T(t) = 25 + 300e^{t \, \ln(2/3)}$.

Finally, we want to determine $T(4)$: $T(4) = 25 + 300e^{4 \, \ln(2/3)}$

$$= 25 + 300e^{\ln(2/3)^4}$$
$$= 25 + 300\left(\frac{2}{3}\right)^4 \approx 84.26°\text{F}$$

**67.** If $P(t)$ is the number of bacteria present at time $t$, then the model for this problem is:

$$\frac{dP}{dt} = kP; \quad P(0) = 100, \quad P(1) = 140, \quad k > 0.$$
$$\frac{dP}{P} = k \, dt$$
$$\int \frac{dP}{P} = \int k \, dt$$
$$\ln|P| = kt + C$$
$$P = e^{kt+C}$$
$$P = Ae^{kt}. \quad \textit{General solution}$$

We use the conditions $P(0) = 100$ and $P(1) = 140$ to evaluate the constants $A$ and $k$.

$$P(0) = 100 = Ae^0, \quad A = 100.$$

Thus, $P(t) = 100e^{kt}$

$$P(1) = 140 = 100e^k$$
$$e^k = 1.4$$
$$k = \ln(1.4).$$

Therefore,
$$P(t) = 100e^{(\ln 1.4)t}. \qquad \textit{Particular solution}$$

(A) When $t = 5$,
$$P(5) = 100e^{\ln(1.4)5}$$
$$\approx 538 \text{ bacteria.}$$

(B) When $P = 1000$,
$$1000 = 100e^{\ln(1.4)t}$$
$$e^{\ln(1.4)t} = 10$$
$$\ln(1.4)t = \ln 10$$
$$t = \frac{\ln 10}{\ln(1.4)} \approx 6.8 \text{ hours.}$$

**69.** If $P(t)$ is the number of people infected at time $t$, then the model for this problem is:
$$\frac{dP}{dt} = kP(50{,}000 - P); \quad P(0) = 100, \quad P(10) = 500.$$

$$\frac{dP}{P(50{,}000 - P)} = k\ dt$$

$$\int \frac{dP}{P(50{,}000 - P)} = \int k\ dt$$

$$\frac{1}{50{,}000} \int \left[ \frac{1}{P} + \frac{1}{50{,}000 - P} \right] dP = kt + C$$

$$\frac{1}{50{,}000} [\ln P - \ln(50{,}000 - P)] = kt + C$$

$$\ln \left[ \frac{P}{50{,}000 - P} \right] = 50{,}000(kt + C)$$

$$\frac{P}{50{,}000 - P} = e^{50{,}000kt + 50{,}000C}$$

$$= e^{50{,}000C} e^{50{,}000kt}$$

$$= Ae^{50{,}000kt}$$

Solving this equation for $P$, we obtain
$$P = (50{,}000 - P)Ae^{50{,}000kt}$$
$$P = \frac{50{,}000Ae^{50{,}000kt}}{1 + Ae^{50{,}000kt}},$$

which can be written
$$P(t) = \frac{50{,}000}{1 + Be^{-50{,}000kt}}, \quad B = \frac{1}{A}.$$

Using the conditions $P(0) = 100$ and $P(10) = 500$ to evaluate the constants $B$ and $k$, we obtain
$$P(0) = 100 = \frac{50{,}000}{1 + Be^0}$$
$$100(1 + B) = 50{,}000$$
$$1 + B = 500$$
$$B = 499.$$

$$P(t) = \frac{50{,}000}{1 + 499e^{-50{,}000kt}}.$$

$$P(10) = 500 = \frac{50{,}000}{1 + 499e^{-500{,}000k}}$$

$$1 + 499e^{-500{,}000k} = 100$$

$$499e^{-500{,}000k} = 99$$

$$-500{,}000k = \ln\left(\frac{99}{499}\right)$$

$$k = -\frac{1}{500{,}000}\ln\left(\frac{99}{499}\right).$$

Therefore, $P(t) = \dfrac{50{,}000}{1 + 499e^{0.1\ln(99/499)t}}.$ *Particular solution*

(A) When $t = 20$,

$$P(20) = \frac{50{,}000}{1 + 499e^{2\ln(99/499)}}$$

$$\approx 2422 \text{ people.}$$

(B) When $P = 25{,}000$,

$$25{,}000 = \frac{50{,}000}{1 + 499e^{0.1\ln(99/499)t}}$$

$$1 + 499e^{0.1\ln(99/499)t} = 2$$

$$e^{0.1\ln(99/499)t} = \frac{1}{499}$$

$$0.1\ln\left(\frac{99}{499}\right)t = \ln\left(\frac{1}{499}\right)$$

$$t = \frac{10\ln\left(\frac{1}{499}\right)}{\ln\left(\frac{99}{499}\right)} \approx 38.4 \text{ days.}$$

71. Let $y$ be the body temperature and let us assume that the time body was found (1:00 a.m.) as 0. Then $y(0) = 75°$ and $y(1) = 70°$ (2:00 a.m. is 1 hour after body was found).
Note that

$$\frac{dy}{dt} = k(y - C) = k(y - 50)$$

or

$$\frac{dy}{y - 50} = kt$$

or

$$\ln(y - 50) = kt + C_1$$

or

$$y = 50 + Ae^{kt}, \text{ where } A \text{ and } k \text{ are constants.}$$

$$y(0) = 75 = 50 + Ae^{k(0)} = 50 + A \quad \text{or} \quad A = 25$$

$$y(1) = 70 = 50 + 25e^{k(1)} \quad \text{or} \quad e^k = \frac{20}{25} = 0.8 \quad \text{or} \quad k = \ln(0.8).$$

Thus, $y(t) = 50 + 25e^{t\ln(0.8)}$.

To find the time of death we have to solve the following equation for $t$:

$$98.6 = 50 + 25e^{t \ln(0.8)}$$

Observe that

$$e^{t \ln(0.8)} = \frac{98.6 - 50}{25}$$

or

$$t = \frac{\ln\left(\frac{48.6}{25}\right)}{\ln(0.8)} \approx -3; \text{ i.e. 3 hours before body was found.}$$

So, the time of death will be 10:00 p.m. since the body was found at 1:00 a.m.

73. $\dfrac{dI}{ds} = k\dfrac{I}{s}$, $I > 0$, $s > 0$.

$$\frac{dI}{I} = \frac{k}{s}ds$$

$$\int \frac{dI}{I} = \int \frac{k}{s}ds$$

$$\ln I = k \ln s + C$$

$$\qquad = \ln s^k + C \qquad (k \ln s = \ln s^k)$$

$$I = e^{\ln s^k + C}$$

$$\qquad = e^C e^{\ln s^k}.$$

Therefore,

$$I = As^k. \qquad (\underline{\text{Note}}: e^{\ln s^k} = s^k.)$$

75. If $P(t)$ is the number of people who have heard the rumor at time $t$, then the model for this problem is:

$$\frac{dP}{dt} = kP(1000 - P); \; P(0) = 5, \; P(1) = 10.$$

$$\frac{dP}{P(1000 - P)} = k\,dt$$

$$\int \frac{dP}{P(1000 - P)} = \int k\,dt$$

$$\frac{1}{1000}\int\left[\frac{1}{P} + \frac{1}{1000 - P}\right]dP = kt + C$$

$$\frac{1}{1000}[\ln P - \ln(1000 - P)] = kt + C$$

$$\ln\left(\frac{P}{1000 - P}\right) = 1000(kt + C)$$

$$\frac{P}{1000 - P} = e^{1000kt + 1000C}$$

$$\qquad = e^{1000C}e^{1000kt}$$

$$\qquad = Ae^{1000kt}$$

Solving this equation for $P$, we obtain

$$P = \frac{1000Ae^{1000kt}}{1 + Ae^{1000kt}} \quad \text{or} \quad P = \frac{1000}{1 + Be^{-1000kt}} \left(B = \frac{1}{A}\right). \quad \textit{General solution}$$

Using the conditions $P(0) = 5$ and $P(1) = 10$ to evaluate the constants $B$ and $k$, we have

$$P(0) = 5 = \frac{1000}{1 + Be^0}$$
$$5(1 + B) = 1000$$
$$1 + B = 200$$
$$B = 199.$$

Thus,

$$P(t) = \frac{1000}{1 + 199e^{-1000kt}}$$
$$P(1) = 10 = \frac{1000}{1 + 199e^{-1000k}}$$
$$1 + 199e^{-1000k} = 100$$
$$199e^{-1000k} = 99$$
$$e^{-1000k} = \frac{99}{199}$$
$$-1000k = \ln\left(\frac{99}{199}\right)$$
$$k = -\frac{1}{1000}\ln\left(\frac{99}{199}\right).$$

Therefore,

$$P(t) = \frac{1000}{1 + 199e^{\ln(99/199)t}}. \quad \textit{Particular solution}$$

(A) When $t = 7$,
$$P(7) = \frac{1000}{1 + 199e^{\ln(99/199)7}} \approx 400 \text{ people.}$$

(B) When $P = 850$,
$$850 = \frac{1000}{1 + 199e^{\ln(99/199)t}}$$
$$1 + 199e^{\ln(99/199)t} = \frac{1000}{850} = \frac{20}{17}$$
$$e^{\ln(99/199)t} = \frac{3}{3383}$$
$$\ln\left(\frac{99}{199}\right)t = \ln\left(\frac{3}{3383}\right)$$
$$t = \frac{\ln\left(\dfrac{3}{3383}\right)}{\ln\left(\dfrac{99}{199}\right)} \approx 10 \text{ days.}$$

**1.** $y' = ky$

$y' - ky = 0;$ $f(x) = -k,$ $g(x) = 0$

The differential equation is first-order linear.

**3.** $y' = ky(M - y) = kMy - ky^2$

The equation has a $y^2$ term so it is **not** first-order linear.

**5.** $3x^2;$ $\int (x^3 y' + 3x^2 y)\,dx = x^3 y$

**7.** $-3e^{-3x};$ $\int (e^{-3x} y' - 3e^{-3x} y)\,dx = e^{-3x} y$

**9.** $x^4;$ $\int (x^4 y' + 4x^3 y)\,dx = x^4 y$

**11.** $e^{-0.5x};$ $\int (e^{-0.5x} y' - 0.5e^{-0.5x} y)\,dx = e^{-0.5x} y$

**13.** $y' + 2y = 4;$ $y(0) = 1$

Step 1: The equation is in standard form.

Step 2: Find the integrating factor.

$$f(x) = 2 \quad \text{and} \quad I(x) = e^{\int f(x)\,dx} = e^{\int 2\,dx} = e^{2x}$$

Step 3: Multiply both sides of the standard form by the integrating factor.

$$e^{2x}(y' + 2y) = e^{2x}(4)$$
$$e^{2x} y' + 2e^{2x} y = 4e^{2x}$$
$$[e^{2x} y]' = 4e^{2x}$$

Step 4: Integrate both sides.

$$\int [e^{2x} y]'\,dx = \int 4e^{2x}\,dx$$
$$e^{2x} y = 2e^{2x} + C$$

Step 5: Solve for $y$.

$$y = \frac{1}{e^{2x}}(2e^{2x} + C) = 2 + Ce^{-2x} \quad \textit{General solution}$$

To find the particular solution satisfying the initial condition $y(0) = 1$, substitute $x = 0$, $y = 1$ in the general solution:

$$1 = 2 + Ce^0 = 2 + C.$$

Thus, $C = -1$ and the particular solution is

$$y = 2 - e^{-2x}$$

**15.** $y' + y = e^{-2x};$ $y(0) = 3$

Step 1: The equation is in standard form.

Step 2: Find the integrating factor.

$$f(x) = 1 \quad \text{and} \quad I(x) = e^{\int f(x)\,dx} = e^{\int 1\,dx} = e^x$$

<u>Step 3</u>: Multiply both sides of the standard form by the integrating factor.

$$e^x [y' + y] = e^x \cdot e^{-2x}$$

$$e^x y' + e^x y = e^{-x}$$

$$[e^x y]' = e^{-x}$$

<u>Step 4</u>: Integrate both sides.

$$\int [e^x y]' \, dx = \int e^{-x} \, dx$$

$$e^x y = \frac{e^{-x}}{-1} + C = -e^{-x} + C$$

<u>Step 5</u>: Solve for $y$.

$$y = \frac{1}{e^x} [-e^{-x} + C] = -e^{-2x} + Ce^{-x} \quad \textit{General solution}$$

To find the particular solution satisfying the initial condition $y(0) = 3$, substitute $x = 0$, $y = 3$ in the general solution:

$$3 = -e^0 + Ce^0 = -1 + C$$

Thus, $C = 4$ and the particular solution is

$$y = -e^{-2x} + 4e^{-x}$$

**17.** $y' - y = 2e^x$; $y(0) = -4$

<u>Step 1</u>: The equation is in standard form.

<u>Step 2</u>: Find the integrating factor.

$$f(x) = -1 \quad \text{and} \quad I(x) = e^{\int f(x)\,dx} = e^{\int (-1)\,dx} = e^{-x}$$

<u>Step 3</u>: Multiply both sides of the standard form by the integrating factor.

$$e^{-x} [y' - y] = e^{-x}(2e^x)$$

$$e^{-x} y' - e^{-x} y = 2$$

$$[e^{-x} y]' = 2$$

<u>Step 4</u>: Integrate both sides.

$$\int [e^{-x} y]' \, dx = \int 2 \, dx$$

$$e^{-x} y = 2x + C$$

<u>Step 5</u>: Solve for $y$.

$$y = \frac{1}{e^{-x}} [2x + C] = 2xe^x + Ce^x \quad \textit{General solution}$$

To find the particular solution satisfying the initial condition $y(0) = -4$, substitute $x = 0$, $y = -4$ in the general solution:

$$-4 = 2(0)e^0 + Ce^0 = C$$

Thus, $C = -4$ and the particular solution is

$$y = 2xe^x - 4e^x$$

**19.** $y' + y = 9x^2 e^{-x}$; $y(0) = 2$

Step 1: The equation is in standard form.

Step 2: Find the integrating factor.

$$f(x) = 1 \quad \text{and} \quad I(x) = e^{\int f(x)\,dx} = e^{\int 1\,dx} = e^x$$

Step 3: Multiply both sides of the standard form by the integrating factor.

$$e^x [y' + y] = e^x (9x^2 e^{-x})$$
$$e^x y' + e^x y = 9x^2$$
$$[e^x y]' = 9x^2$$

Step 4: Integrate both sides.

$$\int [e^x y]'\,dx = \int 9x^2\,dx$$
$$e^x y = 3x^3 + C$$

Step 5: Solve for $y$.

$$y = \frac{1}{e^x}(3x^3 + C) = 3x^3 e^{-x} + Ce^{-x} \quad \textit{General solution}$$

To find the particular solution satisfying the initial condition $y(0) = 2$, substitute $x = 0$, $y = 2$ in the general solution:

$$2 = 3(0)^3 e^0 + Ce^0 = C$$

Thus, $C = 2$ and the particular solution is
$$y = 3x^3 e^{-x} + 2e^{-x}$$

**21.** $xy' + y = 2x$; $y(1) = 1$

Step 1: Write the differential equation in standard form.

Multiply both sides by $\dfrac{1}{x}$ to obtain:

$$y' + \frac{1}{x}y = 2$$

Step 2: Find the integrating factor.

$$f(x) = \frac{1}{x} \quad \text{and} \quad I(x) = e^{\int f(x)\,dx} = e^{\int (1/x)\,dx} = e^{\ln x} = x$$

Step 3: Multiply both sides of the standard form by the integrating factor.

$$x\left(y' + \frac{1}{x}\,y\right) = x(2)$$
$$xy' + y = 2x$$
$$[xy]' = 2x$$

Step 4: Integrate both sides.

$$\int [xy]'\,dx = \int 2x\,dx$$
$$xy = x^2 + C$$

Step 5: Solve for $y$.

$$y = \frac{1}{x}(x^2 + C) = x + \frac{C}{x} \qquad \textit{General solution}$$

To find the particular solution satisfying the initial condition $y(1) = 1$, substitute $x = 1$, $y = 1$ in the general solution:

$$1 = 1 + \frac{C}{1} = 1 + C$$

Thus, $C = 0$ and the particular solution is

$$y = x$$

**23.** $xy' + 2y = 10x^3$; $y(2) = 8$

Step 1: Write the differential equation in standard form.

Multiply both sides by $\frac{1}{x}$ to obtain:

$$y' + \frac{2}{x}y = 10x^2$$

Step 2: Find the integrating factor.

$$f(x) = \frac{2}{x} \quad \text{and} \quad I(x) = e^{\int f(x)\,dx} = e^{\int (2/x)\,dx} = e^{2\ln x} = e^{\ln x^2} = x^2$$

Step 3: Multiply both sides of the standard form by the integrating factor.

$$x^2\left(y' + \frac{2}{x}\,y\right) = x^2(10x^2)$$
$$x^2 y' + 2xy = 10x^4$$
$$[x^2 y]' = 10x^4$$

Step 4: Integrate both sides.

$$\int [x^2 y]'\,dx = \int 10x^4\,dx$$
$$x^2 y = 2x^5 + C$$

Step 5: Solve for $y$.

$$y = \frac{1}{x^2}(2x^5 + C) = 2x^3 + \frac{C}{x^2} \qquad \textit{General solution}$$

To find the particular solution satisfying the initial condition $y(2) = 8$, substitute $x = 2$, $y = 8$ in the general solution:

$$8 = 2(2)^3 + \frac{C}{2^2} = 16 + \frac{C}{4}$$

and $\quad -8 = \frac{C}{4}$.

Thus, $C = -32$ and the particular solution is

$$y = 2x^3 - \frac{32}{x^2}$$

**25.** $y' + xy = 5x$

$\quad I(x) = e^{\int f(x)\,dx} = e^{\int x\,dx} = e^{x^2/2}$ *Integrating factor*

$\quad y = \dfrac{1}{I(x)}\int I(x)\,g(x)\,dx$

$\qquad = \dfrac{1}{e^{x^2/2}}\int e^{x^2/2}5x\,dx$

$\qquad = 5e^{-x^2/2}\int xe^{x^2/2}\,dx \qquad \left(u = \dfrac{x^2}{2},\ du = x\,dx\right)$

$\qquad = 5e^{-x^2/2}[e^{x^2/2} + C] = 5 + 5Ce^{-x^2/2}$

$\quad y = 5 + Ae^{-x^2/2}.$ *General solution*

**27.** $y' - 2y = 4x$

$\quad I(x) = e^{\int(-2)\,dx} = e^{-2x} \qquad$ *Integrating factor*

$\quad y = \dfrac{1}{I(x)}\int I(x)\,g(x)\,dx = \dfrac{1}{e^{-2x}}\int e^{-2x}4x\,dx$

$\qquad = 4e^{2x}\int xe^{-2x}\,dx \qquad$ (Integrate by parts)

$\qquad = 4e^{2x}\left[-\dfrac{1}{2}xe^{-2x} - \dfrac{1}{4}e^{-2x} + C\right]$

$\qquad = -2x - 1 + 4Ce^{2x} \qquad (A = 4C)$

$\quad y = -2x - 1 + Ae^{2x}.$ *General solution*

**29.** $xy' + y = xe^x$

First write the equation in standard form: $y' + \dfrac{1}{x}y = e^x$.

Then,

$\quad I(x) = e^{\int f(x)\,dx} = e^{\int(1/x)\,dx} = e^{\ln x} = x \qquad$ *Integrating factor*

$\quad y = \dfrac{1}{I(x)}\int I(x)\,g(x)\,dx$

$\qquad = \dfrac{1}{x}\int xe^x\,dx \qquad$ (Integrate by parts)

$\qquad = \dfrac{1}{x}[xe^x - e^x + C]$

$\quad y = e^x - \dfrac{e^x}{x} + \dfrac{C}{x}.$ *General solution*

**31.** $xy' + y = x \ln x$

First write the equation in standard form: $y' + \dfrac{1}{x}y = \ln x$

Then,

$y' + \dfrac{1}{x}y = \ln x$

$I(x) = e^{\int f(x)\,dx} = e^{\int (1/x)\,dx} = e^{\ln x} = x$     *Integrating factor*

$y = \dfrac{1}{I(x)}\int I(x)g(x)\,dx$

$\phantom{y} = \dfrac{1}{x}\int x \ln x\,dx$          (Integrate by parts)

$\phantom{y} = \dfrac{1}{x}\left[\dfrac{x^2}{2}\ln x - \dfrac{1}{4}x^2 + C\right]$

$\phantom{y} = \dfrac{1}{2}x \ln x - \dfrac{1}{4}x + \dfrac{C}{x}.$     *General solution*


**33.** $2xy' + 3y = 20x$

First write the equation in standard form

$\quad y' + \dfrac{3}{2x}y = 10$

Then,

$I(x) = e^{\int f(x)\,dx} = e^{\int (3/2x)\,dx} = e^{(3/2)\ln x}$

$\phantom{I(x) = e^{\int f(x)\,dx} = e^{\int (3/2x)\,dx}} = e^{\ln x^{3/2}} = x^{3/2}$     *Integrating factor*

$y = \dfrac{1}{I(x)}\int I(x)g(x)\,dx = \dfrac{1}{x^{3/2}}\int x^{3/2}10\,dx$

$\phantom{y} = 10x^{-3/2}\int x^{3/2}\,dx$

$\phantom{y} = 10x^{-3/2}\left[\dfrac{2}{5}x^{5/2} + C\right]$

$\phantom{y} = 4x + 10Cx^{-3/2}$   $(A = 10C)$

$\phantom{y} = 4x + Ax^{-3/2}$          *General solution*


**35.** (A) Substitute $y = \dfrac{1}{3}(x + 1)^3 + C$; $y' = (x + 1)^2$ into the differential equation to determine whether these substitutions reduce the equation to an identity.

(B) The solution is wrong:

$\quad \dfrac{1}{x}\int x(x + 1)^2\,dx \neq \int (x + 1)^2\,dx$

(C) $y = \frac{1}{x}\int x(x + 1)^2\,dx = \frac{1}{x}\int(x^3 + 2x^2 + x)\,dx$

$$= \frac{1}{x}\left[\frac{1}{4}x^4 + \frac{2}{3}x^3 + \frac{1}{2}x^2 + C\right]$$

$$= \frac{1}{4}x^3 + \frac{2}{3}x^2 + \frac{1}{2}x + \frac{C}{x}$$

Thus, $y = \frac{1}{4}x^3 + \frac{2}{3}x^2 + \frac{1}{2}x + \frac{C}{x}$      *General solution*

Substituting $y$ and $y' = \frac{3}{4}x^2 + \frac{4}{3}x + \frac{1}{2} - \frac{C}{x^2}$

into the differential equation yields:

$$\frac{3}{4}x^2 + \frac{4}{3}x + \frac{1}{2} - \frac{C}{x^2} + \frac{1}{x}\left(\frac{1}{4}x^3 + \frac{2}{3}x^2 + \frac{1}{2}x + \frac{C}{x}\right) = x^2 + 2x + 1$$

$$= (x + 1)^2$$

37. (A) Substitute $y = \frac{1}{2}e^{-x}$, $y' = -\frac{1}{2}e^{-x}$ into the differential equation to determine whether these substitutions reduce the equation to an identity.

(B) $y = \frac{1}{2}e^{-x}$ is **not** the general solution since the constant of integration has been omitted. However, $y = \frac{1}{2}e^{-x}$ is a particular solution.

(C) $y = \frac{1}{e^{3x}}\int e^{2x}\,dx = \frac{1}{e^{3x}}\left[\frac{1}{2}e^{2x} + C\right]$

$$= \frac{1}{2}e^{-x} + Ce^{-3x}$$

Thus, $y = \frac{1}{2}e^{-x} + Ce^{-3x}$      *General solution*

Substituting $y$ and $y' = -\frac{1}{2}e^{-x} - 3Ce^{-3x}$ into the differential equation yields:

$$-\frac{1}{2}e^{-x} - 3Ce^{-3x} + 3\left(\frac{1}{2}e^{-x} + Ce^{-3x}\right) = e^{-x}$$

39. $y' = \frac{1 - y}{x}$

This equation can be rewritten as a first-order linear differential equation in the standard form.

(A) $y' + \dfrac{1}{x}y = \dfrac{1}{x}$.

Then $f(x) = \dfrac{1}{x}$ and the integrating factor is

$I(x) = e^{\int (1/x)\,dx} = e^{\ln x} = x.$

Thus, $y = \dfrac{1}{x}\int x \cdot \dfrac{1}{x}\,dx = \dfrac{1}{x}\int dx = \dfrac{1}{x}(x + C) = 1 + \dfrac{C}{x}.$

(B) Using separation of variables on the original equation, we have:

$$\dfrac{y'}{1 - y} = \dfrac{1}{x}$$

$$\int \dfrac{dy}{1 - y} = \int \dfrac{1}{x}\,dx$$

$-\ln(1 - y) = \ln(x) + C \qquad$ (assuming $1 - y > 0$ and $x > 0$)

$\ln(1 - y) = -\ln(x) - C$

$1 - y = e^{-\ln x - C}$

$1 - y = e^{-C}e^{\ln x^{-1}}$

$1 - y = \dfrac{K}{x} \qquad\qquad \left( e^{\ln x^{-1}} = \dfrac{1}{x} \right)$

$y = 1 + \dfrac{K}{x}$

**41.** $y' = \dfrac{2x + 2xy}{1 + x^2}$

This equation can be rewritten as a first-order linear differential equation in the standard form.

(A) $y' - \dfrac{2x + 2xy}{1 + x^2}y = \dfrac{2x}{1 + x^2}$.

Then, $f(x) = \dfrac{-2x}{1 + x^2}$ and the integrating factor is:

$I(x) = e^{\int (-2x/1 + x^2)\,dx} = e^{-\ln(1 + x^2)} = \dfrac{1}{1 + x^2}$

Thus, $y = \dfrac{1}{\dfrac{1}{1 + x^2}}\int \dfrac{1}{1 + x^2} \cdot \dfrac{2x}{1 + x^2}\,dx$

$= (1 + x^2)\int \dfrac{2x}{(1 + x^2)^2}\,dx = (1 + x^2)\left[ \dfrac{-1}{1 + x^2} + C \right]$

and $y = -1 + C(1 + x^2).$

(B) Using separation of variables on the original equation, we have:

$$\frac{y'}{1 + y} = \frac{2x}{1 + x^2}$$

$$\int \frac{dy}{1 + y} = \int \frac{2x}{1 + x^2}\, dx$$

$$\ln|1 + y| = \ln(1 + x^2) + C$$

$$1 + y = e^{\ln(1+x^2)+C}$$

$$1 + y = e^C e^{\ln(1+x^2)}$$

$$1 + y = K(1 + x^2) \qquad (e^{\ln(1+x^2)} = 1 + x^2)$$

$$y = -1 + K(1 + x^2)$$

**43.** $y' = 2x(y + 1)$

This equation can be rewritten as a first-order linear differential equation in the standard form.

(A) $y' - 2xy = 2x$

Then, $f(x) = -2x$ and the integrating factor is:

$$I(x) = e^{\int -2x\, dx} = e^{-x^2}$$

Thus, $y = \dfrac{1}{e^{-x^2}}\int e^{-x^2}\, 2x\, dx = e^{x^2}\int 2xe^{-x^2}\, dx = e^{x^2}[-e^{-x^2} + C]$

and $\quad y = -1 + Ce^{x^2}$.

(B) Using separation of variables on the original equation, we have:

$$\frac{y'}{y + 1} = 2x$$

$$\int \frac{dy}{1 + y} = \int 2x\, dx$$

$$\ln|y + 1| = x^2 + C$$

$$y + 1 = e^{x^2+C}$$

$$y + 1 = e^C e^{x^2}$$

$$= Ke^{x^2}$$

Thus, $y = -1 + Ke^{x^2}$.

**45.** $\dfrac{dy}{dt} = ky$.

This equation can be rewritten as

$$\frac{dy}{dt} - ky = 0$$

Here $f(t) = -k$ and the integrating factor is:

$I(t) = e^{\int -k\, dt} = e^{-kt}$

Thus, $y = \dfrac{1}{e^{-kt}}\int e^{-kt} \cdot 0\, dt = e^{kt}\int 0\, dt = e^{kt}C$

and $\quad y = Ce^{kt}$.

**47.** If $f(x) = a \neq 0$ and $g(x) = b$, $a,b$ constant, then $y' + f(x)y = g(x)$ becomes $y' + ay = b$.

The integrating factor is: $I(x) = e^{\int a\,dx} = e^{ax}$, and

$$e^{ax}y' + ae^{ax}y = be^{ax}$$

$$(e^{ax}y)' = be^{ax}$$

$$\int (e^{ax}y)'\,dx = be^{ax}dx$$

$$e^{ax}y = \frac{b}{a}e^{ax} + C$$

$$y = \frac{b}{a} + Ce^{-ax} \qquad \text{(General solution)}$$

**49.** The amount $A$ in the account at any time $t$ must satisfy

$$\frac{dA}{dt} - 0.048A = -3000.$$

Now $f(t) = -0.048$ and the integrating factor is:

$$I(t) = e^{\int (-0.048)\,dt} = e^{-0.048t}$$

Thus, $A = \dfrac{1}{e^{-0.048t}}\displaystyle\int -3000e^{-0.048t}\,dt$

$$= e^{0.048t}\left[\frac{-3000e^{-0.048t}}{-0.048} + C\right]$$

$$A = 62{,}500 + Ce^{0.048t} \qquad \textit{General solution}$$

Applying the initial condition $A(0) = 20{,}000$ yields:

$$A(0) = 62{,}500 + Ce^0 = 20{,}000 \quad \text{and} \quad C = -42{,}500$$

Thus, the amount in the account at any time $t$ is:

$$A(t) = 62{,}500 - 42{,}500e^{0.048t}$$

To determine when the amount in the account is 0, we must solve $A(t) = 0$ for $t$:

$$62{,}500 - 42{,}500e^{0.048t} = 0$$

$$42{,}500e^{0.048t} = 62{,}500$$

$$e^{0.048t} = 1.4706$$

$$t = \frac{\ln 1.4706}{0.048} \approx 8.035$$

Thus, the account is depleted after 8.035 years. The total amount withdrawn from the account is:

$$3000(8.035) = \$24{,}105.$$

**51.** The amount in the account at any time $t$ must satisfy

$$\frac{dA}{dt} - 0.0475A = -1900.$$

Now $f(t) = -0.0475$ and the integrating factor is:

$$I(t) = e^{\int (-0.0475)\,dt} = e^{-0.0475t}$$

Thus,

$$A = \frac{1}{e^{-0.0475t}} \int -1900 e^{-0.0475t}\,dt = e^{0.0475t} \left[ \frac{-1900 e^{-0.0475t}}{-0.0475} + C \right]$$

$$= 40{,}000 + Ce^{0.0475t} \qquad \textit{General solution}$$

Applying the initial condition $A(0) = P$ yields:

$$40{,}000 + C = P$$
$$C = P - 40{,}000$$

Thus, the amount in the account at any time $t$ is:

$$A(t) = 40{,}000 + (P - 40{,}000)e^{0.0475t}$$

Since $A(10) = 0$, we have:

$$0 = 40{,}000 + (P - 40{,}000)e^{0.0475(10)}$$
$$(P - 40{,}000)e^{0.475} = -40{,}000$$

Solving for the initial deposit $P$ yields:

$$P = \frac{-40{,}000}{e^{0.475}} + 40{,}000 = 40{,}000(1 - e^{-0.475}) \approx 15{,}124.60$$

Thus, the initial deposit was \$15,124.60.

**53.** The amount in the account at any time $t$ must satisfy

$$\frac{dA}{dt} - 0.07A = 2100.$$

Now $f(t) = -0.07$ and the integrating factor is

$$I(t) = e^{\int (-0.07)\,dt} = e^{-0.07t}$$

Thus,

$$A = \frac{1}{e^{-0.07t}} \int 2100 e^{-0.07t}\,dt = e^{0.07t} \left[ \frac{2100 e^{-0.07t}}{-0.07} + C \right]$$

$$= -30{,}000 + Ce^{0.07t}$$

Applying the initial condition $A(0) = 5000$ yields:

$$5000 = -30{,}000 + C$$

and $\quad C = 35{,}000$

Thus, the amount in the account at any time $t$ is:

$$A(t) = 35{,}000 e^{0.07t} - 30{,}000$$

After 5 years, the amount in the account is

$$A(5) = 35{,}000 e^{0.07(5)} - 30{,}000 = 35{,}000 e^{0.35} - 30{,}000$$
$$\approx \$19{,}667.36$$

**55.** Let $r$ (expressed as a decimal) be the interest rate. Then the amount in the account at any time $t$ must satisfy

$$\frac{dA}{dt} - rA = 1{,}000$$

Now, $f(t) = -r$ and the integrating factor is:

$$I(t) = e^{\int -r\, dt} = e^{-rt}$$

Thus,

$$A(t) = \frac{1}{e^{-rt}}\int 1{,}000 e^{-rt}\, dt$$

$$= \frac{1{,}000}{e^{-rt}}\left[-\frac{1}{r}e^{-rt} + C\right]$$

$$= Ce^{rt} - \frac{1{,}000}{r}$$

Since $A(0) = 10{,}000$, we have

$$10{,}000 = Ce^{0} - \frac{1{,}000}{r}$$

and

$$C = 10{,}000 + \frac{1{,}000}{r}$$

Therefore,

$$A(t) = 10{,}000 e^{rt} + \frac{1{,}000}{r}(e^{rt} - 1)$$

Now, at $t = 10$, we have

$$30{,}000 = 10{,}000 e^{10r} + \frac{1{,}000}{r}(e^{10r} - 1)$$

Solving this equation for $r$ using a graphing utility, we find that $r \approx 0.0523$. Expressed as a percentage, $r = 5.23\%$.

**57.** The equilibrium price at time $t$ is the solution of the equation

$$95 - 5p(t) + 2p'(t) = 35 - 2p(t) + 3p'(t)$$

which satisfies the initial condition $p(0) = 30$.

The equation simplifies to

$$p'(t) + 3p(t) = 60,$$

a first-order linear equation. The integrating factor is:

$$I(t) = e^{\int 3\, dt} = e^{3t}$$

Thus,

$$p(t) = \frac{1}{e^{3t}}\int 60 e^{3t}\, dt = e^{-3t}\left[\frac{60 e^{3t}}{3} + C\right]$$

$$= 20 + Ce^{-3t} \qquad \textit{General solution}$$

Applying the initial condition yields

$$p(0) = 20 + C = 30$$
$$C = 10$$

Thus, the equilibrium price at time $t$ is:

$$p(t) = 20 + 10e^{-3t}$$

The long-range equilibrium price is:

$$\overline{p} = \lim_{t \to \infty}(20 + 10e^{-3t}) = 20$$

**59.** Let $p(t)$ be the amount of pollutants in the tank at time $t$. The initial amount of pollutants in the tank is $p(0) = 2 \cdot 200 = 400$ pounds.

Pollutants are entering the tank at the constant rate of $3 \cdot 75 = 225$ pounds per hour.

The amount of water in the tank at time $t$ is $200 + 25t$.

The amount of pollutants in each gallon of water at time $t$ is $\dfrac{p(t)}{200 + 25t}$.

The rate at which pollutants are leaving the tank is
$$\frac{50p(t)}{200 + 25t} = \frac{2p(t)}{8 + t}.$$
Thus, the model for this problem is
$$p'(t) = 225 - \frac{2p(t)}{8 + t}; \ p(0) = 400 \text{ or } p'(t) + \frac{2p(t)}{8 + t} = 225; \ p(0) = 400.$$

Now $f(t) = \dfrac{2}{8 + t}$ and the integrating factor is:
$$I(t) = e^{\int (2/8+t))dt} = e^{2 \ln(8+t)} = e^{\ln(8+t)^2} = (8 + t)^2$$
Thus,
$$p(t) = \frac{1}{(8 + t)^2} \int 225(8 + t)^2 dt = \frac{1}{(8 + t)^2}\left[225 \frac{(8 + t)^3}{3} + C\right]$$
$$p(t) = 75(8 + t) + \frac{C}{(8 + t)^2}. \quad \textit{General solution}$$

We use the initial condition $p(0) = 400$ to evaluate the constant $C$.
$$p(0) = 400 = 75(8) + \frac{C}{8^2}$$
$$\frac{C}{64} = -200$$
$$C = -12,800$$
$$p(t) = 75(8 + t) - \frac{12,800}{(8 + t)^2}. \quad \textit{Particular solution}$$

To find the total amount of pollutants in the tank after two hours, we evaluate $p(2)$:
$$p(2) = 75(10) - \frac{12,800}{(10)^2} = 750 - 128 = 622$$

After two hours, the tank contains 250 gallons of water. Thus, the rate at which pollutants are being released is
$$\frac{622}{250} \approx 2.5 \text{ pounds per gallon.}$$

**61.** Let $p(t)$ be the amount of pollutants in the tank at time $t$. The initial amount of pollutants in the tank is $p(0) = 2 \cdot 200 = 400$ pounds.

Pollutants are entering the tank at the constant rate of $3(50) = 150$ pounds per hour.

Since water is entering and leaving the tank at the same rate, the amount of water in the tank at all times $t$ is 200 gallons.

The amount of pollutants in each gallon of water at time $t$ is $\dfrac{p(t)}{200}$.

The rate at which pollutants are leaving the tank is $\dfrac{50p(t)}{200} = \dfrac{p(t)}{4}$.

Thus, the model for this problem is:

$$p'(t) = 150 - \frac{p(t)}{4}; \quad p(0) = 400 \quad \text{or} \quad p'(t) + \frac{1}{4}p(t) = 150; \quad p(0) = 400$$

Now $f(t) = \dfrac{1}{4}$ and the integrating factor is:

$$I(t) = e^{\int (1/4)\, dt} = e^{t/4}$$

Thus, $p(t) = \dfrac{1}{e^{t/4}} \int e^{t/4}(150)\, dt$

$$= 150e^{-t/4} \int e^{t/4}\, dt = 150e^{-t/4}[4e^{t/4} + C]$$

$$= 600 + 150Ce^{-t/4} \qquad (A = 150C)$$

$$p(t) = 600 + Ae^{-t/4}. \textit{General solution}$$

We use the initial condition $p(0) = 400$ to evaluate the constant $A$.

$$p(0) = 400 = 600 + Ae^0, \quad A = -200$$

Therefore, $p(t) = 600 - 200e^{-t/4}$.  *Particular solution*

To find the amount of pollutants in the tank after two hours, we evaluate $p(2)$:

$$p(2) = 600 - 200e^{-1/2} \approx 479 \text{ pounds}$$

The rate at which pollutants are being released after two hours is

$$\frac{600 - 200e^{-1/2}}{200} = 3 - e^{-1/2} \approx 2.4 \text{ pounds per gallon.}$$

**63.** From Problem 59, the amount of pollutants in the tank at time $t$ is given by:

$$p(t) = 75(8 + t) - \frac{12,800}{(8 + t)^2}$$

To find the time $t$ when the tank contains 1,000 pounds of pollutants, we solve the equation:

$$75(8 + t) - \frac{12,800}{(8 + t)^2} = 1,000$$

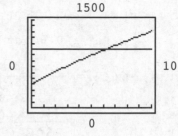

Using a graphing utility, we find that
$$t \approx 6.2 \text{ hrs.}$$
The graphs of $p(t)$ and $p = 1,000$ are shown above.

**65.** The model for this problem is:

$$\frac{dw}{dt} + 0.005w = \frac{2100}{3500} \quad \text{or} \quad \frac{dw}{dt} + 0.005w = \frac{3}{5}$$

Now $f(t) = 0.005$, and the integrating factor is:
$$I(t) = e^{\int 0.005 dt} = e^{0.005t}$$

Thus, $w(t) = \dfrac{1}{e^{0.005t}} \displaystyle\int \frac{3}{5} e^{0.005t} \, dt$

$$= \frac{3}{5} e^{-0.005t} \left[ \frac{e^{0.005t}}{0.005} + k \right]$$

$$= 120 + \frac{3}{5} k e^{-0.005t}$$

$$= 120 + A e^{-0.005t} \qquad \textit{General solution}$$

Applying the initial condition $w(0) = 160$, we have:
$$160 = 120 + Ae^0$$
$$A = 40$$

Thus, $w(t) = 120 + 40e^{-0.005t}$.

How much will a person weigh after 30 days on the diet?
$$w(30) = 120 + 40e^{(-0.005)(30)} \approx 154 \text{ pounds}$$

Now, we want to find $t$ such that $w(t) = 150$.
$$150 = 120 + 40e^{-0.005t}$$
$$40e^{-0.005t} = 30$$
$$e^{-0.005t} = \frac{3}{4}$$
$$-0.005t = \ln\left(\frac{3}{4}\right)$$
$$t = \frac{-\ln\left(\dfrac{3}{4}\right)}{0.005} \approx 58$$

Thus, it will take 58 days to lose 10 pounds.

Finally, $\displaystyle\lim_{t \to \infty} w(t) = \lim_{t \to \infty} (120 + 40e^{-0.005t}) = 120$, since $\displaystyle\lim_{t \to \infty} e^{-0.005t} = 0$.

Therefore, the person's weight will approach 120 pounds if this diet is maintained for a long period.

**67.** The model for this problem is
$$\frac{dw}{dt} + 0.005w = \frac{1}{3500}C,$$
where $C$ is to be determined.

Now, $f(t) = 0.005$, and the integrating factor is:
$$I(t) = e^{\int 0.005\, dt} = e^{0.005t}$$

Thus, $w(t) = \dfrac{1}{e^{0.005t}} \displaystyle\int \frac{C}{3500} e^{0.005t}\, dt$

$$= \frac{C}{3500} e^{-0.005t}\left[\frac{e^{0.005t}}{0.005} + k\right]$$

$$= \frac{C}{17.5} + \frac{Ck}{3500} e^{-0.005t}$$

$$= \frac{C}{17.5} + Ae^{-0.005t} \qquad \text{\textit{General solution}}$$

Applying the initial condition $w(0) = 130$, we have
$$130 = \frac{C}{17.5} + Ae^0$$

$$A = 130 - \frac{C}{17.5}$$

and
$$w(t) = \frac{C}{17.5} + \left(130 - \frac{C}{17.5}\right)e^{-0.005t}$$

Now, we want to determine $C$ such that $w(30) = 125$.

$$125 = \frac{C}{17.5} + \left(130 - \frac{C}{17.5}\right)e^{-0.005(30)}$$

$$125 = \frac{C}{17.5} + \left(130 - \frac{C}{17.5}\right)e^{-0.15}$$

$$\frac{C}{17.5}(1 - e^{-0.15}) = 125 - 130e^{-0.15}$$

$$C = \frac{17.5(125 - 130e^{-0.15})}{1 - e^{-0.15}} \approx 1{,}647$$

Thus, the person should consume 1,647 calories per day.

**69.** The model for this problem is
$$\frac{dk}{dt} + \ell k = \lambda\ell$$

where $\ell$ and $\lambda$ are constants.

Now, $f(t) = \ell$, and the integrating factor is:
$$I(t) = e^{\int \ell\, dt} = e^{\ell t}$$
Thus,

$$k(t) = \frac{1}{e^{\ell t}}\int \lambda\ell e^{\ell t}\, dt = e^{-\ell t}\left[\ell\lambda \frac{e^{\ell t}}{\ell} + C\right] = \lambda + \frac{C}{\ell} e^{-\ell t}$$

$$= \lambda + Me^{-\ell t} \quad \text{\textit{General solution}}$$

For Student $A$, $\ell = 0.8$ and $\lambda = 0.9$. Thus,

$$k(t) = 0.9 + Me^{-0.8t}.$$

Applying the initial condition $k(0) = 0.1$ yields:

$$0.1 = 0.9 + Me^0 \quad \text{or} \quad M = -0.8$$

and $k(t) = 0.9 - 0.8e^{-0.8t}$.

When $t = 6$, we have:

$$k(6) = 0.9 - 0.8e^{-0.8(6)} = 0.9 - 0.8e^{-4.8} \approx 0.8934 \text{ or } 89.34\%$$

For Student $B$, $\ell = 0.8$ and $\lambda = 0.7$. Thus,

$$k(t) = 0.7 + Me^{-0.8t}.$$

Applying the initial condition $k(0) = 0.4$ yields:

$$0.4 = 0.7 + Me^0 \quad \text{or} \quad M = -0.3$$

and $k(t) = 0.7 - 0.3e^{-0.8t}$

When $t = 6$, we have:

$$k(6) = 0.7 - 0.3e^{-0.8(6)} = 0.7 - 0.3e^{-4.8} \approx 0.6975 \text{ or } 69.75\%$$

## CHAPTER 1 REVIEW

1. Substitute $y = C\sqrt{x}$, $y' = \dfrac{C}{2\sqrt{x}}$ into the given differential equation:

$$2x\left(\frac{C}{2\sqrt{x}}\right) = C\sqrt{x}$$
$$C\sqrt{x} = C\sqrt{x}$$

Thus, $y = C\sqrt{x}$ is the general solution. (1-1)

2. Substitute $y = 1 + Ce^{-x/3}$, $y' = -\dfrac{C}{3}e^{-x/3}$ into the given differential equation:

$$3\left(-\frac{C}{3}e^{-x/3}\right) + 1 + Ce^{-x/3} = 1$$
$$-Ce^{-x/3} + 1 + Ce^{-x/3} = 1$$
$$1 = 1$$

Thus, $y = 1 + Ce^{-x/3}$ is the general solution. (1-1)

3. (B) (1-1)    4. (A) (1-1)

5.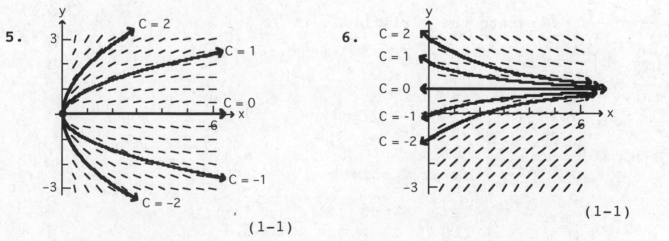

(1-1)

6.

(1-1)

**7.** $\dfrac{dy}{dt} = -k(y - 5)$, $k < 0$    (1-2)      **8.** $\dfrac{dy}{dt} = ky$, $k > 0$    (1-2)

**9.** A single person began the spread of a disease; the disease is spreading at a rate that is proportional to the product of the number of people who have the disease and the number who don't have it. (1-2)

**10.** There are 100 grams of a radioactive material at the instant a nuclear accident occurs, and the material is decaying at a rate proportional to the amount present.    (1-2)

**11.** $y' + 3x^2 y + 9 = e^x(1 - x + y) = e^x - xe^x + ye^x$
$y' + (3x^2 - e^x)y = (1 - x)e^x - 9$
The differential equation is first-order linear.    (1-2, 1-3)

**12.** $y' + 10y + 25 = 5xy^2$
The $5xy^2$ term implies that the equation is not first-order linear.
Also, the equation cannot be written in the form $f(y)y' = g(x)$. (1-2, 1-3)

**13.** $y' + 3y = xy + 2x - 6$
$\qquad y' = (x - 3)y + 2(x - 3) = (y + 2)(x - 3)$
Separation of variables form: $\dfrac{1}{y + 2}y' = x - 3$.

Also, the equation can be written in first-order linear form:
$y' + (3 - x)y = 2x - 6$    (1-2, 1-3)

**14.** $y^2 y' = 4xe^{x-y} = 4xe^x e^{-y}$
Separation of variables form: $y^2 e^y y' = 4xe^{-x}$
The equation is **not** first-order linear.    (1-2, 1-3)

**15.** $y' = -\dfrac{4y}{x}$

$\dfrac{y'}{y} = \dfrac{-4}{x}$     Separate the variables

$\displaystyle\int \dfrac{dy}{y} = \int \dfrac{-4}{x}\,dx$

$\ln|y| = -4\ln|x| + C = \ln x^{-4} + C$

$\qquad y = e^{\ln x^{-4} + C}$

$\qquad\quad = e^C e^{\ln x^{-4}}$

$\qquad y = Ax^{-4} = \dfrac{A}{x^4}$     *General solution*    (1-2)

**16.** $y' = \dfrac{-4y}{x} + x$

$y' + \dfrac{4}{x}y = x$   First-order linear equation

Integrating factor: $I(x) = e^{\int f(x)\,dx} = e^{\int (4/x)\,dx} = e^{4\ln x} = e^{\ln x^4} = x^4$
Therefore,

$y = \dfrac{1}{I(x)}\int I(x)\,g(x)\,dx, \quad g(x) = x$

$\quad = \dfrac{1}{x^4}\int x^4 x\ dx = \dfrac{1}{x^4}\int x^5\ dx = \dfrac{1}{x^4}\left[\dfrac{x^6}{6} + C\right]$

$y = \dfrac{x^2}{6} + \dfrac{C}{x^4}$   *General solution*                          (1-3)

**17.** $y' = 3x^2 y^2$

$\dfrac{y'}{y^2} = 3x^2$     Separate the variables

$\int\dfrac{dy}{y^2} = \int 3x^2\ dx$

$-\dfrac{1}{y} = x^3 + C$

$y = \dfrac{-1}{x^3 + C}$   *General solution*                          (1-2)

**18.** $y' = 2y - e^x$

$y' - 2y = -e^x$     First-order linear equation

Integrating factor: $I(x) = e^{\int (-2)\,dx} = e^{-2x}$

$y = \dfrac{1}{I(x)}\int I(x)\,g(x)\,dx, \quad g(x) = -e^x$

$\quad = \dfrac{1}{e^{-2x}}\int e^{-2x}(-e^x)\,dx = e^{2x}\int -e^{-x}\ dx = e^{2x}[e^{-x} + C]$

$y = e^x + Ce^{2x}$   *General solution*                          (1-3)

**19.** $y' = \dfrac{5}{x}y + x^6$

$y' - \dfrac{5}{x}y = x^6$     First-order linear equation

Integrating factor: $I(x) = e^{\int (-5/x)\,dx} = e^{-5\ln x} = e^{\ln x^{-5}} = x^{-5}$

$y = \dfrac{1}{I(x)}\int I(x)\,g(x)\,dx, \quad g(x) = x^6$

$\quad = \dfrac{1}{x^{-5}}\int x^{-5}x^6\ dx = x^5\int x\ dx = x^5$

$y = \dfrac{x^7}{2} + Cx^5$   *General solution*                          (1-3)

**20.** $y' = \dfrac{3 + y}{2 + x}$

$\dfrac{y'}{3 + y} = \dfrac{1}{2 + x}$  Separate the variables

$\displaystyle\int\dfrac{dy}{3 + y} = \int\dfrac{dx}{2 + x}$

$\ln|3 + y| = \ln(2 + x) + C$  (<u>Note</u>: $2 + x > 0$.)

$3 + y = e^{\ln(2+x)+C} = e^C(2 + x) = A(2 + x)$

$y = A(2 + x) - 3$  *General solution*  (1-2)

**21.** $y' = 10 - y;\ y(0) = 0$

$\dfrac{y'}{10 - y} = 1$  Separate the variables

$\displaystyle\int\dfrac{dy}{10 - y} = \int dx$

$-\ln|10 - y| = x + C$

$\ln|10 - y| = -x - C$

$10 - y = e^{-x-C}$

$\qquad = e^{-C}e^{-x} = Ae^{-x}$

$y = 10 - Ae^{-x}$. *General solution*

Applying the initial condition $y(0) = 0$, we have:

$0 = 10 - Ae^0$, $A = 10$

Thus, $y = 10 - 10e^{-x}$.  *Particular solution*  (1-2 or 1-3)

**22.** $y' + y = x;\ y(0) = 0$

Integrating factor: $I(x) = e^{\int 1\ dx} = e^x$

Thus, $y = \dfrac{1}{I(x)}\displaystyle\int I(x)g(x)\,dx,\ g(x) = x$

$\qquad = \dfrac{1}{e^x}\displaystyle\int e^x x\ dx = e^{-x}\int xe^x\ dx = e^{-x}[xe^x - e^x + C]$

$y = x - 1 + Ce^{-x}$.  *General solution*

Applying the initial condition $y(0) = 0$, we have:

$0 = 0 - 1 + Ce^0$, $C = 1$

Therefore, $y = x - 1 + e^{-x}$. *Particular solution*  (1-3)

**23.** $y' = 2ye^{-x};\ y(0) = 1$

$\dfrac{y'}{y} = 2e^{-x}$  Separate the variables

$\displaystyle\int\dfrac{dy}{y} = \int 2e^{-x}\ dx$

$\ln|y| = -2e^{-x} + C$

$y = e^{-2e^{-x}+C} = e^0 e^{-2e^{-x}}$

$y = Ae^{-2e^{-x}}$  *General solution*

Applying the initial condition $y(0) = 1$, we have:

$1 = Ae^{-2e^0} = Ae^{-2}$, $A = e^2$

Thus, $y = e^2 e^{-2e^{-x}}$.  *Particular solution* (1-2)

24. $y' = \dfrac{2x - y}{x + 4}$; $y(0) = 1$

$y' + \dfrac{1}{x + 4}y = \dfrac{2x}{x + 4}$    First-order linear equation

Integrating factor: $I(x) = e^{\int (1/(x+4))\,dx} = e^{\ln(x+4)} = x + 4$

Thus, $y = \dfrac{1}{I(x)}\int I(x)g(x)\,dx$, $g(x) = \dfrac{2x}{x + 4}$

$\qquad = \dfrac{1}{x + 4}\int (x + 4)\dfrac{2x}{x + 4}\,dx = \dfrac{1}{x + 4}\int 2x\,dx = \dfrac{1}{x + 4}[x^2 + C]$

$\qquad y = \dfrac{x^2}{x + 4} + \dfrac{C}{x + 4}$.        *General solution*

Applying the initial condition $y(0) = 1$, we have:

$1 = 0 + \dfrac{C}{4}$, $C = 4$

Therefore, $y = \dfrac{x^2}{x + 4} + \dfrac{4}{x + 4} = \dfrac{x^2 + 4}{x + 4}$.  *Particular solution*    (1-3)

25. $y' = \dfrac{x}{y + 4}$; $y(0) = 0$

$(y + 4)y' = x$    Separate the variables

$\int (y + 4)\,dy = \int x\,dx$

$\qquad \dfrac{(y + 4)^2}{2} = \dfrac{x^2}{2} + C$

$\qquad (y + 4)^2 = x^2 + 2C$

or $(y + 4)^2 = x^2 + A$.        *General solution, implicit form*

Solving for $y$ and applying the initial condition $y(0) = 0$, we have:

$y + 4 = \sqrt{x^2 + A}$

$0 + 4 = \sqrt{0 + A}$

$\quad A = 16$

Therefore, $y = \sqrt{x^2 + 16} - 4$.    *Particular solution*    (1-2)

**26.** $y' + \dfrac{2}{x}y = \ln x;\ y(1) = 2$

Integrating factor: $I(x) = e^{\int (2/x)\,dx} = e^{2\,\ln x} = e^{\ln x^2} = x^2$

Thus, $y = \dfrac{1}{I(x)}\displaystyle\int I(x)\,g(x)\,dx,\ g(x) = \ln x$

$\qquad = \dfrac{1}{x^2}\displaystyle\int x^2 \ln x\ dx$   (Integrate by parts)

$\qquad = \dfrac{1}{x^2}\left[\dfrac{1}{3}x^3 \ln x - \dfrac{x^3}{9} + C\right]$

$\quad y = \dfrac{1}{3}x \ln x - \dfrac{x}{9} + \dfrac{C}{x^2}.$   *General solution*

Applying the initial condition $y(1) = 2$, we have:

$2 = \dfrac{1}{3}(1)\ln 1 - \dfrac{1}{9} + C,\ C = \dfrac{19}{9}$

Therefore, $y = \dfrac{1}{3}x \ln x - \dfrac{x}{9} + \dfrac{19}{9x^2}.$ *Particular solution*     (1-3)

**27.** $yy' = \dfrac{x(1 + y^2)}{1 + x^2};\ y(0) = 1$

$\dfrac{yy'}{1 + y^2} = \dfrac{x}{1 + x^2}$     Separate the variables

$\displaystyle\int \dfrac{y\,dy}{1 + y^2} = \int \dfrac{x\,dx}{1 + x^2}$

$\dfrac{1}{2}\ln(1 + y^2) = \dfrac{1}{2}\ln(1 + x^2) + C$

$\ln(1 + y^2) = \ln(1 + x^2) + 2C$

$\qquad 1 + y^2 = e^{\ln(1+x^2)\ +\ 2C}$

$\qquad\qquad = e^{2C}e^{\ln(1+x^2)}$

$\qquad\qquad = A(1 + x^2)$

$\qquad\quad y^2 = A(1 + x^2) - 1.$   *General solution, implicit form*

Solving this equation for $y$ and applying the initial condition $y(0) = 1$, we have:

$\qquad y = \sqrt{A(1 + x^2) - 1}$

$\qquad 1 = \sqrt{A - 1}$

$A - 1 = 1$

$\quad A = 2$

Thus, $y = \sqrt{2(1 + x^2) - 1} = \sqrt{1 + 2x^2}.$     *Particular solution*     (1-2)

**28.** $y' + 2xy = 2e^{-x^2}$; $y(0) = 1$

Integrating factor: $I(x) = e^{\int 2x\, dx} = e^{x^2}$

Thus, $y = \dfrac{1}{I(x)}\int I(x)g(x)\,dx$, $g(x) = 2e^{-x^2}$

$$= \frac{1}{e^{x^2}}\int e^{x^2} 2e^{-x^2}\, dx = e^{-x^2}\int 2\ dx = e^{-x^2}[2x + C]$$

$$y = 2xe^{-x^2} + Ce^{-x^2}. \qquad\qquad\qquad \textit{General solution}$$

Applying the initial condition $y(0) = 1$, we have:

$1 = 2 \cdot 0 \cdot e^0 + Ce^0$, $C = 1$

Therefore, $y = 2xe^{-x^2} + e^{-x^2} = (2x + 1)e^{-x^2}$.  *Particular solution*

(1-3)

**29.** $xy' - 4y = 8$

(A) <u>Solution using an integrating factor</u>:

<u>Step 1</u>: Write the differential equation in standard form:

Multiply both sides of the equation by $\dfrac{1}{x}$

$$y' - \frac{4}{x}y = \frac{8}{x}$$

<u>Step 2</u>: Find the integrating factor:

$$f(x) = -\frac{4}{x},\ I(x) = e^{\int f(x)\,dx} = e^{\int (-4/x)\,dx}$$

$$= e^{-4\ \ln x}$$

$$= e^{\ln(x^{-4})}$$

$$= x^{-4}$$

<u>Step 3</u>: Multiply both sides of the equation by the integrating factor:

$$x^{-4}(y' - 4x^{-1}y) = (8x^{-1})x^{-4}$$
$$x^{-4}y' - 4x^{-5}y = 8x^{-5}$$
$$[x^{-4}y]' = 8x^{-5}$$

<u>Step 4</u>: Integrate both sides:

$$\int[x^{-4}y]'\,dx = \int 8x^{-5}\,dx$$
$$x^{-4}y = -2x^{-4} + C$$

<u>Step 5</u>: Solve for $y$:

$$y = -2 + Cx^4 \qquad \textit{General solution}$$

(B) <u>Solution using separation of variables</u>

$$xy' - 4y = 8$$

$$xy' = 4y + 8$$

$$\frac{1}{4y + 8} y' = \frac{1}{x}$$

$$\int \frac{1}{4y + 8} \, dy = \int \frac{1}{x} \, dx$$

$$\frac{1}{4} \ln|4y + 8| = \ln|x| + K$$

$$\ln|4y + 8| = 4 \ln|x| + L \qquad (L = 4K)$$

$$\ln|4y + 8| = \ln x^4 + L$$

$$|4y + 8| = e^{\ln x^4 + L} = e^L x^4$$

$$4y + 8 = Mx^4 \qquad\qquad (M = \pm e^L)$$

$$y = Cx^4 - 2 \qquad\qquad (C = M/4) \qquad\qquad (1\text{-}2,\ 1\text{-}3)$$

30. The equation $y' + y = x$ can be solved using an integrating factor; it cannot be solved by separating the variables.

Integrating factor:

$$f(x) = 1, \quad I(x) = e^{\int dx} = e^x$$

Multiply by $e^x$:

$$e^x y' + e^x y = xe^x$$

$$[e^x y]' = xe^x$$

Integrate:

$$e^x y = xe^x - e^x + C$$

Solve for $y$:

$$y = Ce^{-x} + x - 1 \qquad\qquad (1\text{-}3)$$

31. The equation $y' = xy^2$ can be solved by separating the variables; it cannot be solved using an integrating factor

$$y' = xy^2$$

$$\frac{1}{y^2} y' = x$$

$$\int \frac{1}{y^2} \, dy = \int x \, dx$$

$$-\frac{1}{y} = \frac{1}{2} x^2 + K$$

$$y = \frac{-1}{\frac{1}{2} x^2 + K} = \frac{-2}{x^2 + 2K}$$

$$y = \frac{-2}{x^2 + C} \qquad (C = 2K) \qquad\qquad (1\text{-}2)$$

**32.** $xy' - 5y = -10$

Step 1: Write the differential equation in standard form:
$$y' - \frac{5}{x}y = -\frac{10}{x}$$

Step 2: Find an integrating factor:
$$f(x) = -\frac{5}{x} \; ; \; I(x) = e^{\int (-5/x)\, dx} = e^{-5 \ln x}$$
$$= e^{\ln x^{-5}} = x^{-5}$$

Step 3: Multiply by the integrating factor:
$$x^{-5}(y' - 5x^{-1}y) = x^{-5}(-10x^{-1})$$
$$x^{-5}y' - 5x^{-6}y = -10x^{-6}$$
$$(x^{-5}y)' = -10x^{-6}$$

Step 4: Integrate both sides:
$$\int (x^{-5}y)'\, dx = \int -10x^{-6}\, dx$$
$$x^{-5}y = 2x^{-5} + C$$

Step 5: Solve for $y$:
$$y = 2 + Cx^5 \qquad \textit{General solution}$$

(A) Applying the initial condition $y(0) = 2$, we have:
$$2 = 2 + 0 = 2$$
Thus, $y = 2 + Cx^5$ satisfies $y(0) = 2$ for all values of $C$.

(B) Applying the initial condition $y(0) = 0$, we have:
$$0 = 2 + 0 \quad \text{or} \quad 2 = 0$$
Thus, there is **no** particular solution that satisfies $y(0) = 0$.

(C) Applying the initial condition $y(1) = 1$, we have:
$$1 = 2 + C(1) \quad \text{and} \quad C = -1$$
Thus, $y = 2 - x^5$ is the particular solution that satisfies $y(1) = 1$. $\hspace{2cm}$ (1-3)

**33.**
$$yy' = x$$
$$\int y\, dy = \int x\, dx$$
$$\frac{1}{2}y^2 = \frac{1}{2}x^2 + K$$

or $\hspace{1cm} y^2 = x^2 + C \quad (C = 2K) \qquad \textit{General solution}$

Applying the initial condition $y(0) = 4$, we have:
$$16 = 0^2 + C \quad \text{and} \quad C = 16$$
Thus, $y^2 = x^2 + 16$ and $y = \sqrt{x^2 + 16}$ $\qquad$ *Particular solution* $\qquad$ (1-2)

**34.** (A)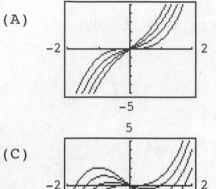

(B) The graphs are increasing and cross the $x$ axis only at $x = 0$.

(C)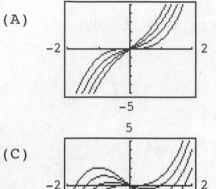

(D) The graphs have a local maximum, a local minimum, and cross the $x$ axis three times.

(1-1)

**35.** $y = M(1 - e^{-kt})$, $M > 0$, $k > 0$

(A) At $t = 2$, $4 = M(1 - e^{-2k})$ so $M = \dfrac{4}{1 - e^{-2k}}$

At $t = 5$, $7 = M(1 - e^{-5k})$ so $M = \dfrac{7}{1 - e^{-5k}}$

(B) Using a graphing utility to graph the two equations in part (A), we find that $M \approx 9.2$. (1-2)

**36.** Let $V(t)$ denote the value of the refrigerator at time $t$. Then the model for this problem is: $\dfrac{dV}{dt} = kV$; $V(0) = 500$, $V(20) = 25$

$$\frac{dV}{dt} - kV = 0$$

Integrating factor: $I(t) = e^{\int (-k)\,dt} = e^{-kt}$

Thus, $V = \dfrac{1}{I(t)}\int I(t)g(t)\,dt$, $g(t) = 0$

$$= \frac{1}{e^{-kt}}\int e^{-kt} \cdot 0 \; dt = e^{kt}\int 0 \; dt$$

$V = Ce^{kt}$.  *General solution*

Applying the conditions $V(0) = 500$ and $V(20) = 25$, we have:

$V(0) = 500 = Ce^0$, $C = 500$

Therefore, $V(t) = 500e^{kt}$.

$V(20) = 25 = 500e^{k(20)}$

$$e^{20k} = \frac{25}{500} = \frac{1}{20}$$

$$20k = \ln\left(\frac{1}{20}\right)$$

$$k = \frac{1}{20}\ln\left(\frac{1}{20}\right)$$

Therefore, $V(t) = 500e^{(1/20)\ln(1/20)t}$.

Finally, we want to calculate $V$ when $t = 5$:

$V(5) = 500e^{(1/20)\ln(1/20)5} = 500e^{(1/4)\ln(1/20)} = 500e^{-0.25\ln 20} \approx \$236.44$

(1-2 or 1-3)

**37.** The model for this problem is:

$$\frac{ds}{dt} = k(200{,}000 - s); \quad s(0) = 0, \; s(1) = 50{,}000, \; k > 0$$

(A) $\dfrac{ds}{200{,}000 - s} = k \, dt$      Separate the variables

$$\int \frac{ds}{200{,}000 - s} = \int k \, dt$$

$-\ln(200{,}000 - s) = kt + C$      $(0 < s < 200{,}000)$

$$200{,}000 - s = e^{-kt - C}$$
$$= e^{-C} e^{-kt}$$
$$= A e^{-kt}$$
$$s = 200{,}000 - A e^{-kt} \quad \textit{General solution}$$

We use the conditions $s(0) = 0$ and $s(1) = 50{,}000$ to evaluate the constants $A$ and $k$.

$$s(0) = 0 = 200{,}000 - A e^{0}$$
$$A = 200{,}000$$

Thus, $\quad s = 200{,}000 - 200{,}000 e^{-kt}$.

$$s(1) = 50{,}000 = 200{,}000 - 200{,}000 e^{-k}$$
$$200{,}000 e^{-k} = 150{,}000$$
$$e^{-k} = \frac{150{,}000}{200{,}000} = \frac{3}{4}$$
$$-k = \ln\!\left(\frac{3}{4}\right)$$

Therefore, $s = 200{,}000 - 200{,}000 e^{\ln(3/4)t}$.

Finally, we determine $t$ such that $s(t) = 150{,}000$.

$$150{,}000 = 200{,}000 - 200{,}000 e^{\ln(3/4)t}$$
$$200{,}000 e^{\ln(3/4)t} = 50{,}000$$
$$e^{\ln(3/4)t} = \frac{50{,}000}{200{,}000} = \frac{1}{4}$$
$$\ln\!\left(\frac{3}{4}\right) t = \ln\!\left(\frac{1}{4}\right)$$
$$t = \frac{\ln(1/4)}{\ln(3/4)} \approx 5 \text{ years}$$

(B) From part (A), we know that

$$S(t) = 200{,}000 - 200{,}000 e^{-kt}$$

Since we want the sales to be \$150,000 after 3 years, we have

$$150{,}000 = 200{,}000 - 200{,}000 e^{-3k}$$
$$200{,}000 e^{-3k} = 50{,}000$$
$$e^{-3k} = 0.25$$
$$-3k = \ln 0.25$$
$$k = -\frac{\ln(0.25)}{3}$$

Thus, $s(t) = 200,000 - 200,000e^{(t/3)\ln 0.25}$

Now, $s(1) = 200,000 - 200,000e^{(1/3)\ln 0.25} \approx 74,000$

Therefore, the sales in the first year should be \$74,000 to ensure that the sales after 3 years will be \$150,000.                        (1-2)

38. (A) The equilibrium price $p(t)$ at time $t$ satisfies
$S = D$; $p(0) = 75$. Thus, $100 + p + p' = 200 - p' - p$
$$2p' + 2p = 100$$
$$p' + p = 50$$

Integrating factor: $I(t) = e^{\int f(t)\,dt} = e^{\int 1\,dt} = e^t$

Thus, $p = \dfrac{1}{I(t)}\int I(t)g(t)\,dt$,   $g(t) = 50$

$$= \dfrac{1}{e^t}\int e^t \cdot 50\ dt = 50e^{-t}[e^t + C]$$

$$= 50 + 50Ce^{-t}\qquad (A = 50C)$$

$$p = 50 + Ae^{-t}. \qquad \textit{General solution}$$

(B) The equilibrium price $\bar{p}$ is given by
$$\bar{p} = \lim_{t\to\infty}(50 + 25e^{-t}) = 50 + 25\lim_{t\to\infty}e^{-t} = 50.$$

(C) Applying the initial condition $p(0) = 75$, we have:
$$75 = 50 + Ae^0 \quad \text{and} \quad A = 25$$
Therefore, $p_1 = 50 + 25e^{-t}$   *Particular solution*

Applying the initial condition $p(0) = 25$, we have
$$25 = 50 + Ae^0 \quad \text{and} \quad A = -25$$

Therefore, $p_2 = 50 - 25e^{-t}$   *Particular solution*

The graphs of $p_1$ and $p_2$ are shown at the right.

(D) For an initial price above the equilibrium price $\bar{p} = 50$, the price of the commodity decreases toward $\bar{p}$. For an initial price below the equilibrium price $\bar{p}$, the price of the commodity increases toward $\bar{p}$. (1-1, 1-2)

39. The amount in the account at any time $t$ must satisfy
$$\dfrac{dA}{dt} - 0.05A = -5000.$$

Now, $f(t) = -0.05$, and the integrating factor is:
$$I(t) = e^{\int(-0.05)dt} = e^{-0.05t}$$

Thus, $A = \dfrac{1}{e^{-0.05t}}\int -5000e^{-0.05t}\,dt = e^{0.05t}\left[-5000\,\dfrac{e^{-0.05t}}{-0.05} + C\right]$

$$A = 100,000 + Ce^{0.05t} \quad \textit{General solution}$$

Applying the initial condition $A(0) = 60,000$ yields:

$A(0) = 100,000 + Ce^0 = 60,000$

$$C = -40,000$$

Thus, the amount in the account at any time $t$ is:

$A(t) = 100,000 - 40,000e^{0.05t}$

To determine when the amount in the account is 0, we must solve $A(t) = 0$ for $t$:

$$100,000 - 40,000e^{0.05t} = 0$$

$$e^{0.05t} = \frac{100,000}{40,000} = \frac{5}{2}$$

$$t = \frac{\ln\left(\frac{5}{2}\right)}{0.05} \approx 18.326$$

Thus, the account will be depleted after 18.326 years. The total amount withdrawn from the account is: $5000(18.326) = \$91,630$      (1-3)

40. Let $r$ be the continuous compound rate of interest (expressed in decimal form). Then the amount in the account at any time $t$ is given by:

$$\frac{dA}{dt} = rA + 2,000$$

or    $\dfrac{dA}{dt} - rA = 2,000$

(Note: the method of separation of variables could also have been used to solve the equation.)

Now, $f(t) = -r$ and the integrating factor is:

$$I(t) = e^{\int -r\, dt} = e^{-rt}$$

Thus,

$$A = \frac{1}{e^{-rt}}\int 2,000e^{-rt}\, dt$$

$$= e^{rt}\left[-\frac{2,000}{r}e^{-rt} + C\right]$$

and    $A = Ce^{rt} - \dfrac{2,000}{r}$      *General solution*

Applying the initial condition $A(0) = 15,000$ yields:

$$A(0) = 15,000 = Ce^0 - \frac{2,000}{r}$$

$$C = 15,000 + \frac{2,000}{r}$$

Thus, the amount in the account at any time $t$ is:

$$A(t) = \left(15,000 + \frac{2,000}{r}\right)e^{rt} - \frac{2,000}{r}$$

Now, at $t = 10$, $A(10) = 50,000$. Therefore, we have

$$50,000 = \left(15,000 + \frac{2,000}{r}\right)e^{10r} - \frac{2,000}{r}$$

Using a graphing utility to solve this equation for $r$, we find that $r = 0.0482$. Expressed as a percentage, $r = 4.82\%$.      (1-2 or 1-3)

**41.** $\dfrac{dy}{dt} = 100 + e^{-t} - y; \quad y(0) = 0$

$\dfrac{dy}{dt} + y = 100 + e^{-t}$

Integrating factor: $I(t) = e^{\int 1\,dt} = e^{t}$

Thus, $y = \dfrac{1}{I(t)}\int I(t)g(t)\,dt, \quad g(t) = 100 + e^{-t}$

$\qquad = \dfrac{1}{e^{t}}\int e^{t}(100 + e^{-t})\,dt = e^{-t}\int(100e^{t} + 1)\,dt = e^{-t}[100e^{t} + t + C]$

$\qquad y = 100 + te^{-t} + Ce^{-t}. \quad$ *General solution*

Applying the initial condition, we have:

$0 = 100 + 0e^{0} + Ce^{0}, \quad C = -100$

Therefore, $y = 100 + te^{-t} - 100e^{-t}. \quad$ *Particular solution* $\hspace{2em}$ (1-3)

**42.** Let $p(t)$ be the amount of pollutants in the tank at time $t$. At $t = 0$, we have $p(0) = 0$.

Pollutants are entering the tank at the constant rate $2(75) = 150$ pounds per hour. The amount of water in the tank at time $t$ is:

$\qquad 100 + 75t - 50t = 100 + 25t$

The amount of pollutants in each gallon of water at time $t$ is:

$\qquad \dfrac{p(t)}{100 + 25t}$

The rate at which pollutants are leaving the tank at time $t$ is:

$\qquad 50\!\left(\dfrac{p(t)}{100 + 25t}\right) = \dfrac{2p(t)}{4 + t}$

The mathematical model for this problem is:

$\qquad \dfrac{dp}{dt} = 150 - \dfrac{2p}{4 + t}; \quad p(0) = 0$

(A) In standard form, the differential equation is:

$\qquad \dfrac{dp}{dt} + \dfrac{2}{4 + t}p = 150$

Now, $f(t) = \dfrac{2}{4 + t}$ and $I(t) = e^{\int [2/(4+t)]\,dt} = e^{2\ln(4 + t)}$

$\hspace{18em} = e^{\ln(4 + t)^{2}}$

$\hspace{18em} = (4 + t)^{2}$

Thus, $p(t) = \dfrac{1}{I(t)}\int I(t)g(t)\,dt, \quad g(t) = 150$

$\qquad = \dfrac{1}{(4 + t)^{2}}\int(4 + t)^{2}\,150\,dt$

$\qquad = \dfrac{150}{(4 + t)^{2}}\left[\dfrac{(4 + t)^{3}}{3} + K\right]$

$\qquad = 50(4 + t) + \dfrac{C}{(4 + t)^{2}} \quad (C = 150K) \quad$ *General solution*

Applying the initial condition $p(0) = 0$, we have

$$0 = 50(4) + \frac{C}{(4)^2} \quad \text{and} \quad C = -3{,}200$$

Therefore, $p(t) = 50(4 + t) - \dfrac{3{,}200}{(4 + t)^2}$    *Particular solution*

Now, at $t = 2$,

$$p(2) = 50(6) - \frac{3{,}200}{6^2} \approx 211.1$$

There are approximately 211.1 pounds of pollutants in the tank after 2 hours.

(B) To find how long it will take for the tank to contain 700 pounds of pollutants, we solve the equation

$$50(4 + t) - \frac{3{,}200}{(4 + t)^2} = 700$$

for $t$ using a graphing utility. The result is $t \approx 10.3$ hours. (1-3)

**43.** Let $p(t)$ be the bird population at time $t$. Then $\dfrac{dp}{dt} = k(p - 200)$, $k < 0$  constant.

(Note: this equation can be solved either by separating the variables or as a first order linear equation; we'll illustrate the latter.)

(A) $\dfrac{dp}{dt} = kp - 200k, \quad p(0) = 500$

In standard form, the differential equation is:

$$\frac{dp}{dt} - kp = -200k, \quad k < 0 \text{ constant}$$

Now, $f(t) = -k$ and $I(t) = e^{\int -k\,dt} = e^{-kt}$

Thus, $p(t) = \dfrac{1}{e^{-kt}} \displaystyle\int e^{-kt}(-200k)\,dt$

$$= e^{kt}[200e^{-kt} + C]$$

and    $p(t) = 200 + Ce^{kt}$   *General solution*

Applying the initial condition $p(0) = 500$, yields

$$500 = 200 + C \quad \text{and} \quad C = 300$$

Therefore, $p(t) = 200 + 300e^{kt}$

Now, at $t = -5$, $p(-5) = 1{,}000$ and

$$1000 = 200 + 300e^{-5k}$$

$$e^{-5k} = \frac{8}{3}$$

$$-5k = \ln\left(\frac{8}{3}\right)$$

$$k = \frac{\ln(8/3)}{-5}$$

Thus, the bird population at any time $t$ is:

$$p(t) = 200 + 300e^{-(t/5)\ln(8/3)}$$

The bird population 4 years from now will be:

$$p(4) = 200 + 300e^{-(4/5)\ln(8/3)} \approx 337 \text{ birds}$$

(B) From part (A), the general solution of

$$\frac{dp}{dt} = k(p - M), \; k < 0 \text{ constant}$$

is $p(t) = M + Ce^{kt}$

Applying the initial condition $p(0) = 500$, we have

$$500 = M + C \quad \text{and} \quad C = 500 - M$$

Therefore,

$$p(t) = M + (500 - M)e^{kt}$$

Now, at $t = -5$, $p(-5) = 1{,}000$ and

$$1{,}000 = M + (500 - M)e^{-5k}$$

$$e^{-5k} = \frac{1{,}000 - M}{500 - M}$$

$$-5k = \ln\left(\frac{1{,}000 - M}{500 - M}\right)$$

$$k = -\frac{1}{5}\ln\left(\frac{1{,}000 - M}{500 - M}\right)$$

Thus, the bird population at any time $t$ is

$$p(t) = M + (500 - M)e^{-(t/5)\ln[(1000 - M)/(500 - M)]}$$

$$= M + (500 - M)\left(\frac{1{,}000 - M}{500 - M}\right)^{-t/5}$$

Now, at $t = 4$, $p(4) = 400$. Using a graphing calculator to solve

$$400 = M + (500 - M)\left(\frac{1{,}000 - M}{500 - M}\right)^{-4/5}$$

we get $\quad M \approx 357$ birds $\hfill$ (1-2 or 1-3)

44. Let $p(t)$ denote the number of people who have heard the rumor at time $t$. Then the model for this problem is:

$$\frac{dp}{dt} = k(200 - p); \; p(0) = 1, \; p(2) = 10$$

$$\frac{dp}{200 - p} = k \, dt \qquad \text{Separating the variables}$$

$$\int \frac{dp}{200 - p} = \int k \, dt$$

$$-\ln(200 - p) = kt + C$$

$$\ln(200 - p) = -kt - C$$

$$200 - p = e^{-kt-C}$$

$$= e^{-C}e^{-kt}$$

$$= Ae^{-kt}$$

$$p = 200 - Ae^{-kt} \qquad \textit{General solution}$$

Apply the conditions $p(0) = 1$ and $p(2) = 10$ to evaluate the constants $A$ and $k$:  $p(0) = 1 = 200 - Ae^0$, $A = 199$

Thus, $p = 200 - 199e^{-kt}$.

$$p(2) = 10 = 200 - 199e^{-2k}$$

$$199e^{-2k} = 190$$

$$e^{-2k} = \frac{190}{199}$$

$$-2k = \ln\left(\frac{190}{199}\right)$$

$$k = \frac{-1}{2}\ln\left(\frac{190}{199}\right)$$

Therefore, $p = 200 - 199e^{(1/2)\ln(190/199)t}$.     *Particular solution*

(A) When $t = 5$,

$$p(5) = 200 - 199e^{(1/2)\ln(190/199)5}$$

$$= 200 - 199e^{(5/2)\ln(190/199)} \approx 23 \text{ people}$$

(B) Find $t$ such that $p(t) = 100$.

$$100 = 200 - 199e^{(1/2)\ln(190/199)t}$$

$$199e^{(1/2)\ln(190/199)t} = 100$$

$$e^{(1/2)\ln(190/199)t} = \frac{100}{199}$$

$$\frac{1}{2}\ln\left(\frac{190}{199}\right)t = \ln\left(\frac{100}{199}\right)$$

$$t = \frac{2\ln\left(\dfrac{100}{199}\right)}{\ln\left(\dfrac{190}{199}\right)} \approx 30 \text{ days} \qquad (1\text{-}2)$$

1.  $f(x) = \dfrac{1}{x} = x^{-1}$

$f'(x) = -x^{-2}$  (Using Power Rule)

$f''(x) = 2x^{-3}$

$f^{(3)}(x) = -6x^{-4} = -\dfrac{6}{x^4}$

3.  $f(x) = e^{-x}$

$f'(x) = -e^{-x}$

$f''(x) = e^{-x}$

$f^{(3)}(x) = -e^{-x}$

5.  $f(x) = \ln(1 + 3x)$

$f'(x) = \dfrac{3}{1 + 3x} = 3(1 + 3x)^{-1}$

$f''(x) = 3(-1)(3)(1 + 3x)^{-2} = -9(1 + 3x)^{-2}$  (Using Power Rule)

$f^{(3)}(x) = (-9)(-2)(3)(1 + 3x)^{-3} = 54(1 + 3x)^{-3}$

$f^{(4)}(x) = (54)(-3)(3)(1 + 3x)^{-4} = -486(1 + 3x)^{-4} = -\dfrac{486}{(1 + 3x)^4}$

7.  $f(x) = \sqrt{1 + x} = (1 + x)^{1/2}$

$f'(x) = \dfrac{1}{2}(1 + x)^{-1/2}$   (Using Power Rule)

$f''(x) = \left(\dfrac{1}{2}\right)\left(-\dfrac{1}{2}\right)(1 + x)^{-3/2} = -\dfrac{1}{4}(1 + x)^{-3/2}$

$f^{(3)}(x) = \left(-\dfrac{1}{4}\right)\left(-\dfrac{3}{2}\right)(1 + x)^{-5/2} = \dfrac{3}{8}(1 + x)^{-5/2}$

$f^{(4)}(x) = \left(\dfrac{3}{8}\right)\left(-\dfrac{5}{2}\right)(1 + x)^{-7/2} = -\dfrac{15}{16}(1 + x)^{-7/2}$

9.  $f(x) = e^{-x}$         $f(0) = e^{-0} = 1$

$f'(x) = -e^{-x}$         $f'(0) = -e^{-0} = -1$

$f''(x) = e^{-x}$         $f''(0) = e^{-0} = 1$

$f^{(3)}(x) = -e^{-x}$         $f^{(3)}(0) = -e^{-0} = -1$

$f^{(4)}(x) = e^{-x}$         $f^{(4)}(0) = e^{-0} = 1$

Using $\underline{2}$,

$$p_4(x) = f(0) + f'(0)x + \dfrac{f''(0)}{2!}x^2 + \dfrac{f^{(3)}(0)}{3!}x^3 + \dfrac{f^{(4)}(0)}{4!}x^4$$

Thus,

$$p_4(x) = 1 - x + \dfrac{1}{2!}x^2 - \dfrac{1}{3!}x^3 + \dfrac{1}{4!}x^4 = 1 - x + \dfrac{1}{2}x^2 - \dfrac{1}{6}x^3 + \dfrac{1}{24}x^4$$

**11.**
$$f(x) = (x + 1)^3, \qquad f(0) = 1$$
$$f'(x) = 3(x + 1)^2, \qquad f'(0) = 3$$
$$f''(x) = 6(x + 1), \qquad f''(0) = 6$$
$$f^{(3)}(x) = 6, \qquad f^{(3)}(0) = 6$$
$$f^{(4)}(x) = 0 \qquad f^{(4)}(0) = 0$$
$$p_4(x) = 1 + 3x + \frac{6}{2!}x^2 + \frac{6}{3!}x^3 = 1 + 3x + 3x^2 + x^3$$

**13.**
$$f(x) = \ln(1 + 2x) \qquad f(0) = \ln(1) = 0$$
$$f'(x) = \frac{1}{1 + 2x}(2) = \frac{2}{1 + 2x} \qquad f'(0) = \frac{2}{1 + 2 \cdot 0} = 2$$
$$f''(x) = -2(1 + 2x)^{-2}(2) = \frac{-4}{(1 + 2x)^2} \qquad f''(0) = -4$$
$$f^{(3)}(x) = 8(1 + 2x)^{-3}(2) = \frac{16}{(1 + 2x)^3} \qquad f^{(3)}(0) = 16$$

Using $\underline{2}$, $p_3(x) = f(0) + f'(0)x + \frac{f''(0)}{2!}x^2 + \frac{f^{(3)}(0)}{3!}x^3$

Thus, $p_3(x) = 0 + 2x - \frac{4}{2!}x^2 + \frac{16}{3!}x^3 = 2x - 2x^2 + \frac{8}{3}x^3$

**15.**
$$f(x) = \sqrt[3]{x + 1} = (x + 1)^{1/3} \qquad f(0) = \sqrt[3]{1} = 1$$
$$f'(x) = \frac{1}{3(x + 1)^{2/3}} \qquad f'(0) = \frac{1}{3}$$
$$f''(x) = \frac{-2}{9(x + 1)^{5/3}} \qquad f''(0) = -\frac{2}{9}$$
$$f^{(3)}(x) = \frac{10}{27(x + 1)^{8/3}} \qquad f^{(3)}(0) = \frac{10}{27}$$

$$p_3(x) = f(0) + f'(0)x + \frac{f''(0)}{2!}x^2 + \frac{f^{(3)}(0)}{3!}x^3$$

Thus, $p_3(x) = 1 + \frac{1}{3}x - \frac{2}{9 \cdot 2!}x^2 + \frac{10}{27 \cdot 3!}x^3 = 1 + \frac{1}{3}x - \frac{1}{9}x^2 + \frac{5}{81}x^3$

**17.** (A)
$$f(x) = x^4 - 1 \qquad f(0) = -1$$
$$f'(x) = 4x^3 \qquad f'(0) = 0$$
$$f''(x) = 12x^2 \qquad f''(0) = 0$$
$$f^{(3)}(x) = 24x \qquad f^{(3)}(0) = 0$$

Using $\underline{2}$, $p_3(x) = f(0) + f'(0)x + \frac{f''(0)}{2!}x^2 + \frac{f^{(3)}(0)}{3!}x^3$

Thus, $p_3(x) = -1$

Now $|p_3(x) - f(x)| = |-1 - (x^4 - 1)| = |x^4| = |x|^4$

and $|x|^4 < 0.1$ implies $|x| < (0.1)^{1/4} \approx 0.562$

Therefore, $-0.562 < x < 0.562$

(B) From part (A),

$$f^{(4)}(x) = 24 \qquad f^{(4)}(0) = 24$$

Using $\underline{2}$, $p_4(x) = f(0) + f'(0)x + \dfrac{f''(0)}{2!}x^2 + \dfrac{f^{(3)}(0)}{3!}x^3 + \dfrac{f^{(4)}(0)}{4!}x^4$

Thus, $p_4(x) = -1 + \dfrac{24}{4!}x^4 = -1 + x^4 = x^4 - 1 = f(x)$

and $|p_4(x) - f(x)| = 0 < 0.1$ for all $x$.

**19.**
$$\begin{aligned} f(x) &= e^{x-1} & f(1) &= 1 \\ f'(x) &= e^{x-1} & f'(1) &= 1 \\ f''(x) &= e^{x-1} & f''(1) &= 1 \\ f^{(3)}(x) &= e^{x-1} & f^{(3)}(1) &= 1 \\ f^{(4)}(x) &= e^{x-1} & f^{(4)}(1) &= 1 \end{aligned}$$

Thus, using $\underline{3}$,

$$p_4(x) = f(1) + f'(1)(x-1) + \dfrac{f''(1)}{2!}(x-1)^2 + \dfrac{f^{(3)}(1)}{3!}(x-1)^3 + \dfrac{f^{(4)}(1)}{4!}(x-1)^4$$

$$= 1 + (x-1) + \dfrac{1}{2!}(x-1)^2 + \dfrac{1}{3!}(x-1)^3 + \dfrac{1}{4!}(x-1)^4$$

$$= 1 + (x-1) + \dfrac{1}{2}(x-1)^2 + \dfrac{1}{6}(x-1)^3 + \dfrac{1}{24}(x-1)^4$$

**21.**
$$\begin{aligned} f(x) &= x^3 & f(1) &= 1 \\ f'(x) &= 3x^2 & f'(1) &= 3 \\ f''(x) &= 6x & f''(1) &= 6 \\ f^{(3)}(x) &= 6 & f^{(3)}(1) &= 6 \end{aligned}$$

$$p_3(x) = 1 + 3(x-1) + \dfrac{6}{2!}(x-1)^2 + \dfrac{6}{3!}(x-1)^3$$

$$= 1 + 3(x-1) + 3(x-1)^2 + (x-1)^3$$

**23.**
$$\begin{aligned} f(x) &= \ln 3x & f\left(\tfrac{1}{3}\right) &= \ln 1 = 0 \\[2mm] f'(x) &= \dfrac{3}{3x} = \dfrac{1}{x} & f'\left(\tfrac{1}{3}\right) &= 3 \\[2mm] f''(x) &= -\dfrac{1}{x^2} & f''\left(\tfrac{1}{3}\right) &= -9 \\[2mm] f^{(3)}(x) &= \dfrac{2}{x^3} & f^{(3)}\left(\tfrac{1}{3}\right) &= 54 \end{aligned}$$

Thus, using $\underline{3}$,

$$p_3(x) = f\left(\tfrac{1}{3}\right) + f'\left(\tfrac{1}{3}\right)\left(x - \tfrac{1}{3}\right) + \dfrac{f''\left(\tfrac{1}{3}\right)}{2!}\left(x - \tfrac{1}{3}\right)^2 + \dfrac{f^{(3)}\left(\tfrac{1}{3}\right)}{3!}\left(x - \tfrac{1}{3}\right)^3$$

$$= 0 + 3\left(x - \tfrac{1}{3}\right) - \dfrac{9}{2!}\left(x - \tfrac{1}{3}\right)^2 + \dfrac{54}{3!}\left(x - \tfrac{1}{3}\right)^3$$

$$= 3\left(x - \tfrac{1}{3}\right) - \dfrac{9}{2}\left(x - \tfrac{1}{3}\right)^2 + 9\left(x - \tfrac{1}{3}\right)^3$$

**25.**

$$f(x) = e^{-2x} \qquad\qquad f(0) = 1$$
$$f'(x) = -2e^{-2x} \qquad\qquad f'(0) = -2$$
$$f''(x) = 4e^{-2x} \qquad\qquad f''(0) = 4$$
$$f^{(3)}(x) = -8e^{-2x} \qquad\qquad f^{(3)}(0) = -8$$

Using $\underline{2}$, $p_3(x) = f(0) + f'(0)x + \dfrac{f''(0)}{2!}x^2 + \dfrac{f^{(3)}(0)}{3!}x^3$

Thus, $p_3(x) = 1 - 2x + \dfrac{4}{2!}x^2 - \dfrac{8}{3!}x^3 = 1 - 2x + 2x^2 - \dfrac{4}{3}x^3.$

Now, $e^{-0.5} = e^{-2(0.25)} = f(0.25) \approx p_3(0.25)$

$$= 1 - 2(0.25) + 2(0.25)^2 - \dfrac{4}{3}(0.25)^3 = 0.60416667.$$

**27.**

$$f(x) = \sqrt{x + 16} = (x + 16)^{1/2} \qquad\qquad f(0) = 4$$
$$f'(x) = \frac{1}{2}(x + 16)^{-1/2} = \frac{1}{2\sqrt{x + 16}} \qquad\qquad f'(0) = \frac{1}{8}$$
$$f''(x) = \frac{-1}{4}(x + 16)^{-3/2} = \frac{-1}{4(x + 16)^{3/2}} \qquad\qquad f''(0) = \frac{-1}{256}$$

$p_2(x) = f(0) + f'(0) + \dfrac{f''(0)}{2!}x^2 = 4 + \dfrac{1}{8}x - \dfrac{1}{256 \cdot 2!}x^2 = 4 + \dfrac{1}{8}x - \dfrac{1}{512}x^2.$

Now, $\sqrt{17} = \sqrt{1 + 16} = f(1) \approx p_2(1) = 4 + \dfrac{1}{8}(1) - \dfrac{1}{512}(1)^2 = 4.1230469.$

**29.**

$$f(x) = \sqrt{x} = x^{1/2} \qquad\qquad f(1) = 1$$
$$f'(x) = \frac{1}{2}x^{-1/2} \qquad\qquad f'(1) = \frac{1}{2}$$
$$f''(x) = -\frac{1}{4}x^{-3/2} \qquad\qquad f''(1) = -\frac{1}{4}$$
$$f^{(3)}(x) = \frac{3}{8}x^{-5/2} \qquad\qquad f^{(3)}(1) = \frac{3}{8}$$
$$f^{(4)}(x) = -\frac{15}{16}x^{-7/2} \qquad\qquad f^{(4)}(1) = -\frac{15}{16}$$

Thus,

$p_4(x) = f(1) + f'(1)(x - 1) + \dfrac{f''(1)}{2!}(x - 1)^2 + \dfrac{f^{(3)}(1)}{3!}(x - 1)^3 + \dfrac{f^{(4)}(1)}{4!}(x - 1)^4$

$\qquad = 1 + \dfrac{1}{2}(x - 1) - \dfrac{1}{4 \cdot 2!}(x - 1)^2 + \dfrac{3}{8 \cdot 3!}(x - 1)^3 - \dfrac{15}{16 \cdot 4!}(x - 1)^4$

$\qquad = 1 + \dfrac{1}{2}(x - 1) - \dfrac{1}{8}(x - 1)^2 + \dfrac{1}{16}(x - 1)^3 - \dfrac{5}{128}(x - 1)^4.$

Now,

$\sqrt{1.2} = f(1.2) \approx p_4(1.2)$

$\qquad = 1 + \dfrac{1}{2}(1.2 - 1) - \dfrac{1}{8}(1.2 - 1)^2 + \dfrac{1}{16}(1.2 - 1)^3 - \dfrac{5}{128}(1.2 - 1)^4$

$\qquad = 1 + \dfrac{1}{2}(0.2) - \dfrac{1}{8}(0.2)^2 + \dfrac{1}{16}(0.2)^3 - \dfrac{5}{128}(0.2)^4 = 1.0954375.$

**31.** 
$$f(x) = \frac{1}{4-x} = (4-x)^{-1}$$
$$f'(x) = -1(4-x)^{-2}(-1) = (4-x)^{-2}$$
$$f''(x) = -2(4-x)^{-3}(-1) = 1\cdot2(4-x)^{-3}$$
$$f^{(3)}(x) = 2(-3)(4-x)^{-4}(-1) = 1\cdot2\cdot3(1-x)^{-4}$$
$$f^{(4)}(x) = 2\cdot3(-4)(4-x)^{-5}(-1) = 1\cdot2\cdot3\cdot4(1-x)^{-5}$$
$$\vdots$$
$$f^{(n)}(x) = n!(4-x)^{-(n+1)}$$

**33.** 
$$f(x) = e^{3x}$$
$$f'(x) = e^{3x}(3) = 3e^{3x}$$
$$f''(x) = 3e^{3x}(3) = 9e^{3x} = 3^2e^{3x}$$
$$f^{(3)}(x) = 9e^{3x}(3) = 27e^{3x} = 3^3e^{3x}$$
$$f^{(4)}(x) = 27e^{3x}(3) = 81e^{3x} = 3^4e^{3x}$$
$$\vdots$$
$$f^{(n)}(x) = 3^ne^{3x}$$

**35.** 
$$f(x) = \ln(6-x)$$
$$f'(x) = \frac{1}{6-x}(-1) = -\frac{1}{6-x} = -(6-x)^{-1}$$
$$f''(x) = (6-x)^{-2}(-1) = -(6-x)^{-2}$$
$$f^{(3)}(x) = 2(6-x)^{-3}(-1) = -1\cdot2(6-x)^{-3}$$
$$f^{(4)}(x) = 1\cdot2\cdot3(6-x)^{-4}(-1) = -1\cdot2\cdot3(6-x)^{-4}$$
$$\vdots$$
$$f^{(n)}(x) = -(n-1)!(6-x)^{-n}$$

**37.** From Problem 31,
$$f(0) = \frac{1}{4}, \ f'(0) = \frac{1}{4^2}, \ f''(0) = \frac{2!}{4^3}, \ f^{(3)}(0) = \frac{3!}{4^4}, \ \ldots, \ f^{(n)}(0) = \frac{n!}{4^{n+1}}$$
Thus,
$$p_n(x) = \frac{1}{4} + \frac{1}{4^2}x + \frac{1}{4^3}x^2 + \frac{1}{4^4}x^3 + \ldots + \frac{1}{4^{n+1}}x^n.$$

**39.** From Problem 33,
$$f(0) = e^0 = 1, \ f'(0) = 3e^0 = 3, \ f''(0) = 3^2e^0 = 3^2,$$
$$f^{(3)}(0) = 3^3e^0 = 3^3, \ \ldots, \ f^{(n)}(0) = 3^ne^0 = 3^n$$
Thus,
$$p_n(x) = 1 + 3x + \frac{3^2}{2!}x^2 + \frac{3^3}{3!}x^3 + \ldots + \frac{3^n}{n!}x^n.$$

**41.** From Problem 35,

$$f(0) = \ln 6, \quad f'(0) = -\frac{1}{6}, \quad f''(0) = -\frac{1}{6^2}, \quad f^{(3)}(0) = -\frac{2!}{6^3}, \quad \dots, \quad f^{(n)}(0) = -\frac{(n-1)!}{6^n}$$

Thus,

$$p_n(x) = \ln 6 - \frac{1}{6}x - \frac{1}{2 \cdot 6^2}x^2 - \frac{1}{3 \cdot 6^3}x^3 - \dots - \frac{1}{n \cdot 6^n}x^n.$$

**43.** Step 1.              Step 2.              Step 3.

$$f(x) = \frac{1}{x} \qquad f(-1) = -1 \qquad a_0 = f(-1) = -1$$

$$f'(x) = -\frac{1}{x^2} \qquad f'(-1) = -1 \qquad a_1 = f'(-1) = -1$$

$$f''(x) = \frac{2}{x^3} \qquad f''(-1) = -2! \qquad a_2 = \frac{f''(-1)}{2!} = -\frac{2!}{2!} = -1$$

$$f^{(3)}(x) = -\frac{3!}{x^4} \qquad f^{(3)}(-1) = -3! \qquad a_3 = \frac{f^{(3)}(-1)}{3!} = -\frac{3!}{3!} = -1$$

$$f^{(4)}(x) = \frac{4!}{x^5} \qquad f^{(4)}(-1) = -4! \qquad a_4 = \frac{f^{(4)}(-1)}{4!} = -\frac{4!}{4!} = -1$$

$$\vdots \qquad\qquad \vdots \qquad\qquad \vdots$$

$$f^{(n)} = \frac{(-1)^n n!}{x^n} \qquad f^{(n)}(-1) = -n! \qquad a_n = \frac{f^{(n)}(-1)}{n!} = -\frac{n!}{n!} = -1$$

Step 4. The $n$th degree Taylor polynomial is:

$$p_n(x) = -1 - (x+1) - (x+1)^2 - (x+1)^3 - (x+1)^4 - \dots - (x+1)^n.$$

**45.** Step 1.              Step 2.              Step 3.

$$f(x) = \ln x \qquad f(1) = 0 \qquad a_0 = f(1) = 0$$

$$f'(x) = \frac{1}{x} \qquad f'(1) = 1 \qquad a_1 = f'(1) = 1$$

$$f''(x) = -\frac{1}{x^2} \qquad f''(1) = -1 \qquad a_2 = \frac{f''(1)}{2!} = -\frac{1}{2!} = -\frac{1}{2}$$

$$f^{(3)}(x) = \frac{2}{x^3} \qquad f^{(3)}(1) = 2 \qquad a_3 = \frac{f^{(3)}(1)}{3!} = \frac{2}{3!} = \frac{1}{3}$$

$$f^{(4)}(x) = -\frac{2 \cdot 3}{x^4} \qquad f^{(4)}(1) = -3! \qquad a_4 = \frac{f^{(4)}(1)}{4!} = -\frac{3!}{4!} = -\frac{1}{4}$$

$$\vdots \qquad\qquad \vdots \qquad\qquad \vdots$$

$$f^{(n)}(x) = \qquad f^{(n)}(1) = \qquad a_n =$$

$$\frac{(-1)^{n+1}(n-1)!}{x^n} \qquad (-1)^{n+1}(n-1)! \qquad \frac{f^{(n)}(1)}{n!} = \frac{(-1)^{n+1}(n-1)!}{n!} = \frac{(-1)^{n+1}}{n}$$

Step 4. The $n$th degree Taylor polynomial is:

$$p_n(x) = (x-1) - \frac{1}{2}(x-1)^2 + \frac{1}{3}(x-1)^3 - \frac{1}{4}(x-1)^4 + \dots + \frac{(-1)^{n+1}}{n}(x-1)^n.$$

**47.** Step 1.

$f(x) = e^{-x}$

$f'(x) = -e^{-x}$

$f''(x) = e^{-x}$

$f^{(3)}(x) = -e^{-x}$

$f^{(4)}(x) = e^{-x}$

$\vdots$

$f^{(n)}(x) = (-1)^n e^{-x}$

Step 2.

$f(2) = \dfrac{1}{e^2}$

$f'(2) = -\dfrac{1}{e^2}$

$f''(2) = \dfrac{1}{e^2}$

$f^{(3)}(2) = -\dfrac{1}{e^2}$

$f^{(4)}(2) = \dfrac{1}{e^2}$

$\vdots$

$f^{(n)}(2) = \dfrac{(-1)^n}{e^2}$

Step 3.

$a_0 = f(2) = \dfrac{1}{e^2}$

$a_1 = f'(2) = -\dfrac{1}{e^2}$

$a_2 = \dfrac{f''(2)}{2!} = \dfrac{1}{2!\,e^2}$

$a_3 = \dfrac{f^{(3)}(2)}{3!} = \dfrac{-1}{3!\,e^2}$

$a_4 = \dfrac{f^{(4)}(2)}{4!} = \dfrac{1}{4!\,e^2}$

$\vdots$

$a_n = \dfrac{f^{(n)}(2)}{n!} = \dfrac{(-1)^n}{n!\,e^2}$

Step 4. The $n$th degree Taylor polynomial is:

$$p_n(x) = \frac{1}{e^2} - \frac{1}{e^2}(x-2) + \frac{1}{2!\,e^2}(x-2)^2 - \frac{1}{3!\,e^2}(x-2)^3$$

$$+ \frac{1}{4!\,e^2}(x-2)^4 - \dots + \frac{(-1)^n}{n!\,e^2}(x-2)^n$$

**49.** Step 1.

$f(x) = x \ln x$

$f'(x) = 1 + \ln x$

$f''(x) = \dfrac{1}{x}$

$f^{(3)}(x) = -\dfrac{1}{x^2}$

$f^{(4)}(x) = \dfrac{2}{x^3}$

$\vdots$

$f^{(n)}(x) = \dfrac{(-1)^n(n-2)!}{x^{n-1}}$

Step 2.

$f(3) = 3 \ln 3$

$f'(3) = 1 + \ln 3$

$f''(3) = \dfrac{1}{3}$

$f^{(3)}(3) = -\dfrac{1}{3^2}$

$f^{(4)}(3) = \dfrac{2}{3^3}$

$\vdots$

$f^{(n)}(3) = \dfrac{(-1)^n(n-2)!}{3^{n-1}}$

Step 3.

$a_0 = 3 \ln 3$

$a_1 = 1 + \ln 3$

$a_2 = \dfrac{f''(3)}{2!} = \dfrac{1}{3 \cdot 2}$

$a_3 = \dfrac{f^{(3)}(3)}{3!} = \dfrac{-1}{3^2 \cdot 2 \cdot 3}$

$a_4 = \dfrac{f^{(4)}(3)}{4!} = \dfrac{1}{3^3 \cdot 3 \cdot 4}$

$\vdots$

$a_n = \dfrac{f^{(n)}(3)}{n!} = \dfrac{(-1)^n}{3^{n-1}(n-1)n}$

Step 4. The $n$th degree Taylor polynomial is:

$$p_n(x) = 3 \ln 3 + (1 + \ln 3)(x-3) + \frac{1}{3 \cdot 2}(x-3)^2 - \frac{1}{3^2 \cdot 2 \cdot 3}(x-3)^3$$

$$+ \frac{1}{3^3 \cdot 3 \cdot 4}(x-3)^4 - \dots + \frac{(-1)^n}{3^{n-1}(n-1)n}(x-3)^n.$$

**51.** $f(x) = x^4 + 5x^2 + 1$

(A) Third-degree Taylor polynomial $p_3(x)$ for $f$ at 0 is:

$$p_3(x) = f(0) + \frac{f'(0)}{1!}x + \frac{f''(0)}{2!}x^2 + \frac{f^{(3)}(0)}{3!}x^3$$

$$f(0) = 1$$
$$f'(x) = 4x^3 + 10x; \quad f'(0) = 0$$
$$f''(x) = 12x^2 + 10; \quad f''(0) = 10$$
$$f^{(3)}(x) = 24x \quad\quad; \quad f^{(3)}(0) = 0$$

Thus,

$$p_3(x) = 1 + \frac{10}{2!}x^2 = 1 + 5x^2 = 5x^2 + 1$$

(B) The degree of the polynomial is 2.

**53.** $f(x) = x^6 + 2x^3 + 1$

$$f(0) = 1$$
$$f'(x) = 6x^5 + 6x^2 \quad\quad; \quad f'(0) = 0$$
$$f''(x) = 30x^4 + 12x \quad\quad; \quad f''(0) = 0$$
$$f^{(3)}(x) = 120x^3 + 12 \quad\quad; \quad f^{(3)}(0) = 12$$
$$f^{(4)}(x) = 360x^2 \quad\quad; \quad f^{(4)}(0) = 0$$
$$f^{(5)}(x) = 720x \quad\quad; \quad f^{(5)}(0) = 0$$
$$f^{(6)}(x) = 720 \quad\quad; \quad f^{(6)}(0) = 720$$
$$f^{(n)}(x) = 0 \quad \text{for} \quad n \geq 7.$$

Thus, for $n = 0$, 3 and 6, the $n$th-degree Taylor polynomial for $f$ at 0 has degree $n$.

**55.**
$$f(x) = \ln(1 + x) \quad\quad\quad f(0) = 0$$
$$f'(x) = \frac{1}{1 + x} \quad\quad\quad f'(0) = 1$$
$$f''(x) = \frac{-1}{(1 + x)^2} \quad\quad\quad f''(0) = -1$$
$$f^{(3)}(x) = \frac{2}{(1 + x)^3} \quad\quad\quad f^{(3)}(0) = 2$$

Thus,

$$p_1(x) = x, \; p_2(x) = x - \frac{x^2}{2}, \; p_3(x) = x - \frac{x^2}{2} + \frac{x^3}{3}.$$

| $x$ | $p_1(x)$ | $p_2(x)$ | $p_3(x)$ | $f(x)$ |
|---|---|---|---|---|
| −0.2 | −0.2 | −0.22 | −0.222667 | −0.223144 |
| −0.1 | −0.1 | −0.105 | −0.105333 | −0.105361 |
| 0 | 0 | 0 | 0 | 0 |
| 0.1 | 0.1 | 0.095 | 0.095333 | 0.09531 |
| 0.2 | 0.2 | 0.18 | 0.182667 | 0.182322 |

| $x$ | $\left\|p_1(x) - f(x)\right\|$ | $\left\|p_2(x) - f(x)\right\|$ | $\left\|p_3(x) - f(x)\right\|$ |
|---|---|---|---|
| -0.2 | 0.023144 | 0.003144 | 0.000477 |
| -0.1 | 0.005361 | 0.000361 | 0.000028 |
| 0 | 0 | 0 | 0 |
| 0.1 | 0.00469 | 0.00031 | 0.000023 |
| 0.2 | 0.017678 | 0.002322 | 0.000345 |

**57.** From Problem 55, the first three Taylor polynomials for $f(x) = \ln(1 + x)$ are:

$$p_1(x) = x, \qquad p_2(x) = x - \frac{x^2}{2},$$

$$p_3(x) = x - \frac{x^2}{2} + \frac{x^3}{3}$$

The graphs of $f$, $p_1$, $p_2$, and $p_3$ are shown at the right.

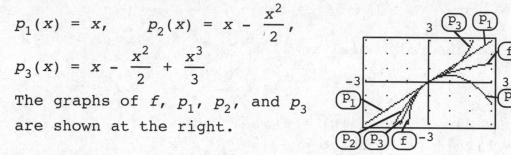

**59.** From the text we know that

$$p_3(x) = 1 + x + \frac{1}{2}x^2 + \frac{1}{6}x^3$$

Using a graphing utility, we find that

$$\left|p_3(x) - e^x\right| < 0.1 \quad \text{for} \quad -1.323 < x < 1.168$$

**61.**

$$\begin{array}{ll} f(x) = e^x & f(a) = e^a \\ f'(x) = e^x & f'(a) = e^a \\ f''(x) = e^x & f''(a) = e^a \\ f^{(3)}(x) = e^x & f^{(3)}(a) = e^a \\ \vdots & \vdots \\ f^{(n)}(x) = e^x & f^{(n)}(a) = e^a \end{array}$$

Thus,

$$p_n(x) = f(a) + f'(a)(x - a) + \frac{f''(a)}{2!}(x - a)^2 + \frac{f^{(3)}(a)}{3!}(x - a)^3$$

$$+ \cdots + \frac{f^{(n)}(a)}{n!}(x - a)^n$$

$$= e^a + e^a(x - a) + \frac{e^a}{2!}(x - a)^2 + \frac{e^a}{3!}(x - a)^3 + \cdots + \frac{e^a}{n!}(x - a)^n$$

$$= e^a\left[1 + (x - a) + \frac{1}{2!}(x - a)^2 + \frac{1}{3!}(x - a)^3 + \cdots + \frac{1}{n!}(x - a)^n\right]$$

$$= e^a \sum_{k=0}^{n} \frac{1}{k!}(x - a)^k$$

**63.**

$$f(x) = \frac{1}{x} \qquad\qquad\qquad f(a) = \frac{1}{a}$$

$$f'(x) = \frac{-1}{x^2} \qquad\qquad\qquad f'(a) = \frac{-1}{a^2}$$

$$f''(x) = \frac{2}{x^3} \qquad\qquad\qquad f''(a) = \frac{2}{a^3}$$

$$f^{(3)}(x) = \frac{-2 \cdot 3}{x^4} = \frac{-3!}{x^4} \qquad\qquad f^{(3)}(a) = \frac{-3!}{a^4}$$

$$\vdots \qquad\qquad\qquad\qquad\qquad \vdots$$

$$f^{(n)}(x) = \frac{(-1)^n n!}{x^{n+1}} \qquad\qquad f^{(n)}(a) = \frac{(-1)^n n!}{a^{n+1}}$$

Thus,

$$p_n(x) = f(a) + f'(a)(x - a) + \frac{f''(a)}{2!}(x - a)^2 + \frac{f^{(3)}(a)}{3!}(x - a)^3$$
$$+ \dots + \frac{f^{(n)}(a)}{n!}(x - a)^n$$

$$= \frac{1}{a} - \frac{1}{a^2}(x - a) + \frac{2}{a^3} \cdot \frac{1}{2!}(x - a)^2 - \frac{3!}{a^4} \cdot \frac{1}{3!}(x - a)^3$$
$$+ \dots + \frac{(-1)^n n!}{a^{n+1}} \cdot \frac{1}{n!}(x - a)^n$$

$$= \frac{1}{a} - \frac{1}{a^2}(x - a) + \frac{1}{a^3}(x - a)^2 - \frac{1}{a^4}(x - a)^3 + \dots + \frac{(-1)^n}{a^{n+1}}(x - a)^n$$

$$= \sum_{k=0}^{n} \frac{(-1)^k}{a^{k+1}}(x - a)^k$$

**65.** Let $f(x)$ be a polynomial of degree $k$, $k \geq 0$. Then

$$f(x) = a_0 + a_1 x + a_2 x^2 + a_3 x^3 + \dots + a_k x^k \qquad f(0) = a_0$$

$$f'(x) = a_1 + 2a_2 x + 3a_3 x^2 + \dots + k a_k x^{k-1} \qquad f'(0) = a_1$$

$$f''(x) = 2a_2 + 6a_3 x + \dots + k(k - 1)a_k x^{k-2} \qquad f''(0) = 2a_2 + 2!a_2$$

$$f^{(3)}(x) = 6a_3 + \dots + k(k - 1)(k - 2)a_k x^{k-3} \qquad f^{(3)}(0) = 6a_3 = 3!a_3$$

$$\vdots$$

In general,

$$f^{(m)}(0) = \begin{cases} m!\, a_m & \text{for } m = 0, 1, 2, \dots k \\ 0 & \text{for } m > k \end{cases}$$

Since $p_n(x) = f(0) + f'(0)x + \dfrac{f''(0)}{2!}x^2 + \dfrac{f^{(3)}(0)}{3!}x^3 + \dots + \dfrac{f^{(n)}(0)}{n!}x^n$

it follows that $p_n(x) = f(x)$ for all $n \geq k$.

**67.** $f(x) = e^x$

$f(0) = 1$

$f'(x) = e^x$ ; $f'(0) = 1$

$f''(x) = e^x$ ; $f''(0) = 1$

$\vdots$

$f^{(k)}(x) = e^x$ ; $f^{(k)}(0) = 1$

Therefore,

$$p_{10}(x) = 1 + \frac{1}{1!}x + \frac{1}{2!}x^2 + \dots + \frac{1}{10!}x^{10}$$

$$p_{11}(x) = 1 + \frac{1}{1!}x + \frac{1}{2!}x^2 + \dots + \frac{1}{10!}x^{10} + \frac{1}{11!}x^{11}$$

For $x > 0$, $e^x > p_{11}(x)$ and hence

$$e^x - p_{10}(x) > p_{11}(x) - p_{10}(x) = \frac{1}{11!}x^{11}$$

Take $x = 2(11!)^{1/11}$, then

$$e^x - p_{10}(x) > \frac{1}{11!}(2(11!)^{1/11})^{11} = 2^{11} = 2048.$$

So, there exist values of $x$ for which

$$|p_{10}(x) - e^x| = |e^x - p_{10}(x)| \geq 100.$$

**69.** $D(x) = \dfrac{1}{10}\sqrt{10,000 - x^2} = \dfrac{1}{10}(10,000 - x^2)^{1/2}$ $\qquad D(0) = \dfrac{1}{10}(100) = 10$

$D'(x) = \dfrac{1}{10}\left(\dfrac{1}{2}\right)(10,000 - x^2)^{-1/2}(-2x)$

$\qquad = \dfrac{-x}{10}(10,000 - x^2)^{-1/2}$ $\qquad\qquad\qquad\qquad D'(0) = 0$

$D''(x) = -\dfrac{1}{10}(10,000 - x^2)^{-1/2} + \dfrac{x}{20}(10,000 - x^2)^{-3/2}(-2x)$

$\qquad = -\dfrac{1}{10}(10,000 - x^2)^{-1/2} - \dfrac{x^2}{10}(10,000 - x^2)^{-3/2}$ $\quad D''(0) = \dfrac{-1}{10(100)}$

$\qquad = \dfrac{-1}{10(10,000 - x^2)^{1/2}} - \dfrac{x^2}{10(10,000 - x^2)^{3/2}}$ $\qquad\qquad = \dfrac{-1}{1000}$

Thus,

$$p_2(x) = D(0) + D'(0)x + \frac{D''(0)}{2!}x^2 = 10 - \frac{1}{2000}x^2 = 10 - 0.0005x^2.$$

Average price on $[0, 30]$:

$$\frac{1}{30 - 0}\int_0^{30} D(x)\,dx \approx \frac{1}{30}\int_0^{30}(10 - 0.0005x^2)\,dx$$

$$= \frac{1}{30}\left[\left(10x - \frac{0.0005x^3}{3}\right)\Big|_0^{30}\right] = \frac{1}{30}\left[10(30) - \frac{0.0005(30)^3}{3}\right]$$

$$= \frac{1}{30}[300 - 4.5] = \$9.85$$

**71.** $\quad D(x) = \dfrac{1}{10}\sqrt{10,000 - x^2}$

$\qquad\quad = \dfrac{1}{10}(10,000 - x^2)^{1/2}$

$\quad D'(x) = \dfrac{1}{10}\left(\dfrac{1}{2}\right)(10,000 - x^2)^{-1/2}(-2x)$

$\qquad\quad = \dfrac{1}{10}(-x)(10,000 - x^2)^{-1/2}$

$\qquad\quad = \dfrac{-x}{10(10,000 - x^2)^{1/2}}$

$\quad D''(x) = -\dfrac{1}{10}(10,000 - x^2)^{-1/2}$

$\qquad\qquad + \dfrac{1}{20}x(10,000 - x^2)^{-3/2}(-2x)$

$\qquad\quad = \dfrac{-1}{10(10,000 - x^2)^{1/2}} - \dfrac{x^2}{10(10,000 - x^2)^{3/2}}$

$D(60) = \dfrac{1}{10}\sqrt{10,000 - (60)^2}$

$\qquad\quad = \dfrac{1}{10}\sqrt{6400} = 8$

$D'(60) = \dfrac{-60}{10\cdot 80} = \dfrac{-3}{40}$

$D''(60) = \dfrac{-1}{10\cdot 80} - \dfrac{3600}{10(80)^3}$

$\qquad\quad = \dfrac{-1}{512}$

Thus,

$p_2(x) = D(60) + D'(60)(x - 60) + \dfrac{D''(60)}{2!}(x - 60)^2$

$\qquad\quad = 8 - \dfrac{3}{40}(x - 60) - \dfrac{1}{1024}(x - 60)^2.$

Average price on [60, 80]:

$\dfrac{1}{80 - 60}\displaystyle\int_{60}^{80} D(x)\,dx \approx \dfrac{1}{20}\int_{60}^{80}\left[8 - \dfrac{3}{40}(x - 60) - \dfrac{1}{1024}(x - 60)^2\right]dx$

$\qquad\qquad\qquad = \dfrac{1}{20}\left[8x - \dfrac{3}{40}\cdot\dfrac{(x - 60)^2}{2} - \dfrac{1}{1024}\cdot\dfrac{(x - 60)^3}{3}\right]\Big|_{60}^{80}$

$\qquad\qquad\qquad = \dfrac{1}{20}\left[640 - \dfrac{3}{40}\cdot\dfrac{400}{2} - \dfrac{1}{1024}\cdot\dfrac{8000}{3}\right] - \dfrac{1}{20}(480)$

$\qquad\qquad\qquad = \$7.12$

**73.** $\quad R(t) = 2 + 8e^{-0.1t^2}$

$\qquad R'(t) = -1.6te^{-0.1t^2}$

$\qquad R''(t) = -1.6e^{-0.1t^2} + 0.32t^2e^{-0.1t^2}$

$\qquad R(0) = 10$

$\qquad R'(0) = 0$

$\qquad R''(0) = -1.6$

Thus,

$p_2(t) = R(0) + R'(0)t + \dfrac{R''(0)}{2!}t^2$

$\qquad\quad = 10 - 0.8t^2$

Now, the total production during the first two years is

$P = \displaystyle\int_0^2 R(t)\,dt \approx \int_0^2 (10 - 0.8t^2)\,dt$

$\quad = \left(10t - \dfrac{0.8t^3}{3}\right)\Big|_0^2$

$\quad \approx 17.87$ or \$18 million

**75.** Let $A'(t) = F(t) = \dfrac{-75}{t^2 + 25}$ $\qquad\qquad F(0) = -3$

$\qquad F'(t) = \dfrac{75(2t)}{(t^2 + 25)^2} = \dfrac{150t}{(t^2 + 25)^2}$ $\qquad F'(0) = 0$

$\qquad F''(t) = \dfrac{(t + 25)^2(150) - 150t(2)(t^2 + 25)(2t)}{(t + 25)^4}$ $\qquad F''(0) = \dfrac{6}{25}$

$\qquad\qquad = \dfrac{150(t^2 + 25) - 600t^2}{(t^2 + 25)^3}$

Thus, $p_2(t) = F(0) + F'(0)t + \dfrac{F''(0)}{2!}t^2 = -3 + \dfrac{3}{25}t^2.$

Therefore,

$A(t) = \int A'(t)\,dt \approx \int p_2(t)\,dt = \int\left[-3 + \dfrac{3}{25}t^2\right]dt = -3t + \dfrac{1}{25}t^3 + C.$

Now, $A(0) = 12.$  Therefore, $C = 12$ and $A(t) \approx -3t + \dfrac{1}{25}t^3 + 12$;

thus, $A(2) \approx -6 + \dfrac{8}{25} + 12 = 6.32$ square centimeters.

**77.** $P(x) = 20\sqrt{3x^2 + 25} - 80$ $\qquad\qquad P(5) = 20(3\cdot 5^2 + 25)^{1/2} - 80$

$\qquad\quad = 20(3x^2 + 25)^{1/2} - 80$ $\qquad\qquad\qquad = 120$

$P'(x) = 20\left(\dfrac{1}{2}\right)(3x^2 + 25)^{-1/2}(6x)$ $\qquad P'(5) = \dfrac{60(5)}{10} = 30$

$\qquad\quad = 60x(3x^2 + 25)^{-1/2} = \dfrac{60x}{(3x^2 + 25)^{1/2}}$

$P''(x) = 60(3x^2 + 25)^{-1/2}$ $\qquad\qquad P''(5) = \dfrac{60}{10} - \dfrac{180(5)^2}{1000}$

$\qquad\quad + 60x\left(\dfrac{1}{2}\right)(3x^2 + 25)^{-3/2}(6x)$ $\qquad\qquad = 6 - \dfrac{9}{2} = \dfrac{3}{2}$

$\qquad = \dfrac{60}{(3x^2 + 25)^{1/2}} - \dfrac{180x^2}{(3x^2 + 25)^{3/2}}$

Thus, $p_2(x) = P(5) + P'(5)(x - 5) + \dfrac{P''(5)}{2!}(x - 5)^2$

$\qquad\qquad = 120 + 30(x - 5) + \dfrac{3}{4}(x - 5)^2$

Average pollution:

$\dfrac{1}{10 - 0}\displaystyle\int_0^{10} P(x)\,dx \approx \dfrac{1}{10}\int_0^{10}\left[120 + 30(x - 5) + \dfrac{3}{4}(x - 5)^2\right]dx$

$\qquad\qquad = \dfrac{1}{10}\left[120x + 15(x - 5)^2 + \dfrac{1}{4}(x - 5)^3\right]\Big|_0^{10}$

$\qquad\qquad = \dfrac{1}{10}\left[1200 + 375 + \dfrac{125}{4}\right] - \dfrac{1}{10}\left[375 - \dfrac{125}{4}\right]$

$\qquad\qquad = 120 + 6.25 = 126.25$ parts per million

**79.** Let $N'(t) = F(t) = 6e^{-0.01t^2}$                                                  $F(0) = 6$

$\quad F'(t) = 6e^{-0.01t^2}(-0.02t) = -0.12te^{-0.01t^2}$        $F'(0) = 0$

$\quad F''(t) = -0.12e^{-0.01t^2} - 0.12te^{-0.01t^2}(-0.02t)$     $F''(0) = -0.12$

$\quad\quad\quad = -0.12e^{-0.01t^2} + 0.0024t^2e^{-0.01t^2}$

Thus,

$$p_2(t) = F(0) + F'(0)t + \frac{F''(0)}{2!}t^2 = 6 - 0.06t^2.$$

Therefore,

$$N(t) = \int N'(t)\,dt \approx \int p_2(t)\,dt = \int(6 - 0.06t^2)\,dt = 6t - 0.02t^3 + C.$$

Now, $N(0) = 40$. Therefore, $C = 40$ and
$N(t) \approx 6t - 0.02t^3 + 40$.

Thus, $N(5) \approx 30 - 0.02(5)^3 + 40 = 67.5$ words per minute and the improvement is $N(5) - N(0) = 67.5 - 40 = 27.5$ words per minute.

**81.** From Problem 69,

$\quad D(x) = \dfrac{1}{10}\sqrt{10,000 - x^2}$ and

$\quad p_2(x) = 10 - 0.0005x^2$.

The graphs of $D$ and $p_2$ are shown at the right.

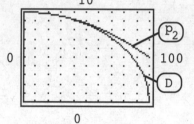

**83.** From Problem 73, $R(t) = 2 + 8e^{-0.1t^2}$ and $p_2(t) = 10 - 0.8t^2$. The graphs of $R$ and $p_2$ are shown at the right.

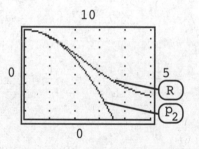

**85.** From Problem 77,

$\quad P(x) = 20\sqrt{3x^2 + 25} - 80$ and

$\quad p_2(x) = 120 + 30(x - 5) + \dfrac{3}{4}(x - 5)^2$.

The graphs of $P$ and $p_2$ are shown at the right.

**87.** From Problem 79,

$\quad N'(t) = 6e^{-0.01t^2}$ and $p_2(t) = 6 - 0.06t^2$.

The graphs of $N'$ and $p_2$ are shown at the right.

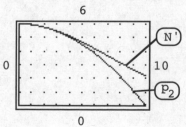

**1.** $\dfrac{4}{1-x} = 4 + 4x + 4x^2 + \ldots + 4x^n + \ldots$

$\qquad a_n = 4$

$\qquad a_{n+1} = 4$

$\qquad \dfrac{a_{n+1}}{a_n} = 1, \quad$ so $\quad \displaystyle\lim_{n \to \infty} \left| \dfrac{a_{n+1}}{a_n} \right| = 1.$

Thus, $R = \dfrac{1}{L} = 1$ and the series converges for $|x| < 1$ or $-1 < x < 1$.

**3.** $\dfrac{1}{1+7x} = 1 - 7x + 7^2 x^2 - \ldots + (-1)^n 7^n x^n + \ldots$

$\qquad a_n = (-1)^n 7^n$

$\qquad a_{n+1} = (-1)^{n+1} 7^{n+1}$

$\qquad \dfrac{a_{n+1}}{a_n} = \dfrac{(-1)^{n+1} 7^{n+1}}{(-1)^n 7^n} = -7$

$\displaystyle\lim_{n \to \infty} \left| \dfrac{a_{n+1}}{a_n} \right| = 7,$ thus $R = \dfrac{1}{L} = \dfrac{1}{7}$ and the series converges for $|x| < \dfrac{1}{7}$

or $-\dfrac{1}{7} < x < \dfrac{1}{7}.$

**5.** $\dfrac{1}{6-x} = \dfrac{1}{6} + \dfrac{1}{6^2}x + \dfrac{1}{6^3}x^2 + \ldots + \dfrac{1}{6^{n+1}}x^n + \ldots$

$\qquad a_n = \dfrac{1}{6^{n+1}}$

$\qquad a_{n+1} = \dfrac{1}{6^{n+2}}$

$\qquad \dfrac{a_{n+1}}{a_n} = \dfrac{\frac{1}{6^{n+2}}}{\frac{1}{6^{n+1}}} = \dfrac{1}{6}$

$\displaystyle\lim_{n \to \infty} \left| \dfrac{a_{n+1}}{a_n} \right| = \dfrac{1}{6},$ thus $R = \dfrac{1}{L} = 6$ and the series converges for $|x| < 6$

or $-6 < x < 6.$

**7.** $\ln(1 + x) = x - \dfrac{1}{2}x^2 + \dfrac{1}{3}x^3 - \cdots + \dfrac{(-1)^{n-1}}{n}x^n + \cdots$

We have $a = 0$ and

$$a_n = \dfrac{(-1)^{n-1}}{n}$$

$$a_{n+1} = \dfrac{(-1)^n}{n+1}$$

Thus, $\dfrac{a_{n+1}}{a_n} = \dfrac{\dfrac{(-1)^n}{n+1}}{\dfrac{(-1)^{n-1}}{n}} = -\dfrac{n}{n+1}$ and $\displaystyle\lim_{n\to\infty}\left|\dfrac{a_{n+1}}{a_n}\right| = \lim_{n\to\infty}\left|\dfrac{-n}{n+1}\right| = 1.$

Therefore, Case (a) of $\underline{2}$ applies. We have $L = 1$, $R = \dfrac{1}{1} = 1$, and the series converges for $|x - 0| < 1$. Thus, the interval of convergence is:

$|x| < 1$ or $-1 < x < 1$

**9.** $\dfrac{1}{(1+x)^2} = 1 - 2x + 3x^2 - \cdots + (-1)^n(n+1)x^n + \cdots$

We have $a = 0$ and

$$a_n = (-1)^n(n+1)$$

$$a_{n+1} = (-1)^{n+1}(n+2)$$

Thus, $\dfrac{a_{n+1}}{a_n} = \dfrac{(-1)^{n+1}(n+2)}{(-1)^n(n+1)} = -\dfrac{(n+2)}{n+1}$ and $\displaystyle\lim_{n\to\infty}\left|\dfrac{a_{n+1}}{a_n}\right| = \lim_{n\to\infty}\left|\dfrac{-(n+2)}{(n+1)}\right| = 1.$

Therefore, Case (a) of $\underline{2}$ applies. We have $L = 1$, $R = \dfrac{1}{1} = 1$, and the series converges for $|x - 0| < 1$. Thus, the interval of convergence is:

$|x| < 1$ or $-1 < x < 1$

**11.** $e^x = 1 + x + \dfrac{1}{2!}x^2 + \cdots + \dfrac{1}{n!}x^n + \cdots$

We have $a = 0$ and $a_n = \dfrac{1}{n!}$

$$a_{n+1} = \dfrac{1}{(n+1)!}$$

Thus, $\dfrac{a_{n+1}}{a_n} = \dfrac{\dfrac{1}{(n+1)!}}{\dfrac{1}{n!}} = \dfrac{n!}{(n+1)!} = \dfrac{1}{n+1}$ and $\displaystyle\lim_{n\to\infty}\left|\dfrac{a_{n+1}}{a_n}\right| = \lim_{n\to\infty}\left|\dfrac{1}{n+1}\right| = 0.$

Therefore, Case (b) of $\underline{2}$ applies and the series converges for all $x$. The interval of convergence is:

$-\infty < x < \infty$.

**13.** (A) Using a calculator, $e^2 \approx 7.3891$.
Now

$$p_1(x) = 1 + x \qquad\qquad\qquad p_1(2) = 3$$

$$p_2(x) = 1 + x + \frac{1}{2!}x^2 \qquad\qquad p_2(2) = 5$$

$$p_3(x) = 1 + x + \frac{1}{2!}x^2 + \frac{1}{3!}x^3 \qquad p_3(2) \approx 6.3333$$

$$p_4(x) = 1 + x + \frac{1}{2!}x^2 + \frac{1}{3!}x^3 + \frac{1}{4!}x^4 \qquad p_4(2) = 7$$

$$p_5(x) = 1 + x + \frac{1}{2!}x^2 + \ldots + \frac{1}{5!}x^5 \qquad p_5(2) \approx 7.2667$$

$$p_6(x) = 1 + x + \frac{1}{2!}x^2 + \ldots + \frac{1}{6!}x^6 \qquad p_6(2) \approx 7.3556$$

Thus, the smallest $n$ such that $|p_n(2) - e^2| < 0.1$ is $n = 6$.

(B) Since the interval of convergence for $e^x$ at 0 is $-\infty < x < \infty$, it follows from $\underline{3}$ that

$$\lim_{n \to \infty} p_n(x) = e^x$$

for every $x$. Therefore,

$$\lim_{n \to \infty} p_n(100) = e^{100}$$

and there does exist a positive integer $n$ such that $|p_n(100) - e^{100}| < 0.1$

**15.** $\ln(4 + x) = (x + 3) - \frac{1}{2}(x + 3)^2 + \frac{1}{3}(x + 3)^3 - \ldots + \frac{(-1)^{n+1}}{n}(x + 3)^n + \ldots$

$$a_n = \frac{(-1)^{n+1}}{n}, \quad a_{n+1} = \frac{(-1)^{n+2}}{n + 1}$$

$$\frac{a_{n+1}}{a_n} = \frac{\dfrac{(-1)^{n+2}}{n + 1}}{\dfrac{(-1)^{n+1}}{n}} = \frac{(-1)^{n+2}}{n + 1} \cdot \frac{n}{(-1)^{n+1}} = -\frac{n}{n + 1}$$

$$\lim_{n \to \infty} \left| \frac{a_{n+1}}{a_n} \right| = \lim_{n \to \infty} \left| \frac{-n}{n + 1} \right| = \lim_{n \to \infty} \frac{n}{n + 1} = \lim_{n \to \infty} \frac{1}{1 + \dfrac{1}{n}} = 1$$

Thus, $L = 1$, $R = \frac{1}{L} = \frac{1}{1} = 1$, and the series converges for $|x + 3| < 1$. The interval of convergence is: $|x + 3| < 1$ or $-4 < x < -2$.

**17.** $e^{2x} = 1 + 2x + \dfrac{2^2}{2!}x^2 + \dfrac{2^3}{3!}x^3 + \ldots + \dfrac{2^n}{n!}x^n + \ldots$

$a_n = \dfrac{2^n}{n!}, \quad a_{n+1} = \dfrac{2^{n+1}}{(n+1)!}$

$\dfrac{a_{n+1}}{a_n} = \dfrac{\dfrac{2^{n+1}}{(n+1)!}}{\dfrac{2^n}{n!}} = \dfrac{2^{n+1}}{(n+1)!} \cdot \dfrac{n!}{2^n} = \dfrac{2}{n+1}$

$\lim\limits_{n \to \infty} \left| \dfrac{a_{n+1}}{a_n} \right| = \lim\limits_{n \to \infty} \left| \dfrac{2}{n+1} \right| = \lim\limits_{n \to \infty} \dfrac{2}{n+1} = 0$

The series converges for all values of $x$. The interval of convergence is: $-\infty < x < \infty$.

**19.** $\dfrac{1}{6-x} = \dfrac{1}{4} + \dfrac{1}{4^2}(x-2) + \dfrac{1}{4^3}(x-2)^2 + \ldots + \dfrac{1}{4^{n+1}}(x-2)^n + \ldots$

$a_n = \dfrac{1}{4^{n+1}}, \quad a_{n+1} = \dfrac{1}{4^{n+2}}$

$\dfrac{a_{n+1}}{a_n} = \dfrac{\dfrac{1}{4^{n+2}}}{\dfrac{1}{4^{n+1}}} = \dfrac{1}{4^{n+2}} \cdot \dfrac{4^{n+1}}{1} = \dfrac{1}{4}$

$\lim\limits_{n \to \infty} \left| \dfrac{a_{n+1}}{a_n} \right| = \lim\limits_{n \to \infty} \left| \dfrac{1}{4} \right| = \lim\limits_{n \to \infty} \dfrac{1}{4} = \dfrac{1}{4}$

Thus, $L = \dfrac{1}{4}$, $R = \dfrac{1}{L} = 4$, and the interval of convergence is:
$|x-2| < 4$ or $-2 < x < 6$.

**21.** $\dfrac{1}{6-x} = -\dfrac{1}{2} + \dfrac{1}{2^2}(x-8) - \dfrac{1}{2^3}(x-8)^2 + \ldots + \dfrac{(-1)^n}{2^{n+1}}(x-8)^n$

$a_n = \dfrac{(-1)^n}{2^{n+1}}, \quad a_{n+1} = \dfrac{(-1)^{n+1}}{2^{n+2}}$

$\dfrac{a_{n+1}}{a_n} = \dfrac{\dfrac{(-1)^{n+1}}{2^{n+2}}}{\dfrac{(-1)^n}{2^{n+1}}} = \dfrac{(-1)^{n+1}}{2^{n+2}} \cdot \dfrac{2^{n+1}}{(-1)^n} = -\dfrac{1}{2}$

$\lim\limits_{n \to \infty} \left| \dfrac{a_{n+1}}{a_n} \right| = \lim\limits_{n \to \infty} \left| -\dfrac{1}{2} \right| = \dfrac{1}{2}$

Thus, $L = \dfrac{1}{2}$, $R = 2$, and the interval of convergence is:
$|x-8| < 2$ or $6 < x < 10$.

**23. (A)**

Based on these graphs it appears that the interval of convergence will be $-\infty < x < \infty$.

(B) The Taylor series at 0 for $f(x) = e^x$ is:
$$e^x = 1 + x + \frac{1}{2!}x^2 + \ldots + \frac{1}{n!}x^n + \ldots$$

By Problem 11, the interval of convergence is $-\infty < x < \infty$.

**25.** $f(x) = e^{4x}$

(A)

| <u>Step 1:</u> | <u>Step 2:</u> | <u>Step 3:</u> |
|---|---|---|
| $f(x) = e^{4x}$ | $f(0) = 1$ | $a_0 = f(0) = 1$ |
| $f'(x) = 4e^{4x}$ | $f'(0) = 4$ | $a_1 = f'(0) = 4$ |
| $f''(x) = 4^2 e^{4x}$ | $f''(0) = 4^2$ | $a_2 = \dfrac{f''(0)}{2!} = \dfrac{4^2}{2}$ |
| $\vdots$ | $\vdots$ | $\vdots$ |
| $f^{(n)}(x) = 4^n e^{4x}$ | $f^{(n)}(0) = 4^n$ | $a_n = \dfrac{f^{(n)}(0)}{n!} = \dfrac{4^n}{n!}$ |

<u>Step 4:</u> The $n$th-degree Taylor polynomial at $a = 0$ for $f(x) = e^{4x}$ is:
$$p_n(x) = 1 + 4x + \frac{4^2}{2}x^2 + \ldots + \frac{4^n}{n!}x^n$$

(B) The Taylor series at $a = 0$ for $f$ is:
$$\lim_{n \to \infty} p_n(x) = 1 + 4x + \frac{4^2}{2!}x^2 + \ldots + \frac{4^n}{n!}x^n + \ldots$$

(C) From part (A):
$$a_n = \frac{4^n}{n!}$$
$$a_{n+1} = \frac{4^{n+1}}{(n+1)!}$$

Thus, $\dfrac{a_{n+1}}{a_n} = \dfrac{\dfrac{4^{n+1}}{(n+1)!}}{\dfrac{4^n}{n!}} = \dfrac{4^{n+1}}{4^n} \cdot \dfrac{n!}{(n+1)!} = \dfrac{4}{n+1}$ and

$$\lim_{n \to \infty} \left| \frac{a_{n+1}}{a_n} \right| = \lim_{n \to \infty} \left| \frac{4}{n+1} \right| = 0.$$

Therefore, the series converges for all $x$, and the interval of convergence is: $-\infty < x < \infty$.

**27.** $f(x) = \ln(1 + 2x)$

(A) Step 1:
$$f(x) = \ln(1 + 2x)$$
$$f'(x) = \frac{1}{1 + 2x}(2) = 2(1 + 2x)^{-1}$$
$$f''(x) = -2(1 + 2x)^{-2}(2) = -2^2(1 + 2x)^{-2}$$
$$f^{(3)}(x) = -2^2(-2)(1 + 2x)^{-3}(2) = 2^3(2)(1 + 2x)^{-3}$$
$$f^{(4)}(x) = 2^3(2)(-3)(1 + 2x)^{-4}(2) = -2^4(3 \cdot 2)(1 + 2x)^{-4}$$
$$\vdots$$
$$f^{(n)}(x) = (-1)^{n-1}2^n(n - 1)!(1 + 2x)^{-n}$$

Step 2:
$$f(0) = \ln 1 = 0$$
$$f'(0) = 2$$
$$f''(0) = -2^2$$
$$f^{(3)}(0) = 2^3 \cdot 2$$
$$f^{(4)}(0) = -2^4 \cdot 3!$$
$$\vdots$$
$$f^{(n)}(0) = (-1)^{n-1}2^n(n - 1)!$$

Step 3:
$$a_0 = f(0) = 0$$
$$a_1 = f'(0) = 2$$
$$a_2 = \frac{f''(0)}{2!} = \frac{-2^2}{2}$$
$$a_3 = \frac{f^{(3)}(0)}{3!} = \frac{2^3}{3}$$
$$a_4 = \frac{f^{(4)}(0)}{4!} = \frac{-2^4}{4}$$
$$\vdots$$
$$a_n = \frac{f^{(n)}(0)}{n!} = \frac{(-1)^{n-1}2^n}{n}$$

Step 4: The $n$th-degree Taylor polynomial at $a = 0$ for
$f(x) = \ln(1 + 2x)$ is:
$$p_n(x) = 2x - \frac{2^2}{2}x^2 + \frac{2^3}{3}x^3 - \frac{2^4}{4}x^4 + \cdots + \frac{(-1)^{n-1}2^n}{n}x^n$$

(B) The Taylor series at $a = 0$ for $f$ is
$$\lim_{n \to \infty} p_n(x) = 2x - \frac{2^2}{2}x^2 + \frac{2^3}{3}x^3 - \frac{2^4}{4}x^4 + \cdots + \frac{(-1)^{n-1}2^n}{n}x^n + \cdots$$

(C) From part (A):
$$a_n = \frac{(-1)^{n-1}2^n}{n}$$
$$a_{n+1} = \frac{(-1)^n2^{n+1}}{n + 1}$$

Thus, $\dfrac{a_{n+1}}{a_n} = \dfrac{\frac{(-1)^n2^{n+1}}{n + 1}}{\frac{(-1)^{n+1}2^n}{n}} = \dfrac{-2n}{n + 1}$ and $\lim\limits_{n \to \infty}\left|\dfrac{a_{n+1}}{a_n}\right| = \lim\limits_{n \to \infty}\left|\dfrac{-2n}{n + 1}\right| = 2.$

Therefore, $L = 2$, $R = \dfrac{1}{2}$, and the series converges for
$$|x| < \frac{1}{2} \quad \text{or} \quad -\frac{1}{2} < x < \frac{1}{2}.$$

**29.** $f(x) = \dfrac{2}{2-x}$

(A)    <u>Step 1:</u>                              <u>Step 2:</u>      <u>Step 3:</u>

$$f(x) = \frac{2}{2-x} = 2(2-x)^{-1} \qquad f(0) = 1 \qquad a_0 = f(0) = 1$$

$$f'(x) = 2(-1)(2-x)^{-2}(-1) \qquad f'(0) = \frac{2}{2^2} = \frac{1}{2} \qquad a_1 = f'(0) = \frac{1}{2}$$

$$= 2(2-x)^{-2}$$

$$f''(x) = 2(-2)(2-x)^{-3}(-1) \qquad f''(0) = \frac{2!}{2^2} = \frac{1}{2} \qquad a_2 = \frac{f''(0)}{2!} = \frac{1}{2^2}$$

$$= 2(2)(2-x)^{-3}$$

$$f^{(3)}(x) = 2(2)(-3)(2-x)^{-4}(-1) \quad f^{(3)}(0) = \frac{3!}{2^3} \qquad a_3 = \frac{f^{(3)}(0)}{3!} = \frac{1}{2^3}$$

$$= 2(3 \cdot 2)(2-x)^{-4}$$

$$\vdots \qquad\qquad\qquad\qquad\qquad \vdots \qquad\qquad\qquad\qquad \vdots$$

$$f^{(n)}(x) = 2(n!)(2-x)^{-(n+1)} \qquad f^{(n)}(0) = \frac{n!}{2^n} \qquad a_n = \frac{f^{(n)}(0)}{n!} = \frac{1}{2^n}$$

<u>Step 4:</u> The $n$th-degree Taylor polynomial at $a = 0$ for

$$f(x) = \frac{2}{2-x} \text{ is:}$$

$$p_n(x) = 1 + \frac{1}{2}x + \frac{1}{2^2}x^2 + \cdots + \frac{1}{2^n}x^n$$

(B) The Taylor series at $a = 0$ for $f$ is:

$$\lim_{n \to \infty} p_n(x) = 1 + \frac{1}{2}x + \frac{1}{2^2}x^2 + \cdots + \frac{1}{2^n}x^n + \cdots$$

(C) From part (A):

$$a_n = \frac{1}{2^n}$$

$$a_{n+1} = \frac{1}{2^{n+1}}$$

Thus, $\dfrac{a_{n+1}}{a_n} = \dfrac{\frac{1}{2^{n+1}}}{\frac{1}{2^n}} = \dfrac{1}{2}$    and    $\displaystyle\lim_{n \to \infty} \left| \dfrac{a_{n+1}}{a_n} \right| = \lim_{n \to \infty} \left| \dfrac{1}{2} \right| = \dfrac{1}{2}.$

Therefore, $L = \dfrac{1}{2}$, $R = \dfrac{1}{\frac{1}{2}} = 2$, and the series converges for:

$$|x| < 2 \quad \text{or} \quad -2 < x < 2.$$

**31.** $f(x) = \ln(2 - x)$, $a = 1$.

(A) <u>Step 1:</u>

$$f(x) = \ln(2 - x)$$

$$f'(x) = \frac{1}{2-x}(-1)$$

$$= -(2 - x)^{-1}$$

$$f''(x) = (2 - x)^{-2}(-1)$$

$$= -(2 - x)^{-2}$$

$$f^{(3)}(x) = 2(2 - x)^{-3}(-1)$$

$$= -2(2 - x)^{-3}$$

$$f^{(4)}(x) = 3 \cdot 2(2 - x)^{-4}(-1)$$

$$= -3!(2 - x)^{-4}$$

$$\vdots$$

$$f^{(n)}(x) = -(n - 1)!(2 - x)^{-n}$$

<u>Step 2:</u>

$$f(1) = \ln 1 = 0$$

$$f'(1) = -1$$

$$f''(1) = -1$$

$$f^{(3)}(1) = -2$$

$$f^{(4)}(1) = -3!$$

$$\vdots$$

$$f^{(n)}(1) = -(n - 1)!$$

<u>Step 3:</u>

$$a_0 = f(1) = 0$$

$$a_1 = f'(1) = -1$$

$$a_2 = \frac{f''(1)}{2!} = -\frac{1}{2}$$

$$a_3 = \frac{f^{(3)}(1)}{3!} = -\frac{1}{3}$$

$$a_4 = \frac{f^{(4)}(1)}{4!} = -\frac{1}{4}$$

$$\vdots$$

$$a_n = \frac{f^{(n)}(1)}{n!} = -\frac{1}{n}$$

<u>Step 4:</u> The $n$th-degree Taylor polynomial at $a = 1$ for
$f(x) = \ln(2 - x)$ is:
$$p_n(x) = -(x - 1) - \frac{1}{2}(x - 1)^2 - \frac{1}{3}(x - 1)^3 - \cdots - \frac{1}{n}(x - 1)^n$$

(B) The Taylor series at $a = 1$ for $f$ is:
$$\lim_{n \to \infty} p_n(x) = -(x - 1) - \frac{1}{2}(x - 1)^2 - \frac{1}{3}(x - 1)^3 - \cdots - \frac{1}{n}(x - 1)^n - \cdots$$

(C) From part (A):
$$a_n = -\frac{1}{n}$$

$$a_{n+1} = -\frac{1}{n + 1}$$

Thus, $\dfrac{a_{n+1}}{a_n} = \dfrac{-\dfrac{1}{n+1}}{-\dfrac{1}{n}} = \dfrac{n}{n + 1}$ and $\displaystyle\lim_{n \to \infty} \left|\dfrac{a_{n+1}}{a_n}\right| = \lim_{n \to \infty} \left|\dfrac{n}{n + 1}\right| = 1$.

Therefore, $L = 1$, $R = \dfrac{1}{1} = 1$, and the series converges for:
$$|x - 1| < 1 \quad \text{or} \quad 0 < x < 2.$$

**33.** $f(x) = \dfrac{1}{1 - x}$, $a = 2$.

(A) 

Step 1:

$$f(x) = \frac{1}{1 - x}$$
$$= (1 - x)^{-1}$$
$$f'(x) = -(1 - x)^{-2}(-1)$$
$$= (1 - x)^{-2}$$
$$f''(x) = -2(1 - x)^{-3}(-1)$$
$$= 2(1 - x)^{-3}$$
$$f^{(3)}(x) = 2(-3)(1 - x)^{-4}(-1)$$
$$= 3!(1 - x)^{-4}$$
$$\vdots$$
$$f^{(n)}(x) = n!(1 - x)^{-(n+1)}$$

Step 2:

$$f(2) = -1$$

$$f'(2) = 1$$

$$f''(2) = -2$$

$$f^{(3)}(2) = 3!$$

$$\vdots$$

$$f^{(n)}(2) = (-1)^{n-1}n!$$

Step 3:

$$a_0 = f(2) = -1$$

$$a_1 = f'(2) = 1$$

$$a_2 = \frac{f''(2)}{2!} = -1$$

$$a_3 = \frac{f^{(3)}(2)}{3!} = 1$$

$$\vdots$$

$$a_n = \frac{f^{(n)}(2)}{n!}$$
$$= (-1)^{n-1}$$

Step 4: The $n$th-degree Taylor polynomial at $a = 2$ for $f(x) = \dfrac{1}{1 - x}$ is:

$$p_n(x) = -1 + (x - 2) - (x - 2)^2 + (x - 2)^3 - \cdots + (-1)^{n-1}(x - 2)^n$$

(B) The Taylor series at $a = 2$ for $f$ is:

$$\lim_{n\to\infty} p_n(x) = -1 + (x - 2) - (x - 2)^2 + (x - 2)^3 - \cdots$$
$$+ (-1)^{n-1}(x - 2)^n + \cdots$$

(C) From part (A): $a_n = (-1)^{n-1}$

$$a_{n+1} = (-1)^n$$

Thus, $\dfrac{a_{n+1}}{a_n} = \dfrac{(-1)^n}{(-1)^{n-1}} = -1$ and $\lim\limits_{n\to\infty} \left| \dfrac{a_{n+1}}{a_n} \right| = \lim\limits_{n\to\infty} |-1| = 1$.

Therefore, $L = 1$, $R = \dfrac{1}{1} = 1$, and the series converges for:

$|x - 2| < 1$ or $1 < x < 3$.

**35.** (A) $\dfrac{1}{4 - x} = \dfrac{1}{4} + \dfrac{1}{4^2}x + \dfrac{1}{4^3}x^2 + \cdots + \dfrac{1}{4^{n+1}}x^n + \cdots$

We have $a = 0$ and

$$a_n = \frac{1}{4^{n+1}}; \quad a_{n+1} = \frac{1}{4^{n+2}}$$

Thus, $\dfrac{a_{n+1}}{a_n} = \dfrac{\frac{1}{4^{n+2}}}{\frac{1}{4^{n+1}}} = \dfrac{4^{n+1}}{4^{n+2}} = \dfrac{1}{4}$ and

$$\lim_{n \to \infty} \left| \frac{a_{n+1}}{a_n} \right| = \lim_{n \to \infty} \left| \frac{1}{4} \right| = \frac{1}{4}.$$

Therefore, $L = \frac{1}{4}$, $R = 4$ and the interval of convergence

is $|x - 0| < 4$; that is, $|x| < 4$ or $-4 < x < 4$.

(B) $\dfrac{1}{4 - x^2} = \dfrac{1}{4} + \dfrac{1}{4^2}x^2 + \dfrac{1}{4^3}x^4 + \dots + \dfrac{1}{4^{n+1}}x^{2n} + \dots$

Theorem 1 is not directly applicable because the coefficients of the odd powers of $x$ are 0.

(C) If we let $t = x^2$, then the Taylor series in (B) becomes

$$\frac{1}{4 - x^2} = \frac{1}{4 - t} = \frac{1}{4} + \frac{1}{4^2}t + \frac{1}{4^3}t^2 + \dots + \frac{1}{4^{n+1}}t^n + \dots$$

From the result in part (A), this series converges for $|t| < 4$.
Expressed in terms of $x$, we have $|x^2| < 4$ which implies $|x| < 2$.
Thus, the interval of convergence is $|x| < 2$ or $-2 < x < 2$.

**37.** (A) $\dfrac{1}{1 - x} + \dfrac{1}{1 - x^2} = 2 + x + 2x^2 + x^3 + 2x^4 + x^5 + \dots$

We have $a = 0$ and

$$a_n = \begin{cases} 2 & \text{if } n \text{ is even} \\ 1 & \text{if } n \text{ is odd} \end{cases}$$

Therefore, $\dfrac{a_{n+1}}{a_n} = \begin{cases} \dfrac{1}{2} & \text{if } n \text{ is even} \\ 2 & \text{if } n \text{ is odd} \end{cases}$

and $\lim\limits_{n \to \infty} \left| \dfrac{a_{n+1}}{a_n} \right|$ does not exist. Thus, Theorem 1 does not apply.

(B) $\dfrac{1}{1 - x} = 1 + x + x^2 + x^3 + \dots + x^n + \dots$

We have $a = 0$ and $a_n = a_{n+1} = 1$.

Thus, $\dfrac{a_{n+1}}{a_n} = 1$ and $\lim\limits_{n \to \infty} \left| \dfrac{a_{n+1}}{a_n} \right| = 1$.

Therefore, $L = R = 1$ and the interval of convergence
is $|x| < 1$ or $-1 < x < 1$.

$$\frac{1}{1 - x^2} = 1 + x^2 + x^4 + x^6 + \dots + x^{2n} + \dots$$

If we let $t = x^2$, then

$$\frac{1}{1 - x^2} = \frac{1}{1 - t} = 1 + t + t^2 + t^3 + \dots + t^n + \dots$$

and the interval of convergence is $|t| < 1$. In terms of $x$, the interval of convergence is $|x^2| < 1$ which implies $|x| < 1$ or $-1 < x < 1$. Thus, both series have the same interval of convergence, $-1 < x < 1$, and so this is the interval of convergence of their sum.

**1.** $f(x) = \dfrac{1}{1-x} + \dfrac{1}{1+x}$

$\dfrac{1}{1-x} = 1 + x + x^2 + \dots + x^n + \dots \qquad ; \quad -1 < x < 1$

$\dfrac{1}{1+x} = 1 - x + x^2 - \dots + (-1)^n x^n + \dots \; ; \quad -1 < x < 1$

$\dfrac{1}{1-x} + \dfrac{1}{1+x} = 2 + 2x^2 + 2x^4 + \dots + 2x^{2n} + \dots$

$a_n = 2; \; a_{n+1} = 2$

$\lim\limits_{n \to \infty} \left| \dfrac{a_{n+1}}{a_n} \right| = 1$, thus $R = \dfrac{1}{L} = 1$ and hence $-1 < x < 1$.

Note: Since the interval of convergence for each series is $(-1, 1)$, we could have concluded that the interval of convergence for the sum of them is also $(-1, 1)$.

**3.** $f(x) = \dfrac{1}{1-x} + e^{-x}$

$\dfrac{1}{1-x} = 1 + x + x^2 + \dots + x^n + \dots \qquad ; \quad -1 < x < 1$

$e^{-x} = 1 - x + \dfrac{1}{2!}x^2 - \dots + \dfrac{(-1)^n}{n!}x^n + \dots \; ; \quad -\infty < x < \infty$

$\dfrac{1}{1-x} + e^{-x} = 2 + \dfrac{3}{2}x^2 + \dfrac{1}{12}x^4 + \dots + \left[1 + \dfrac{(-1)^n}{n!}\right]x^n + \dots$

$a_n = 1 + \dfrac{(-1)^n}{n!}$

$a_{n+1} = 1 + \dfrac{(-1)^{n+1}}{(n+1)!}$

$\lim\limits_{n \to \infty} \left| \dfrac{a_{n+1}}{a_n} \right| = 1$ and hence $R = 1$ or $(-1, 1)$ is the interval of convergence.

<u>Note</u>: We could have concluded that the interval of convergence of the sum of the two series is the intersection of the two interval of convergence, which is $(-1, 1)$.

**5.** $f(x) = \dfrac{x^3}{1+x}$

$\dfrac{1}{1+x} = 1 - x + x^2 - \dots + (-1)^n x^n + \dots \; ; \quad -1 < x < 1$

$\dfrac{x^3}{1+x} = x^3 - x^4 + x^5 - \dots + (-1)^n x^{n+3} + \dots$

$a_n = (-1)^n; \; a_{n+1} = (-1)^{n+1}$

$\lim\limits_{n \to \infty} \left| \dfrac{a_{n+1}}{a_n} \right| = 1$ and hence the series is convergent for $|x| < 1$ or $-1 < x < 1$.

**7.** $f(x) = x^2 \ln(1 - x)$

$$\ln(1 - x) = -x - \frac{x^2}{2} - \frac{x^3}{3} - \dots - \frac{x^n}{n} - \dots \quad , \quad -1 < x < 1$$

$$x^2 \ln(1 - x) = -x^3 - \frac{x^4}{2} - \frac{x^5}{5} - \dots - \frac{x^{n+2}}{n} - \dots$$

$$a_n = -\frac{1}{n}$$

$$a_{n+1} = -\frac{1}{n + 1}$$

$\lim\limits_{n \to \infty} \left| \dfrac{a_{n+1}}{a_n} \right| = \lim\limits_{n \to \infty} \dfrac{n}{n + 1} = 1.$ Thus the series is convergent for $|x| < 1$ or $-1 < x < 1.$

**9.** $f(x) = e^{x^2}$

$$e^u = 1 + u + \frac{1}{2!}u^2 + \dots + \frac{1}{n!}u^n + \dots \quad ; \quad -\infty < u < \infty$$

Let $u = x^2$, then

$$e^{x^2} = 1 + x^2 + \frac{1}{2!}x^4 + \dots + \frac{1}{n!}x^{2n} + \dots \quad ; \quad -\infty < x < \infty$$

**11.** $f(x) = \ln(1 + 3x)$

$$\ln(1 + u) = u - \frac{1}{2}u^2 + \dots + \frac{(-1)^{n-1}}{n}u^n + \dots \quad ; \quad -1 < u < 1$$

Let $u = 3x$, then

$$\ln(1 + 3x) = 3x - \frac{9}{2}x^2 + 9x^3 - \dots + (-1)^{n-1}\frac{3^n}{n}x^n + \dots$$

Note that

$$-1 < 3x < 1 \quad \text{or} \quad -\frac{1}{3} < x < \frac{1}{3},$$

and hence the interval of convergence is $\left( -\dfrac{1}{3}, \dfrac{1}{3} \right).$

**13.** From Table 1,

$$\frac{1}{1 - x} = 1 + x + x^2 + \dots + x^n + \dots \quad (-1 < x < 1).$$

Now, $f(x) = \dfrac{1}{2 - x} = \dfrac{1}{2\left(1 - \dfrac{x}{2}\right)} = \dfrac{1}{2} \cdot \dfrac{1}{1 - \dfrac{x}{2}}.$

Substituting $\dfrac{x}{2}$ for $x$ in the series for $\dfrac{1}{1 - x}$, and using Property $\underline{2}$, we have

$$f(x) = \frac{1}{2-x} = \frac{1}{2}\left(\frac{1}{1-\frac{x}{2}}\right) = \frac{1}{2}\left(1 + \frac{x}{2} + \left(\frac{x}{2}\right)^2 + \cdots + \left(\frac{x}{2}\right)^n + \cdots\right)$$

$$= \frac{1}{2}\left(1 + \frac{x}{2} + \frac{x^2}{2^2} + \cdots + \frac{x^n}{2^n} + \cdots\right)$$

$$= \frac{1}{2} + \frac{x}{2^2} + \frac{x^2}{2^3} + \cdots + \frac{x^n}{2^{n+1}} + \cdots$$

The interval of convergence is $-1 < \dfrac{x}{2} < 1$ or $-2 < x < 2$.

**15.** From Table 1,

$$\frac{1}{1-x} = 1 + x + x^2 + \cdots + x^n + \cdots \quad (-1 < x < 1).$$

Substituting $8x^3$ for $x$, we have

$$f(x) = \frac{1}{1 - 8x^3} = 1 + 8x^3 + [8x^3]^2 + \cdots + [8x^3]^n + \cdots$$

$$= 1 + 8x^3 + 8^2 x^6 + \cdots + 8^n x^{3n} + \cdots.$$

The interval of convergence is $-1 < 8x^3 < 1$, which is

$\dfrac{-1}{8} < x^3 < \dfrac{1}{8}$ or $\dfrac{-1}{2} < x < \dfrac{1}{2}$.

**17.** $f(x) = 10^x = e^{x \ln 10}$

From Table 1,

$$e^x = 1 + x + \frac{1}{2!}x^2 + \cdots + \frac{1}{n!}x^n + \cdots \quad (-\infty < x < \infty)$$

Substituting $x \ln 10$ for $x$, we have

$$10^x = e^{x \ln 10} = 1 + (\ln 10)x + \frac{(\ln 10)^2}{2!}x^2 + \frac{(\ln 10)^3}{3!}x^3 + \cdots$$

$$+ \frac{(\ln 10)^n}{n!}x^n + \cdots.$$

The interval of convergence is $-\infty < x < \infty$.

**19.** $f(x) = \log_2(1 - x) = \dfrac{1}{\ln 2}\ln(1 - x)$

From Table 1,

$$\ln(1 - x) = -x - \frac{1}{2}x^2 - \frac{1}{3}x^3 - \cdots - \frac{1}{n}x^n - \cdots, \quad -1 < x < 1.$$

Therefore, the Taylor series for $f(x) = \log_2(1 - x)$ is:

$$\log_2(1 - x) = \frac{-1}{\ln 2}x - \frac{1}{2\ln 2}x^2 - \frac{1}{3\ln 2}x^3 - \frac{1}{n\ln 2}x^n - \cdots, \quad -1 < x < 1.$$

**21.** From Table 1,

$$\frac{1}{1 + x} = 1 - x + x^2 - x^3 + \cdots + (-1)^n x^n + \cdots \quad (-1 < x < 1).$$

Now, $f(x) = \dfrac{1}{4 + x^2} = \dfrac{1}{4\left(1 + \dfrac{x^2}{4}\right)} = \dfrac{1}{4} \cdot \dfrac{1}{1 + \left(\dfrac{x}{2}\right)^2}.$

Thus, substituting $\left(\dfrac{x}{2}\right)^2 = \dfrac{x^2}{4}$ for $x$ in the series for $\dfrac{1}{1 + x}$, we have

$$f(x) = \frac{1}{4 + x^2} = \frac{1}{4}\left[\frac{1}{1 + x^2/4}\right] = \frac{1}{4}\left(1 - \frac{x^2}{4} + \left[\frac{x^2}{4}\right]^2 - \cdots + (-1)^n \left[\frac{x^2}{4}\right]^n + \cdots\right)$$

$$= \frac{1}{4}\left(1 - \frac{x^2}{4} + \frac{x^4}{4^2} - \cdots + (-1)^n \frac{x^{2n}}{4^n} + \cdots\right)$$

$$= \frac{1}{4} - \frac{x^2}{4^2} + \frac{x^4}{4^3} - \cdots + (-1)^n \frac{x^{2n}}{4^{n+1}} + \cdots.$$

The interval of convergence is $-1 < \dfrac{x^2}{4} < 1$ or $-4 < x^2 < 4$, which is $-2 < x < 2$.

**23.** (A) The Taylor series at 0 for $e^x$ converges for all $x$. Since the odd powers of $\sqrt{x} = x^{1/2}$ are defined only for $x \geq 0$, the formula
$$e^{\sqrt{x}} = 1 + x^{1/2} + \frac{1}{2!}x + \frac{1}{3!}x^{3/2} + \cdots + \frac{1}{n!}x^{n/2} + \cdots$$
is valid only for $x \geq 0$.

(B) The formula for $e^{\sqrt{x}}$ is **not** a Taylor series at 0 because some (in fact, half) of the powers of $x$ are not non-negative integers.

**25.** (A) From Table 1,
$$\frac{1}{1 - x} = 1 + x + x^2 + x^3 + \cdots + x^n + \cdots \ (-1 < x < 1).$$
Substituting $x^2$ for $x$, we have
$$f(x) = \frac{1}{1 - x^2} = 1 + x^2 + (x^2)^2 + (x^2)^3 + \cdots + (x^2)^n + \cdots$$
$$= 1 + x^2 + x^4 + x^6 + \cdots + x^{2n} + \cdots \quad (-1 < x < 1).$$

(B) Using Property $\underline{3}$,
$$f'(x) = 2x + 4x^3 + 6x^5 + \cdots + 2nx^{2n-1} + \cdots \quad (-1 < x < 1).$$
On the other hand, differentiating $f(x) = \dfrac{1}{1 - x^2}$, we have

$$f'(x) = \frac{2x}{(1 - x^2)^2}.$$

Thus, the Taylor series at 0 for $g(x) = \dfrac{2x}{(1 - x^2)^2}$ is

$$g(x) = 2x + 4x^3 + 6x^5 + \cdots + 2nx^{2n-1} + \cdots \quad (-1 < x < 1).$$

**27.** From Table 1,

$$\frac{1}{1 + x} = 1 - x + x^2 - x^3 + \cdots + (-1)^n x^n + \cdots \qquad (-1 < x < 1).$$

Thus,

$$f'(x) = \frac{1}{1 + x^2} = 1 - x^2 + x^4 - x^6 + \cdots + (-1)^n x^{2n} + \cdots \qquad (-1 < x < 1),$$

substituting $x^2$ for $x$ in the series for $\dfrac{1}{1 + x}$. Now, using Property $\underline{4}$,

$$f(x) = \int f(x)\,dx = \int [1 - x^2 + x^4 - x^6 + \cdots + (-1)^n x^{2n} + \cdots]\,dx$$

$$= C + x - \frac{x^3}{3} + \frac{x^5}{5} - \frac{x^7}{7} + \cdots + \frac{(-1)^n}{2n + 1}x^{2n+1} + \cdots .$$

Applying the initial condition $f(0) = 0$, we have

$$0 = C + 0 - \frac{0^3}{3} + \cdots + \frac{(-1)^n}{2n + 1}0^{2n+1} + \cdots ,$$

$$C = 0.$$

Thus,

$$f(x) = x - \frac{x^3}{3} + \frac{x^5}{5} - \frac{x^7}{7} + \cdots + \frac{(-1)^n}{2n + 1}x^{2n+1} + \cdots \qquad (-1 < x < 1).$$

**29.** From Table 1

$$\ln(1 - x) = -x - \frac{1}{2}x^2 - \frac{1}{3}x^3 - \cdots - \frac{1}{n}x^n - \cdots \qquad -1 < x < 1$$

Therefore,

$$f'(x) = x^2 \left[ -x - \frac{1}{2}x^2 - \frac{1}{3}x^3 - \cdots - \frac{1}{n}x^n - \cdots \right]$$

$$= -x^3 - \frac{1}{2}x^4 - \frac{1}{3}x^5 - \cdots - \frac{1}{n}x^{n+2} - \cdots , \qquad -1 < x < 1$$

and

$$f(x) = \int f'(x)\,dx = \int \left[ -x^3 - \frac{1}{2}x^4 - \frac{1}{3}x^5 - \cdots - \frac{1}{n}x^{n+2} - \cdots \right] dx$$

$$= -\int x^3\,dx - \frac{1}{2}\int x^4\,dx - \frac{1}{3}\int x^5\,dx - \cdots - \frac{1}{n}\int x^{n+2}\,dx - \cdots$$

$$= C - \frac{1}{4}x^4 - \frac{1}{10}x^5 - \frac{1}{18}x^6 - \cdots - \frac{1}{n(n + 3)}x^{n+3} - \cdots$$

$$= C - \frac{1}{4}x^4 - \frac{1}{10}x^5 - \frac{1}{18}x^6 - \cdots - \frac{1}{n(n - 3)}x^n - \cdots$$

Since $f(0) = 5$, we have

$$f(0) = 5 = C - 0 - 0 - \cdots - 0 - \cdots$$

$$= C$$

Thus

$$f(x) = 5 - \frac{1}{4}x^4 - \frac{1}{10}x^5 - \frac{1}{18}x^6 - \cdots - \frac{1}{n(n - 3)}x^n - \cdots$$

Since the series for $f'$ converges for $-1 < x < 1$, the series for $f$ converges for $-1 < x < 1$.

**31.** (A)

(B) $\sinh x = x + \dfrac{1}{3!}x^3 + \dfrac{1}{5!}x^5 + \cdots + \dfrac{1}{(2n+1)!}x^{2n+1} + \cdots,\ -\infty < x < \infty.$

Since $\dfrac{d^2}{dx^2}(x) = 0,\ \dfrac{d^2}{dx^2}\left(\dfrac{1}{3!}x^3\right) = x,$

$\dfrac{d^2}{dx^2}\left(\dfrac{1}{5!}x^5\right) = \dfrac{1}{3!}x^3,$ and, in general,

$\dfrac{d^2}{dx^2}\left[\dfrac{1}{(2k+1)!}x^{2k+1}\right] = \dfrac{1}{(2k-1)!}x^{2k-1},\ k = 1,\ 2,\ \cdots,$

it follows that $\dfrac{d^2}{dx^2}(\sinh x) = \sinh x.$

**33.** Let $t = x - 3$. Then $x = t + 3$ and

$\dfrac{1}{4-x} = \dfrac{1}{4-(t+3)} = \dfrac{1}{1-t} = f(t).$

From Table 1,

$\dfrac{1}{1-t} = 1 + t + t^2 + t^3 + \cdots + t^n + \cdots \quad (-1 < t < 1).$

Substituting $x - 3$ for $t$ in this series, we have

$f(x) = \dfrac{1}{4-x} = 1 + (x-3) + (x-3)^2 + (x-3)^3 + \cdots + (x-3)^n + \cdots.$

The interval of convergence is $-1 < x - 3 < 1$ or $2 < x < 4$.

**35.** Let $t = x - 1$. Then $x = 1 + t$ and

$\ln x = \ln(1+t) = t - \dfrac{1}{2}t^2 + \dfrac{1}{3}t^3 - \cdots + \dfrac{(-1)^{n-1}}{n}t^n + \cdots \quad (-1 < t < 1)$

from Table 1. Substituting $x - 1$ for $t$ in this series, we have

$f(x) = \ln x = (x-1) - \dfrac{1}{2}(x-1)^2 + \dfrac{1}{3}(x-1)^3 - \cdots + \dfrac{(-1)^{n-1}}{n}(x-1)^n + \cdots.$

The interval of convergence is $-1 < x - 1 < 1$ or $0 < x < 2$.

**37.** Let $t = x - 1$. Then $x = t + 1$ and

$\dfrac{1}{4-3x} = \dfrac{1}{4-3(t+1)} = \dfrac{1}{1-3t}.$

Now, from Table 1,

$\dfrac{1}{1-u} = 1 + u + u^2 + u^3 + \cdots + u^n + \cdots \quad (-1 < u < 1).$

Substituting $3t$ for $u$ in this series, we have

$$\frac{1}{1 - 3t} = 1 + (3t) + (3t)^2 + (3t)^3 + \cdots + (3t)^n + \cdots$$

$$= 1 + 3t + 3^2 t^2 + 3^3 t^3 + \cdots + 3^n t^n + \cdots ,$$

and the interval of convergence is $-1 < 3t < 1$ or $-\frac{1}{3} < t < \frac{1}{3}$.

Finally, replacing $t$ by $x - 1$, we obtain

$$\frac{1}{4 - 3x} = 1 + 3(x - 1) + 3^2(x - 1)^2 + 3^3(x - 1)^3 + \cdots + 3^n(x - 1)^n + \cdots$$

and the interval of convergence is $-\frac{1}{3} < x - 1 < \frac{1}{3}$ or $\frac{2}{3} < x < \frac{4}{3}$.

**39.** From Table 1,

$$F(x) = \frac{1}{1 - x} = 1 + x + x^2 + x^3 + x^4 + \cdots + x^n + \cdots \qquad (-1 < x < 1).$$

Now, using Property 3,

$$F'(x) = \frac{1}{(1 - x)^2} = 1 + 2x + 3x^2 + 4x^3 + \cdots + nx^{n-1} + \cdots \qquad (-1 < x < 1)$$

and

$$F''(x) = \frac{2}{(1 - x)^3} = 2 + 3 \cdot 2x + 4 \cdot 3x^2 + \cdots + n(n - 1)x^{n-2} + \cdots \quad (-1 < x < 1).$$

Thus,

$$f(x) = \frac{1}{(1 - x)^3} = \frac{1}{2} \cdot \frac{2}{(1 - x)^3}$$

$$= \frac{1}{2}[2 + 3 \cdot 2x + 4 \cdot 3x^2 + \cdots + n(n - 1)x^{n-2} + \cdots]$$

$$= 1 + 3x + 6x^2 + \cdots + \frac{n(n - 1)}{2}x^{n-2} + \cdots \quad (-1 < x < 1).$$

**41.** $f(x) = x + \frac{1}{3}x^3 + \frac{1}{5}x^5 + \cdots + \frac{1}{2n + 1}x^{2n + 1} + \cdots, \ -1 < x < 1.$

(A) Since $\dfrac{d}{dx}\left[\dfrac{1}{2k + 1}x^{2k+1}\right] = x^{2k}$ for $k = 1, 2, \cdots$,

it follows from Property 3 that

$f'(x) = 1 + x^2 + x^4 + \cdots + x^{2n} + \cdots, \ -1 < x < 1.$

(B) From Table 1,

$$\frac{1}{1 - x} = 1 + x + x^2 + \cdots + x^n + \cdots, \ -1 < x < 1$$

Substituting $x^2$ for $x$, the Taylor series for $\dfrac{1}{1 - x^2}$ at $0$ is:

$$\frac{1}{1 - x^2} = 1 + x^2 + x^4 + \cdots + x^{2n} + \cdots, \ -1 < x < 1.$$

Therefore, $f'(x)$ and $\dfrac{1}{1 - x^2}$ have the same Taylor series at $0$ and so

$$f'(x) = \frac{1}{1 - x^2} \text{ for } -1 < x < 1.$$

**43.** Using Table 1 and Property $\underline{2}$, we have

$$\frac{1}{1-x} = 1 + x + x^2 + x^3 + \cdots + x^n + \cdots \quad (-1 < x < 1)$$

$$\frac{x}{1-x} = x\left(\frac{1}{1-x}\right) = x + x^2 + x^3 + x^4 + \cdots + x^{n+1} + \cdots \quad (-1 < x < 1).$$

Thus, by Property $\underline{1}$,

$$\frac{1+x}{1-x} = \frac{1}{1-x} + \frac{x}{1-x} = 1 + 2x + 2x^2 + 2x^3 + \cdots + 2x^n + \cdots$$

and the interval of convergence is $-1 < x < 1$.

**45.** Using Table 1 and Property $\underline{2}$, we have

$$\ln(1+x) = x - \frac{1}{2}x^2 + \frac{1}{3}x^3 - \cdots + \frac{(-1)^{n+1}}{n}x^n + \cdots \quad (-1 < x < 1)$$

$$-\ln(1-x) = x + \frac{1}{2}x^2 + \frac{1}{3}x^3 + \cdots + \frac{1}{n}x^n + \cdots \quad (-1 < x < 1).$$

Thus, by Property $\underline{1}$,

$$\ln\frac{1+x}{1-x} = \ln(1+x) - \ln(1-x) = 2x + \frac{2}{3}x^3 + \cdots + \frac{2}{2n+1}x^{2n+1} + \cdots$$

$$(-1 < x < 1),$$

so, by Property $\underline{2}$,

$$\frac{1}{2}\ln\frac{1+x}{1-x} = x + \frac{1}{3}x^3 + \frac{1}{5}x^5 + \cdots + \frac{1}{2n+1}x^{2n+1} + \cdots \quad (-1 < x < 1).$$

**47.** From Table 1,

$$e^x = 1 + x + \frac{x^2}{2!} + \frac{x^3}{3!} + \cdots + \frac{x^n}{n!} + \cdots \quad (-\infty < x < \infty)$$

and

$$e^{-x} = 1 - x + \frac{x^2}{2!} - \frac{x^3}{3!} + \cdots + \frac{(-1)^n}{n!}x^n + \cdots \quad (-\infty < x < \infty).$$

Thus, using Properties $\underline{1}$ and $\underline{2}$,

$$f(x) = \frac{e^x + e^{-x}}{2} = \frac{1}{2}(e^x + e^{-x}) = \frac{1}{2}\left(2 + \frac{2}{2!}x^2 + \frac{2}{4!}x^4 + \cdots + \frac{2}{(2n)!}x^{2n} + \cdots\right)$$

and

$$f(x) = 1 + \frac{1}{2!}x^2 + \frac{1}{4!}x^4 + \cdots + \frac{1}{(2n)!}x^{2n} + \cdots \quad (-\infty < x < \infty).$$

**49.** From Table 1,

$$f''(x) = \ln(1+x) = x - \frac{1}{2}x^2 + \frac{1}{3}x^3 - \cdots + \frac{(-1)^{n-1}}{n}x^n + \cdots \quad (-1 < x < 1)$$

Therefore

$$f'(x) = \int f''(x)\,dx = \int\left[x - \frac{1}{2}x^2 + \frac{1}{3}x^3 - \cdots + \frac{(-1)^{n-1}}{n}x^n + \cdots\right]dx$$

$$= \int x\,dx - \frac{1}{2}\int x^2\,dx + \frac{1}{3}\int x^3\,dx - \cdots + \frac{(-1)^{n-1}}{n}\int x^n\,dx + \cdots$$

$$= C_1 + \frac{1}{2}x^2 - \frac{1}{6}x^3 + \frac{1}{12}x^4 - \cdots + \frac{(-1)^{n-1}}{n(n+1)}x^{n+1} + \cdots$$

$$= C_1 + \frac{1}{2}x^2 - \frac{1}{6}x^3 + \frac{1}{12}x^4 - \cdots + \frac{(-1)^n}{n(n-1)}x^n + \cdots$$

Now,

$$f'(0) = 3 = C_1 + 0 - 0 + 0 - \cdots$$
$$= C_1$$

Thus

$$f'(x) = 3 + \frac{1}{2}x^2 - \frac{1}{6}x^3 + \frac{1}{12}x^4 - \cdots + \frac{(-1)^n}{n(n-1)}x^n + \cdots \quad (-1 < x < 1)$$

Continuing,

$$f(x) = \int f'(x)\,dx = \int \left[ 3 + \frac{1}{2}x^2 - \frac{1}{6}x^3 + \frac{1}{12}x^4 - \cdots + \frac{(-1)^n}{n(n-1)}x^n + \cdots \right] dx$$

$$= \int 3\,dx + \frac{1}{2}\int x^2\,dx - \frac{1}{6}\int x^3\,dx + \cdots + \frac{(-1)^n}{n(n-1)}\int x^n\,dx + \cdots$$

$$= C_2 + 3x + \frac{1}{6}x^3 - \frac{1}{24}x^4 + \cdots + \frac{(-1)^n}{n(n-1)(n+1)}x^{n+1} + \cdots$$

$$= C_2 + 3x + \frac{1}{6}x^3 - \frac{1}{24}x^4 + \cdots + \frac{(-1)^{n+1}}{n(n-1)(n-2)}x^n + \cdots$$

Now

$$f(0) = -2 = C_2 + 0 + 0 - 0 + \cdots$$
$$= C_2$$

Thus,

$$f(x) = -2 + 3x + \frac{1}{6}x^3 - \frac{1}{24}x^4 + \cdots + \frac{(-1)^{n+1}}{n(n-1)(n-2)}x^n + \cdots, \quad -1 < x < 1.$$

**51.** From Table 1,

$$\frac{1}{1+t} = 1 - t + t^2 - t^3 + \cdots + (-1)^n t^n + \cdots \quad (-1 < t < 1).$$

Now, $\dfrac{1}{x} = \dfrac{1}{a + (x-a)} = \dfrac{1}{a\left[1 + \dfrac{x-a}{a}\right]} = \dfrac{1}{a} \cdot \dfrac{1}{1 + \left(\dfrac{x-a}{a}\right)}.$

Substituting $\dfrac{x-a}{a}$ for $t$ in the series for $\dfrac{1}{1+t}$ and using Property 2, we have

$$\frac{1}{x} = \frac{1}{a}\left[ 1 - \frac{x-a}{a} + \left(\frac{x-a}{a}\right)^2 - \left(\frac{x-a}{a}\right)^3 + \cdots + (-1)^n\left(\frac{x-a}{a}\right)^n + \cdots \right]$$

$$= \frac{1}{a} - \frac{1}{a^2}(x-a) + \frac{1}{a^3}(x-a)^2 - \frac{1}{a^4}(x-a)^3 + \cdots + \frac{(-1)^n}{a^{n+1}}(x-a)^n + \cdots,$$

and the interval of convergence is $-1 < \dfrac{x-a}{a} < 1$ or $-a < x - a < a$, which is $0 < x < 2a$.

Now, $\ln x = \int \frac{1}{x}\, dx =$

$$\int\left(\frac{1}{a} - \frac{1}{a^2}(x - a) + \frac{1}{a^3}(x - a)^2 - \ldots + \frac{(-1)^n}{a^{n+1}}(x - a)^n + \ldots\right) dx$$

$$= C + \frac{1}{a}(x - a) - \frac{1}{2a^2}(x - a)^2 + \frac{1}{3a^3}(x - a)^3 - \ldots$$

$$+ \frac{(-1)^n}{(n + 1)a^{n+1}}(x - a)^{n+1} + \ldots \quad (0 < x < 2a).$$

Evaluating this equation at $x = a$ gives $\ln a = C$.

Therefore, $\ln x = \ln a + \frac{1}{a}(x - a) - \frac{1}{2a^2}(x - a)^2 + \frac{1}{3a^3}(x - a)^3 - \ldots$

$$+ \frac{(-1)^n}{(n + 1)a^{n+1}}(x - a)^{n+1} + \ldots \quad (0 < x < 2a).$$

Equivalently, this can be written:

$\ln x = \ln a + \frac{1}{a}(x - a) - \frac{1}{2a^2}(x - a)^2 + \frac{1}{3a^3}(x - a)^3 - \ldots$

$$+ \frac{(-1)^{n-1}}{na^n}(x - a)^n + \ldots \quad (0 < x < 2a).$$

**53.** From Table 1,

$$\frac{1}{1 - x} = 1 + x + x^2 + x^3 + \ldots + x^n + \ldots \qquad -1 < x < 1$$

Therefore, if $\alpha$ and $\beta$ are constants, then:

$$\frac{1}{\alpha - \beta x} = \frac{1}{\alpha\left[1 - \frac{\beta}{\alpha}x\right]} = \frac{1}{\alpha}\left[1 + \frac{\beta}{\alpha}x + \frac{\beta^2}{\alpha^2}x^2 + \frac{\beta^3}{\alpha^3}x^3 + \ldots + \frac{\beta^n}{\alpha^n}x^n + \ldots\right]$$

$$= \frac{1}{\alpha} + \frac{\beta}{\alpha^2}x + \frac{\beta^2}{\alpha^3}x^2 + \frac{\beta^3}{\alpha^4}x^3 + \ldots + \frac{\beta^n}{\alpha^{n+1}}x^n + \ldots$$

$$\left|\frac{\beta}{\alpha}x\right| < 1 \quad \text{or} \quad |x| < \left|\frac{\alpha}{\beta}\right|$$

Finally, replacing $x$ by $x - a$, we have

$$\frac{1}{\alpha - \beta(x - a)} = \frac{1}{\alpha} + \frac{\beta}{\alpha^2}(x - a) + \frac{\beta^2}{\alpha^3}(x - a)^2$$

$$+ \ldots + \frac{\beta^n}{\alpha^{n+1}}(x - a)^n + \ldots ,$$

$$|x - a| < \left|\frac{\alpha}{\beta}\right|$$

Now, $f(x) = \frac{1}{1 - bx} = \frac{1}{1 - ab - b(x - a)}$, $ab \neq 1$

$$= \frac{1}{1 - ab} + \frac{b}{(1 - ab)^2}x + \frac{b^2}{(1 - ab)^3}x^2 + \ldots + \frac{b^n}{(1 - ab)^{n+1}}x^n + \ldots ,$$

$$|x - a| < \left|\frac{1 - ab}{b}\right| \text{ by our result above.}$$

**1.** Using $\underline{3}$ with $n = 1$, we have
$e^{-0.6} \approx 1 - 0.6 = 0.4$
and $|R_1(0.6)| < 0.18$.

**3.** Using $\underline{3}$ with $n = 3$, we have
$e^{-0.6} \approx 1 - 0.6 + 0.18 - 0.036 = 0.544$
and $|R_3(0.6)| < 0.0054$.

**5.** Using $\underline{3}$ with $n = 2$, we have
$\ln 1.3 \approx 0.3 - 0.045 = 0.255$
and $|R_2(0.3)| < 0.009$.

**7.** Using $\underline{3}$ with $n = 4$, we have
$\ln 1.3 \approx 0.3 - 0.045 + 0.009 - 0.00203$
$\qquad = 0.261975$
and $|R_4(0.3)| < 0.000486$.

*For Problems 9 and 11:* $e^{-x} = 1 - x + \dfrac{x^2}{2!} - \dfrac{x^3}{3!} + \dfrac{x^4}{4!} - \cdots$

*and* $\qquad p_3(x) = 1 - x + \dfrac{x^2}{2!} - \dfrac{x^3}{3!}.$

**9.** Let $x = 0.2$. Then

$e^{-0.2} \approx p_3(0.2) = 1 - 0.2 + \dfrac{(0.2)^2}{2!} - \dfrac{(0.2)^3}{3!}$

$\qquad\qquad = 1 - 0.2 + 0.02 - 0.001333$

$\qquad\qquad = 0.818667$

and $|R_3(0.2)| < \left|\dfrac{(0.2)^4}{4!}\right| = 0.000067.$

**11.** Let $x = 0.03$. Then

$e^{-0.03} \approx p_3(0.03) = 1 - 0.03 + \dfrac{(0.03)^2}{2!} - \dfrac{(0.03)^3}{3!}$

$\qquad\qquad = 1 - 0.03 + 0.00045 - 0.000005$

$\qquad\qquad = 0.970446$

and $|R_3(0.03)| < \left|\dfrac{(0.03)^4}{4!}\right| = 0.000000033.$

*For Problems 13 and 15:* $\ln(1 + x) = x - \dfrac{x^2}{2} + \dfrac{x^3}{3} - \dfrac{x^4}{4} + \cdots$

$\qquad\qquad p_3(x) = x - \dfrac{x^2}{2} + \dfrac{x^3}{3}.$

**13.** Let $x = 0.6$. Then

$\ln(1.6) \approx p_3(0.6) = 0.6 - \dfrac{(0.6)^2}{2} + \dfrac{(0.6)^3}{3}$

$\qquad\qquad = 0.6 - 0.18 + 0.072 = 0.492$

and $|R_3(0.6)| < \left|\dfrac{-(0.6)^4}{4}\right| = 0.0324.$

**15.** Let $x = 0.06$. Then

$$\ln(1.06) \approx p_3(0.06) = 0.06 - \frac{(0.06)^2}{2} + \frac{(0.06)^3}{3}$$

$$= 0.06 - 0.0018 + 0.000072 = 0.058272$$

and $\left| R_3(0.06) \right| < \left| \dfrac{-(0.06)^4}{4} \right| = 0.00000324.$

**17.** $e^{-0.1}$

Let $f(x) = e^{-x}$, then

$f'(x) = -e^{-x}$

$f''(x) = e^{-x}$

$\vdots$

$f^{(n)}(x) = (-1)^n e^{-x}$

$\vdots$

$R_n(x) = \dfrac{f^{(n+1)}(t)x^{n+1}}{(n+1)!}$ for some $t$ between $0$ and $x$.

Note that $f^{(n+1)}(t) = (-1)^{n+1}e^{-t}$ and hence

$\left| f^{(n+1)}(t) \right| = \left| (-1)^{n+1}e^{-t} \right| = e^{-t} < 1$ for $t > 0$.

Therefore, $\left| R_n(x) \right| = \left| \dfrac{f^{(n+1)}(t)x^{n+1}}{(n+1)!} \right| < \dfrac{|x|^{n+1}}{(n+1)!}$ and $\left| R_n(0.1) \right| < \dfrac{(0.1)^{n+1}}{(n+1)!}.$

For $n = 3$, $\dfrac{(0.1)^4}{4!} < 0.0000042 < 0.000005$,

and hence $\left| R_3(0.1) \right| < 0.000005$. Thus, the polynomial of the lowest degree is

$$p_3(x) = 1 - x + \frac{1}{2!}x^2 - \frac{1}{3!}x^3 \text{ which has degree 3.}$$

$$e^{-0.1} \approx p_3(0.1) = 1 - 0.1 + \frac{1}{2}(0.1)^2 - \frac{1}{6}(0.1)^3 = 0.904833.$$

**19.** $e^{-0.01}$

From Problem 17 we have

$$\left| R_n(x) \right| < \frac{|x|^{n+1}}{(n+1)!}$$

For $x = 0.01$,

$$\left| R_n(0.01) \right| < \frac{(0.01)^{n+1}}{(n+1)!}$$

Now, for $n = 2$, $\dfrac{(0.01)^3}{3!} < 0.000002 < 0.000005$, and hence

$\left| R_2(0.01) \right| < 0.000005$. Thus, the polynomial with the lowest degree is

$$p_2(x) = 1 - x + \frac{1}{2!}x^2 \text{ which has degree 2.}$$

$$e^{-0.01} \approx p_2(0.01) = 1 - 0.01 + \frac{1}{2}(0.01)^2 = 0.990050.$$

**21.** ln 1.2

Let $f(x) = \ln(1 + x)$

$$f'(x) = \frac{1}{1 + x} = (1 + x)^{-1}$$

$$f''(x) = -(1 + x)^{-2}$$

$$f^{(3)}(x) = 2(1 + x)^{-3}$$

$\vdots$

$$f^{(n)}(x) = (n - 1)!(-1)^{n-1}(1 + x)^{-n}$$

$\vdots$

$$R_n(x) = \frac{f^{(n+1)}(t)x^{n+1}}{(n + 1)!} \quad \text{for some } t \text{ between 0 and } x.$$

Note that $f^{(n+1)}(t) = n!(-1)^n(1 + t)^{-(n+1)}$ and hence

$$\left|f^{(n+1)}(t)\right| = \left|n!(-1)^n(1 + t)^{-(n+1)}\right| = n!(1 + t)^{-(n+1)}$$
$$< n! \text{ for } t > 0.$$

Therefore,

$$\left|R_n(x)\right| = \left|\frac{f^{(n+1)}(t)x^{n+1}}{(n + 1)!}\right| < \frac{n!\,|x|^{n+1}}{(n + 1)!} = \frac{|x|^{n+1}}{n + 1}, \text{ and } R_n(0.2) < \frac{(0.2)^{n+1}}{n + 1}.$$

For $n = 6$, $\left|R_6(0.2)\right| < \dfrac{(0.2)^7}{8} < 0.000002 < 0.000005$,

and hence the polynomial with the lowest degree is

$$p_6(x) = x - \frac{1}{2}x^2 + \frac{1}{3}x^3 - \frac{1}{4}x^4 + \frac{1}{5}x^5 - \frac{1}{6}x^6$$

which has degree 6.

$$\ln(1.2) \approx p_6(0.2) = 0.2 - \frac{1}{2}(0.2)^2 + \frac{1}{3}(0.2)^3 - \frac{1}{4}(0.2)^4$$
$$+ \frac{1}{5}(0.2)^5 - \frac{1}{6}(0.2)^6 = 0.182320.$$

**23.** ln(1.02)

From Problem 21 we have

$$\left|R_n(x)\right| < \frac{|x|^{n+1}}{n + 1}$$

For $x = 0.02$,

$$\left|R_n(0.02)\right| < \frac{(0.02)^{n+1}}{n + 1}$$

Now, for $n = 2$, $\dfrac{(0.02)^3}{3} < 0.000003 < 0.000005$, and hence

$\left|R_2(0.02)\right| < 0.000005$. Thus, the polynomial with the lowest degree is

$p_2(x) = x - \dfrac{1}{2}x^2$ which has degree 2.

$$\ln(1.02) \approx p_2(0.02) = 0.02 - \frac{1}{2}(0.02)^2 = 0.019800.$$

**25.** The Taylor series at 0 for $f(x) = \dfrac{1}{1 + x^2}$ is

$$\frac{1}{1 + x^2} = 1 - x^2 + x^4 - x^6 + \cdots + (-1)^n x^{2n} + \cdots \qquad (-1 < x < 1).$$

[<u>Note</u>: Use the Taylor series at 0 for $\dfrac{1}{1 + x^2}$ and substitute $x^2$ for $x$.]

Thus,

$$\int \frac{1}{1 + x^2}\, dx = C + x - \frac{x^3}{3} + \frac{x^5}{5} - \frac{x^7}{7} + \cdots + \frac{(-1)^n}{2n + 1} x^{2n+1} + \cdots \ (-1 < x < 1)$$

and

$$\int_0^{0.2} \frac{1}{1 + x^2}\, dx = \left[ x - \frac{x^3}{3} + \frac{x^5}{5} - \frac{x^7}{7} + \cdots + \frac{(-1)^n}{2n + 1} x^{2n+1} + \cdots \right]\Bigg|_0^{0.2}$$

$$= 0.2 - \frac{(0.2)^3}{3} + \frac{(0.2)^5}{5} - \frac{(0.2)^7}{7} + \cdots$$

$$= 0.2 - 0.00267 + 0.000064 - \cdots.$$

Now, <u>3</u> applies and the error in using the first two terms in this series will be less than 0.000064. Thus,

$$\int_0^{0.2} \frac{1}{1 + x^2}\, dx \approx 0.19733.$$

**27.** The Taylor series at 0 for $f(x) = \ln(1 + x^2)$ is

$$\ln(1 + x^2) = x^2 - \frac{1}{2}x^4 + \frac{1}{3}x^6 - \cdots + \frac{(-1)^{n+1}}{n} x^{2n} + \cdots \qquad (-1 < x < 1).$$

[<u>Note</u>: Use the Taylor series at 0 for $\ln(1 + x)$ and substitute $x^2$ for $x$.]

Thus,

$$\int \ln(1 + x^2)\, dx = C + \frac{x^3}{3} - \frac{x^5}{10} + \frac{x^7}{21} - \cdots + \frac{(-1)^{n+1}}{n(2n + 1)} x^{2n+1} + \cdots$$

and

$$\int_0^{0.6} \ln(1 + x^2)\, dx = \left[ \frac{x^3}{3} - \frac{x^5}{10} + \frac{x^7}{21} - \cdots + \frac{(-1)^{n+1}}{n(2n + 1)} x^{2n+1} + \cdots \right]\Bigg|_0^{0.6}$$

$$= 0.072 - 0.00778 + 0.00133 - 0.00028 + \cdots$$

Now, <u>3</u> applies and the error in using the first three terms in this series will be less than 0.00028. Therefore,

$$\int_0^{0.6} \ln(1 + x^2)\, dx \approx 0.0656.$$

**29.** The Taylor series at 0 for $f(x) = x^2 e^{-x^2}$ is

$$x^2 e^{-x^2} = x^2 - x^4 + \frac{1}{2!}x^6 - \cdots + \frac{(-1)^n}{n!} x^{2n+2} + \cdots \qquad (-\infty < x < \infty).$$

[<u>Note</u>: Use the Taylor series at 0 for $e^{-x}$, substitute $x^2$ for $x$, and multiply the resulting series by $x^2$.]

Thus,

$$\int x^2 e^{-x^2}\, dx = C + \frac{x^3}{3} - \frac{x^5}{5} + \frac{1}{7 \cdot 2!}x^7 - \cdots + \frac{(-1)^n}{(2n+3)n!}x^{2n+3} + \cdots$$

and

$$\int_0^{0.4} x^2 e^{-x^2}\, dx = \left[ \frac{x^3}{3} - \frac{x^5}{5} + \frac{1}{14}x^7 - \cdots + \frac{(-1)^n}{(2n+3)n!}x^{2n+3} + \cdots \right]\Bigg|_0^{0.4}$$

$$= 0.02133 - 0.00205 + 0.00012 - \cdots .$$

Now, $\underline{3}$ applies and the error in using the first two terms in this series will be less than 0.00012. Therefore,

$$\int_0^{0.4} x^2 e^{-x^2}\, dx \approx 0.01928.$$

**31.** From Taylor's formula for the remainder,

$$R_n(2) = \frac{f^{(n+1)}(t)}{(n+1)!}2^{n+1}$$

where $t$ is some number between 0 and 2. Since $|f^{n+1}(t)| \le 1$, we have

$$|R_n(2)| = \left| \frac{f^{(n+1)}(t)}{(n+1)!}2^{n+1} \right| = \frac{|f^{(n+1)}(t)|}{(n+1)!}2^{n+1} \le \frac{2^{n+1}}{(n+1)!}$$

Now, we want $|R_n(2)| < 0.001$ which will be true if

$$\frac{2^{n+1}}{(n+1)!} < 0.001$$

which implies $(n+1)! > \frac{1}{0.001}2^{n+1} = 1000(2^{n+1})$.

The smallest $n$ that satisfies this inequality is $n = 9$.

**33.** From Taylor's formula for the remainder,

$$R_5(x) = \frac{f^{(6)}(t)}{6!}x^6$$

where $t$ is between 0 and $x$. Now

$$|R_6(x)| = \left| \frac{f^{(6)}(t)}{6!}x^6 \right| \le \frac{|x^6|}{6!} = \frac{|x|^6}{720} < 0.05$$

implies $\quad |x|^6 < 720(0.05) = 36$

Therefore, $|x| < (36)^{1/6} \approx 1.817$.

The error in the approximation of $f(x)$ by $P_5(x)$ will be less than 0.05 if $|x| < 1.817$.

**35.** Since $xe^x = x\left(1 + x + \dfrac{1}{2!}x^2 + \ldots + \dfrac{1}{n!}x^n + \ldots\right)$

$$= x + x^2 + \frac{1}{2!}x^3 + \ldots + \frac{1}{n!}x^{n+1} + \ldots, \quad -\infty < x < \infty,$$

it follows that

$$p_3(x) = x + x^2 + \frac{1}{2!}x^3$$

Using a graphing utility to graph

$$\left|p_3(x) - xe^x\right|$$

we find that

$$\left|p_3(x) - xe^x\right| < 0.005$$

for $-0.427 < x < 0.405$

**37.** From Section 2-3, Table 1, the Taylor series at 0 for $f(x) = e^x$ is

$$e^x = 1 + x + \frac{x^2}{2!} + \frac{x^3}{3!} + \ldots + \frac{x^n}{n!} + \ldots \quad (-\infty < x < \infty).$$

Thus, the second-degree polynomial at 0 is

$$p_2(x) = 1 + x + \frac{x^2}{2!} = 1 + x + \frac{x^2}{2}$$

and

$$e^{-0.3} = f(-0.3) \approx p_2(-0.3) = 1 - 0.3 + \frac{(-0.3)^2}{2} = 0.745.$$

To estimate the error, we use the formula in 2:

$$R_2(-0.3) = \frac{f^{(3)}(t)}{3!}(-0.3)^3, \text{ where } -0.3 < t < 0.$$

Now,
$$f(x) = e^x$$
$$f'(x) = e^x$$
$$f''(x) = e^x$$
$$f^{(3)}(x) = e^x.$$

Therefore, $R_2(-0.3) = \dfrac{e^t}{3!}(-0.3)^3$, $-0.3 < t < 0$, and the error $\left|R_2(-0.3)\right|$

satisfies $\left|R_2(-0.3)\right| = \left|\dfrac{e^t}{3!}(-0.3)^3\right| = \dfrac{e^t}{6}(.027) < \dfrac{1}{6}(.027) = 0.0045.$

[Note: Here we used $e^t \le e^0 = 1$, $-0.3 < t < 0$.]

**39.** The second-degree Taylor polynomial at 0 for $f(x) = e^x$ is

$$p_2(x) = 1 + x + \frac{x^2}{2!} = 1 + x + \frac{x^2}{2}.$$

Now,

$$e^{0.05} = f(0.05) \approx p_2(0.05) = 1 + 0.05 + \frac{(0.05)^2}{2} = 1.05125.$$

The remainder $R_2(0.05)$ is given by

$$R_2(0.05) = \frac{f^{(3)}(t)}{3!}(0.05)^3, \text{ where } 0 < t < 0.05.$$

Since $f^{(3)}(x) = e^x$, we have

$$R_2(0.05) = \frac{e^t}{3!}(0.05)^3, \text{ where } 0 < t < 0.05,$$

and the error $\left|R_2(0.05)\right|$ satisfies

$$\left|R_2(0.05)\right| = \left|\frac{e^t}{3!}(0.05)^3\right| = \frac{e^t}{6}(0.000125) < \frac{3}{6}(0.000125) = 0.0000625.$$

[<u>Note</u>: Here we used $e^t < e = 2.71828\ldots < 3$, $0 < t < 0.05$.]

**41.** $f(x) = \sqrt{16 + x} = (16 + x)^{1/2}$

$$p_2(x) = f(0) + f'(0)x + \frac{f''(0)}{2!}x^2.$$

Now,

$$f(x) = (16 + x)^{1/2} \qquad\qquad f(0) = 16^{1/2} = 4$$

$$f'(x) = \frac{1}{2}(16 + x)^{-1/2} = \frac{1}{2(16 + x)^{1/2}} \qquad f'(0) = \frac{1}{2 \cdot 16^{1/2}} = \frac{1}{8}$$

$$f''(x) = -\frac{1}{4}(16 + x)^{-3/2} = \frac{-1}{4(16 + x)^{3/2}} \qquad f''(0) = \frac{-1}{4 \cdot 16^{3/2}} = \frac{-1}{256}$$

Thus, $p_2(x) = 4 + \dfrac{1}{8}x - \dfrac{1}{2!}\cdot\dfrac{1}{256}x^2 = 4 + \dfrac{1}{8}x - \dfrac{1}{512}x^2$

and $\sqrt{17} = f(1) \approx P_2(1) = 4 + \dfrac{1}{8} - \dfrac{1}{512} = 4 + .125 - 0.00195 = 4.12305.$

The error in this approximation is

$$\left|R_2(1)\right| = \left|\frac{f^{(3)}(t)}{3!}(1)^3\right| \qquad 0 < t < 1$$

$$= \frac{\left|f^{(3)}(t)\right|}{3!}.$$

Now, $f^{(3)}(x) = \dfrac{3}{8}(16 + x)^{-5/2} = \dfrac{3}{8(16 + x)^{5/2}}.$

Since the maximum value of $\dfrac{3}{8(16 + t)^{5/2}}$ on $[0, 1]$ occurs at $t = 0$ and

is $\dfrac{3}{8(16)^{5/2}} = \dfrac{3}{8 \cdot 4^5}$,

we have

$$\left|R_2(1)\right| = \frac{\left|f^{(3)}(t)\right|}{3!} \le \frac{\frac{3}{8 \cdot 4^5}}{3!} = \frac{1}{16 \cdot 4^5} = \frac{1}{4^7} = \frac{1}{16,384} = 0.000061.$$

**43.** The Taylor series for $f(x) = \dfrac{1}{1-x}$ at 0

is $\dfrac{1}{1-x} = 1 + x + x^2 + \cdots + x^n + \cdots$,

with interval of convergence $-1 < x < 1$.
Since the lower limit of integration, $-2$, lies outside the interval of convergence, the Taylor series cannot be used to approximate the integral $\displaystyle\int_{-2}^{0} \dfrac{1}{1-x}\,dx$.
The second computation:

$$\int_{-2}^{0} \dfrac{1}{1-x}\,dx = -\ln(1-x)\,\Big|_{-2}^{0} = -\ln 1 + \ln 3 = \ln 3 \approx 1.099$$

is correct.

**45.** (A) The Taylor series at 0 for $f(x) = e^x$ is:

$$e^x = 1 + x + \dfrac{x^2}{2!} + \dfrac{x^3}{3!} + \cdots + \dfrac{x^n}{n!} + \cdots \quad (-\infty < x < \infty)$$

Thus, $p_2(x) = 1 + x + \dfrac{1}{2!}x^2$

The quadratic regression polynomial $q_2(x)$ for the points

(−0.1, $e^{-0.1}$), (0, $e^0$), (0.1, $e^{0.1}$) is shown at the right.

```
QuadReg
 y=ax²+bx+c
 a=.5004168056
 b=1.0016675
 c=1
 R²=1
```

(B) Max $|f(x) - p_2(x)| < 0.000171$
Max $|f(x) - q_2(x)| < 0.000065$

(C) $q_2(x)$ will give a better approximation to $\displaystyle\int_{-0.1}^{0.1} e^x\,dx$.

**47.** Index of income concentration $= 2\displaystyle\int_0^1 [x - f(x)]\,dx = 2\int_0^1 \left[x - \dfrac{5x^6}{4 + x^2}\right]dx$

$$= \int_0^1 2x\,dx - \int_0^1 \dfrac{10x^6}{4 + x^2}\,dx = x^2\,\Big|_0^1 - \int_0^1 \dfrac{10x^6}{4 + x^2}\,dx = 1 - \int_0^1 \dfrac{10x^6}{4 + x^2}\,dx.$$

Step 1: Find the Taylor series for the integrand.

$$\dfrac{10x^6}{4 + x^2} = \dfrac{10x^6}{4} \cdot \dfrac{1}{1 + \dfrac{x^2}{4}}$$

$$= \dfrac{5x^6}{2}\left[1 - \dfrac{x^2}{4} + \dfrac{x^4}{16} - \dfrac{x^6}{64} + \cdots + \dfrac{(-1)^n x^{2n}}{4^n} + \cdots\right]$$

[substitute $x^2/4$ into the series for $1/{1+x}$]

$$= 5\left[\dfrac{x^6}{2} - \dfrac{x^8}{8} + \dfrac{x^{10}}{32} - \dfrac{x^{12}}{128} + \cdots + \dfrac{(-1)^n x^{2n+6}}{2 \cdot 4^n} + \cdots\right].$$

This series converges for $\left|\dfrac{x^2}{4}\right| < 1$ or $-2 < x < 2.$

<u>Step 2</u>: Find the Taylor series for the antiderivative.

$$\int \frac{10x^6}{4 + x^2}\, dx = \int 5\left[\frac{x^6}{2} - \frac{x^8}{8} + \frac{x^{10}}{32} - \frac{x^{12}}{128} + \dots + \frac{(-1)^n x^{2n+6}}{2 \cdot 4^n} + \dots\right] dx$$

$$= C + 5\left[\frac{x^7}{14} - \frac{x^9}{72} + \frac{x^{11}}{352} - \frac{x^{13}}{1664} + \dots + \frac{(-1)^n x^{2n+7}}{2 \cdot 4^n (2n + 7)} + \dots\right].$$

<u>Step 3</u>: Approximate the definite integral.
Choosing $C = 0$ in the antiderivative, we have

$$\int_0^1 \frac{10x^6}{4 + x^2}\, dx = 5\left[\frac{x^7}{14} - \frac{x^9}{72} + \frac{x^{11}}{352} - \frac{x^{13}}{1664} + \dots + \frac{(-1)^n x^{2n+7}}{2 \cdot 4^n (2n + 7)} + \dots\right]\Bigg|_0^1$$

$$= \frac{5}{14} - \frac{5}{72} + \frac{5}{352} - \frac{5}{1664} + \dots$$

$$= 0.3571 - 0.0694 + 0.0142 - 0.003 + \dots .$$

Since the terms are alternating in sign and decreasing in absolute value, and since the fourth term is less than 0.005, we can use the first three terms of this series to approximate the definite integral to the specified accuracy:

$$\int_0^1 \frac{10x^6}{4 + x^2}\, dx \approx 0.3571 - 0.0694 + 0.0142 = 0.3019 \approx 0.302$$

Finally, the index of income concentration is:

$$2\int_0^1 \left[x - \frac{10x^6}{4 + x^2}\right] dx \approx 1 - 0.302 = 0.698$$

**49.** Total Sales $= \int_0^4 S'(t)\, dt = \int_0^4 [10 - 10e^{-0.01t^2}]\, dt$

<u>Step 1</u>: Find the fourth-degree Taylor polynomial at 0 for its integrand. Substituting $0.01t^2 = \left(\frac{t}{10}\right)^2$ into the Taylor series at 0 for $e^{-x}$, we have:

$$e^{-0.01t^2} = 1 - \left(\frac{t}{10}\right)^2 + \frac{1}{2!}\left(\frac{t}{10}\right)^4 - \frac{1}{3!}\left(\frac{t}{10}\right)^6 + \dots + \frac{(-1)^n}{n!}\left(\frac{t}{10}\right)^{2n} + \dots$$

Thus,

$$10 - 10e^{-0.01t^2} = 10 - 10\left[1 - \frac{t^2}{(10)^2} + \frac{1}{2!}\frac{t^4}{(10)^4} - \frac{1}{3!}\frac{t^6}{(10)^6} + \dots \right.$$

$$\left. + \frac{(-1)^n}{n!}\frac{t^{2n}}{(10)^{2n}} + \dots\right]$$

$$= \frac{t^2}{10} - \frac{t^4}{2!\,(10)^3} + \frac{t^6}{3!\,(10)^5} - \dots + \frac{(-1)^{n+1} t^{2n}}{n!\,(10)^{2n-1}} + \dots$$

and $p_4(t) = \dfrac{t^2}{10} - \dfrac{t^4}{2(10)^3} = 0.1t^2 - 0.0005t^4.$

<u>Step 2</u>: Approximate the definite integral.

$$\int_0^4 S'(t)\,dt \approx \int_0^4 p_4(t)\,dt = \int_0^4 (0.1t^2 - 0.0005t^4)\,dt$$

$$= \left(\frac{0.1t^3}{3} - 0.0001t^5\right)\Big|_0^4$$

$$= \frac{0.1(4)^3}{3} - 0.0001(4^5) = 2.1333 - 0.1024$$

$$= 2.0309 \approx 2.031$$

Therefore, $S(4) \approx 2.031$ and total sales during the first four months are approximately \$2,031,000.

<u>Step 3</u>: Estimate the error in this approximation.
From Steps 1 and 2, the Taylor series at 0 for $S(t)$ is:

$$S(t) = \frac{t^3}{3(10)} - \frac{t^5}{5 \cdot 2!\,(10)^3} + \frac{t^7}{7 \cdot 3!\,(10)^5} - \cdots + \frac{(-1)^n t^{2n+1}}{(2n+1)n!\,(10)^{2n-1}} + \cdots$$

and

$$S(4) = \frac{(4)^3}{3(10)} - \frac{(4)^5}{5 \cdot 2!\,(10)^3} + \frac{(4)^7}{7 \cdot 3!\,(10)^5} - \cdots + \frac{(-1)^n (4)^{2n+1}}{(2n+1)n!\,(10)^{2n-1}} + \cdots .$$

Since the terms alternate in sign and decrease in absolute value, the error in this approximation is less than

$$\left|\frac{(4)^7}{7 \cdot 3!\,(10)^5}\right| \approx 0.004.$$

Thus, $S(4) = 2.031$ to within $\pm 0.004$ or $S(4) = \$2,031,000$ to within $\pm\$4000$.

**51.** To find the useful life, set $C'(t) = R'(t)$ and solve for $t$:

$$4 = \frac{80}{16 + t^2}$$

$$4t^2 + 64 = 80$$

$$t^2 = 4$$

and $\quad t = 2$ or $-2$

Thus, the useful life of the system is 2 years. The total profit of the system is given by the definite integral

Total Profit $= \int_0^2 [R'(t) - C'(t)]\,dt$

<u>Step 1</u>: Find the Taylor series for the integrand.

$$R'(t) = \frac{80}{16 + t^2} = \frac{80}{16} \cdot \frac{1}{1 + \dfrac{t^2}{16}} = 5 \cdot \frac{1}{1 + \left(\dfrac{t}{4}\right)^2}$$

$$= 5\left[1 - \frac{t^2}{4^2} + \frac{t^4}{4^4} - \frac{t^6}{4^6} + \cdots + (-1)^n \frac{t^{2n}}{4^{2n}} + \cdots\right]$$

(substitute $\left(\dfrac{t}{4}\right)^2$ into the series for $\dfrac{1}{1+x}$)

Thus,

$$R'(t) - C'(t) = 5\left[1 - \frac{t^2}{4^2} + \frac{t^4}{4^4} - \frac{t^6}{4^6} + \dots + (-1)^n \frac{t^{2n}}{4^{2n}} + \dots\right] - 4$$

$$= 1 - \frac{5t^2}{4^2} + \frac{5t^4}{t^4} - \frac{5t^6}{4^6} + \dots + (-1)^n \frac{5t^{2n}}{4^{2n}} + \dots$$

Step 2: Find the Taylor series for the antiderivative.

$$\int\left[1 - \frac{5t^2}{4^2} + \frac{5t^4}{4^4} - \frac{5t^6}{4^6} + \dots + (-1)^n \frac{5t^{2n}}{4^{2n}} + \dots\right] dt$$

$$= C + t - \frac{5t^3}{3 \cdot 4^2} + \frac{t^5}{4^4} - \frac{5t^7}{7 \cdot 4^6} + \dots + (-1)^n \frac{5t^{2n+1}}{(2n+1)4^{2n}} + \dots$$

Step 3: Approximate the definite integral. Let $C = 0$. Then

$$\int_0^2 [R'(t) - C'(t)]dt = \left[ t - \frac{5t^3}{3 \cdot 4^2} + \frac{t^5}{4^4} - \frac{5t^7}{7 \cdot 4^6} + \dots \right.$$

$$\left. + (-1)^n \frac{5t^{2n+1}}{(2n+1)4^{2n}} + \dots \right]\Big|_0^2$$

$$= 2 - \frac{5 \cdot 2^3}{3 \cdot 4^2} + \frac{2^5}{4^4} - \frac{5 \cdot 2^7}{7 \cdot 4^6} + \frac{5 \cdot 2^9}{9 \cdot 4^8} - \dots$$

$$= 2 - \frac{5}{6} + \frac{1}{8} - \frac{5}{224} + \frac{5}{1152} - \dots$$

Since the terms are alternating in sign and decreasing in absolute value, and since $\frac{5}{1152} \approx 0.0043$ is less than 0.005, we can use the first four terms of this series. Thus,

Total Profit $\approx 2 - \frac{5}{6} + \frac{1}{8} - \frac{5}{224} \approx 1.270$ thousand or \$1,270.

**53.** $C(t) = 20 + 800 \ln\left(1 + \frac{t^2}{100}\right)$, $0 \le t \le 2$.

Average Temperature $= \frac{1}{2-0}\int_0^2 C(t)\,dt = \frac{1}{2}\int_0^2 \left[20 + 800 \ln\left(1 + \frac{t^2}{100}\right)\right] dt.$

Step 1: Determine the Taylor series at 0 for the integrand.

Substituting $\frac{t^2}{100}$ into the Taylor series at 0 for $\ln(1 + x)$, we have:

$$\ln\left(1 + \frac{t^2}{100}\right) = \frac{t^2}{100} - \frac{1}{2}\left(\frac{t^2}{100}\right)^2 + \frac{1}{3}\left(\frac{t^2}{100}\right)^3 - \dots + (-1)^{n+1}\frac{1}{n}\left(\frac{t^2}{100}\right)^n + \dots$$

$$= \frac{t^2}{100} - \frac{1}{2 \cdot 10^4}t^4 + \frac{1}{3 \cdot 10^6}t^6 - \dots$$

$$+ (-1)^{n+1}\frac{1}{n \cdot 10^{2n}}t^{2n} + \dots \ .$$

Thus,

$$20 + 8 \cdot 10^2 \ln\left(1 + \frac{t^2}{100}\right) = 20 + 8t^2 - \frac{8}{2 \cdot 10^2}t^4 + \frac{8}{3 \cdot 10^4}t^6 - \cdots$$

$$+ (-1)^{n+1}\frac{8}{n \cdot 10^{2n-2}}t^{2n} + \cdots .$$

Step 2: Calculate the definite integral.

Average Temperature $= \dfrac{1}{2}\displaystyle\int_0^2 C(t)\,dt$

$$= \frac{1}{2}\int_0^2 20 \; dt + \frac{1}{2}\int\left[8t^2 - \frac{8}{2 \cdot 10^2}t^4 + \frac{8}{3 \cdot 10^4}t^6 - \cdots\right.$$

$$\left. + \frac{(-1)^{n+1}8}{n \cdot 10^{2n-2}}t^{2n} + \cdots\right]dt$$

$$= 10t\Big|_0^2 + \frac{1}{2}\left[\frac{8}{3}t^3 - \frac{8}{2 \cdot 5 \cdot 10^2}t^5 + \frac{8}{3 \cdot 7 \cdot 10^4}t^7 - \cdots\right.$$

$$\left. + \frac{(-1)^{n+1}8}{n(2n+1)10^{2n-2}}t^{2n+1} + \cdots\right]\Big|_0^2$$

$$= 20 + \frac{1}{2}\left[\frac{8}{3}(2)^3 - \frac{8}{2 \cdot 5 \cdot 10^2}(2)^5 + \frac{8}{3 \cdot 7 \cdot 10^4}(2)^7 - \cdots\right.$$

$$\left. + \frac{(-1)^{n+1}8}{n(2n+1)10^{2n-2}}(2)^{2n+1} + \cdots\right]$$

$$= 20 + \frac{1}{2}\left[\frac{2^6}{3} - \frac{2^8}{2 \cdot 5 \cdot 10^2} + \frac{2^{10}}{3 \cdot 7 \cdot 10^4} - \cdots + \frac{(-1)^{n+1}2^{2n+4}}{n(2n+1)10^{2n-2}} + \cdots\right]$$

$$= 20 + \frac{2^5}{3} - \frac{2^7}{2 \cdot 5 \cdot 10^2} + \frac{2^9}{3 \cdot 7 \cdot 10^4} - \cdots + \frac{(-1)^{n+1}2^{2n+3}}{n(2n+1)10^{2n-2}} + \cdots$$

Step 3: Approximate the definite integral to within ±0.005.

Since the terms are alternating in sign and decreasing in absolute value, and since the fourth term

$$\frac{2^9}{3 \cdot 7 \cdot 10^4} = \frac{512}{210,000} = 0.0024 < 0.005,$$

the approximate average temperature is:

$$20 + \frac{2^5}{3} - \frac{2^7}{2 \cdot 5 \cdot 10^2} = 30.539°C$$

**55.** $A'(t) = \dfrac{-75}{t^2 + 25}$

Step 1: Find the second-degree Taylor polynomial at 0 for
$$A'(t) = \frac{-75}{t^2 + 25} = -3\frac{1}{1 + \left(\dfrac{t}{5}\right)^2}$$

Substituting $\dfrac{t^2}{25}$ into the Taylor series at 0 for $\dfrac{1}{1 + x}$, we have:

$$-3\frac{1}{1 + \left(\dfrac{t}{5}\right)^2} = -3\left[1 - \left(\frac{t}{5}\right)^2 + \left(\frac{t}{5}\right)^4 - \left(\frac{t}{5}\right)^6 + \ldots + (-1)^n\left(\frac{t}{5}\right)^{2n} + \ldots\right]$$

Thus, $p_2(t) = -3 + \dfrac{3t^2}{25}$.

Step 2: Approximate $A(t) = \int A'(t)\,dt$, and find $A(2)$.

$$A(t) = \int A'(t)\,dt \approx \int p_2(t)\,dt = \int\left[-3 + \frac{3t^2}{25}\right]dt = -3t + \frac{t^3}{25} + C.$$

Since $A(0) = 12$, we have:

$$A(t) \approx -3t + \frac{t^3}{25} + 12$$

Thus, $A(2) \approx -3\cdot2 + \dfrac{2^3}{25} + 12 = 6.32$ square inches.

Step 3: Estimate the error in this approximation. From Steps 1 and 2, the Taylor series at 0 for $A(t)$ is

$$A(t) = 12 - 3t + \frac{t^3}{5^2} - \frac{3t^5}{5 \cdot 5^4} + \frac{3t^7}{7 \cdot 5^6} - \ldots$$

and
$$A(2) = 12 - 6 + \frac{8}{25} - \frac{3 \cdot 2^5}{5^5} + \frac{3 \cdot 2^7}{7 \cdot 5^6} - \ldots.$$

Since the terms are alternating in sign and decreasing in absolute value, the error in the approximation is less than

$$\left|\frac{-3 \cdot 2^5}{5^5}\right| = 0.03072 \approx 0.031.$$

Thus, $A(2) = 6.32$ square cm to within $\pm0.031$.

**57.** $N'(t) = 6e^{-0.01t^2}$

Step 1: Find the second-degree polynomial at 0 for the integrand.

Substituting $0.01t^2 = \left(\dfrac{t}{10}\right)^2$ into the Taylor series at 0 for $e^{-x}$, we have:

$$e^{-0.01t^2} = 1 - \left(\frac{t}{10}\right)^2 + \frac{1}{2!}\left(\frac{t}{10}\right)^4 - \cdots + \frac{(-1)^n}{n!}\left(\frac{t}{10}\right)^{2n} + \cdots$$

Thus, $6e^{-0.01t^2} = 6 - \dfrac{6t^2}{(10)^2} + \dfrac{6t^4}{2!(10)^4} - \cdots + \dfrac{(-1)^n 6t^{2n}}{n!(10)^{2n}} + \cdots$

and

$$P_2(t) = 6 - \frac{6t^2}{100} = 6 - 0.06t^2.$$

Step 2: Approximate $N(t) = \int N'(t)\,dt$ and calculate $N(5)$.

$$N(t) = \int 6e^{-0.01t^2}\,dt \approx \int P_2(t)\,dt = \int [6 - 0.06t^2]\,dt$$

$$= 6t - \frac{0.06t^3}{3} + C = 6t - 0.02t^3 + C$$

Since $N(0) = 40$, we have:

$N(t) \approx 40 + 6t - 0.02t^3$.

Thus, $N(5) \approx 40 + 6(5) - 0.02(5)^3 = 67.5$ words per minute.

Step 3: Estimate the error in the approximation.

From Steps 1 and 2, the Taylor series at 0 for $N(t)$ is:

$$N(t) = 40 + 6t - 0.02t^3 + \frac{6t^5}{5 \cdot 2!(10)^4} - \cdots$$

$$+ \frac{(-1)^n 6t^{2n+1}}{(2n+1)n!(10)^{2n}} + \cdots$$

and

$$N(5) = 40 + 6(5) - 0.02(5)^3 + \frac{6(5)^5}{5 \cdot 2!(10)^4} - \cdots .$$

Since the terms are alternating in sign and decreasing in absolute value, the error in the approximation is less than

$$\left| \frac{6(5)^5}{5 \cdot 2!(10)^4} \right| = 0.1875.$$

Thus, $N(5) = 67.5$ words per minute to within $\pm 0.1875$ and the improvement is $N(5) - N(0) = 67.5 - 40 = 27.5$ words per minute to within $\pm 0.1875$.

**1.** $f(x) = \ln(x + 5)$

$f'(x) = \dfrac{1}{x + 5}(1) = (x + 5)^{-1}$

$f''(x) = -1(x + 5)^{-2}(1) = -(x + 5)^{-2} = \dfrac{-1}{(x + 5)^2}$

$f^{(3)}(x) = -(-2)(x + 5)^{-3}(1) = 2(x + 5)^{-3} = \dfrac{2}{(x + 5)^3}$

$f^{(4)}(x) = 2(-3)(x + 5)^{-4}(1) = -6(x + 5)^{-4} = \dfrac{-6}{(x + 5)^4}$ (2-1)

**2.** $f(x) = \sqrt[3]{1 + x} = (1 + x)^{1/3}$ $\qquad f(0) = 1$

$f'(x) = \dfrac{1}{3}(1 + x)^{-2/3} = \dfrac{1}{3(1 + x)^{2/3}}$ $\qquad f'(0) = \dfrac{1}{3}$

$f''(x) = \dfrac{-2}{9}(1 + x)^{-5/3} = \dfrac{-2}{9(1 + x)^{5/3}}$ $\qquad f''(0) = \dfrac{-2}{9}$

$f^{(3)}(x) = \dfrac{10}{27}(1 + x)^{-8/3} = \dfrac{10}{27(1 + x)^{8/3}}$ $\qquad f^{(3)}(0) = \dfrac{10}{27}$

Thus,

$p_3(x) = f(0) + f'(0)x + \dfrac{f''(0)}{2!}x^2 + \dfrac{f^{(3)}(0)}{3!}x^3$

$= 1 + \dfrac{1}{3}x - \dfrac{2}{9 \cdot 2!}x^2 + \dfrac{10}{27 \cdot 3!}x^3 = 1 + \dfrac{1}{3}x - \dfrac{1}{9}x^2 + \dfrac{5}{81}x^3.$

Now,

$\sqrt[3]{1.01} = f(0.01) \approx p_3(0.01) = 1 + \dfrac{1}{3}(0.01) - \dfrac{1}{9}(0.01)^2 + \dfrac{5}{81}(0.01)^3$

$\approx 1.003322.$ (2-1)

**3.** $f(x) = \sqrt{1 + x} = (1 + x)^{1/2}$ $\qquad f(3) = \sqrt{4} = 2$

$f'(x) = \dfrac{1}{2}(1 + x)^{-1/2}(1) = \dfrac{1}{2(1 + x)^{1/2}}$ $\qquad f'(3) = \dfrac{1}{4}$

$f''(x) = -\dfrac{1}{4}(1 + x)^{-3/2}(1) = \dfrac{-1}{4(1 + x)^{3/2}}$ $\qquad f''(3) = -\dfrac{1}{32}$

$f^{(3)}(x) = \dfrac{3}{8}(1 + x)^{-5/2}(1) = \dfrac{3}{8(1 + x)^{5/2}}$ $\qquad f^{(3)}(3) = \dfrac{3}{256}$

Thus,

$p_3(x) = f(3) + f'(3)(x - 3) + \dfrac{f''(3)}{2!}(x - 3)^2 + \dfrac{f^{(3)}(3)}{3!}(x - 3)^3$

$= 2 + \dfrac{1}{4}(x - 3) - \dfrac{1}{32 \cdot 2!}(x - 3)^2 + \dfrac{3}{256 \cdot 3!}(x - 3)^3$

$= 2 + \dfrac{1}{4}(x - 3) - \dfrac{1}{64}(x - 3)^2 + \dfrac{1}{512}(x - 3)^3.$

Now, $\sqrt{3.9} = f(2.9) \approx p_3(2.9) = 2 + \dfrac{1}{4}(-0.1) - \dfrac{1}{64}(-0.1)^2 + \dfrac{1}{512}(-0.1)^3$

$= 1.974842.$ (2-1)

**4.** $f(x) = \sqrt{9 + x^2} = (9 + x^2)^{1/2}$  $\qquad\qquad\qquad$ $f(0) = 3$

$f'(x) = \dfrac{1}{2}(9 + x^2)^{-1/2}(2x) = x(9 + x^2)^{-1/2}$ $\qquad$ $f'(0) = 0$

$f''(x) = (9 + x^2)^{-1/2} - \dfrac{1}{2}x(9 + x^2)^{-3/2}(2x)$

$\qquad = (9 + x^2)^{-1/2} - x^2(9 + x^2)^{-3/2}$ $\qquad\qquad$ $f''(0) = \dfrac{1}{3}$

Thus,

$$p_2(x) = f(0) + f'(0)x + \dfrac{f''(0)}{2!}x^2 = 3 + \dfrac{\frac{1}{3}}{2!}x^2 = 3 + \dfrac{1}{6}x^2.$$

Now, $\sqrt{9.1} = f(0.1) \approx p_2(0.1) = 3 + \dfrac{1}{6}(0.1)^2 = 3.001667.$ $\qquad\qquad$ (2-1)

**5.** $\dfrac{1}{1 - 4x} = 1 + 4x + 4^2 x^2 + \cdots + 4^n x^n + \cdots$

We have $a = 0$ and

$a_n = 4^n$

$a_{n+1} = 4^{n+1}$

Thus, $\dfrac{a_{n+1}}{a_n} = \dfrac{4^{n+1}}{4^n} = 4$ and $\displaystyle\lim_{n \to \infty}\left|\dfrac{a_{n+1}}{a_n}\right| = \lim_{n \to \infty}|4| = 4.$

Therefore, $L = 4$, $R = \dfrac{1}{4}$, and the series converges for $|x - 0| < \dfrac{1}{4}$.

The interval of convergence is:

$|x| < \dfrac{1}{4}$ or $-\dfrac{1}{4} < x < \dfrac{1}{4}$ $\qquad\qquad\qquad\qquad\qquad$ (2-2)

**6.** $\dfrac{5}{x - 1} = 1 - \dfrac{1}{5}(x - 6) + \dfrac{1}{5^2}(x - 6)^2 - \cdots + \dfrac{(-1)^n}{5^n}(x - 6)^n + \cdots$

We have $a = 6$, and

$a_n = \dfrac{(-1)^n}{5^n}$

$a_{n+1} = \dfrac{(-1)^{n+1}}{5^{n+1}}$

Thus, $\dfrac{a_{n+1}}{a_n} = \dfrac{\frac{(-1)^{n+1}}{5^{n+1}}}{\frac{(-1)^n}{5^n}} = \dfrac{-5^n}{5^{n+1}} = \dfrac{-1}{5}$ and $\displaystyle\lim_{n \to \infty}\left|\dfrac{a_{n+1}}{a_n}\right| = \lim_{n \to \infty}\left|\dfrac{-1}{5}\right| = \dfrac{1}{5}.$

Therefore, $L = \dfrac{1}{5}$, $R = \dfrac{1}{\frac{1}{5}} = 5$, and the series converges for $|x - 6| < 5$.

The interval of convergence is:

$|x - 6| < 5$ or $1 < x < 11$ $\qquad\qquad\qquad\qquad\qquad\qquad$ (2-2)

**7.** $\dfrac{2x}{(1-x)^3} = 1 \cdot 2x + 2 \cdot 3x^2 + 3 \cdot 4x^3 + \cdots + n(n+1)x^n + \cdots$

We have $a = 0$, and

$a_n = n(n+1)$

$a_{n+1} = (n+1)(n+2)$

Thus, $\dfrac{a_{n+1}}{a_n} = \dfrac{(n+1)(n+2)}{n(n+1)} = \dfrac{n+2}{n}$ and $\lim\limits_{n \to \infty} \left| \dfrac{a_{n+1}}{a_n} \right| = \lim\limits_{n \to \infty} \left| \dfrac{n+2}{n} \right| = 1.$

Therefore, $L = 1$, $R = 1$, and the series converges for $|x - 0| < 1$. The interval of convergence is:

$|x| < 1$ or $-1 < x < 1.$

(2-2)

**8.** $e^{10x} = 1 + 10x + \dfrac{10^2}{2!}x^2 + \cdots + \dfrac{10^n}{n!}x^n + \cdots$

We have $a = 0$ and

$a_n = \dfrac{10^n}{n!}$

$a_{n+1} = \dfrac{10^{n+1}}{(n+1)!}$

Thus, $\dfrac{a_{n+1}}{a_n} = \dfrac{\dfrac{10^{n+1}}{(n+1)!}}{\dfrac{10^n}{n!}} = \dfrac{10^{n+1}}{(n+1)!} \dfrac{n!}{10^n} = \dfrac{10}{n+1}$

and $\lim\limits_{n \to \infty} \left| \dfrac{a_{n+1}}{a_n} \right| = \lim\limits_{n \to \infty} \left| \dfrac{10}{n+1} \right| = 0.$

Therefore, the series converges for all $x$. The interval of convergence is: $-\infty < x < \infty.$

(2-2)

**9.** $f(x) = e^{-9x}$

$f'(x) = e^{-9x}(-9) = -9e^{-9x}$

$f''(x) = -9e^{-9x}(-9) = 9^2 e^{-9x}$

$f^{(3)}(x) = 9^2 e^{-9x}(-9) = -9^3 e^{-9x}$

$\vdots$

$f^{(n)}(x) = (-1)^n 9^n e^{-9x}$

(2-1)

**10.** $f(x) = \dfrac{1}{7-x} = (7-x)^{-1}$ $\qquad f(0) = \dfrac{1}{7}$

$f'(x) = -(7-x)^{-2}(-1) = (7-x)^{-2}$ $\qquad f'(0) = \dfrac{1}{7^2}$

$f''(x) = -2(7-x)^{-3}(-1) = 2(7-x)^{-3}$ $\qquad f''(0) = \dfrac{2}{7^3}$

$f^{(3)}(x) = -3(2)(7-x)^{-4}(-1) = 3!(7-x)^{-4}$ $\qquad f^{(3)}(0) = \dfrac{3!}{7^4}$

$\vdots$ $\qquad\qquad\qquad\qquad\qquad\qquad \vdots$

$f^{(n)}(x) = n!(7-x)^{-(n+1)}$ $\qquad f^{(n)}(0) = \dfrac{n!}{7^{n+1}}$

$$a_0 = f(0) = \frac{1}{7}$$

$$a_1 = f'(0) = \frac{1}{7^2}$$

$$a_2 = \frac{f''(0)}{2!} = \frac{2}{7^3 \cdot 2!} = \frac{1}{7^3}$$

$$a_3 = \frac{f^{(3)}(0)}{3!} = \frac{3!}{7^4 \cdot 3!} = \frac{1}{7^4}$$

$$\vdots$$

$$a_n = \frac{f^{(n)}(0)}{n!} = \frac{n!}{7^{n+1} n!} = \frac{1}{7^{n+1}}$$

The Taylor series at 0 for $f$ is:

$$f(x) = f(0) + f'(0)x + \frac{f''(0)}{2!}x^2 + \frac{f^{(3)}(0)}{3!}x^3 + \cdots + \frac{f^{(n)}(0)}{n!}x^n + \cdots$$

$$= \frac{1}{7} + \frac{1}{7^2}x + \frac{1}{7^3}x^2 + \cdots + \frac{1}{7^{n+1}}x^n + \cdots$$

Now, $a_n = \dfrac{1}{7^{n+1}}$, $a_{n+1} = \dfrac{1}{7^{n+2}}$, and $\dfrac{a_{n+1}}{a_n} = \dfrac{\frac{1}{7^{n+2}}}{\frac{1}{7^{n+1}}} = \dfrac{7^{n+1}}{7^{n+2}} = \dfrac{1}{7}$.

Thus, $\displaystyle\lim_{n \to \infty}\left|\frac{a_{n+1}}{a_n}\right| = \lim_{n \to \infty}\left|\frac{1}{7}\right| = \frac{1}{7}$. Therefore, $L = \dfrac{1}{7}$, $R = \dfrac{1}{\frac{1}{7}} = 7$, and the

series converges for $|x| < 7$  or  $-7 < x < 7$. (2-2)

11.   $\begin{aligned}f(x) &= \ln x\end{aligned}$           $f(2) = \ln 2$

$$f'(x) = \frac{1}{x} = x^{-1} \qquad\qquad f'(2) = \frac{1}{2}$$

$$f''(x) = -x^{-2} \qquad\qquad f''(2) = -\frac{1}{4} = -\frac{1}{2^2}$$

$$f^{(3)}(x) = 2x^{-3} \qquad\qquad f^{(3)}(2) = \frac{2!}{2^3}$$

$$f^{(4)}(x) = -3!\,x^{-4} \qquad\qquad f^{(4)}(2) = \frac{-3!}{2^4}$$

$$\vdots \qquad\qquad\qquad\qquad \vdots$$

$$f^{(n)}(x) = (-1)^{n-1}(n-1)!\,x^{-n} \qquad f^{(n)}(2) = \frac{(-1)^{n-1}(n-1)!}{2^n}$$

$$a_0 = f(2) = \ln 2$$

$$a_1 = f'(2) = \frac{1}{2}$$

$$a_2 = \frac{f''(2)}{2!} = \frac{-1}{2^2 \cdot 2!}$$

$$a_3 = \frac{f^{(3)}(2)}{3!} = \frac{2}{2^3 \cdot 3!} = \frac{1}{3 \cdot 2^3}$$

$$a_4 = \frac{f^{(4)}(2)}{4!} = \frac{-3!}{2^4 \cdot 4!} = \frac{-1}{4 \cdot 2^4}$$

$$\vdots$$

$$a_n = \frac{f^{(n)}(2)}{n!} = \frac{(-1)^{n-1}(n-1)!}{2^n \cdot n!} = \frac{(-1)^{n-1}}{n \cdot 2^n}$$

The Taylor series at 2 for $f$ is:

$$f(x) = f(2) + f'(2)(x-2) + \frac{f''(2)}{2!}(x-2)^2 + \cdots + \frac{f^{(n)}(2)}{n!}(x-2)^n + \cdots$$

$$= \ln 2 + \frac{1}{2}(x-2) - \frac{1}{2 \cdot 2^2}(x-2)^2 + \cdots + \frac{(-1)^{n-1}}{n \cdot 2^n}(x-2)^n + \cdots$$

$$= \ln 2 + \frac{1}{2}(x-2) - \frac{1}{8}(x-2)^2 + \cdots + \frac{(-1)^{n-1}}{n \cdot 2^n}(x-2)^n + \cdots .$$

Now, $a_n = \dfrac{(-1)^{n-1}}{n \cdot 2^n}$, $a_{n+1} = \dfrac{(-1)^n}{(n+1)2^{n+1}}$

and

$$\frac{a_{n+1}}{a_n} = \frac{\dfrac{(-1)^n}{(n+1)2^{n+1}}}{\dfrac{(-1)^{n-1}}{n \cdot 2^n}} = \frac{-n \cdot 2^n}{(n+1)2^{n+1}} = \frac{-n}{2(n+1)}.$$

Thus, $\displaystyle\lim_{n \to \infty}\left|\frac{a_{n+1}}{a_n}\right| = \lim_{n \to \infty}\left|\frac{-n}{2(n+1)}\right| = \frac{1}{2}$

Therefore, $L = \dfrac{1}{2}$, $R = \dfrac{1}{\frac{1}{2}} = 2$, and the series converges for

$|x-2| < 2$ or $0 < x < 4$. $\qquad\qquad$ (2-2)

**12.** $f(x) = \dfrac{1}{10+x} = \dfrac{1}{10\left(1 + \dfrac{x}{10}\right)} = \dfrac{1}{10} \cdot \dfrac{1}{1 + \dfrac{x}{10}}$

From Table 1, the Taylor series at 0 for $\dfrac{1}{1+x}$ is

$$\frac{1}{1+x} = 1 - x + x^2 - x^3 + \cdots + (-1)^n x^n + \cdots \qquad (-1 < x < 1).$$

Replacing $x$ by $\dfrac{x}{10}$, we have

$$\frac{1}{10} \cdot \frac{1}{1 + \dfrac{x}{10}} = \frac{1}{10}\left[1 - \frac{x}{10} + \frac{x^2}{10^2} - \frac{x^3}{10^3} + \cdots + \frac{(-1)^n}{10^n}x^n + \cdots\right]$$

$$= \frac{1}{10} - \frac{x}{10^2} + \frac{x^2}{10^3} - \frac{x^3}{10^4} + \cdots + \frac{(-1)^n}{10^{n+1}}x^n + \cdots .$$

This series converges for $-1 < \dfrac{x}{10} < 1$ or $-10 < x < 10$. $\qquad$ (2-3)

**13.** $f(x) = \dfrac{x^2}{4 - x^2} = \dfrac{x^2}{4\left(1 - \dfrac{x^2}{4}\right)} = \dfrac{x^2}{4} \cdot \dfrac{1}{1 - \dfrac{x^2}{4}}$

From Table 1, the Taylor series at 0 for $\dfrac{1}{1 - x}$ is

$$\frac{1}{1 - x} = 1 + x + x^2 + x^3 + \cdots + x^n + \cdots \quad (-1 < x < 1).$$

Replacing $x$ by $\dfrac{x^2}{4}$, we have

$$\frac{x^2}{4} \cdot \frac{1}{1 - \dfrac{x^2}{4}} = \frac{x^2}{4}\left[1 + \frac{x^2}{4} + \frac{x^4}{4^2} + \frac{x^6}{4^3} + \cdots + \frac{x^{2n}}{4^n} + \cdots\right]$$

$$= \frac{x^2}{4} + \frac{x^4}{4^2} + \frac{x^6}{4^3} + \frac{x^8}{4^4} + \cdots + \frac{x^{2n+2}}{4^{n+1}} + \cdots$$

The interval of convergence is $-1 < \dfrac{x^2}{4} < 1$ or $-2 < x < 2$. $\qquad$ (2-3)

**14.** $f(x) = x^2 e^{3x}$

From Table 1, the Taylor series at 0 for $e^x$ is

$$e^x = 1 + x + \frac{x^2}{2!} + \frac{x^3}{3!} + \cdots + \frac{x^n}{n!} + \cdots \quad (-\infty < x < \infty).$$

Replacing $x$ by $3x$, we have

$$x^2 e^{3x} = x^2\left[1 + 3x + \frac{(3x)^2}{2!} + \frac{(3x)^3}{3!} + \cdots + \frac{(3x)^n}{n!} + \cdots\right]$$

$$= x^2 + 3x^3 + \frac{3^2}{2!}x^4 + \frac{3^3}{3!}x^5 + \cdots + \frac{3^n}{n!}x^{n+2} + \cdots .$$

The interval of convergence for this series is $-\infty < x < \infty$. $\qquad$ (2-3)

**15.** $f(x) = x \ln(e + x) = x\left[\ln e\left(1 + \dfrac{x}{e}\right)\right]$

$$= x\left[\ln e + \ln\left(1 + \frac{x}{e}\right)\right] = \left[1 + \ln\left(1 + \frac{x}{e}\right)\right].$$

The Taylor series at 0 for $\ln(1 + x)$ is

$$\ln(1 + x) = x - \frac{x^2}{2} + \frac{x^3}{3} - \frac{x^4}{4} + \cdots + \frac{(-1)^{n-1}}{n}x^n + \cdots \quad (-1 < x < 1).$$

Replacing $x$ by $\dfrac{x}{e}$, we have

$$x \ln(e + x) = x\left[1 + \frac{1}{e}x - \frac{1}{2e^2}x^2 + \frac{1}{3e^3}x^3 - \cdots + \frac{(-1)^{n-1}}{ne^n}x^n + \cdots\right]$$

$$= x + \frac{1}{e}x^2 - \frac{1}{2e^2}x^3 + \frac{1}{3e^3}x^4 - \cdots + \frac{(-1)^{n-1}}{ne^n}x^{n+1} + \cdots .$$

The interval of convergence is $-1 < \dfrac{x}{e} < 1$ or $-e < x < e$. $\qquad$ (2-3)

**16.** $f(x) = \dfrac{1}{4 - x} = \dfrac{1}{2 - (x - 2)} = \dfrac{1}{2} \cdot \dfrac{1}{1 - \dfrac{x - 2}{2}}$

The Taylor series at 0 for $\dfrac{1}{1 - x}$ is

$$\frac{1}{1 - x} = 1 + x + x^2 + x^3 + \cdots + x^n + \cdots \quad (-1 < x < 1).$$

Replacing $x$ by $\dfrac{x - 2}{2}$, we have

$$\frac{1}{4 - x} = \frac{1}{2} \cdot \frac{1}{1 - \dfrac{x - 2}{2}}$$

$$= \frac{1}{2}\left[1 + \frac{x - 2}{2} + \frac{(x - 2)^2}{2^2} + \frac{(x - 2)^3}{2^3} + \cdots + \frac{(x - 2)^n}{2^n} + \cdots\right]$$

$$= \frac{1}{2} + \frac{x - 2}{2^2} + \frac{(x - 2)^2}{2^3} + \frac{(x - 2)^3}{2^4} + \cdots + \frac{(x - 2)^n}{2^{n+1}} + \cdots.$$

The interval of convergence is $-1 < \dfrac{x - 2}{2} < 1$ or $-2 < x - 2 < 2$ or

$0 < x < 4.$

$(2-3)$

**17.** (A) Theorem 1 of Section 2-2 is not applicable because the coefficients of the odd powers of $x$ are 0.

(B) If we let $t = 5x^2$, then we have
$$\ln(1 - 5x^2) = \ln(1 - t) = -t - \frac{1}{2}t^2 - \frac{1}{3}t^3 - \cdots - \frac{1}{n}t^n - \cdots, \quad -1 < t < 1,$$
from Table 1, Section 2-3.

Thus, the interval of convergence for the given Taylor series is $-1 < 5x^2 < 1$ which is the same as $-\dfrac{1}{\sqrt{5}} < x < \dfrac{1}{\sqrt{5}}$ or $-\dfrac{\sqrt{5}}{5} < x < \dfrac{\sqrt{5}}{5}$.

$(2-2, \ 2-3)$

**18.** (A) The Taylor series at 0 for $\dfrac{1}{1 + x}$ converges for $-1 < x < 1$.

Since the odd powers of $\sqrt{x} = x^{1/2}$ are defined only for $x \geq 0$, the formula $\dfrac{1}{1 + \sqrt{x}} = 1 - x^{1/2} + x - x^{3/2} + \cdots + (-1)^n x^{n/2} + \cdots$ is valid only for $0 \leq x < 1$.

(B) The formula for $\dfrac{1}{1 + \sqrt{x}}$ is **not** a Taylor series at 0 because some (in fact, half) of the powers of $x$ are not non-negative integers.

$(2-2, \ 2-3)$

**19.** $f(x) = \dfrac{1}{2-x} = \dfrac{1}{2} \cdot \dfrac{1}{1-\dfrac{x}{2}}$

Use the Taylor series at 0 for $\dfrac{1}{1-x}$ and replace $x$ by $\dfrac{x}{2}$:

$$\frac{1}{2-x} = \frac{1}{2} \cdot \frac{1}{1-\dfrac{x}{2}} = \frac{1}{2}\left[1 + \frac{x}{2} + \frac{x^2}{2^2} + \cdots + \frac{x^n}{2^n} + \cdots\right]$$

$$= \frac{1}{2} + \frac{x}{2^2} + \frac{x^2}{2^3} + \cdots + \frac{x^n}{2^{n+1}} + \cdots.$$

The interval of convergence is $-1 < \dfrac{x}{2} < 1$ or $-2 < x < 2$.

Now, $f'(x) = \dfrac{1}{(2-x)^2} = g(x)$.

Thus, the Taylor series at 0 for $g(x)$ can be obtained by differentiating the Taylor series for $f(x)$. We have

$$\frac{1}{(2-x)^2} = f'(x) = \frac{1}{2^2} + \frac{2x}{2^3} + \frac{3x^2}{2^4} + \cdots + \frac{nx^{n-1}}{2^{n+1}} + \cdots$$

$$= \frac{1}{4} + \frac{1}{4}x + \frac{3}{16}x^2 + \cdots + \frac{n}{2^{n+1}}x^{n-1} + \cdots.$$

The interval of convergence is $-2 < x < 2$. $\hspace{2cm}$ (2-3)

**20.** $f(x) = \dfrac{x^2}{1+x^2}$

Use the Taylor series at 0 for $\dfrac{1}{1+x}$, replace $x$ by $x^2$, and multiply by $x^2$. This yields

$$\frac{x^2}{1+x^2} = x^2 \cdot \frac{1}{1+x^2} = x^2[1 - x^2 + x^4 - x^6 + \cdots + (-1)^n x^{2n} + \cdots]$$

$$= x^2 - x^4 + x^6 - x^8 + \cdots + (-1)^n x^{2n+2} + \cdots.$$

The interval of convergence is $-1 < x^2 < 1$ or $-1 < x < 1$.
Now,

$$f'(x) = \frac{2x}{(1+x^2)^2} = g(x).$$

Thus, we differentiate the series for $f(x)$ to obtain the series for $g(x)$.

$$\frac{2x}{(1+x^2)^2} = f'(x) = 2x - 4x^3 + 6x^5 - 8x^7 + \cdots + (-1)^n(2n+2)x^{2n+1} + \cdots.$$

The interval of convergence is $-1 < x < 1$. $\hspace{2cm}$ (2-3)

**21.** $f(x) = \int_0^x \dfrac{t^2}{9 + t^2}\, dt$

$$\frac{t^2}{9 + t^2} = \frac{t^2}{9} \cdot \frac{1}{1 + \dfrac{t^2}{9}}$$

Use the Taylor series at 0 for $\dfrac{1}{1 + x}$, replace $x$ by $\dfrac{t^2}{9}$, and multiply the result by $\dfrac{t^2}{9}$. We obtain

$$\frac{t^2}{t^2 + 9} = \frac{t^2}{9} \cdot \frac{1}{1 + \dfrac{t^2}{9}} = \frac{t^2}{9}\left[1 - \frac{t^2}{9} + \frac{t^4}{9^2} - \frac{t^6}{9^3} + \ldots + \frac{(-1)^n}{9^n}\, t^{2n} + \ldots\right]$$

$$= \frac{t^2}{9} - \frac{t^4}{9^2} + \frac{t^6}{9^3} - \ldots + \frac{(-1)^n}{9^{n+1}}\, t^{2n+2} + \ldots .$$

The interval of convergence for this series is $-1 < \dfrac{t^2}{9} < 1$ or $-3 < t < 3$.

Now,

$$f(x) = \int_0^x \frac{t^2}{9 + t^2}\, dt$$

$$= \left[\frac{t^3}{3 \cdot 9} - \frac{t^5}{5 \cdot 9^2} + \frac{t^7}{7 \cdot 9^3} - \ldots + \frac{(-1)^n}{(2n+3)9^{n+1}}\, t^{2n+3} + \ldots\right]\Bigg|_0^x$$

$$= \frac{1}{3 \cdot 9}\, x^3 - \frac{1}{5 \cdot 9^2}\, x^5 + \frac{1}{7 \cdot 9^3}\, x^7 - \ldots + \frac{(-1)^n}{(2n+3)9^{n+1}}\, x^{2n+3} + \ldots$$

and the interval of convergence is $-3 < x < 3$. $\hspace{2cm}$ (2-3)

**22.** $f(x) = \int_0^x \dfrac{t^4}{16 - t^2}\, dt$

$$\frac{t^4}{16 - t^2} = \frac{t^4}{16} \cdot \frac{1}{1 - \dfrac{t^2}{16}}$$

Thus, we use the Taylor series at 0 for $\dfrac{1}{1 - x}$, replace $x$ by $\dfrac{t^2}{16}$, and multiply the result by $\dfrac{t^4}{16}$. We obtain

$$\frac{t^4}{16 - t^2} = \frac{t^4}{16} \cdot \frac{1}{1 - \dfrac{t^2}{16}} = \frac{t^4}{16}\left[1 + \frac{t^2}{16} + \frac{t^4}{16^2} + \frac{t^6}{16^3} + \ldots + \frac{t^{2n}}{16^n} + \ldots\right]$$

$$= \frac{1}{16}\, t^4 + \frac{1}{16^2}\, t^6 + \frac{1}{16^3}\, t^8 + \ldots + \frac{1}{16^{n+1}}\, t^{2n+4} + \ldots .$$

The interval of convergence is $-1 < \dfrac{t^2}{16} < 1$ or $-4 < t < 4$.

Now,

$$f(x) = \int_0^x \frac{t^4}{16 - t^2}\, dt$$

$$= \left[ \frac{1}{5 \cdot 16} t^5 + \frac{1}{7 \cdot 16^2} t^7 + \frac{1}{9 \cdot 16^3} t^9 + \dots + \frac{1}{(2n + 5)16^{n+1}} t^{2n+5} + \dots \right] \Big|_0^x$$

$$= \frac{1}{5 \cdot 16} x^5 + \frac{1}{7 \cdot 16^2} x^7 + \frac{1}{9 \cdot 16^3} x^9 + \dots + \frac{1}{(2n + 5)16^{n+1}} x^{2n+5} + \dots$$

and the interval of convergence is $-4 < x < 4$. (2-3)

**23.** Let $f(x) = x^3 - 3x^2 + 4$

(A) At $a = 1$:

$$f(x) = x^3 - 3x^2 + 4 \qquad\qquad f(1) = 2$$
$$f'(x) = 3x^2 - 6x \qquad\qquad\qquad f'(1) = -3$$
$$f''(x) = 6x - 6 \qquad\qquad\qquad f''(1) = 0$$
$$f^{(3)}(x) = 6 \qquad\qquad\qquad f^{(3)}(1) = 6$$
$$f^{(4)}(x) = f^{(5)}(x) = \dots = 0 \qquad f^{(4)}(1) = f^{(5)}(1) = \dots = 0$$

Thus,

$$f(x) = f(1) + f'(1)(x - 1) + \frac{f''(1)}{2!}(x - 1)^2 + \frac{f^{(3)}(1)}{3!}(x - 1)^3$$

$$= 2 - 3(x - 1) + (x - 1)^3$$

At $a = -1$:

$$f(-1) = 0$$
$$f'(-1) = 9$$
$$f''(-1) = -12$$
$$f^{(3)}(-1) = 6$$
$$f^{(4)}(-1) = f^{(5)}(-1) = \dots = 0$$

Thus,

$$f(x) = f(-1) + f'(-1)(x + 1) + \frac{f''(1)}{2!}(x + 1)^2 + \frac{f^{(3)}(-1)}{3!}(x + 1)^3$$

$$= 9(x + 1) - 6(x + 1)^2 + (x + 1)^3$$

(B) Each series represents $f(x) = x^3 - 3x^2 + 4$ for $-\infty < x < \infty$. (2-2)

**24.** (A) $f(x) = |x|$ does not have any Taylor polynomials at 0 because $f$ is not differentiable at 0.

(B) $f(x) = |x| = \begin{cases} x & \text{if} \quad x \geq 0 \\ -x & \text{if} \quad x < 0 \end{cases}$

and $f'(x) = \begin{cases} 1 & \text{if} \quad x > 0 \\ -1 & \text{if} \quad x < 0 \end{cases}$

If $a > 0$, then $p_n(x) = a + (x - a) = x$, $n \geq 1$

If $a < 0$, then $p_n(x) = -a - (x - a) = -x$, $n \geq 1$ (2-2)

*For Problems 25 and 26:*

$$e^x = 1 + x + \frac{x^2}{2!} + \frac{x^3}{3!} + \cdots,$$

$$p_2(x) = 1 + x + \frac{x^2}{2!} = 1 + x + \frac{1}{2}x^2, \text{ and}$$

$$R_2(x) = \frac{f^{(3)}(t)}{3!}x^3 = \frac{e^t}{3!}x^3, \text{ where } 0 < t < x.$$

**25.** Let $x = 0.6$. Then

$$e^{0.6} \approx p_2(0.6) = 1 + 0.6 + \frac{(0.6)^2}{2} = 1.78$$

and $\left| R_2(x) \right| = \left| \frac{e^t}{3!}(0.6)^3 \right| < \frac{3(0.6)^3}{3!} = 0.108.$  (2-4)

**26.** Let $x = 0.06$. Then

$$e^{0.06} \approx p_2(0.06) = 1 + 0.06 + \frac{(0.06)^2}{2} = 1.0618$$

and $\left| R_2(x) \right| = \left| \frac{e^t(0.06)^3}{3!} \right| < \frac{3(0.06)^3}{3!} = 0.000108.$  (2-4)

*For Problems 27 and 28:*

$$\ln(1 + x) = x - \frac{x^2}{2} + \frac{x^3}{3} - \frac{x^4}{4} + \frac{x^5}{5} - \cdots$$

**27.** Let $x = 0.3$. Then

$$\ln(1.3) = \ln(1 + 0.3) = 0.3 - \frac{(0.3)^2}{2} + \frac{(0.3)^3}{3} - \frac{(0.3)^4}{4} + \frac{(0.3)^5}{5} - \cdots$$
$$= 0.3 - 0.045 + 0.009 - 0.002025 + 0.000486 \approx 0.261975.$$

The terms are alternating in sign and decreasing in absolute value. If we use the Taylor polynomial of degree 4, then the error will be less than $0.000486 < 0.0005$. Thus, $\ln(1.3) \approx 0.261975$ to within $\pm 0.0005$.

(2-4)

**28.** Let $x = 0.03$. Then

$$\ln(1.03) = \ln(1 + 0.03) = 0.03 - \frac{(0.03)^2}{2} + \frac{(0.03)^3}{3} - \cdots$$
$$= 0.03 - 0.00045 + 0.000009 - \cdots \approx 0.03.$$

The terms are alternating in sign and decreasing in absolute value. If we use the Taylor polynomial of degree 1, then the error will be less than $0.00045 < 0.0005$. Thus, $\ln(1.03) \approx 0.03$ to within $\pm 0.0005$.

(2-4)

**29.** $f'(x) = x \ln(1 - x)$. From Table 1, Section 2-3,

$$\ln(1 - x) = -x - \frac{1}{2}x^2 - \frac{1}{3}x^3 - \ldots - \frac{1}{n}x^n - \ldots, \quad -1 < x < 1$$

Therefore

$$f'(x) = -x^2 - \frac{1}{2}x^3 - \frac{1}{4}x^4 - \ldots - \frac{1}{n}x^{n+1} - \ldots$$

and

$$f(x) = \int f'(x)\,dx = \int \left( -x^2 - \frac{1}{2}x^3 - \frac{1}{4}x^4 - \ldots - \frac{1}{n}x^{n+1} - \ldots \right) dx$$

$$= C - \frac{1}{3}x^3 - \frac{1}{8}x^4 - \frac{1}{20}x^5 - \ldots - \frac{1}{n(n+2)}x^{n+2} - \ldots$$

Since $f(0) = 5$, we have

$$5 = C - 0 - 0 - \ldots \quad \text{so} \quad C = 5$$

Thus,

$$f(x) = 5 - \frac{1}{3}x^3 - \frac{1}{8}x^4 - \frac{1}{20}x^5 - \ldots - \frac{1}{n(n+2)}x^{n+2} - \ldots$$

$$= 5 - \frac{1}{3}x^3 - \frac{1}{8}x^4 - \frac{1}{20}x^5 - \ldots - \frac{1}{(n-2)n}x^n - \ldots, \quad -1 < x < 1.$$

$$(2\text{-}3)$$

**30.** $f''(x) = xe^{-x}$. From Table 1, Section 2-3,

$$e^{-x} = 1 - x + \frac{1}{2!}x^2 - \frac{1}{3!}x^3 + \ldots + \frac{(-1)^n}{n!}x^n + \ldots, \quad -\infty < x < \infty.$$

Therefore

$$f''(x) = x - x^2 + \frac{1}{2!}x^3 - \frac{1}{3!}x^4 + \ldots + \frac{(-1)^n}{n!}x^{n+1} + \ldots$$

and

$$f'(x) = \int f''(x)\,dx = \int \left( x - x^2 + \frac{1}{2!}x^3 - \frac{1}{3!}x^4 + \ldots + \frac{(-1)^n}{n!}x^{n+1} + \ldots \right) dx$$

$$= C_1 + \frac{1}{2}x^2 - \frac{1}{3}x^3 + \frac{1}{4(2!)}x^4 - \frac{1}{5(3!)}x^5 + \ldots + \frac{(-1)^n}{(n+2)n!}x^{n+2} + \ldots$$

Since $f'(0) = -4$, we have

$$-4 = C_1 + 0 - 0 + 0 \ldots = C_1 \quad \text{so} \quad C_1 = -4$$

and

$$f'(x) = -4 + \frac{1}{2}x^2 - \frac{1}{3}x^3 + \frac{1}{4(2!)}x^4 - \ldots + \frac{(-1)^n}{(n+2)n!}x^{n+2} + \ldots$$

Now

$$f(x) = \int f'(x)\,dx = \int \left( -4 + \frac{1}{2}x^2 - \frac{1}{3}x^3 + \frac{1}{4(2!)}x^4 - \ldots + \frac{(-1)^n}{(n+2)n!}x^{n+2} + \ldots \right) dx$$

$$= C_2 - 4x + \frac{1}{6}x^3 - \frac{1}{12}x^4 + \frac{1}{4 \cdot 5(2!)}x^5 - \ldots + \frac{(-1)^n}{(n+2)(n+3)n!}x^{n+3} + \ldots$$

Since $f(0) = 3$, we have

$$3 = C_2 - 0 + 0 - 0 + \ldots = C_2 \quad \text{so} \quad C_2 = 3$$

and

$$f(x) = 3 - 4x + \frac{1}{6}x^3 - \frac{1}{12}x^4 + \ldots + \frac{(-1)^n}{(n+2)(n+3)n!}x^{n+3} + \ldots$$

$$= 3 - 4x + \frac{1}{6}x^3 - \frac{1}{12}x^4 + \ldots + \frac{(-1)^{n-3}}{(n-3)!(n-1)n}x^n + \ldots$$

$$\text{(replacing } n \text{ by } n-3\text{)}$$

$$= 3 - 4x + \frac{1}{6}x^3 - \frac{1}{12}x^4 + \ldots + \frac{(-1)^{n+1}}{(n-3)!(n-1)n}x^n + \ldots , \quad -\infty < x < \infty.$$

$$(2\text{-}3)$$

**31.** Using the Taylor series at 0 for $\dfrac{1}{1+x}$, we have

$$\frac{1}{16+x^2} = \frac{1}{16} \cdot \frac{1}{1+\dfrac{x^2}{16}}$$

$$= \frac{1}{16}\left[1 - \frac{x^2}{16} + \frac{x^4}{16^2} - \frac{x^6}{16^3} + \ldots + \frac{(-1)^n}{16^n}x^{2n} + \ldots\right]$$

$$= \frac{1}{16} - \frac{x^2}{16^2} + \frac{x^4}{16^3} - \frac{x^6}{16^4} + \ldots + \frac{(-1)^n}{16^{n+1}}x^{2n} + \ldots$$

The interval of convergence is $-1 < \dfrac{x^2}{16} < 1$ or $-4 < x < 4$. Thus,

$$\int_0^1 \frac{1}{16+x^2}\,dx = \left(\frac{1}{16}x - \frac{1}{3\cdot 16^2}x^3 + \frac{1}{5\cdot 16^3}x^5 - \ldots \right.$$

$$\left. + \frac{(-1)^n}{(2n+1)16^{n+1}}x^{2n+1} + \ldots\right)\Big|_0^1$$

$$= \frac{1}{16} - \frac{1}{3\cdot 16^2} + \frac{1}{5\cdot 16^3} - \ldots$$

$$= 0.0625 - 0.00130 + 0.00005 - \ldots .$$

Since the terms in the series are alternating in sign and decreasing in absolute value, the error in using the first two terms in this series will be less than 0.00005. Thus,

$$\int_0^1 \frac{1}{16+x^2}\,dx \approx 0.0625 - 0.00130 = 0.0612. \qquad (2\text{-}4)$$

**32.** Using the Taylor series at 0 for $f(x) = e^{-x}$, we obtain

$$x^2 e^{-(0.1)x^2} = x^2\left[1 - \frac{x^2}{10} + \frac{1}{2!\,10^2}x^4 - \frac{1}{3!\,10^3}x^6 + \ldots\right]$$

$$= x^2 - \frac{x^4}{10} + \frac{1}{2!\,10^2}x^6 - \frac{1}{3!\,10^3}x^8 + \ldots .$$

Thus,

$$\int_0^1 x^2 e^{-(0.1)x^2} dx = \left( \frac{1}{3} x^3 - \frac{1}{50} x^5 + \frac{1}{7 \cdot 2! \, 10^2} x^7 - \frac{1}{9 \cdot 3! \, 10^3} x^9 + \cdots \right) \Big|_0^1$$

$$= \frac{1}{3} - \frac{1}{50} + \frac{1}{1400} - \frac{1}{54,000} + \cdots .$$

Since the terms in the series are alternating in sign and decreasing in absolute value, the error in using the first three terms in this series will be less than $\frac{1}{54,000} = 0.000019 < 0.0005$.

Thus,

$$\int_0^1 x^2 e^{-(0.1)x^2} dx \approx 0.33333 - 0.02 + 0.0071 = 0.31404. \qquad (2\text{-}4)$$

33. From Taylor's formula for the remainder,

$$R_n(0.5) = \frac{f^{(n+1)}(t)}{(n+1)!} (0.5)^{n+1}$$

where $t$ is some number between 0 and 0.5.
Since $\left| f^{(n+1)}(t) \right| \leq 10$,

$$\left| R_n(0.5) \right| = \left| \frac{f^{(n+1)}(t)}{(n+1)!} (0.5)^{n+1} \right| = \frac{\left| f^{(n+1)}(t) \right|}{(n+1)!} (0.5)^{n+1} \leq \frac{10}{(n+1)!} (0.5)^{n+1}$$

Now, we want $\left| R_n(0.5) \right| < 10^{-6}$ which will be true if

$$\frac{10}{(n+1)!} (0.5)^{n+1} < 10^{-6}$$

which implies $(n+1)! > \dfrac{10(0.5)^{n+1}}{10^{-6}} - 10^7 (0.5)^{n+1} = \dfrac{10^7}{2^{n+1}}$.

The smallest $n$ that satisfies this inequality is $n = 7$. $\qquad (2\text{-}4)$

34. From Taylor's formula for the remainder,

$$R_8(x) = \frac{f^{(9)}(t)}{9!} x^9$$

where $t$ is between 0 and $x$.
Now

$$\left| R_8(x) \right| = \left| \frac{f^{(9)}(t)}{9!} x^9 \right| \leq \frac{10}{9!} |x|^9 < 10^{-6}$$

implies

$$|x|^9 < 9! (10)^{-7} = \frac{9!}{10^7} = 0.036288$$

Therefore

$$|x| < (0.036288)^{1/9} \approx 0.6918.$$

The error in the approximation of $f(x)$ by $P_8(x)$ will be less than $10^{-6}$ if $\quad |x| < 0.6918$. $\qquad (2\text{-}4)$

**35.** $e^x = 1 + x + \dfrac{1}{2!}x^2 + \cdots + \dfrac{1}{n!}x^n + \cdots, \quad -\infty < x < \infty$

$e^{x^2} = 1 + x^2 + \dfrac{1}{2!}x^4 + \cdots + \dfrac{1}{n!}x^{2n} + \cdots, \quad -\infty < x < \infty.$

Thus, the fourth-degree Taylor polynomial for $f(x) = e^{x^2}$ at 0 is:

$p_4(x) = 1 + x^2 + \dfrac{1}{2}x^4$

Using a graphing utility, we find that

$\left| p_4(x) - e^{x^2} \right| < 0.01 \quad \text{for} \quad -0.616 < x < 0.616.$ $\qquad$ (2-4)

**36.** From Table 1, Section 2-3,

$\ln(1 - x) = -x - \dfrac{1}{2}x^2 - \dfrac{1}{3}x^3 - \cdots - \dfrac{1}{n}x^n - \cdots, \quad -1 < x < 1$

Substituting $x^2$ for $x$, we have

$f(x) = \ln(1 - x^2) = -x^2 - \dfrac{1}{2}x^4 - \dfrac{1}{3}x^6 - \cdots - \dfrac{1}{n}x^{2n} - \cdots, \quad -1 < x < 1$

Thus, the sixth-degree Taylor polynomial for $f$ at 0 is:

$p_6(x) = -x^2 - \dfrac{1}{2}x^4 - \dfrac{1}{3}x^6$

Using a graphing utility, we find that

$\left| p_6(x) - \ln(1 - x^2) \right| < 0.001 \quad \text{for} \quad -0.488 < x < 0.488.$ $\qquad$ (2-4)

**37.** $D(x) = \dfrac{1}{10}\sqrt{2500 - x^2} = \dfrac{1}{10}(2500 - x^2)^{1/2}, \qquad\qquad D(0) = 5$

$D'(x) = \dfrac{1}{20}(2500 - x^2)^{-1/2}(-2x) = -\dfrac{1}{10}x(2500 - x^2)^{-1/2} \qquad D'(0) = 0$

$D''(x) = -\dfrac{1}{10}(2500 - x^2)^{-1/2} + \dfrac{1}{20}x(2500 - x^2)^{-3/2}(-2x) \qquad D''(0) = -\dfrac{1}{500}$

$\qquad\; = -\dfrac{1}{10}(2500 - x^2)^{-1/2} - \dfrac{1}{10}x^2(2500 - x^2)^{-3/2}$

Now, $p_2(x) = D(0) + D'(0)x + \dfrac{D''(0)}{2!}x^2 = 5 - \dfrac{1}{1000}x^2$. Therefore,

Average price $= \dfrac{1}{15 - 0}\displaystyle\int_0^{15} D(x)\,dx \approx \dfrac{1}{15}\int_0^{15} p_2(x)\,dx = \dfrac{1}{15}\int_0^{15}\left(5 - \dfrac{1}{1000}x^2\right)dx$

$\qquad\qquad\quad = \dfrac{1}{15}\left[5x - \dfrac{1}{3000}x^3\right]\Big|_0^{15} = \dfrac{1}{15}\left(75 - \dfrac{3375}{3000}\right) = 4.925 \text{ or } \$4.93.$

$\qquad$ (2-4)

**38.** $R(t) = 6 + 3e^{-0.01t^2}$

Substituting $-0.01t^2$ into the Taylor series at 0 for $e^x$, we have

$$e^{-0.01t^2} = 1 - (0.01)t^2 + \frac{(0.01)^2}{2!}t^4 - \frac{(0.01)^3}{3!}t^6 + \dots$$

Thus, the second degree Taylor polynomial for $R$ is:

$$p_2(t) = 6 + 3(1 - 0.01t^2) = 9 - 0.03t^2$$

The total production during the first 10 years of operation of the well is:

$$T = \int_0^{10} R(t)\,dt \approx \int_0^{10} p_2(t)\,dt = \int_0^{10} (9 - 0.03t^2)\,dt = (9t - 0.01t^3)\Big|_0^{10} = 80$$

or \$80,000                                                                                      (2-4)

**39.** Index of income concentration $= 2\int_0^1 [x - f(x)]\,dx = 2\int_0^1 \left(x - \frac{9x^3}{8 + x^2}\right)dx$

$$= \int_0^1 2x\,dx - \int_0^1 \frac{18x^3}{8 + x^2}\,dx = x^2\Big|_0^1 - \int_0^1 \frac{18x^3}{8 + x^2}\,dx$$

$$= 1 - \int_0^1 \frac{18x^3}{8 + x^2}\,dx$$

Step 1: Find the Taylor series for the integrand.

$$\frac{18x^3}{8 + x^2} = \frac{18x^3}{8} \cdot \frac{1}{1 + \dfrac{x^2}{8}}$$

$$= \frac{9x^3}{4}\left[1 - \frac{x^2}{8} + \frac{x^4}{8^2} - \frac{x^6}{8^3} + \dots + \frac{(-1)^n}{8^n}x^{2n} + \dots\right]$$

$$\left[\text{Substitute } \frac{x^2}{8} \text{ into the series for } \frac{1}{1 + x}\right]$$

$$= \frac{9}{4}\left[x^3 - \frac{x^5}{8} + \frac{x^7}{8^2} - \frac{x^9}{8^3} + \dots + \frac{(-1)^n}{8^n}x^{2n+3} + \dots\right]$$

This series converges for $\left|\dfrac{x^2}{8}\right| < 1$ or $-\sqrt{8} < x < \sqrt{8}$.

Step 2: Find the Taylor series for the antiderivative.

$$\int \frac{18x^3}{8 + x^2}\,dx = \int \frac{9}{4}\left[x^3 - \frac{x^5}{8} + \frac{x^7}{8^2} - \frac{x^9}{8^3} + \dots + \frac{(-1)^n}{8^n}x^{2n+3} + \dots\right]dx$$

$$= C + \frac{9}{4}\left[\frac{x^4}{4} - \frac{x^6}{6 \cdot 8} + \frac{x^8}{8 \cdot 8^2} - \frac{x^{10}}{10 \cdot 8^3} + \dots + \frac{(-1)^n}{(2n + 4)8^n}x^{2n+4} + \dots\right]$$

<u>Step 3</u>: Approximate the definite integral.

Choosing $C = 0$ in the antiderivative, we have:

$$\int \frac{18x^3}{8 + x^2}\,dx = \frac{9}{4}\left[\frac{x^4}{4} - \frac{x^6}{6 \cdot 8} + \frac{x^8}{8 \cdot 8^2} - \frac{x^{10}}{10 \cdot 8^3} + \cdots\right.$$

$$\left. + \frac{(-1)^n}{(2n + 4)8^n}\,x^{2n+4} + \cdots\right]\Bigg|_0^1$$

$$= \frac{9}{4}\left(\frac{1}{4} - \frac{1}{48} + \frac{1}{512} - \frac{1}{5120} + \cdots\right)$$

$$= 0.5625 - 0.046875 + 0.004395 - \cdots$$

Since the terms are alternating in sign and decreasing in absolute value, and since the third term is less than 0.005, we can use the first two terms of this series to approximate the definite integral within the specified accuracy.

$$\int \frac{18x^3}{8 + x^2}\,dx \approx 0.5625 - 0.046875 = 0.515625 \approx 0.516$$

Finally, the index of income concentration is:

$$2\int_0^1\left(x - \frac{9x^3}{8 + x^2}\right)dx \approx 1 - 0.516 = 0.484. \tag{2-4}$$

**40.** Total sales $= \int_0^8 S'(t)\,dt = \int_0^8 [20 - 20e^{-0.001t^2}]\,dt..$

<u>Step 1</u>: Find the fourth-degree polynomial at 0 for the integrand.

Substituting $0.001t^2 = \dfrac{t^2}{10^3}$ into the Taylor series at 0 for $e^{-x}$, we have:

$$e^{-0.001t^2} = 1 - \frac{t^2}{10^3} + \frac{1}{2!}\cdot\frac{t^4}{10^6} - \frac{1}{3!}\cdot\frac{t^6}{10^9} + \cdots$$

Thus,

$$20 - 20e^{-0.001t^2} = \frac{20t^2}{10^3} - 20\cdot\frac{1}{2!}\cdot\frac{t^4}{10^6} + 20\cdot\frac{1}{3!}\cdot\frac{t^6}{10^9} + \cdots$$

$$= \frac{t^2}{50} - \frac{t^4}{10^5} + \frac{1}{3\cdot10^8}t^6 - \cdots$$

and $p_4(t) = \dfrac{t^2}{50} - \dfrac{t^4}{10^5}$.

<u>Step 2</u>: Approximate the definite integral.

$$\int_0^8 S'(t)\,dt \approx \int_0^8 p_4(t)\,dt = \int_0^8\left(\frac{t^2}{50} - \frac{t^4}{10^5}\right)dt = \left(\frac{t^3}{150} - \frac{t^5}{5\cdot10^5}\right)\Bigg|_0^8$$

$$= \frac{8^3}{150} - \frac{8^5}{5\cdot10^5} \approx 3.3478$$

Therefore, $S(8) \approx 3.348$, and the total sales during the first eight months are approximately \$3,348,000.

Step 3: Estimate the error in this approximation.

From Steps 1 and 2, the Taylor series at 0 for $S(t)$ is:

$$S(t) = \frac{t^3}{150} - \frac{t^5}{5 \cdot 10^5} + \frac{t^7}{3 \cdot 7 \cdot 10^8} - \cdots \text{, and}$$

$$S(8) = \frac{8^3}{150} - \frac{8^5}{5 \cdot 10^5} + \frac{8^7}{21 \cdot 10^8} - \cdots .$$

Since the terms alternate in sign and decrease in absolute value, the error in this approximation is less than

$$\left| \frac{8^7}{21 \cdot 10^8} \right| = 0.00099 < 0.001.$$

Thus, $S(8) = 3.348$ to within $\pm 0.001$ or $S(8) = \$3,348,000$ to within $\pm\$1000$. (2-4)

**41.** $A'(t) = \dfrac{-100}{t^2 + 40}$

Step 1: Find the second-degree Taylor polynomial at 0 for

$$A'(t) = \frac{-100}{t^2 + 40} = \frac{-100}{40} \cdot \frac{1}{1 + \dfrac{t^2}{40}} = \frac{-5}{2} \cdot \frac{1}{1 + \dfrac{t^2}{40}}$$

Substituting $\dfrac{t^2}{40}$ into the Taylor series at 0 for $\dfrac{1}{1 + x}$, we have:

$$-\frac{5}{2} \cdot \frac{1}{1 + \dfrac{t^2}{40}} = -\frac{5}{2}\left[ 1 - \frac{t^2}{40} + \frac{t^4}{(40)^2} - \frac{t^6}{(40)^3} + \cdots \right]$$

Thus, $p_2(t) = -\dfrac{5}{2} + \dfrac{5}{2} \cdot \dfrac{t^2}{40} = -\dfrac{5}{2} + \dfrac{1}{16} t^2$.

Step 2: Approximate $A(t)$ and find $A(2)$.

$$A(t) = \int A'(t)\,dt \approx \int p_2(t)\,dt = \int \left[ -\frac{5}{2} + \frac{1}{16} t^2 \right] dt = -\frac{5}{2} t + \frac{1}{48} t^3 + C$$

Since $A(0) = 15$, we have:

$$A(t) \approx -\frac{5}{2} t + \frac{1}{48} t^3 + 15$$

Thus, $A(2) \approx -\dfrac{5}{2}(2) + \dfrac{1}{48}(2)^3 + 15 \approx 10.16667$.

Step 3: Estimate the error in this approximation.

From Steps 1 and 2, the Taylor series at 0 for $A(t)$ is:

$$A(t) = 15 - \frac{5}{2} t + \frac{t^3}{48} - \frac{t^5}{5(40)^2} + \cdots \text{, and}$$

$$A(2) = 15 - \frac{5}{2}(2) + \frac{2^3}{48} - \frac{2^5}{5(40)^2} + \cdots$$

$$= 15 - 5 + \frac{1}{6} - \frac{32}{8000} + \cdots .$$

Since the terms alternate in sign and decrease in absolute value, the error in this approximation is less than
$$\left| \frac{-32}{8000} \right| = 0.004 < 0.01.$$

Thus, $A(2) = 10.17$ square centimeters to within $\pm 0.01$.    (2-4)

42. The average insulin level for the five-minute interval following the injection is:
$$\frac{1}{5} \int_0^5 \frac{5000t^2}{10,000 + t^4} \, dt$$

The Taylor series at 0 for $\dfrac{5000t^2}{10,000 + t^4} = \dfrac{5000t^2}{10,000} \cdot \dfrac{1}{1 + \dfrac{t^4}{10,000}}$

$$= \frac{t^2}{2} \cdot \frac{1}{1 + \left(\dfrac{t}{10}\right)^4} \quad \text{is:}$$

$$\frac{t^2}{2}\left[ 1 - \left(\frac{t}{10}\right)^4 + \left(\frac{t}{10}\right)^8 - \left(\frac{t}{10}\right)^{12} + \cdots + (-1)^n\left(\frac{t}{10}\right)^{4n} + \cdots \right],$$

where $\left| \left(\dfrac{t}{10}\right)^4 \right| < 1$ or $-10 < t < 10$.

$$\left[ \text{Substitute } \left(\frac{t}{10}\right)^4 \text{ into the Taylor series at 0 for } \frac{1}{1+x}. \right]$$

Therefore, $\dfrac{5000t^2}{10,000 + t^4} = \dfrac{t^2}{2} - \dfrac{t^6}{2 \cdot 10^4} + \dfrac{t^{10}}{2 \cdot 10^8} - \cdots + \dfrac{(-1)^n t^{4n+2}}{2 \cdot 10^{4n}} + \cdots ,$

and $\dfrac{1}{5}\displaystyle\int_0^5 \dfrac{5000t^2}{10,000 + t^4}\,dt = \dfrac{1}{5}\left[ \dfrac{t^3}{6} - \dfrac{t^7}{2 \cdot 7 \cdot 10^4} + \dfrac{t^{11}}{2 \cdot 11 \cdot 10^8} - \cdots \right]\Bigg|_0^5$

$$= \frac{1}{5}\left( \frac{5^3}{6} - \frac{5^7}{2 \cdot 7 \cdot 10^4} + \frac{5^{11}}{2 \cdot 11 \cdot 10^8} - \cdots \right)$$

$$= 4.16667 - 0.11161 + 0.00443 - \cdots$$

Since the terms alternate in sign and decrease in absolute value, the error in using the first two terms in this series will be less than $0.00443 < 0.005$. Thus, average insulin level for the first five minutes is, approximately, $4.05505 \approx 4.06$.    (2-4)

**43.** The average number of voters in the five-year period is
$$\frac{1}{5}\int_0^5 N(t)\,dt = \frac{1}{5}\int_0^5 (10 + 2t - 5e^{-0.01t})\,dt.$$

The Taylor series at 0 for $f(t) = e^{-0.01t^2}$ is

$$e^{-0.01t^2} = 1 - (0.01)t^2 + \frac{(0.01)^2}{2!}t^4 - \frac{(0.01)^3}{3!}t^6 + \cdots .$$

[Note: Use the Taylor series for $e^{-x}$ and replace $x$ by $(0.01)t^2$.]

Thus,

$$N(t) = 10 + 2t - 5\left(1 - (0.01)t^2 + \frac{(0.01)^2}{2}t^4 - \frac{(0.01)^3}{3!}t^6 + \cdots\right)$$

$$= 5 + 2t + (0.05)t^2 - \frac{0.0005}{2}t^4 + \frac{0.000005}{6}t^6 + \cdots .$$

Therefore,

$$\frac{1}{5}\int_0^5 N(t)\,dt = \frac{1}{5}\left[5t + t^2 + \frac{(0.05)}{3}t^3 - \frac{(0.0005)}{10}t^5 + \frac{(0.000005)}{42}t^7 - \cdots\right]\Big|_0^5$$

$$= \frac{1}{5}\left[25 + 25 + \frac{(0.05)}{3}(5)^3 - \frac{(0.0005)}{10}(5)^5 + \frac{(0.000005)}{42}(5)^7 - \cdots\right]$$

$$= \frac{1}{5}(50 + 2.08333 - 0.15625 + 0.00930 - \cdots)$$

$$= 10 + 0.41667 - 0.03125 + 0.00186 - \cdots$$

$$\approx 10.41667 \approx 10.4.$$

Since the terms alternate in sign and decrease in absolute value, the error in approximating this integral with the first two terms in the series will be less than $|-0.03125| = 0.03125 < 0.05$.

Thus, the average number of voters over the interval $[0, 5]$ is 10.4 to within ±0.05.

(2-4)

## EXERCISE 3-1

**1.** $\int_1^\infty \dfrac{dx}{x^3}$

Note that $\int_1^b \dfrac{1}{x^3}\,dx = -\dfrac{1}{2x^2}\Big|_1^b = -\dfrac{1}{2b^2} + \dfrac{1}{2}$ and

$\displaystyle\lim_{b\to\infty}\left(-\dfrac{1}{2b^2} + \dfrac{1}{2}\right) = \dfrac{1}{2}$. Therefore $\int_1^\infty \dfrac{dx}{x^3} = \dfrac{1}{2}$.

**3.** $\int_1^\infty x^2\,dx$

Note that $\int_1^b x^2\,dx = \dfrac{x^3}{3}\Big|_1^b = \dfrac{b^3}{3} - \dfrac{1}{3}$ and

$\displaystyle\lim_{b\to\infty}\left(\dfrac{b^3}{3} - \dfrac{1}{3}\right) = \infty$. Therefore $\int_1^\infty x^2\,dx$ diverges.

**5.** $\int_{-\infty}^0 e^{x/2}\,dx$

Note that $\int_b^0 e^{x/2}\,dx = 2e^{x/2}\Big|_b^0 = 2 - 2e^{b/2}$ and

$\displaystyle\lim_{b\to -\infty}(2 - 2e^{b/2}) = 2$. Therefore

$\int_{-\infty}^0 e^{x/2}\,dx = 2$.

**7.** $\int_4^\infty \dfrac{dx}{100\sqrt{x}}$

Note that $\int_4^b \dfrac{1}{100\sqrt{x}}\,dx = \dfrac{1}{50}x^{1/2}\Big|_4^b = \dfrac{b^{1/2}}{50} - \dfrac{2}{50}$ and

$\displaystyle\lim_{b\to\infty}\left(\dfrac{b^{1/2}}{50} - \dfrac{2}{50}\right) = \infty$. Therefore $\int_4^\infty \dfrac{dx}{100\sqrt{x}}$ diverges.

**9.** $\int_9^\infty \dfrac{dx}{x\sqrt{x}}$

Note that $\int_9^b \dfrac{dx}{x\sqrt{x}} = \int_9^b x^{-3/2}\,dx = -2x^{-1/2}\Big|_9^b = -\dfrac{2}{b^{1/2}} + \dfrac{2}{3}$ and

$\displaystyle\lim_{b\to\infty}\left(-\dfrac{2}{b^{1/2}} + \dfrac{2}{3}\right) = \dfrac{2}{3}$. Therefore

$\int_9^\infty \dfrac{dx}{x\sqrt{x}} = \dfrac{2}{3}$.

**11.** $\displaystyle\int_0^\infty \frac{dx}{(x+1)^{2/3}} = \lim_{b\to\infty}\int_0^b \frac{dx}{(x+1)^{2/3}}$

$\displaystyle\qquad\qquad = \lim_{b\to\infty}\left(3(x+1)^{1/3}\right)\Big|_0^b$  [<u>Note</u>: If $u = x+1$, then $du = dx$.]

$\displaystyle\qquad\qquad = 3\lim_{b\to\infty}(x+1)^{1/3}\Big|_0^b = 3\lim_{b\to\infty}\left((b+1)^{1/3} - 1\right)$

Since $(b+1)^{1/3} \to \infty$ as $b \to \infty$, the limit does not exist; hence, the improper integral *diverges*.

**13.** $\displaystyle\int_1^\infty \frac{dx}{x^{0.99}} = \lim_{b\to\infty}\int_1^b \frac{dx}{x^{0.99}} = \lim_{b\to\infty}\frac{x^{0.01}}{0.01}\Big|_1^b$

$\displaystyle\qquad = 100\lim_{b\to\infty}(x^{0.01})\Big|_1^b = 100\lim_{b\to\infty}(b^{0.01} - 1)$

Since $b^{0.01} \to \infty$ as $b \to \infty$, the limit does not exist; hence, the improper integral *diverges*.

**15.** $\displaystyle 0.3\int_0^\infty e^{-0.3x}\, dx = 0.3\lim_{b\to\infty}\int_0^b e^{-0.3x}\, dx = 0.3\lim_{b\to\infty}\frac{e^{-0.3x}}{-0.3}\Big|_0^b$

$\displaystyle\qquad\qquad = -\lim_{b\to\infty}(e^{-0.3x})\Big|_0^b = -\lim_{b\to\infty}(e^{-0.3b} - 1)$

$\displaystyle\qquad\qquad = -\lim_{b\to\infty}\left(\frac{1}{e^{0.3b}} - 1\right)$

$\displaystyle\qquad\qquad = -(0 - 1)\ \left[\text{Note: }\frac{1}{e^{0.3b}} \to 0 \text{ as } b \to \infty.\right]$

$\displaystyle\qquad\qquad = 1$

Thus, the improper integral *converges*.

**17.** The graph of $f$ is shown at the right. Since $f(x) = 0$ for $x < 0$ and $x > 2$,

$\displaystyle\int_{-\infty}^\infty f(x)\,dx = \int_0^2 f(x)\,dx \qquad (f(x) = 0 \text{ for } x < 0 \text{ and } x > 2)$

$\displaystyle\qquad = \int_0^2 (1 + x^2)\,dx$

$\displaystyle\qquad = \left[\left(x + \frac{1}{3}x^3\right)\Big|_0^2\right]$

$\displaystyle\qquad = \frac{14}{3}$

The integral *converges* and has value $\dfrac{14}{3}$.

**19.** The graph of $f$ is shown at the right. Since $f(x) = 0$ for $x < 0$,

$\displaystyle\int_{-\infty}^{\infty} f(x)\,dx = \int_{0}^{\infty} f(x)$     $(f(x) = 0$ for $x < 0)$

$\displaystyle = \lim_{b \to \infty} \int_{0}^{b} e^{-0.1x}\,dx$

$\displaystyle = \lim_{b \to \infty} \left[ -10 e^{-0.1x} \Big|_{0}^{b} \right]$

$\displaystyle = \lim_{b \to \infty} [-10 e^{-0.1b} + 10] = 10$

The integral *converges* and has value 10.

**21.** The graph of $f$ is shown at the right. Since $f(x) = 0$ for $x < 2$,

$\displaystyle\int_{-\infty}^{\infty} f(x)\,dx = \int_{2}^{\infty} f(x)\,dx$     $(f(x) = 0$ for $x < 2)$

$\displaystyle = \lim_{b \to \infty} \int_{2}^{b} \frac{4}{(x+2)}\,dx$

$\displaystyle = \lim_{b \to \infty} \left[ 4 \ln(x+2) \Big|_{2}^{b} \right]$

$\displaystyle = \lim_{b \to \infty} [4 \ln(b+2) - 4 \ln 4]$

Since $4 \ln(b+2) \to \infty$ as $b \to \infty$, the integral *diverges*.

**23.** Always true. If $f$ is positive, continuous and increasing on $[0, \infty)$, then $f(x) > f(0) > 0$ for all $x > 0$ and

$\displaystyle\int_{0}^{\infty} f(x)\,dx = \lim_{b \to \infty} \int_{0}^{b} f(x)\,dx > \lim_{b \to \infty} \int_{0}^{b} f(0)\,dx$

$\displaystyle = \lim_{b \to \infty} f(0) \int_{0}^{b} dx$

$\displaystyle = f(0) \lim_{b \to \infty} x \Big|_{0}^{b}$

$\displaystyle = f(0) \lim_{b \to \infty} b \to \infty$

as $b \to \infty$

**25.** False. Let $f(x) = \begin{cases} 10 & \text{if } 1 \le x \le 1000 \\ 10 e^{-(x-1000)} & \text{if } x > 1000 \end{cases}$

Then $f(x) > 0$ for all $x \in [1, \infty)$, $f$ is continuous, $f(x) \ge 10$ on $[1, 1000]$, and

$\displaystyle\int_{1}^{\infty} f(x)\,dx = \int_{1}^{1000} f(x)\,dx + \int_{1000}^{\infty} f(x)\,dx$

$\displaystyle = \int_{1}^{1000} 10\,dx + \lim_{b \to \infty} \int_{1000}^{b} 10 e^{-(x-1000)}\,dx$

$\displaystyle = \left[ 10x \right]_{1}^{1000} + \lim_{b \to \infty} \left[ -10 e^{-(x-1000)} \right]_{1000}^{b}$

$\displaystyle = 10{,}000 - 10 + \lim_{b \to \infty} \left[ 10 - e^{-(b-1000)} \right] = 10{,}000.$

The improper integral converges.

**27.** $F(b) = \int_1^b \dfrac{dx}{x^3} = -\dfrac{1}{2b^2} + \dfrac{1}{2}$

$\displaystyle \lim_{b \to \infty} F(b) = \lim_{b \to \infty}\left(-\dfrac{1}{2b^2} + \dfrac{1}{2}\right) = \dfrac{1}{2}.$

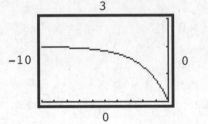

**29.** $F(b) = \int_1^b x^2\, dx = \dfrac{b^3}{3} - \dfrac{1}{3}$

$\displaystyle \lim_{b \to \infty} F(b) = \lim_{b \to \infty}\left(\dfrac{b^3}{3} - \dfrac{1}{3}\right) = \infty.$

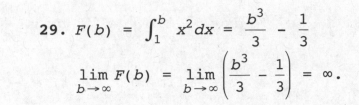

**31.** $F(b) = \int_b^0 e^{x/2}\, dx = 2 - 2e^{b/2}$

$\displaystyle \lim_{b \to -\infty} F(b) = \lim_{b \to -\infty}(2 - 2e^{b/2}) = 2.$

**33.** $F(b) = \int_4^b \dfrac{dx}{100\sqrt{x}} = \dfrac{b^{1/2}}{50} - \dfrac{1}{25}$

$\displaystyle \lim_{b \to \infty} F(b) = \lim_{b \to \infty}\left(\dfrac{b^{1/2}}{50} - \dfrac{1}{25}\right) = \infty$

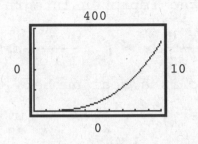

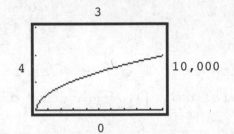

**35.** $F(b) = \int_9^b \dfrac{dx}{x\sqrt{x}} = -\dfrac{2}{b^{1/2}} + \dfrac{2}{3}$

$\displaystyle \lim_{b \to \infty} F(b) = \lim_{b \to \infty}\left(-\dfrac{2}{b^{1/2}} + \dfrac{2}{3}\right) = \dfrac{2}{3}.$

**37.** $f$ is continuous on $[0, \infty)$ and $\int_0^\infty f(x)\,dx = L$ converges. Since

$$L = \int_0^\infty f(x)\,dx = \lim_{b \to \infty}\int_0^b f(x)\,dx$$

$$= \lim_{b \to \infty}\left[\int_0^1 f(x)\,dx + \int_1^b f(x)\,dx\right]$$

$$= \int_0^1 f(x)\,dx + \lim_{b \to \infty}\int_1^b f(x)\,dx,$$

it follows that

$$\lim_{b \to \infty}\int_1^b f(x)\,dx = L - \int_0^1 f(x)\,dx \text{ exists.}$$

Thus, $\int_1^\infty f(x)\,dx$ *converges.*

**39.** $\displaystyle\int_0^\infty \frac{1}{k} e^{-x/k}\, dx = \frac{1}{k} \lim_{b\to\infty} \int_0^b e^{-x/k}\, dx = \frac{1}{k} \lim_{b\to\infty} \frac{e^{-x/k}}{-\dfrac{1}{k}} \bigg|_0^b = -\lim_{b\to\infty} e^{-x/k} \bigg|_0^b$

$$= -\lim_{b\to\infty} (e^{-b/k} - 1) = -\lim_{b\to\infty} \left( \frac{1}{e^{b/k}} - 1 \right) = -(0 - 1) = 1$$

Thus, the improper integral *converges*.

**41.** $\displaystyle\int_{-\infty}^\infty \frac{x}{1 + x^2}\, dx = \int_{-\infty}^c \frac{x}{1 + x^2}\, dx + \int_c^\infty \frac{x}{1 + x^2}\, dx$

where $c$ is a real number, using $\underline{1}$(c).  Let $c = 0$.

Consider $\displaystyle\int_0^\infty \frac{x}{1 + x^2}\, dx = \lim_{b\to\infty} \int_0^b \frac{x}{1 + x^2}\, dx$

$\displaystyle\int \frac{x}{1 + x^2}\, dx = \frac{1}{2} \int \frac{du}{u} = \frac{1}{2} \ln u + C = \frac{1}{2} \ln(1 + x^2) + C$

[Let $u = 1 + x^2$.  Then $du = 2x\, dx$.]

Thus, $\displaystyle\lim_{b\to\infty} \int_0^b \frac{x}{1 + x^2}\, dx = \lim_{b\to\infty} \frac{1}{2} \ln(1 + x^2) \bigg|_0^b$

$$= \lim_{b\to\infty} \frac{1}{2} \ln(1 + b^2) - \frac{1}{2} \ln 1$$

$$= \frac{1}{2} \lim_{b\to\infty} \ln(1 + b^2)$$

Since $\ln(1 + b^2) \to \infty$ as $b \to \infty$, this improper integral *diverges*.
Therefore, $\displaystyle\int_{-\infty}^\infty \frac{x}{1 + x^2}\, dx$ *diverges*.

**43.** $\displaystyle\int_0^\infty (e^{-x} - e^{-2x})\, dx = \lim_{b\to\infty} \int_0^b (e^{-x} - e^{-2x})\, dx = \lim_{b\to\infty} \left( -e^{-x} - \frac{e^{-2x}}{-2} \right) \bigg|_0^b$

$$= \lim_{b\to\infty} \left[ -e^{-b} + \frac{e^{-2b}}{2} - \left( -1 + \frac{1}{2} \right) \right]$$

$$= \frac{1}{2} + \lim_{b\to\infty} \left( -e^{-b} + \frac{e^{-2b}}{2} \right) = \frac{1}{2} + 0 = \frac{1}{2}$$

Thus, the improper integral *converges*.

**45.** $\displaystyle\int_{-\infty}^0 \frac{dx}{\sqrt{1 - x}} = \lim_{a\to-\infty} \int_a^0 \frac{dx}{\sqrt{1 - x}}$     (using $\underline{1}$(b))

$\displaystyle\int \frac{dx}{\sqrt{1 - x}} = \int -\frac{du}{u^{1/2}} = -\int u^{-1/2}\, du$     Substitution: $u = 1 - x$

$$= -2u^{1/2} + C = -2\sqrt{1 - x} + C$$

$du = -dx$

$-du = dx$

Therefore, $\displaystyle\lim_{a\to-\infty} \int_a^0 \frac{dx}{\sqrt{1 - x}} = \lim_{a\to-\infty} (-2\sqrt{1 - x}) \bigg|_a^0 = \lim_{a\to-\infty} \left( -2 + 2\sqrt{1 - a} \right)$

Since $2\sqrt{1 - a} \to \infty$ as $a \to -\infty$, the limit does not exist; hence, the improper integral *diverges*.

**47.** $\displaystyle\int_1^\infty \frac{\ln x}{x}\,dx = \lim_{b\to\infty}\int_1^b \frac{\ln x}{x}\,dx$

Substitution: $u = \ln x$

$du = \dfrac{1}{x}dx$

$\qquad\qquad = \displaystyle\lim_{b\to\infty}\int_0^{\ln b} u\,du$

Limits: $x = 1$ implies $u = 0$

$\qquad\qquad\qquad\qquad\quad x = b$ implies $u = \ln b$

$\qquad\qquad = \displaystyle\lim_{b\to\infty}\frac{u^2}{2}\Big|_0^{\ln b} = \lim_{b\to\infty}\frac{(\ln b)^2}{2}$

Since $\ln b \to \infty$ as $b \to \infty$, the limit does not exist; hence, the improper integral *diverges*.

**49.** $CV = \displaystyle\int_0^\infty 6000e^{-0.05t}\,dt = \lim_{T\to\infty}\int_0^T 6000e^{-0.05t}\,dt$

$\qquad\qquad = \displaystyle\lim_{T\to\infty}\frac{6000}{-0.05}e^{-0.05t}\Big]_0^T = -120{,}000\cdot\lim_{T\to\infty}[e^{-0.05T} - 1]$

$\qquad\qquad\qquad\qquad\qquad\qquad\qquad\qquad = \$120{,}000$

**51.** $f(t) = 1500e^{0.04t}$

$\qquad CV = \displaystyle\int_0^\infty 1500e^{0.04t}\cdot e^{-0.07t}\,dt = \lim_{T\to\infty}\int_0^T 1500e^{-0.03t}\,dt$

$\qquad\qquad\qquad = \displaystyle\lim_{T\to\infty}\left[\frac{1500}{-0.03}e^{-0.03t}\Big|_0^T\right]$

$\qquad\qquad\qquad = -50{,}000\,\displaystyle\lim_{T\to\infty}[e^{-0.03T} - 1]$

$\qquad\qquad\qquad = \$50{,}000$

**53.** At an interest rate of 6%,

$\qquad CV = \displaystyle\int_0^\infty 6000e^{-0.06t}\,dt = \lim_{T\to\infty}\int_0^T 6000e^{-0.06t}\,dt$

$\qquad\qquad = \displaystyle\lim_{T\to\infty}\left[-100{,}000e^{-0.06t}\Big|_0^T\right]$

$\qquad\qquad = \displaystyle\lim_{T\to\infty}[-100{,}000e^{-0.06T} + 100{,}000]$

$\qquad\qquad = \$100{,}000$

Increasing the interest rate to 6% decreases the capital value to $100,000.

At an interest rate of 4%,

$\qquad CV = \displaystyle\int_0^\infty 6000e^{-0.04t}\,dt = \lim_{T\to\infty}\int_0^T 6000e^{-0.04t}\,dt$

$\qquad\qquad = \displaystyle\lim_{T\to\infty}\left[-150{,}000e^{-0.04t}\Big|_0^T\right]$

$\qquad\qquad = \displaystyle\lim_{T\to\infty}[-150{,}000e^{-0.04T} + 150{,}000]$

$\qquad\qquad = \$150{,}000$

Decreasing the interest rate to 4% increases the capital value to $150,000. In general, increasing the interest rate decreases the amount of capital required to establish the income stream.

**55.** $R(t) = 3e^{-0.2t} - 3e^{-0.4t}$

(A) Total production $= \int_0^\infty R(t)\,dt = \lim_{T \to \infty} \int_0^T R(t)\,dt$

$$= \lim_{T \to \infty} \int_0^T (3e^{-0.2t} - 3e^{-0.4t})\,dt$$

$$= \lim_{T \to \infty} \left[ (-15e^{-0.2t} + 7.5e^{-0.4t}) \Big|_0^T \right]$$

$$= \lim_{T \to \infty} [-15e^{-0.2T} + 7.5e^{-0.4T} + 7.5]$$

$$= 7.5 \text{ billion ft}^3$$

(B) 50% of 7.5 billion ft$^3$ is 3.75 billion ft$^3$. Thus, we need to solve

$$\int_0^T R(t)\,dt = \int_0^T (3e^{-0.2t} - 3e^{-0.4t})\,dt = 3.75$$

for $T$.

Now,

$$\int_0^T (3e^{-0.2t} - 3e^{-0.4t})\,dt = \left[ (-15e^{-0.2t} + 7.5e^{-0.4t}) \Big|_0^T \right]$$

$$= -15e^{-0.2T} + 7.5e^{-0.4T} + 7.5 = 3.75$$

and

$$-15e^{-0.2T} + 7.5e^{-0.4T} = -3.75$$

Using a graphing utility, we find that $T \approx 6.14$ years.

**57.** Total seepage $= \int_0^\infty \dfrac{500}{(1 + t)^2}\,dt = \lim_{b \to \infty} 500 \int_0^b \dfrac{1}{(1 + t)^2}\,dt$

$$= 500 \lim_{b \to \infty} \int_0^b \dfrac{1}{(1 + t)^2}\,dt \qquad \begin{array}{l} \text{Substitution: } u = 1 + t \\ \qquad\qquad\quad du = dt \end{array}$$

$$= 500 \lim_{b \to \infty} \int_1^{1+b} u^{-2}\,du \qquad \begin{array}{l} \text{Limits: } t = 0 \text{ implies } u = 1 \\ \qquad\quad t = b \text{ implies } u = 1 + b \end{array}$$

$$= 500 \lim_{b \to \infty} \left( -\dfrac{1}{u} \right) \Big|_1^{1+b}$$

$$= 500 \lim_{b \to \infty} \left( -\dfrac{1}{1 + b} + 1 \right) = 500 \text{ gallons}$$

**59.** Total $= \int_0^\infty R(t)\,dt = \lim_{T \to \infty} \int_0^T R(t)\,dt = \lim_{T \to \infty} \int_0^T \dfrac{400}{(5 + t)^3}\,dt$

$$= \lim_{T \to \infty} \left[ \dfrac{-200}{(5 + t)^2} \Big|_0^T \right]$$

$$= \lim_{T \to \infty} \left[ \dfrac{-200}{(5 + T)^2} + 8 \right] = 8$$

Thus, 8 million immigrants will enter the country under this policy.

**1.** The graph of $f(x)$ is shown at the right. From the graph we see that $f(x) \geq 0$ for $x \in (-\infty, \infty)$.

$$\int_{-\infty}^{\infty} f(x)\,dx = \int_0^4 \frac{1}{8}x\,dx$$

$$= \frac{1}{8} \cdot \frac{x^2}{2}\Big|_0^4$$

$$= \frac{1}{16}(16 - 0)$$

$$= 1$$

| $x$ | $f(x)$ |
|-----|--------|
| 0 | 0 |
| 1 | $\frac{1}{8}$ |
| 4 | $\frac{1}{2}$ |
| $x > 4$ | 0 |

**3.** $f(x) = \begin{cases} 2 + 2x & \text{if } -1 \leq x \leq 0 \\ 0 & \text{otherwise} \end{cases}$

For $f(x)$ to be a probability density function, it has to satisfy the following two conditions:

(i) $f(x) \geq 0$ for all $x$

(ii) $\int_{-1}^0 f(x)\,dx = 1$

Note that for $-1 \leq x \leq 0$, $f(x) = 2 + 2x \geq 0$ and hence (i) is satisfied.

$$\int_{-1}^0 (2 + 2x)\,dx = (2x + x^2)\Big|_{-1}^0 = -(2(-1) + (-1)^2) = -(-2 + 1) = 1.$$

Therefore, $f(x)$ given above is a probability density function.

**5.** $f(x) = \begin{cases} 0.2 & \text{if } 1 \leq x \leq 5 \\ 0 & \text{otherwise} \end{cases}$

Note that (i) of Problem 3 is satisfied, however,

$$\int_1^5 f(x)\,dx = \int_1^5 0.2\,dx = 0.2x\Big|_1^5 = 0.2(5) - 0.2(1) = 1 - 0.2 = 0.8$$

and hence (ii) of Problem 3 is not satisfied. Thus, $f(x)$ given above is not a probability density function.

**7.** (A) Using $\underline{2}$(c),

$$P(1 < X < 3) = \int_1^3 \frac{1}{8}x\,dx = \frac{1}{8} \cdot \frac{x^2}{2}\Big|_1^3$$

$$= \frac{1}{16}(9 - 1) = \frac{8}{16} = \frac{1}{2}.$$

The graph is shown at the right.

(B) $P(X \leq 2) = \int_{-\infty}^2 \frac{1}{8}x\,dx = \int_0^2 \frac{1}{8}x\,dx$  [since $f(x) = 0$ for $x \leq 0$]

$$= \frac{1}{8} \cdot \frac{x^2}{2}\Big|_0^2 = \frac{1}{16}(4) = \frac{1}{4}.$$

The graph is shown at the right.

(C) $P(X > 3) = \int_3^\infty \frac{1}{8} x\, dx = \int_3^4 \frac{1}{8} x\, dx$    [since $f(x) = 0$ for $x > 4$]

$$= \frac{1}{8} \cdot \frac{x^2}{2} \Big|_3^4 = \frac{1}{16}(16 - 9) = \frac{7}{16}.$$

The graph is shown at the right.

**9.** (A) $P(X = 1) = \int_1^1 f(x)\, dx = 0$

(B) $P(X > 5) = \int_5^\infty f(x)\, dx = \int_5^\infty 0\, dx = 0$    [$f(x) = 0$ for $x > 4$]

(C) $P(X < 5) = \int_{-\infty}^5 f(x)\, dx = \int_0^4 \frac{1}{8} x\, dx$ [$f(x) = 0$ when $x < 0$ and when $x > 4$]

$$= \frac{1}{8} \cdot \frac{x^2}{2} \Big|_0^4 = \frac{1}{16}(16) = 1.$$

**11.** If $x < 0$, then

$$F(x) = \int_{-\infty}^x f(t)\, dt = \int_{-\infty}^x 0\, dt = 0$$

If $0 \le x \le 4$, then

$$F(x) = \int_{-\infty}^x f(t)\, dt = \int_{-\infty}^0 f(t)\, dt + \int_0^x f(t)\, dt = 0 + \int_0^x \frac{1}{8} t\, dt$$

$$= \frac{1}{16} t^2 \Big|_0^x = \frac{1}{16} x^2.$$

If $x > 4$, then

$$F(x) = \int_{-\infty}^x f(t)\, dt = \int_{-\infty}^0 f(t)\, dt + \int_0^4 f(t)\, dt + \int_4^x f(t)\, dt$$

$$= 0 + \int_0^4 \frac{1}{8} t\, dt + 0 = \frac{1}{16} t^2 \Big|_0^4 = \frac{1}{16}(16 - 0) = 1.$$

Thus, the cumulative probability distribution function is:

$$F(x) = \begin{cases} 0 & x < 0 \\ \frac{1}{16} x^2 & 0 \le x \le 4 \\ 1 & x > 4 \end{cases}$$

The graph of $F(x)$ is shown at the right.

**13.** Using $\underline{3}$ and the cumulative probability
distribution function $F$ from Problem 11:

(A) $P(2 \le X \le 4) = F(4) - F(2)$      (B) $P(0 < X < 2) = F(2) - F(0)$

$$= \frac{1}{16}(4)^2 - \frac{1}{16}(2)^2 \qquad\qquad = \frac{1}{16}(2)^2 - 0$$

$$= 1 - \frac{1}{4} = \frac{3}{4} \qquad\qquad\qquad\quad = \frac{1}{4}$$

**15.** From Problem 11:

(A) $P(0 \le X \le x) = F(x) - F(0) = \frac{1}{16} x^2 - 0 = \frac{1}{16} x^2$

Now $\frac{1}{16} x^2 = \frac{1}{4}$

$x^2 = 4$

Since $x \ge 0$, the solution is $x = 2$.

(B) $P(0 \le X \le x) = F(x) - F(0) = \dfrac{1}{9}$ implies $\dfrac{1}{16}x^2 = \dfrac{1}{9}$

$$x^2 = \dfrac{16}{9}$$

Since $x \ge 0$, the solution is $x = \dfrac{4}{3}$.

**17.** From the graph of $f(x)$ shown below, we see that $f(x) \ge 0$ for $x \in (-\infty, \infty)$.

| $x$ | $f(x)$ |
|-----|--------|
| 0 | 2 |
| 2 | $\dfrac{2}{27}$ |
| 4 | $\dfrac{2}{125}$ |
| $x < 0$ | 0 |

Also,

$$\int_{-\infty}^{\infty} f(x)\,dx = \int_0^{\infty} \dfrac{2}{(1+x)^3}\,dx = \lim_{R \to \infty} \int_0^R \dfrac{2}{(1+x)^3}\,dx$$

$$= \lim_{R \to \infty} \left(-\dfrac{1}{(1+x)^2}\right)\bigg|_0^R = -\lim_{R \to \infty}\left(\dfrac{1}{(1+R)^2} - 1\right) = -(0 - 1) = 1.$$

**19.** (A) $P(1 \le X \le 4) = \int_1^4 \dfrac{2}{(1+x)^3}\,dx = \dfrac{-1}{(1+x)^2}\bigg|_1^4 = \dfrac{-1}{(1+4)^2} - \dfrac{-1}{(1+1)^2}$

$$= \dfrac{-1}{25} + \dfrac{1}{4} = \dfrac{21}{100} = .21$$

(B) $P(X > 3) = \int_3^{\infty} \dfrac{2}{(1+x)^3}\,dx$

Now, $1 = \int_{-\infty}^{\infty} \dfrac{2}{(1+x)^3}\,dx = \int_{-\infty}^{3} \dfrac{2}{(1+x)^3}\,dx + \int_3^{\infty} \dfrac{2}{(1+x)^3}\,dx$,

so $\int_3^{\infty} \dfrac{2}{(1+x)^3}\,dx = 1 - \int_{-\infty}^{3} \dfrac{2}{(1+x)^3}\,dx = 1 - \int_0^{3} \dfrac{2}{(1+x)^3}\,dx$,

since $f(x) = 0$ for $x < 0$. Thus,

$$\int_3^{\infty} \dfrac{2}{(1+x)^3}\,dx = 1 - \left(\dfrac{-1}{(1+x)^2}\bigg|_0^3\right) = 1 - \left(-\dfrac{1}{16} + 1\right) = \dfrac{1}{16}.$$

(C) $P(X \le 2) = \int_{-\infty}^{2} \dfrac{2}{(1+x)^3}\,dx = \int_0^{2} \dfrac{2}{(1+x)^3}\,dx = \dfrac{-1}{(1+x)^2}\bigg|_0^2 = -\dfrac{1}{9} + 1 = \dfrac{8}{9}$

**21.** If $x < 0$, then

$F(x) = \int_{-\infty}^{x} f(t)\,dt = \int_{-\infty}^{x} 0\,dt = 0.$

If $x \ge 0$, then

$F(x) = \int_{-\infty}^{x} f(t)\,dt = \int_{-\infty}^{0} f(t)\,dt + \int_0^{x} f(t)\,dt = 0 + \int_0^{x} \dfrac{2}{(1+t)^3}\,dt$

$$= \left(-\dfrac{1}{(1+t)^2}\right)\bigg|_0^x = -\dfrac{1}{(1+x)^2} + 1 \quad \text{or} \quad 1 - \dfrac{1}{(1+x)^2}.$$

$$F(x) = \begin{cases} 0 & x < 0 \\ 1 - \dfrac{1}{(1+x)^2} & x \geq 0 \end{cases}$$

The graph of $F(x)$ is shown at the right.

**23.** From Problem 21,

$$F(x) = \begin{cases} 0 & x < 0 \\ 1 - \dfrac{1}{(1+x)^2} & x \geq 0 \end{cases}$$

and $P(0 \leq X \leq x) = F(x) - F(0) = 1 - \dfrac{1}{(1+x)^2}$.

(A) Set $P(0 \leq X \leq x) = \dfrac{3}{4}$ and solve for $x$.

$$1 - \frac{1}{(1+x)^2} = \frac{3}{4}$$
$$\frac{1}{(1+x)^2} = \frac{1}{4}$$
$$(1+x)^2 = 4$$

Since $x \geq 0$, we have $1 + x = 2$, so $x = 1$.

(B) $1 = P(X \leq x) + P(X > x)$. Therefore,

$$P(X > x) = 1 - P(X \leq x) = 1 - F(x) = 1 - \left(1 - \frac{1}{(1+x)^2}\right) = \frac{1}{(1+x)^2}.$$

Set $P(X > x) = \dfrac{1}{16}$ and solve for $x$.

$$\frac{1}{(1+x)^2} = \frac{1}{16}$$
$$(1+x)^2 = 16$$
$$1 + x = 4$$
$$x = 3$$

**25.** $f(x) = \begin{cases} \dfrac{3}{2}x - \dfrac{3}{4}x^2 & 0 \leq x \leq 2 \\ 0 & \text{otherwise} \end{cases}$

If $x < 0$, then
$$F(x) = \int_{-\infty}^{x} f(t)\,dt = \int_{-\infty}^{x} 0\,dt = 0.$$

If $0 \leq x \leq 2$, then
$$F(x) = \int_{-\infty}^{x} f(t)\,dt = \int_{-\infty}^{0} f(t)\,dt + \int_{0}^{x} f(t)\,dt = 0 + \int_{0}^{x}\left(\frac{3}{2}t - \frac{3}{4}t^2\right)dt$$

$$= \left(\frac{3}{2}\cdot\frac{t^2}{2} - \frac{3}{4}\cdot\frac{t^3}{3}\right)\Big|_0^x = \left(\frac{3}{4}t^2 - \frac{1}{4}t^3\right)\Big|_0^x = \frac{3}{4}x^2 - \frac{1}{4}x^3.$$

If $x > 2$, then

$$F(x) = \int_{-\infty}^{x} f(t)\,dt = \int_{-\infty}^{x} 0 \; dt + \int_{0}^{2} f(t)\,dt + \int_{2}^{x} f(t)\,dt$$

$$= 0 + \int_{0}^{2}\left(\frac{3}{2}t - \frac{3}{4}t^2\right)dt + 0 = \left(\frac{3}{4}t^2 - \frac{1}{4}t^3\right)\Big|_{0}^{2} = \frac{3}{4}(2)^2 - \frac{1}{4}(2)^3 = 1$$

Thus, the cumulative probability distribution function is:

$$F(x) = \begin{cases} 0 & x < 0 \\ \frac{3}{4}x^2 - \frac{1}{4}x^3 & 0 \le x \le 2 \\ 1 & x > 2 \end{cases}$$

The graphs of $F(x)$ and $f(x)$ are as follows:

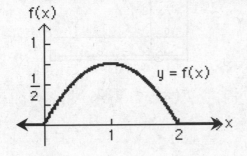
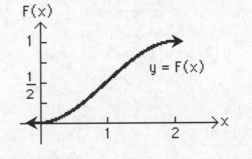

27. $f(x) = \begin{cases} \dfrac{1}{2} + \dfrac{1}{2}x^3 & -1 \le x \le 1 \\ 0 & \text{otherwise} \end{cases}$

If $x < -1$, then

$$F(x) = \int_{-\infty}^{x} f(t)\,dt = \int_{-\infty}^{x} 0 \; dt = 0$$

If $-1 \le x \le 1$, then

$$F(x) = \int_{-\infty}^{x} f(t)\,dt = \int_{-\infty}^{-1} f(t)\,dt + \int_{-1}^{x} f(t)\,dt = 0 + \int_{-1}^{x}\left(\frac{1}{2} + \frac{1}{2}t^3\right)dt$$

$$= \left(\frac{1}{2}t + \frac{1}{8}t^4\right)\Big|_{-1}^{x} = \frac{1}{2}x + \frac{1}{8}x^4 - \left(-\frac{1}{2} + \frac{1}{8}\right)$$

$$= \frac{3}{8} + \frac{1}{2}x + \frac{1}{8}x^4$$

If $x > 1$, then

$$F(x) = \int_{-\infty}^{x} f(t)\,dt = \int_{-\infty}^{-1} f(t)\,dt + \int_{-1}^{1} f(t)\,dt + \int_{1}^{x} f(t)\,dt$$

$$= 0 + \int_{-1}^{1}\left(\frac{1}{2} + \frac{1}{2}t^3\right)dt + 0$$

$$= \left(\frac{1}{2}t + \frac{1}{8}t^4\right)\Big|_{-1}^{1} = \left(\frac{1}{2} + \frac{1}{8}\right) - \left(-\frac{1}{2} + \frac{1}{8}\right)$$

$$= 1$$

Thus, the cumulative probability distribution function is:

$$F(x) = \begin{cases} 0 & x < -1 \\ \frac{3}{8} + \frac{1}{2}x + \frac{1}{8}x^4 & -1 \le x \le 1 \\ 1 & x > 1 \end{cases}$$

The graphs of $f(x)$ and $F(x)$ are:

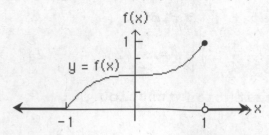

**29.** The graphs of
$$F(x) = \begin{cases} 0 & x < 0 \\ \frac{3}{4}x^2 - \frac{1}{4}x^3 & 0 \le x \le 2 \\ 1 & x > 2 \end{cases}$$
and $y = 0.2$ are shown at the right.

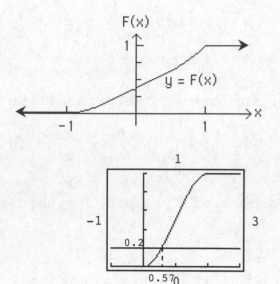

$F(x) = 0.2$ at $x \approx 0.5$

**31.** The graphs of
$$F(x) = \begin{cases} 0 & x < -1 \\ \frac{3}{8} + \frac{1}{2}x + \frac{1}{8}x^4 & -1 \le x \le 1 \\ 1 & x > 1 \end{cases} \quad \text{and } y = 0.6 \text{ are shown below.}$$

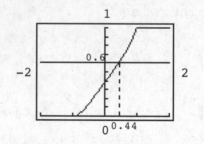

$F(x) = 0.6$ at $x \approx 0.44$

**33.** $f(x) = \begin{cases} e^{-x/3} & \text{if } x \ge 0 \\ 0 & \text{otherwise} \end{cases}$

Note that for $k > 0$, condition (i) of Problem 3 is satisfied. To check the second condition, we observe that

$$\int_0^b e^{-x/3}dx = -3e^{-x/3}\Big|_0^b = -3e^{-b/3} + 3 \text{ and}$$

$$\lim_{b \to \infty}(-3e^{-b/3} + 3) = 3, \text{ therefore } \int_0^\infty e^{-x/3}dx = 3.$$

Now, we have to choose $k$ so that $\int_0^\infty kf(x) = 1$

or $k = \dfrac{1}{\int_0^\infty f(x)dx} = \dfrac{1}{3}$.

**35.** $f(x) = \begin{cases} \dfrac{1}{x^5} & \text{if } x \geq 1 \\ 0 & \text{otherwise} \end{cases}$

Note that for $k > 0$, $kf(x) \geq 0$ for all $x$.

$\displaystyle\int_1^b \frac{1}{x^5}dx = -\frac{1}{4}x^{-4}\Big|_1^b = -\frac{1}{4b^4} + \frac{1}{4}$ and

$\displaystyle\lim_{b\to\infty}\left(-\frac{1}{4b^4} + \frac{1}{4}\right) = 1$, thus $\displaystyle\int_1^\infty \frac{1}{x^5}dx = \frac{1}{4}$.

$\displaystyle\int_1^\infty kf(x)\,dx = 1$ implies that $k = \dfrac{1}{\displaystyle\int_1^\infty \frac{1}{x^5}\,dx} = \dfrac{1}{\frac{1}{4}} = 4$

**37.** No! For example, $F$ does not satisfy (b) and (c) of $\underline{4}$; $F(x) > 1$ on $\left[0, \dfrac{1}{2}\right]$ and $F$ is *decreasing* on $[0, 1]$.

**39.** $f(x) = \dfrac{1}{\sqrt{2\pi}}e^{-0.5x^2}$. The graph of $f$ is shown at the right.

$\displaystyle\int_{-1}^1 f(x)\,dx \approx 0.683$; $\displaystyle\int_{-2}^2 f(x)\,dx \approx 0.954$; $\displaystyle\int_{-3}^3 f(x)\,dx = 0.997$

Based on these calculations, it appears that

$$\int_{-\infty}^\infty f(x)\,dx = 1$$

**41.** The relationship between $f(x)$ and $F(x)$ is $F'(x) = f(x)$. Thus, if

$F(x) = \begin{cases} 0 & x < 0 \\ x^2 & 0 \leq x \leq 1 \\ 1 & x > 1 \end{cases}$

then

$f(x) = \begin{cases} 0 & x < 0 \\ 2x & 0 \leq x < 1 \\ 0 & x > 1 \end{cases}$  or  $f(x) = \begin{cases} 2x & 0 \leq x < 1 \\ 0 & \text{otherwise} \end{cases}$

**43.** $F(x) = \begin{cases} 0 & x < 0 \\ 6x^2 - 8x^3 + 3x^4 & 0 \leq x \leq 1 \\ 1 & x > 1 \end{cases}$

Thus,

$f(x) = \begin{cases} 0 & x < 0 \\ 12x - 24x^2 + 12x^3 & 0 \leq x \leq 1 \\ 0 & \text{otherwise} \end{cases}$  [<u>Note</u>: $f(x) = F'(x)$.]

or

$f(x) = \begin{cases} 12x(1 - 2x + x^2) & 0 \leq x \leq 1 \\ 0 & \text{otherwise} \end{cases}$

**45.** Given $f(x) = \begin{cases} x & 0 \le x \le 1 \\ 2 - x & 1 < x \le 2 \\ 0 & \text{otherwise} \end{cases}$

If $x < 0$, then

$$F(x) = \int_{-\infty}^{x} f(t)\,dt = \int_{-\infty}^{x} 0\,dt = 0.$$

If $0 \le x \le 1$, then

$$F(x) = \int_{-\infty}^{x} f(t)\,dt = \int_{-\infty}^{0} f(t)\,dt + \int_{0}^{x} f(t)\,dt = 0 + \int_{0}^{x} t\,dt = \frac{1}{2}t^2\Big|_0^x = \frac{1}{2}x^2.$$

If $1 < x \le 2$, then

$$F(x) = \int_{-\infty}^{x} f(t)\,dt = \int_{-\infty}^{0} f(t)\,dt + \int_{0}^{1} f(t)\,dt + \int_{1}^{x} f(t)\,dt$$

$$= 0 + \int_{0}^{1} t\,dt + \int_{1}^{x} (2 - t)\,dt = \frac{1}{2}t^2\Big|_0^1 + \left(2t - \frac{1}{2}t^2\right)\Big|_1^x$$

$$= \frac{1}{2} + \left(2x - \frac{1}{2}x^2\right) - \left(2 - \frac{1}{2}\right) = \frac{1}{2} + 2x - \frac{1}{2}x^2 - \frac{3}{2} = 2x - \frac{1}{2}x^2 - 1.$$

If $x > 2$, then

$$F(x) = \int_{-\infty}^{x} f(t)\,dt = \int_{-\infty}^{0} f(t)\,dt + \int_{0}^{1} f(t)\,dt + \int_{1}^{2} f(t)\,dt + \int_{2}^{x} f(t)\,dt$$

$$= 0 + \int_{0}^{1} t\,dt + \int_{1}^{2} (2 - t)\,dt + 0 = \frac{1}{2}t^2\Big|_0^1 + \left(2t - \frac{1}{2}t^2\right)\Big|_1^2$$

$$= \frac{1}{2} + \left[\left(4 - \frac{1}{2}\cdot 4\right) - \left(2 - \frac{1}{2}\right)\right] = 1.$$

Thus, $F(x)$ is given by:

$$F(x) = \begin{cases} 0 & x < 0 \\ \frac{1}{2}x^2 & 0 \le x < 1 \\ 2x - \frac{1}{2}x^2 - 1 & 1 \le x \le 2 \\ 1 & x > 2 \end{cases}$$

**47.** $f(x) = \begin{cases} .2 - .02x & 0 \le x \le 10 \\ 0 & \text{otherwise} \end{cases}$

(A) $\displaystyle\int_{2}^{6} f(x)\,dx = \int_{2}^{6} (0.2 - 0.02x)\,dx$

$$= \left[(0.2x - 0.01x^2)\Big|_2^6\right] = 0.48$$

The probability that the daily demand for electricity is between 2 million and 6 million kilowatt hours is 0.48.

(B) $\displaystyle P(X \le 8) = \int_{-\infty}^{8} f(x)\,dx = \int_{0}^{8} f(x)\,dx = \int_{0}^{8} (0.2 - 0.02x)\,dx$

$$= \left(0.2x - \frac{0.02x^2}{2}\right)\Big|_0^8 = 0.2(8) - 0.01(8)^2 = 1.6 - 0.64 = 0.96$$

(C) $P(X > 5) = \int_5^{\infty} f(x)\,dx = \int_5^{10} f(x)\,dx = \int_5^{10} (0.2 - 0.02x)\,dx$

$$= \left(0.2x - \frac{0.02x^2}{2}\right)\Big|_5^{10}$$

$$= [0.2(10) - 0.01(10)^2] - [0.2(5) - 0.01(5)^2]$$

$$= 2 - 1 - (1 - 0.25) = 0.25$$

**49.** (A) $\int_5^{10} f(x)\,dx = \int_5^{10} \frac{1}{10} e^{-x/10}\,dx$

$$= \left[-e^{-x/10}\Big|_5^{10}\right] = e^{-1/2} - e^{-1} \approx 0.239$$

The probability that it takes the computer more than 5 seconds and less than 10 seconds to respond is 0.239.

(B) $P(0 \le X \le 1) = \int_0^1 f(x)\,dx = \int_0^1 \frac{1}{10} e^{-x/10}\,dx$

$$= \frac{1}{10}\int_0^1 e^{-x/10}\,dx = \frac{1}{10}\cdot\frac{e^{-x/10}}{-\frac{1}{10}}\Big|_0^1$$

$$= -(e^{-1/10} - 1) = 1 - e^{-1/10} \approx 0.0952$$

(C) $P(X > 4) = \int_4^{\infty} \frac{1}{10} e^{-x/10}\,dx = \lim_{R \to \infty} \frac{1}{10}\int_4^R e^{-x/10}\,dx = \lim_{R \to \infty}\left(-e^{-x/10}\Big|_4^R\right)$

$$= -\lim_{R \to \infty}(e^{-R/10} - e^{-4/10}) = -(0 - e^{-2/5}) = e^{-2/5} \approx 0.6703$$

**51.** (A) $P(X > 4) = \int_4^{10} f(x)\,dx$     [Note: 4 stands for 4000.]

$$= \int_4^{10} 0.003x\sqrt{100 - x^2}\,dx$$

$$= \frac{-0.003}{2}\int_4^{10} (100 - x^2)^{1/2}(-2x)\,dx \quad [\text{Note: } d(100 - x^2) = -2x\,dx.]$$

$$= -0.0015 \cdot \frac{2}{3}(100 - x^2)^{3/2}\Big|_4^{10} = -\frac{1}{1000}(100 - x^2)^{3/2}\Big|_4^{10}$$

$$= -\frac{1}{1000}(0 - (84)^{3/2})$$

$$= \frac{(84)^{3/2}}{1000} \approx 0.7699$$

(B) $P(0 \le x \le 8) = \int_0^8 f(x)\,dx = \int_0^8 0.003x\sqrt{100 - x^2}\,dx$

$$= -\frac{1}{1000}(100 - x^2)^{3/2}\Big|_0^8 \quad [\text{refer to part (A)}]$$

$$= -\frac{1}{1000}[(100 - 64)^{3/2} - (100)^{3/2}]$$

$$= -\frac{1}{1000}(36^{3/2} - 100^{3/2}) = \frac{-1}{1000}(216 - 1000) = 0.784$$

(C) We must solve the following for $x$:

$$\int_0^x f(t)\,dt = 0.9$$

$$\int_0^x 0.003\sqrt{100 - t^2}\,dt = 0.9$$

$$-\frac{1}{1000}(100 - t^2)^{3/2}\Big|_0^x = 0.9 \quad [\text{refer to part (A)}]$$

$$(100 - t^2)^{3/2}\Big|_0^x = -900$$

$$(100 - x^2)^{3/2} - (100)^{3/2} = -900$$

$$(100 - x^2)^{3/2} - 1000 = -900$$

$$(100 - x^2)^{3/2} = 100$$

$$100 - x^2 = 100^{2/3}$$

$$x^2 = 100 - 100^{2/3}$$

$$x = \sqrt{100 - 100^{2/3}} \approx 8.858 \quad \text{or} \quad 8858 \text{ pounds}$$

**53.** (A) $P(7 \leq X) = \int_7^\infty f(x)\,dx = \int_7^{10} f(x)\,dx + \int_{10}^\infty f(x)\,dx$

$$= \int_7^{10} \frac{1}{5000}(10x^3 - x^4)\,dx + \int_{10}^\infty 0\,dx = \frac{1}{5000}\left(\frac{10}{4}x^4 - \frac{x^5}{5}\right)\Big|_7^{10} + 0$$

$$= \frac{1}{5000}\left(\left[\frac{5}{2}(10)^4 - \frac{1}{5}(10)^5\right] - \left[\frac{5}{2}(7)^4 - \frac{1}{5}(7)^5\right]\right) \approx 0.47178$$

(B) $P(X \leq 5) = \int_{-\infty}^5 f(x)\,dx = \int_{-\infty}^0 f(x)\,dx + \int_0^5 f(x)\,dx$

$$= \int_{-\infty}^0 0\,dx + \int_0^5 \frac{1}{5000}(10x^3 - x^4)\,dx = 0 + \frac{1}{5000}\left(\frac{5}{2}x^4 - \frac{1}{5}x^5\right)\Big|_0^5$$

$$= \frac{1}{5000}\left[\frac{5}{2}(5)^4 - \frac{1}{5}(5)^5\right] = \frac{3}{16} = 0.1875$$

**55.** (A) $P(X \leq 20) = \int_{-\infty}^{20} f(x)\,dx = \int_0^{20} \frac{800x}{(400 + x^2)^2}\,dx = 400\int_0^{20} \frac{2x}{(400 + x^2)^2}\,dx$

$$= \frac{-400}{400 + x^2}\Big|_0^{20} = \frac{-400}{800} + \frac{400}{400} = 0.5$$

(B) $P(X > 15) = \int_{15}^\infty f(x)\,dx = 1 - \int_{-\infty}^{15} f(x)\,dx = 1 - \int_0^{15} \frac{800x}{(400 + x^2)^2}\,dx$

$$= 1 - \left(\frac{-400}{400 + x^2}\right)\Big|_0^{15} = 1 + \frac{400}{625} - 1 = 0.64$$

(C) We must solve the following for $x$:

$$\int_0^x f(t)\,dt = 0.8$$

$$\int_0^x \frac{800t}{(400 + t^2)^2}\,dt = 0.8$$

$$\frac{-400}{400 + t^2}\Big|_0^x = 0.8$$

CHAPTER 3   PROBABILITY AND CALCULUS

$$\frac{-400}{400 + x^2} + 1 = 0.8$$

$$\frac{-400}{400 + x^2} = -0.2$$

$$-400 = -80 - 0.2x^2$$

$$0.2x^2 = 320$$

$$x^2 = 1600$$

$$x = 40 \text{ days}$$

**57.** (A) $P(X \geq 30) = \int_{30}^{\infty} f(x)\,dx = 1 - \int_{-\infty}^{30} f(x)\,dx = 1 - \int_{0}^{30} \frac{1}{20} e^{-x/20}\,dx$

$\qquad = 1 + e^{-x/20} \Big|_0^{30} = 1 + e^{-30/20} - 1 = e^{-3/2} \approx 0.223$

(B) $P(X \geq 80) = \int_{80}^{\infty} f(x)\,dx = 1 - \int_{-\infty}^{80} f(x)\,dx = 1 - \int_{0}^{80} \frac{1}{20} e^{-x/20}\,dx$

$\qquad = 1 + e^{-x/20} \Big|_0^{80} = 1 + e^{-80/20} - 1 = e^{-4} \approx 0.018$

## EXERCISE 3-3

**1.** $f(x) = \begin{cases} 2 & \text{if } 0 \leq x \leq 0.5 \\ 0 & \text{otherwise} \end{cases}$

$\mu = E(x) = \int_0^{0.5} x f(x)\,dx = \int_0^{0.5} 2x\,dx = x^2 \Big|_0^{0.5} = 0.25 = \dfrac{1}{4}$

$V(x) = \int_0^{0.5} (x - 0.25)^2 (2)\,dx = \int_0^{0.5} [2x^2 - x + 0.125]\,dx$

$\qquad = \left[ \dfrac{2}{3} x^3 - \dfrac{1}{2} x^2 + 0.125x \right]_0^{0.5}$

$\qquad = \dfrac{2}{3}(0.5)^3 - \dfrac{1}{2}(0.5)^2 + 0.125(0.5)$

$\qquad = \dfrac{1}{48}$

$\sigma = \sqrt{V(X)} = \sqrt{\dfrac{1}{48}} = 0.144.$

<u>Note</u>: If the density is constant over an interval $(a, b)$ and zero elsewhere, then the mean is the midpoint of the interval $\dfrac{a + b}{2}$ and the variance is $\dfrac{(b - a)^2}{12}$ and standard deviation is $\dfrac{b - a}{2\sqrt{3}}$.

Use these to check mean, variance, and standard deviation for this problem.

**3.** $f(x) = \begin{cases} \dfrac{1}{8}\,x & \text{if } 0 \le x \le 4 \\ 0 & \text{otherwise} \end{cases}$

$\mu = \displaystyle\int_0^4 x\left(\frac{1}{8}\,x\right)dx = \int_0^4 \frac{1}{8}x^2\,dx = \frac{1}{24}x^3\Big|_0^4 = \frac{4^3}{24} = \frac{8}{3}$

$V(x) = \displaystyle\int_0^4 \left(x - \frac{8}{3}\right)^2\left(\frac{1}{8}\,x\right)dx = \int_0^4 \frac{1}{8}x\left(x^2 - \frac{16}{3}\,x + \frac{64}{9}\right)dx$

$= \dfrac{1}{8}\displaystyle\int_0^4 \left(x^3 - \frac{16}{3}\,x^2 + \frac{64}{9}\,x\right)dx = \frac{1}{8}\left[\frac{1}{4}\,x^4 - \frac{16}{9}\,x^3 + \frac{32}{9}\,x^2\right]_0^4$

$= \dfrac{1}{8}\left[\dfrac{4^4}{4} - \dfrac{16(4^3)}{9} + \dfrac{32(4^2)}{9}\right] = \frac{1}{8}\left[\frac{64}{9}\right] = \frac{8}{9}$

$\sigma = \sqrt{V(X)} = \sqrt{\dfrac{8}{9}} = 0.943.$

**5.** $\mu = E(X) = \displaystyle\int_{-\infty}^{\infty} xf(x)\,dx = \int_1^2 x(4 - 2x)\,dx = \int_1^2 (4x - 2x^2)\,dx = \left(2x^2 - \frac{2}{3}\,x^3\right)\Big|_1^2$

$= 8 - \dfrac{16}{3} - \left(2 - \dfrac{2}{3}\right) = \dfrac{4}{3} \approx 1.333$

Using $\underline{2}$,

$V(X) = \displaystyle\int_{-\infty}^{\infty} x^2 f(x)\,dx - \mu^2 = \int_1^2 x^2(4 - 2x)\,dx - \frac{16}{9} = \int_1^2 (4x^2 - 2x^3)\,dx - \frac{16}{9}$

$= \left(\dfrac{4}{3}\,x^3 - \dfrac{1}{2}\,x^4\right)\Big|_1^2 - \dfrac{16}{9} = \dfrac{32}{3} - 8 - \left(\dfrac{4}{3} - \dfrac{1}{2}\right) - \dfrac{16}{9}$

$= \dfrac{1}{18}$

Thus, $\sigma = \sqrt{V(X)} = \sqrt{\dfrac{1}{18}} \approx 0.236.$

**7.** $f(x) = \begin{cases} 0.2 & \text{if } -2 \le x \le 3 \\ 0 & \text{otherwise} \end{cases}$

Since the density is constant over $[-2, 3]$ and zero elsewhere, the median will be the same as the mean which is $\dfrac{a + b}{2} = \dfrac{-2 + 3}{2} = \dfrac{1}{2}$.

You can find the median directly as follows:

Let $m$ be the median, then $\displaystyle\int_{-2}^{m} f(x)\,dx = \int_{m}^{3} f(x)\,dx = \frac{1}{2}$.

So we need to solve $\displaystyle\int_{-2}^{m} 0.2\,dx = \frac{1}{2}$ or $\displaystyle\int_{m}^{3} 0.2\,dx = \frac{1}{2}$ to find $m$.

$\displaystyle\int_{m}^{3} 0.2\,dx = 0.2x\Big|_m^3 = 0.6 - 0.2m = \frac{1}{2}$ or

$0.2m = 0.1$ or $m = \dfrac{0.1}{0.2} = 0.5.$

**9.** <u>Step 1</u>: Find the cumulative probability distribution function.
If $x < 2$, then $F(x) = 0$. If $2 \leq x \leq 4$, then

$$F(x) = \int_{-\infty}^{x} f(t)\,dt = \int_{2}^{x} \frac{1}{6}t\,dt = \frac{1}{12}t^2 \Big|_{2}^{x} = \frac{x^2}{12} - \frac{1}{3}.$$

If $x > 4$, then

$$F(x) = \int_{-\infty}^{x} f(t)\,dt = \int_{-\infty}^{2} f(t)\,dt + \int_{2}^{4} f(t)\,dt + \int_{4}^{x} f(t)\,dt$$

$$= 0 + \int_{2}^{4} \frac{1}{6}t\,dt + 0 = \frac{1}{12}t^2 \Big|_{2}^{4} = 1.$$

Thus, $F(x) = \begin{cases} 0 & \text{if } x < 2 \\ \dfrac{x^2}{12} - \dfrac{1}{3} & \text{if } 2 \leq x \leq 4 \\ 1 & \text{if } x > 4 \end{cases}$

<u>Step 2</u>: Solve the equation $P(X \leq m) = \dfrac{1}{2}$ for $m$.

$$F(m) = P(X \leq m) = \frac{1}{2}$$

$$\frac{m^2}{12} - \frac{1}{3} = \frac{1}{2}$$

$$\frac{m^2}{12} = \frac{5}{6}$$

$$m^2 = 10$$

$$m = \sqrt{10} \approx 3.162$$

**11.** <u>Step 1</u>: Find the cumulative probability distribution function.
If $x < 0$, then $F(x) = 0$. If $0 \leq x \leq 4$, then

$$F(x) = \int_{-\infty}^{x} f(t)\,dt = \int_{0}^{x} \left(\frac{1}{2} - \frac{1}{8}t\right) dt = \left(\frac{1}{2}t - \frac{1}{16}t^2\right) \Big|_{0}^{x} = \frac{1}{2}x - \frac{1}{16}x^2.$$

If $x > 4$, then

$$F(x) = \int_{-\infty}^{x} f(t)\,dt = \int_{-\infty}^{0} f(t)\,dt + \int_{0}^{4} f(t)\,dt + \int_{4}^{x} f(t)\,dt$$

$$= 0 + \int_{0}^{4} \left(\frac{1}{2} - \frac{1}{8}t\right) dt + 0$$

$$= \left(\frac{1}{2}t - \frac{1}{16}t^2\right) \Big|_{0}^{4} = 2 - 1 = 1$$

Thus, $F(x) = \begin{cases} 0 & \text{if } x < 0 \\ \dfrac{1}{2}x - \dfrac{1}{16}x^2 & \text{if } 0 \leq x \leq 4 \\ 1 & \text{if } x > 4 \end{cases}$

<u>Step 2</u>: Solve the equation $P(X \leq m) = \dfrac{1}{2}$ for $m$.

$$F(m) = P(X \leq m) = \frac{1}{2}$$

$$\frac{1}{2}m - \frac{1}{16}m^2 = \frac{1}{2}$$

$$m^2 - 8m + 8 = 0$$

Now, the roots of the quadratic equation are

$$\frac{8 \pm \sqrt{64 - 32}}{2} = 4 \pm 2\sqrt{2}.$$

Since $m$ must lie in the interval $[0, 4]$, $m = 4 - 2\sqrt{2} \approx 1.172$.

**13.** $f_1(x) = \begin{cases} 0.5 & \text{if } 0 \le x \le 2 \\ 0 & \text{otherwise} \end{cases}$

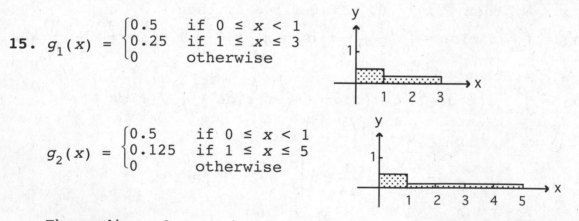

$f_2(x) = \begin{cases} 0.25 & \text{if } -2 \le x \le 2 \\ 0 & \text{otherwise} \end{cases}$

The mean of $X_i$, $i = 1, 2$, is the "balance point" of the region bounded by the graph of $f_i$ and the $x$ axis (see the discussion following explore-discuss 1). From this, we conclude that $\mu_1 = 1$ and $\mu_2 = 0$.

$$\mu_1 = \int_{-\infty}^{\infty} x f_1(x)\,dx = \int_0^2 x(0.5)\,dx = \left.\frac{1}{4}x^2\right|_0^2 = 1$$

$$\mu_2 = \int_{-\infty}^{\infty} x f_2(x)\,dx = \int_{-2}^2 x(0.25)\,dx = \left.\frac{1}{8}x^2\right|_{-2}^2 = \frac{1}{2} - \frac{1}{2} = 0$$

**15.** $g_1(x) = \begin{cases} 0.5 & \text{if } 0 \le x < 1 \\ 0.25 & \text{if } 1 \le x \le 3 \\ 0 & \text{otherwise} \end{cases}$

$g_2(x) = \begin{cases} 0.5 & \text{if } 0 \le x < 1 \\ 0.125 & \text{if } 1 \le x \le 5 \\ 0 & \text{otherwise} \end{cases}$

The median of $X_i$, $i = 1, 2$, is the value of the random variable that divides the area under the graph of $g_i$, $i = 1, 2$, into two equal parts. From the two graphs, it is easy to see that the medians $m_1 = m_2 = 1$.

Since $\int_0^1 g_1(x)\,dx = \int_0^1 g_2(x)\,dx = \int_0^1 \frac{1}{2}\,dx = \left.\frac{1}{2}x\right|_0^1 = \frac{1}{2}$ it follows that $X_1$ and $X_2$ each have median 1.

**17.** $\mu = E(X) = \int_{-\infty}^{\infty} x f(x)\, dx = \int_{1}^{\infty} x \cdot \frac{4}{x^5}\, dx = \int_{1}^{\infty} \frac{4}{x^4}\, dx = \lim_{R \to \infty} \int_{1}^{R} \frac{4}{x^4}\, dx$

$$= \lim_{R \to \infty} \left( -\frac{4}{3x^3} \right) \Big|_{1}^{R} = \lim_{R \to \infty} \left( -\frac{4}{3R^3} + \frac{4}{3} \right) = \frac{4}{3}$$

$V(X) = \int_{-\infty}^{\infty} x^2 f(x)\, dx - \mu^2 = \int_{1}^{\infty} x^2 \cdot \frac{4}{x^5}\, dx - \frac{16}{9} = \int_{1}^{\infty} \frac{4}{x^3}\, dx - \frac{16}{9}$

$$= \lim_{R \to \infty} \int_{1}^{R} \frac{4}{x^3}\, dx - \frac{16}{9} = \lim_{R \to \infty} \left( -\frac{2}{x^2} \right) \Big|_{1}^{R} - \frac{16}{9}$$

$$= \lim_{R \to \infty} \left( -\frac{2}{R^2} + 2 \right) - \frac{16}{9} = \frac{2}{9}$$

$\sigma = \sqrt{V(X)} = \sqrt{\frac{2}{9}} = \frac{\sqrt{2}}{3} \approx 0.471$

**19.** $\mu = E(X) = \int_{-\infty}^{\infty} x f(x)\, dx = \int_{2}^{\infty} x \cdot \frac{64}{x^5}\, dx = \int_{2}^{\infty} \frac{64}{x^4}\, dx = \lim_{R \to \infty} \int_{2}^{R} \frac{64}{x^4}\, dx$

$$= \lim_{R \to \infty} \left( \frac{-64}{3x^3} \right) \Big|_{2}^{R} = \lim_{R \to \infty} \left( \frac{-64}{3R^3} + \frac{64}{24} \right) = \frac{8}{3} \approx 2.667$$

$V(X) = \int_{-\infty}^{\infty} x^2 f(x)\, dx - \mu^2 = \int_{2}^{\infty} x^2 \cdot \frac{64}{x^5}\, dx - \frac{64}{9} = \int_{2}^{\infty} \frac{64}{x^3}\, dx - \frac{64}{9}$

$$= \lim_{R \to \infty} \int_{2}^{R} \frac{64}{x^3}\, dx - \frac{64}{9} = \lim_{R \to \infty} \left( \frac{-32}{x^2} \right) \Big|_{2}^{R} - \frac{64}{9}$$

$$= \lim_{R \to \infty} \left( \frac{-32}{R^2} + 8 \right) - \frac{64}{9} = \frac{8}{9}$$

$\sigma = \sqrt{V(X)} = \sqrt{\frac{8}{9}} = \frac{2\sqrt{2}}{3} \approx 0.943$

**21.** $f(x) = \frac{2}{\sqrt{2\pi}} e^{-2(x-1)^2}$

$\mu = \int_{-\infty}^{\infty} x f(x)\, dx \approx \int_{-10}^{10} x f(x)\, dx \approx 1$

$V(X) = \int_{-\infty}^{\infty} x^2 f(x)\, dx - \mu^2 \approx \int_{-10}^{10} x^2 f(x)\, dx - 1 \approx 1.25 - 1 = 0.25$

$\sigma = \sqrt{0.25} = 0.50$

**23.** <u>Step 1</u>: Find the cumulative probability distribution function.
If $x < 1$, then $F(x) = 0$. If $1 \le x \le e$, then

$F(x) = \int_{-\infty}^{x} f(t)\, dt = \int_{-\infty}^{1} f(t)\, dt + \int_{1}^{x} f(t)\, dt$

$$= 0 + \int_{1}^{x} \frac{1}{t}\, dt = \ln t \Big|_{1}^{x} = \ln x - \ln 1 = \ln x.$$

If $x > e$, then

$F(x) = \int_{-\infty}^{x} f(t)\, dt = \int_{-\infty}^{1} f(t)\, dt = \int_{1}^{e} f(t)\, dt = \int_{e}^{x} f(t)\, dt$

$$= 0 + \int_{1}^{e} \frac{1}{t}\, dt + 0 = \ln t \Big|_{1}^{e} = \ln e - \ln 1 = 1.$$

Thus, $F(x) = \begin{cases} 0 & x < 1 \\ \ln x & 1 \le x \le e \\ 1 & x > e \end{cases}$

Step 2: Solve the equation $P(X \leq m) = \dfrac{1}{2}$ for $m$.

$$F(m) = P(X \leq m) = \frac{1}{2}$$

$$\ln m = \frac{1}{2} \qquad (1 \leq x \leq e)$$

Thus, the median $m = e^{1/2} \approx 1.649$.

25. Step 1: Find the cumulative probability distribution function.
If $x < 0$, then $F(x) = 0$. If $0 \leq x \leq 2$, then

$$F(x) = \int_{-\infty}^{x} f(t)\,dt = \int_{2}^{x} \frac{4}{(2+t)^2}\,dt = \frac{-4}{2+t}\Big|_{0}^{x} = \frac{-4}{2+x} + 2.$$

If $x > 2$, then

$$F(x) = \int_{-\infty}^{x} f(t)\,dt = \int_{-\infty}^{0} f(t)\,dt + \int_{0}^{2} f(t)\,dt + \int_{2}^{x} f(t)\,dt$$

$$= 0 + \int_{0}^{2} \frac{4}{(2+t)^2}\,dt + 0 = \frac{-4}{2+t}\Big|_{0}^{2} = -1 + 2 = 1$$

Thus, $F(x) = \begin{cases} 0 & \text{if } x < 0 \\ \dfrac{-4}{2+x} + 2 & \text{if } 0 \leq x \leq 2 \\ 1 & \text{if } x > 2 \end{cases}$

Step 2: Solve the equation $P(X \leq m) = \dfrac{1}{2}$ for $m$.

$$F(m) = P(X \leq m) = \frac{1}{2}$$

$$\frac{-4}{2+m} + 2 = \frac{1}{2}$$

$$\frac{-4}{2+m} = \frac{-3}{2}$$

$$-6 - 3m = -8$$

$$m = \frac{2}{3}$$

27. Step 1: Find $F(x)$.
If $x < 0$, then $F(x) = 0$. If $x \geq 0$, then

$$F(x) = \int_{-\infty}^{x} f(t)\,dt = \int_{-\infty}^{0} f(t)\,dt + \int_{0}^{x} f(t)\,dt = 0 + \int_{0}^{x} \frac{1}{(1+t)^2}\,dt$$

$$= -\frac{1}{(1+t)}\Big|_{0}^{x} = -\left(\frac{1}{1+x} - 1\right) = 1 - \frac{1}{1+x} = \frac{x}{1+x}.$$

Thus,

$$F(x) = \begin{cases} 0 & x < 0 \\ \dfrac{x}{1+x} & x \geq 0 \end{cases}$$

<u>Step 2</u>: Solve $P(X \leq m) = \dfrac{1}{2}$ for $m$.

$$F(m) = P(X \leq m) = \dfrac{1}{2}$$

$$\dfrac{m}{1 + m} = \dfrac{1}{2} \qquad (x \geq 0)$$

$$2m = 1 + m$$

Thus, the median $m = 1$.

**29.** <u>Step 1</u>: Find $F(x)$.

If $x < 0$, then $F(x) = 0$. If $x \geq 0$, then

$$F(x) = \int_{-\infty}^{x} f(t)\,dt = \int_{-\infty}^{0} f(t)\,dt + \int_{0}^{x} f(t)\,dt = 0 + \int_{0}^{x} 2e^{-2t}\,dt$$

$$= -e^{-2t}\Big|_{0}^{x} = -e^{-2x} + 1$$

Thus, $F(x) = \begin{cases} 0 & \text{if } x < 0 \\ 1 - e^{-2x} & \text{if } x \geq 0 \end{cases}$

<u>Step 2</u>: Solve $P(X \leq m) = \dfrac{1}{2}$ for $m$.

$$F(m) = P(X \leq m) = \dfrac{1}{2}$$

$$1 - e^{-2m} = \dfrac{1}{2}$$

$$e^{-2m} = \dfrac{1}{2}$$

$$-2m = \ln \dfrac{1}{2} = -\ln 2$$

$$m = \dfrac{\ln 2}{2} \approx 0.347$$

**31.** $F(x) = \dfrac{1}{2}$ for any $x$ satisfying $0.5 \leq x \leq 2.5$. The median is *not* unique.

**33.** From the graph,

$$f(x) = \begin{cases} 0 & \text{if } x < 0 \\ 1 - x & \text{if } 0 \leq x \leq 1 \\ 0 & \text{if } 1 < x < 2 \\ x - 2 & \text{if } 2 \leq x \leq 3 \\ 0 & \text{if } x > 3 \end{cases}$$

Now $F(x) = \int_{-\infty}^{x} f(t)\,dt$ and:

for $-\infty < x < 0$, $F(x) = \int_{-\infty}^{x} f(t)\,dt = \int_{-\infty}^{x} 0\,dt = 0$;

for $0 \leq x < 1$,

$$F(x) = \int_{-\infty}^{x} f(t)\,dt = \int_{-\infty}^{0} 0\,dt + \int_{0}^{x} (1 - t)\,dt = \left[\left(t - \dfrac{1}{2}t^2\right)\Big|_{0}^{x}\right] = x - \dfrac{1}{2}x^2;$$

for $1 \leq x \leq 2$,

$$F(x) = \int_{-\infty}^{x} f(t)\,dt = \int_{-\infty}^{0} 0\,dt + \int_{0}^{1}(1 - t)\,dt + \int_{1}^{x} 0\,dt = \left[\left(t - \frac{1}{2}t^2\right)\Big|_{0}^{1}\right] = \frac{1}{2};$$

for $2 \leq x \leq 3$,

$$F(x) = \int_{-\infty}^{x} f(t)\,dt = \int_{-\infty}^{0} 0\,dt + \int_{0}^{1}(1 - t)\,dt + \int_{1}^{2} 0\,dt + \int_{2}^{x}(t - 2)\,dt$$

$$= \frac{1}{2} + \left[\left(\frac{1}{2}t^2 - 2t\right)\Big|_{2}^{x}\right]$$

$$= \frac{1}{2} + \frac{1}{2}x^2 - 2x + 2$$

$$= \frac{1}{2}x^2 - 2x + \frac{5}{2};$$

For $x > 3$

$$F(x) = \int_{-\infty}^{x} f(t)\,dt = \int_{-\infty}^{0} 0\,dt + \int_{0}^{1}(1 - t)\,dt + \int_{1}^{2} 0\,dt$$

$$+ \int_{2}^{3}(t - 2)\,dt + \int_{3}^{x} 0\,dt$$

$$= 0 + \frac{1}{2} + 0 + \frac{1}{2} = 1.$$

Thus,

$$F(x) = \begin{cases} 0 & \text{if } x < 0 \\ x - \frac{1}{2}x^2 & \text{if } 0 \leq x < 1 \\ \frac{1}{2} & \text{if } 1 \leq x \leq 2 \\ \frac{1}{2}x^2 - 2x + \frac{5}{2} & \text{if } 2 < x \leq 3 \\ 1 & \text{if } x > 3 \end{cases}$$

Now, $F(x) = \frac{1}{2}$ for any $x$ satisfying $1 \leq x \leq 2$. The median is *not* unique.

35. Since $f$ is a probability density function,
$\int_{-\infty}^{\infty} f(x)\,dx = 1$ and $\int_{-\infty}^{\infty} xf(x)\,dx = \mu$, the mean.

Now,

$$\int_{-\infty}^{\infty}(ax + b)f(x)\,dx = \int_{-\infty}^{\infty} axf(x)\,dx + \int_{-\infty}^{\infty} bf(x)\,dx$$

$$= a\int_{-\infty}^{\infty} xf(x)\,dx + b\int_{-\infty}^{\infty} f(x)\,dx = a\mu + b.$$

37. Step 1: Find $F(x)$.
If $x < 0$, then $F(x) = 0$. If $0 \leq x \leq 2$, then

$$F(x) = \int_{-\infty}^{x} f(t)\,dt = \int_{-\infty}^{0} f(t)\,dt + \int_{0}^{x} f(t)\,dt$$

$$= 0 + \frac{1}{2}\int_{0}^{x} t\,dt = \frac{1}{4}t^2\Big|_{0}^{x} = \frac{1}{4}x^2.$$

If $x > 2$, then

$$F(x) = \int_{-\infty}^{x} f(t)\,dt = \int_{-\infty}^{0} f(t)\,dt + \int_{0}^{2} f(t)\,dt + \int_{2}^{x} f(t)\,dt$$

$$= 0 + \frac{1}{2}\int_{0}^{2} t\,dt + 0 = \frac{1}{4}t^2 \Big|_{0}^{2} = 1.$$

Thus,

$$F(x) = \begin{cases} 0 & x < 0 \\ \frac{1}{4}x^2 & 0 \le x \le 2 \\ 1 & x > 2 \end{cases}$$

Step 2: In order to find the quartile point $x_1$, we solve the following for $x_1$:

$$F(x_1) = P(X \le x_1) = \frac{1}{4}$$

$$\frac{1}{4}x_1^2 = \frac{1}{4}$$

$$x_1^2 = 1$$

$$x_1 = 1$$

For the quartile point $x_2$ (or $m$), we solve the following for $x_2$:

$$F(x_2) = P(X \le x_2) = \frac{1}{2}$$

$$\frac{1}{4}x_2^2 = \frac{1}{2}$$

$$x_2^2 = 2$$

$$x_2 = \sqrt{2} \approx 1.414$$

For the quartile point $x_3$, we solve the following for $x_3$:

$$F(x_3) = P(X \le x_3) = \frac{3}{4}$$

$$\frac{1}{4}x_3^2 = \frac{3}{4}$$

$$x_3^2 = 3$$

$$x_3 = \sqrt{3} \approx 1.732$$

39. Step 1: Find $F(x)$.

If $x < 0$, then $F(x) = 0$. If $x \ge 0$, then

$$F(x) = \int_{-\infty}^{x} f(t)\,dt = \int_{-\infty}^{0} f(t)\,dt + \int_{0}^{x} f(t)\,dt = 0 + \int_{0}^{x} \frac{3}{(3+t)^2}\,dt$$

$$= \frac{-3}{3+t} \Big|_{0}^{x} = \frac{-3}{3+x} + 1.$$

Thus, $F(x) = \begin{cases} 0 & \text{if } x < 0 \\ 1 - \dfrac{3}{3+x} & \text{if } x \ge 0 \end{cases}$

Step 2: For the quartile point $x_1$, we solve:

$$F(x_1) = P(X \leq x_1) = \frac{1}{4}$$

$$1 - \frac{3}{3 + x_1} = \frac{1}{4}$$

$$-\frac{3}{3 + x_1} = -\frac{3}{4}$$

$$3 + x_1 = 4$$

$$x_1 = 1$$

For the quartile point $x_2$, we solve:

$$F(x_2) = P(X \leq x_2) = \frac{1}{2}$$

$$1 - \frac{3}{3 + x_2} = \frac{1}{2}$$

$$-\frac{3}{3 + x_2} = -\frac{1}{2}$$

$$3 + x_2 = 6$$

$$x_2 = 3$$

For the quartile point $x_3$, we solve:

$$F(x_3) = P(X \leq x_3) = \frac{3}{4}$$

$$1 - \frac{3}{3 + x_3} = \frac{3}{4}$$

$$-\frac{3}{3 + x_3} = -\frac{1}{4}$$

$$3 + x_3 = 12$$

$$x_3 = 9$$

**41.** Given $f(x) = \begin{cases} 4x - 4x^3 & 0 \leq x \leq 1 \\ 0 & \text{otherwise} \end{cases}$

Step 1. Find $F(x)$.

If $x < 0$, then $F(x) = 0$. If $0 \leq x \leq 1$, then

$$F(x) = \int_{-\infty}^{x} f(t)\,dt = \int_{-\infty}^{0} f(t)\,dt + \int_{0}^{x} f(t)\,dt$$

$$= 0 + \int_{0}^{x} (4t - 4t^3)\,dt = (2t^2 - t^4)\Big|_{0}^{x}$$

$$= 2x^2 - x^4$$

If $x > 1$, then

$$F(x) = \int_{-\infty}^{x} f(t)\,dt = \int_{-\infty}^{0} f(t)\,dt + \int_{0}^{1} f(t)\,dt + \int_{1}^{x} f(t)\,dt$$

$$= 0 + \int_{0}^{1} (4t - 4t^3)\,dt + 0 = (2t^2 - t^4)\Big|_{0}^{1} = 1$$

Thus, $F(x) = \begin{cases} 0 & x < 0 \\ 2x^2 - x^4 & 0 \le x \le 1 \\ 1 & x > 1 \end{cases}$

<u>Step 2.</u> Solve the equation $F(m) = \dfrac{1}{2}$ for $m$.

The graphs of $F$ and $y = \dfrac{1}{2}$ are shown at the right.

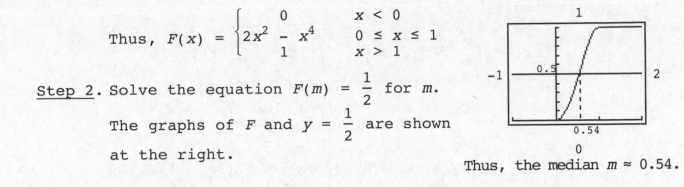

Thus, the median $m \approx 0.54$.

43. $f(x) = \begin{cases} \dfrac{1}{2x^2} + \dfrac{3}{2x^4} & \text{if } x \ge 1 \\ 0 & \text{if } x < 1 \end{cases}$

<u>Step 1</u>: Find $F(x)$.

If $x < 1$, then $F(x) = 0$. If $x \ge 1$, then

$$F(x) = \int_{-\infty}^{x} f(t)\,dt = \int_{-\infty}^{1} 0\,dt + \int_{1}^{x}\left(\frac{1}{2t^2} + \frac{3}{2t^4}\right)dt$$

$$= \left[\left(-\frac{1}{2t} - \frac{3}{6t^3}\right)\Big|_{1}^{x}\right]$$

$$= 1 - \frac{1}{2x} - \frac{1}{2x^3}$$

Thus, $F(x) = \begin{cases} 0 & \text{if } x < 1 \\ 1 - \dfrac{1}{2x} - \dfrac{1}{2x^3} & \text{if } x \ge 1 \end{cases}$

<u>Step 2.</u> Solve the equation $F(m) = \dfrac{1}{2}$ for $m$. The graphs of $F$ and $y = \dfrac{1}{2}$ are shown below.

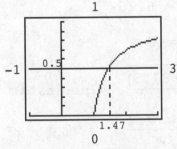

Thus, $m \approx 1.47$.

45. (A) The contractor's expected profit is given by:

$$E(x) = \int_{-\infty}^{\infty} x f(x)\,dx = \int_{15}^{20} x(0.08)(20 - x)\,dx = 0.08\int_{15}^{20}(20x - x^2)\,dx$$

$$= 0.08\left[10x^2 - \frac{1}{3}x^3\right]_{15}^{20}$$

$$= 0.08\left(10(20)^2 - \frac{(20)^3}{3} - \left[10(15)^2 - \frac{(15)^3}{3}\right]\right)$$

$$= 0.08(1333.33 - 1125) = 16.667 \text{ or } \$16,667.$$

(B) <u>Step 1</u>: Find $F(x)$.

If $x < 15$, then $F(x) = 0$. If $15 \leq x \leq 20$, then

$$F(x) = \int_{-\infty}^{x} f(t)\,dt = \int_{-\infty}^{15} f(t)\,dt + \int_{15}^{x} f(t)\,dt$$

$$= 0 + 0.08\int_{15}^{x} (20 - t)\,dt$$

$$= 0.08\left[20t - \frac{1}{2}t^2\right]_{15}^{x}$$

$$= 0.08\left[20x\,\frac{1}{2}\,x^2 - (300 - 112.50)\right]$$

$$= 0.08\left[20x\,\frac{1}{2}\,x^2 - 187.5\right]$$

If $x > 20$, then

$$F(x) = \int_{-\infty}^{x} f(t)\,dt = \int_{-\infty}^{15} f(t)\,dt + \int_{15}^{20} f(t)\,dt + \int_{20}^{x} f(t)\,dt$$

$$= 0 + 0.08\int_{15}^{20} (20 - x)\,dx + 0$$

$$= 0.08\left[20x - \frac{1}{2}x^2\right]_{15}^{20}$$

$$= 0.08[400 - 200 - (300 - 112.5)] = 1.$$

<u>Step 2</u>: To find the median profit, $x_m$, we solve the following:

$$F(x_m) = P(X \leq x_m) = \frac{1}{2}$$

$$0.08\left[20x_m - \frac{1}{2}x_m^2 - 187.5\right] = \frac{1}{2}$$

$$20x_m - \frac{1}{2}x_m^2 - 187.5 = 6.25$$

$$x_m^2 - 40x_m + 387.5 = 0$$

$$x_m = \frac{40 \pm \sqrt{(40)^2 - 4(387.5)}}{2} \quad \text{(quadratic formula)}$$

$$= \frac{40 \pm \sqrt{50}}{2} = \frac{40 \pm 5\sqrt{2}}{2}$$

Since $x_m$ lies in the interval $[15, 20]$,

$$x_m = \frac{40 - 5\sqrt{2}}{2} = 20 - \frac{5}{2}\sqrt{2} \approx 16.464 \text{ or } \$16{,}464$$

**47.** <u>Step 1</u>: Find $F(x)$.

If $x < 0$, then $F(x) = 0$. If $x \geq 0$, then

$$F(x) = \int_{-\infty}^{x} f(t)\,dt = \int_{-\infty}^{0} f(t)\,dt + \int_{0}^{x} f(t)\,dt = 0 + \int_{0}^{x} \frac{1}{3} e^{-t/3}\,dt$$

$$= -e^{-t/3} \Big|_{0}^{x} = -e^{-x/3} + 1.$$

Thus, $F(x) = \begin{cases} 0 & \text{if } x < 0 \\ 1 - e^{-x/3} & \text{if } x \geq 0 \end{cases}$

<u>Step 2</u>: Solve $P(X \leq m) = \dfrac{1}{2}$ for $m$.

$$F(m) = P(X \leq m) = \frac{1}{2}$$

$$1 - e^{-m/3} = \frac{1}{2}$$

$$e^{-m/3} = \frac{1}{2}$$

$$\frac{-m}{3} = \ln \frac{1}{2} = -\ln 2$$

$$m = 3 \ln 2 \approx 2.079 \text{ minutes}$$

**49.** The expected daily consumption is given by:

$$E(X) = \int_{-\infty}^{\infty} x f(x)\,dx = \int_{-\infty}^{\infty} \frac{x}{(1 + x^2)^{3/2}}\,dx$$

$$= \lim_{R \to \infty} \int_{0}^{R} \frac{x}{(1 + x^2)^{3/2}}\,dx \quad [\underline{\text{Note}}: \text{ If } u = 1 + x^2, \text{ then } \frac{du}{2} = x\,dx \text{ and}$$

$$\int \frac{x}{(1 + x^2)^{3/2}}\,dx = \frac{1}{2}\int \frac{du}{u^{3/2}} = \frac{1}{2} \cdot \frac{-2}{u^{1/2}}$$

$$= -\frac{1}{(1 + x^2)^{1/2}} \cdot]$$

$$= -\lim_{R \to \infty} \frac{1}{(1 + x^2)^{1/2}} \Big|_{0}^{R} = -\lim_{R \to \infty} \left( \frac{1}{(1 + R^2)^{1/2}} - \frac{1}{1} \right) = 1 \text{ or } 1 \text{ million gallons}$$

**51.** Mean life expectancy is given by:

$$E(X) = \mu = \int_{-\infty}^{\infty} x f(x)\,dx = \frac{1}{5000} \int_{0}^{10} x(10x^3 - x^4)\,dx = \frac{1}{5000} \left( 10 \cdot \frac{x^5}{5} - \frac{x^6}{6} \right) \Big|_{0}^{10}$$

$$= \frac{1}{5000} \left( 2 \cdot 10^5 - \frac{1}{6} \cdot 10^6 - 0 \right) = \frac{1}{5000} \left( \frac{1}{3} \cdot 10^5 \right) = \frac{20}{3} \approx 6.7 \text{ minutes}$$

**53.** Step 1: Find $F(x)$.

If $x < 0$, $F(x) = 0$. If $x \geq 0$, then

$$F(x) = \int_{-\infty}^{x} f(t)\,dt = \int_{-\infty}^{0} f(t)\,dt + \int_{0}^{x} f(t)\,dt = 0 + \int_{0}^{x} \frac{800t}{(400 + t^2)^2}\,dt$$

$$= \frac{-400}{400 + t^2}\Big|_{0}^{x} = \frac{-400}{400 + x^2} + 1.$$

Thus, $F(x) = \begin{cases} 0 & \text{if } x < 0 \\ 1 - \dfrac{400}{400 + x^2} & \text{if } x \geq 0 \end{cases}$

Step 2: Solve $P(X \leq m) = \dfrac{1}{2}$ for $m$.

$$1 - \frac{400}{400 + m^2} = \frac{1}{2}$$

$$\frac{-400}{400 + m^2} = \frac{-1}{2}$$

$$400 + m^2 = 800$$

$$m^2 = 400$$

$$m = 20 \text{ days}$$

**55.** The expected number of hours to learn the task is given by:

$$E(X) = \mu = \int_{-\infty}^{\infty} x f(x)\,dx = \int_{0}^{3} x\left(\frac{4}{9}x^2 - \frac{4}{27}x^3\right)dx = \left(\frac{4}{9}\cdot\frac{x^4}{4} - \frac{4}{27}\cdot\frac{x^5}{5}\right)\Big|_{0}^{3}$$

$$= \frac{1}{9}(3^4) - \frac{4}{3^3(5)}(3^5) = 9 - \frac{36}{5} = \frac{9}{5} = 1.8 \text{ hours}$$

## EXERCISE 3-4

**1.** Using 1(a) with $[a, b] = [0, 2]$, we have:

$$f(x) = \begin{cases} \dfrac{1}{2 - 0} & 0 \leq x \leq 2 \\ 0 & \text{otherwise} \end{cases} = \begin{cases} \dfrac{1}{2} & 0 \leq x \leq 2 \\ 0 & \text{otherwise} \end{cases}$$

Using 1(b) with $[a, b] = [0, 2]$, we have:

$$F(x) = \begin{cases} 0 & x < 0 \\ \dfrac{x - 0}{2 - 0} & 0 \leq x \leq 2 \\ 1 & x > 2 \end{cases} = \begin{cases} 0 & x < 0 \\ \dfrac{x}{2} & 0 \leq x \leq 2 \\ 1 & x > 2 \end{cases}$$

**3.** Using 2(a) with $\lambda = \dfrac{1}{2}$, we have:

$$f(x) = \begin{cases} \dfrac{1}{1/2} e^{-x/(1/2)} & x \geq 0 \\ 0 & \text{otherwise} \end{cases} = \begin{cases} 2e^{-2x} & x \geq 0 \\ 0 & \text{otherwise} \end{cases}$$

Using 2(b) with $\lambda = \dfrac{1}{2}$, we have:

$$F(x) = \begin{cases} 1 - e^{-x/(1/2)} & x \geq 0 \\ 0 & \text{otherwise} \end{cases} = \begin{cases} 1 - e^{-2x} & x \geq 0 \\ 0 & \text{otherwise} \end{cases}$$

5. Using 1(c), (d), (e), with $[a, b] = [1, 5]$, we have:

Mean: $\mu = \dfrac{1}{2}(a + b) = \dfrac{1}{2}(1 + 5) = 3$

Median: $m = \dfrac{1}{2}(a + b) = \dfrac{1}{2}(1 + 6) = 3$

Standard deviation: $\sigma = \dfrac{1}{\sqrt{12}}(b - a) = \dfrac{1}{\sqrt{12}}(5 - 1) = \dfrac{4}{\sqrt{12}} = \dfrac{2}{\sqrt{3}} \approx 1.155$

7. Using 2(c), (d), (e), with $\lambda = 5$, we have:
Mean: $\mu = 5$
Median: $m = 5 \ln 2 \approx 3.466$
Standard deviation: $\sigma = 5$

9. From Table I, the area under the standard normal curve from 0 to 1.5 is 0.4332.

11. The area under the standard normal curve between −0.72 and 0 is the same as the area between 0 and 0.72 (due to the symmetry of the curve). From Table I, the area under the standard normal curve between 0 and 0.72 is 0.2642.

13. From Table I, the area under the standard normal curve from 0 to 2.01 is 0.4778.

15. The area under the standard normal curve from −1.93 to 0 is the same as the area from 0 to 1.93 which from Table I is 0.4732.

17. The area under a normal curve with mean 100 and standard deviation 10 from 100 to 120 is the same as the area under the standard normal curve from 0 to $\dfrac{120 - 100}{10} = 2$. From Table I, this area is 0.4772.

19. For 95, the area is from $\dfrac{95 - 100}{10} = -\dfrac{1}{2}$ to 0 which is the same as from

0 to $\dfrac{1}{2}$ under the standard normal curve. From Table I for $\dfrac{1}{2} = 0.5$, this area is 0.1915.

21. For 106, the area is from 0 to $\dfrac{106 - 100}{10} = 0.6$ under the standard normal curve. From Table I for 0.6, this area is 0.2257.

23. For 78, the area is from $\dfrac{78 - 100}{10} = -2.2$ to 0 which is the same as from 0 to 2.2 under the standard normal curve. From Table I for 2.2, this area is 0.4861.

**25.** $f_1(x) = \dfrac{1}{\sqrt{2\pi}} e^{-x^2/2}$

$f_2(x) = \begin{cases} 0.4 & \text{if } -1.25 \leq x \leq 1.25 \\ 0 & \text{otherwise} \end{cases}$

From the graphs, it appears that $X_1$ has the greater standard deviation.

$\mu_1 = \displaystyle\int_{-\infty}^{\infty} x f_1(x)\, dx = 0$

$V_1(X) = \displaystyle\int_{-\infty}^{\infty} (x - 0)^2 f_1(x)\, dx = \int_{-\infty}^{\infty} x^2 f_1(x)\, dx = 1;\ \sigma_1 = 1$

$\mu_2 = \displaystyle\int_{-\infty}^{\infty} x f_2(x)\, dx = \int_{-1.25}^{1.25} 0.4x\, dx = 0$

$V_2(X) = \displaystyle\int_{-\infty}^{\infty} (x - 0)^2 f(x)\, dx = \int_{-1.25}^{1.25} 0.4x^2\, dx = 0.4\left.\dfrac{x^3}{3}\right|_{-1.25}^{1.25} \approx 0.5208$

$\sigma_2 = \sqrt{0.5208} = 0.722$

**27.** $g_1(x) = \dfrac{1}{\sqrt{2\pi}} e^{-(x-1)^2/2}$

$g_2(x) = \begin{cases} e^{-x} & \text{if } x \geq 0 \\ 0 & \text{otherwise} \end{cases}$

By symmetry, $X_1$ has median 1. From the graph of $g_2$, it appears that $X_2$ has median less than 1.

$F_2(x) = \displaystyle\int_{-\infty}^{x} g_2(t)\, dt = \int_0^x e^{-t}\, dt = -e^{-t}\Big|_0^x = 1 - e^{-x}$

Now solve the equation $F_2(m) = \dfrac{1}{2}$:

$1 - e^{-m} = \dfrac{1}{2}$

$e^{-m} = \dfrac{1}{2}$

$-m = \ln(1/2)$

$m = 0.693$

**29.** $X$ is uniformly distributed on $[0,\ 4]$. The probability density function is

$f(x) = \begin{cases} \dfrac{1}{4} & 0 \leq x \leq 4 \\ 0 & \text{otherwise} \end{cases}$

and the mean is

$\mu = \dfrac{1}{2}(0 + 4) = 2.$

Thus, $P(X \leq 2) = \displaystyle\int_0^2 f(x)\, dx = \int_0^2 \dfrac{1}{4}\, dx = \dfrac{1}{4}x\Big|_0^2 = \dfrac{1}{2}.$

**31.** $X$ is an exponential random variable with $\mu = 1$. Then

$$f(x) = \begin{cases} e^{-x} & x \geq 0 \\ 0 & \text{otherwise} \end{cases}$$

and the mean is
$\mu = 1$.

Thus, $P(X \leq 1) = \int_0^1 e^{-x}\, dx = -e^{-x}\Big|_0^1 = -e^{-1} + 1 = 1 - e^{-1} \approx 0.632$.

**33.** $X$ is uniformly distributed on $[-5, 5]$. The mean is

$$\mu = \frac{1}{2}(-5 + 5) = 0$$

and the standard deviation is

$$\sigma = \frac{1}{\sqrt{12}}[5 - (-5)] = \frac{10}{\sqrt{12}} = \frac{5}{\sqrt{3}} \approx 2.887.$$

Also,

$$f(x) = \begin{cases} \dfrac{1}{10} & -5 \leq x \leq 5 \\ 0 & \text{otherwise} \end{cases}$$

Thus, $P(\mu - \sigma \leq X \leq \mu + \sigma) = P\left(\dfrac{-5}{\sqrt{3}} \leq X \leq \dfrac{5}{\sqrt{3}}\right) = \displaystyle\int_{-5\sqrt{3}}^{5\sqrt{3}} f(x)\,dx = \int_{-5\sqrt{3}}^{5\sqrt{3}} \frac{1}{10}\,dx$

$$= \frac{1}{10}x\Big|_{-5\sqrt{3}}^{5\sqrt{3}} = \frac{1}{10}\left(\frac{5}{\sqrt{3}}\right) - \frac{1}{10}\left(\frac{-5}{\sqrt{3}}\right)$$

$$= \frac{1}{\sqrt{3}} \approx 0.577.$$

**35.** $X$ is an exponential random variable with median $m = 6 \ln 2$. Since $m = \lambda \ln 2$, we have $\lambda = 6$. Thus, mean $\mu = 6$ and standard deviation $\sigma = 6$. Also,

$$f(x) = \begin{cases} \dfrac{1}{6}e^{-x/6} & x \geq 0 \\ 0 & \text{otherwise} \end{cases}$$

Now, $P(\mu - \sigma \leq X \leq \mu + \sigma) = P(0 \leq X \leq 12) = \displaystyle\int_0^{12} f(x)\,dx = \int_0^{12} \frac{1}{6}e^{-x/6}\,dx$

$$= -e^{-x/6}\Big|_0^{12} = -e^{-2} + 1 \approx 0.865.$$

**37.** $\mu = 70$, $\sigma = 8$

$z$ (for $x = 60$) $= \dfrac{60 - 70}{8} = -1.25$

$z$ (for $x = 80$) $= \dfrac{80 - 70}{8} = 1.25$

Area $A_1 = 0.3944$.  Area $A_2 = 0.3944$.
Total area $= A = A_1 + A_2 = 0.7888$.

**39.** $\mu = 70$, $\sigma = 8$

$z$ (for $x = 62$) $= \dfrac{62 - 70}{8} = -1.00$

$z$ (for $x = 74$) $= \dfrac{74 - 70}{8} = 0.5$

Area $A_1 = 0.3413$.  Area $A_2 = 0.1915$.

Total area $= A = A_1 + A_2 = 0.5328$.

**41.** $\mu = 70$, $\sigma = 8$

$z$ (for $x = 88$) $= \dfrac{88 - 70}{8} = 2.25$

Required area $= 0.5 - \begin{array}{l}\text{(area corresponding to} \\ \quad z = 2.25)\end{array}$

$\qquad\qquad\qquad = 0.5 - 0.4878$

$\qquad\qquad\qquad = 0.0122$

**43.** $\mu = 70$, $\sigma = 8$

$z$ (for $x = 60$) $= \dfrac{60 - 70}{8} = -1.25$

Required area $= 0.5 - \begin{array}{l}\text{(area corresponding to} \\ \quad z = 1.25)\end{array}$

$\qquad\qquad\qquad = 0.5 - 0.3944$

$\qquad\qquad\qquad = 0.1056$

**45.** Normal probability density function with $\mu = 0$:

$$f(x) = \frac{1}{\sigma\sqrt{2\pi}}\, e^{-x^2/2\sigma^2}, \quad \sigma > 0.$$

(A) $\sigma = 0.5$

$$P(-0.5 \le X \le 0.5) = \frac{1}{0.5\sqrt{2\pi}} \int_{-0.5}^{0.5} e^{-x^2/0.5}\, dx$$

Using a graphing utility with a numerical integration routine, we find that

$$P(-0.5 \le X \le 0.5) \approx 0.6827$$

(B) $\sigma = 1$

$$P(-1 \le X \le 1) = \frac{1}{\sqrt{2\pi}} \int_{-1}^{1} e^{-x^2/0.5}\, dx \approx 0.6827$$

(C) $\sigma = 2$

$$P(-2 \le X \le 2) = \frac{1}{2\sqrt{2\pi}} \int_{-2}^{2} e^{-x^2/8}\, dx \approx 0.6827$$

Interpretation: the probability of being within one standard deviation of the mean is always 0.6827.

**47.** Normal probability density function with $\mu = 0$:

$$f(x) = \frac{1}{\sigma\sqrt{2\pi}}\, e^{-x^2/2\sigma^2}, \quad \sigma > 0.$$

(A) $\sigma = 0.5$

$$P(-1.5 \le X \le 1.5) = \frac{1}{0.5\sqrt{2\pi}} \int_{-1.5}^{1.5} e^{-x^2/0.5}\, dx$$

Using a graphing utility with a numerical integration routine, we find that

$$P(-1.5 \le X \le 1.5) \approx 0.9973$$

(B) $\sigma = 1$

$P(-3 \le X \le 3) = \dfrac{1}{\sqrt{2\pi}} \displaystyle\int_{-3}^{3} e^{-x^2/2}\, dx \approx 0.9973$

(C) $\sigma = 2$

$P(-6 \le X \le 6) = \dfrac{1}{2\sqrt{2\pi}} \displaystyle\int_{-6}^{6} e^{-x^2/8}\, dx \approx 0.9973$

Interpretation: the probability of being within 3 standard deviations of the mean is always 0.9973.

**49.** From Section 3-3,

$\mu = \displaystyle\int_{-\infty}^{\infty} xf(x)\, dx,$

where

$f(x) = \begin{cases} \dfrac{1}{b-a} & a \le x \le b \\ 0 & \text{otherwise} \end{cases}$

Thus,

$\mu = \displaystyle\int_{a}^{b} x\left(\dfrac{1}{b-a}\right) dx = \dfrac{1}{b-a}\int_{a}^{b} x\, dx = \dfrac{1}{b-a} \cdot \dfrac{x^2}{2}\Big|_{a}^{b} = \dfrac{1}{b-a}\left(\dfrac{b^2}{2} - \dfrac{a^2}{2}\right)$

$= \dfrac{1}{b-a} \cdot \dfrac{1}{2}(b-a)(b+a) = \dfrac{a+b}{2}.$

**51.** $\displaystyle\int_{-\infty}^{\infty} x^2 f(x)\, dx = \int_{a}^{b} x^2\left(\dfrac{1}{b-a}\right) dx = \dfrac{1}{b-a} \cdot \dfrac{x^3}{3}\Big|_{a}^{b} = \dfrac{1}{3} \cdot \dfrac{1}{b-a}(b^3 - a^3)$

$= \dfrac{1}{3} \cdot \dfrac{1}{b-a}(b-a)(b^2 + ab + a^2) = \dfrac{1}{3}(b^2 + ab + a^2)$

**53.** $f(x) = \begin{cases} 5e^{-5x} & \text{if } x \ge 0 \\ 0 & \text{otherwise} \end{cases}$

Mean: $\mu = \displaystyle\int_{-\infty}^{\infty} xf(x)\, dx = \int_{0}^{\infty} 5xe^{-5x}\, dx$ since $f(x) = 0$ for $x < 0$.

Also, $10\, f(10) \approx 9.644 \times 10^{-21} \approx 0$. Therefore, it is reasonable to replace the integral by

$\displaystyle\int_{0}^{10} 5xe^{-5x}\, dx = 0.2$

Median: $F(x) = \displaystyle\int_{-\infty}^{x} f(t)\, dt = \int_{0}^{x} 5e^{-5t}\, dt, \ x \ge 0$

$= -e^{-5t}\Big|_{0}^{x} \quad 1 - e^{-5x}$

Now solve the equation $F(m) = \dfrac{1}{2}$:

$1 - e^{-5m} = \dfrac{1}{2}$

$e^{-5m} = \dfrac{1}{2}$

$-5m = \ln(1/2)$

$m = 0.139$

Standard Deviation:

$$V(X) = \int_{-\infty}^{\infty} x^2 f(x)dx - (0.2)^2 = \int_0^{\infty} 5x^2 e^{-5x}\, dx - 0.04$$

since $f(x) = 0$ for $x < 0$

Also, $100\, f(10) \approx 9.644 \times 10^{-20} \approx 0$. Therefore, it is reasonable to integrate over the interval $[0, 10]$:

$$\int_0^{10} 5x^2 e^{-5x}\, dx - 0.04 = 0.08 - 0.04 = 0.04;$$

$$\sigma = \sqrt{0.04} = 0.2$$

**55.** $f(x) = \begin{cases} \dfrac{p}{x^{p+1}} & \text{if } x \geq 1 \\ 0 & \text{otherwise} \end{cases} \quad p > 0$

(A) Clearly, $f(x) \geq 0$ for all $x$.

(B) $\displaystyle\int_{-\infty}^{\infty} f(x)dx = \int_1^{\infty} \frac{p}{x^{p+1}}\, dx = \lim_{b \to \infty} \int_1^b \frac{p}{x^{p+1}}\, dx$

$$= \lim_{b \to \infty}\left[ -\frac{1}{x^p}\Big|_1^b \right]$$

$$= \lim_{b \to \infty}\left[ 1 - \frac{1}{b^p} \right] = 1$$

Thus, $f$ is a probability density function.
The mean $\mu$ of $f$ is given by:

$$\mu = \int_{-\infty}^{\infty} xf(x)dx = \int_{-\infty}^{\infty} x\left(\frac{p}{x^{p+1}}\right)dx$$

$$= \int_1^{\infty} \frac{p}{x^p}\, dx = \lim_{b \to \infty} \int_1^b \frac{p}{x^p}\, dx$$

$$= \lim_{b \to \infty}\left[ \frac{px^{-p+1}}{-p+1}\Big|_1^b \right],\ p \neq 1$$

$$= \lim_{b \to \infty}\left[ \frac{pb^{-p+1}}{1-p} - \frac{p}{1-p} \right]$$

$$= \frac{p}{p-1} \text{ provided } p > 1;$$

The limit does not exist if $p < 1$.
If $p = 1$, then

$$\int_1^{\infty} \frac{1}{x}\, dx = \lim_{b \to \infty} \int_1^b \frac{1}{x}\, dx$$

$$= \lim_{b \to \infty}[\ln b]$$

does not exist.
Thus,

$$\mu = \begin{cases} \dfrac{p}{p-1} & \text{if } p > 1 \\ \text{does not exist} & \text{if } 0 < p \leq 1 \end{cases}$$

**57.** The median $m$ is given by
$$P(X \leq m) = \frac{1}{2}$$
Clearly, $m > 1$ since $f(x) = 0$ for $x < 1$.
Now,
$$P(X \leq m) = \int_{-\infty}^{m} f(x)dx = \int_{1}^{m} \frac{p}{x^{p+1}}\, dx$$
$$= \frac{-1}{x^p}\Big|_{1}^{m} = 1 - \frac{1}{m^p}$$

and
$$1 - \frac{1}{m^p} = \frac{1}{2}$$
$$\frac{1}{m^p} = \frac{1}{2}$$
$$m^p = 2$$
$$m = \sqrt[p]{2}$$

**59.** Using the uniform cumulative probability distribution function:
$$P(25 \leq X \leq 40) = F(40) - F(25)$$
$$= \frac{40 - 0}{40 - 0} - \frac{25 - 0}{40 - 0} \quad [\underline{\text{Note}}:\ a = 0,\ b = 40.]$$
$$= 1 - \frac{25}{40} = \frac{15}{40} = \frac{3}{8} = 0.375$$

**61.** We are given that $\mu = 3$. Also $\lambda = \mu$. Thus, $\lambda = 3$, and we have:
$$P(0 \leq X \leq 2) = F(2) - F(0) = (1 - e^{-2/3}) - (1 - e^0)$$
$$= 1 - e^{-2/3} \approx 0.487$$

**63.** $X$ is an exponential random variable and the median $m = 2$. Since $m = \lambda \ln 2$, we have $2 = \lambda \ln 2$ or $\lambda = \dfrac{2}{\ln 2}$. Now $P(X \leq 1) = F(1)$, where $F$ is the cumulative probability distribution:
$$F(x) = \begin{cases} 1 - e^{-x/\lambda} & x \geq 0 \\ 0 & \text{otherwise} \end{cases}$$

Setting $\lambda = \dfrac{2}{\ln 2}$, we have
$$F(x) = \begin{cases} 1 - e^{-(x\ln 2)/2} & x \geq 0 \\ 0 & \text{otherwise} \end{cases}$$
Thus, $P(X \leq 1) = 1 - e^{-(\ln 2)/2} = 1 - e^{-(1/2)\ln 2}$
$$= 1 - e^{\ln(2)^{-1/2}}$$
$$= 1 - 2^{-1/2}$$
$$= 1 - \frac{1}{\sqrt{2}} \approx 0.293.$$

**65.** $\mu = 200{,}000$, $\sigma = 20{,}000$, $x \geq 240{,}000$

$$z \text{ (for } x = 240{,}000) = \frac{240{,}000 - 200{,}000}{20{,}000} = 2.0$$

Fraction of the salesmen who would be expected to make annual sales of \$240,000 or more

$$= \text{Area } A_1$$
$$= 0.5 - (\text{area between } \mu \text{ and } 240{,}000)$$
$$= 0.5 - 0.4772$$
$$= 0.0228$$

Thus, the percentage of salesmen expected to make annual sales of \$240,000 or more is 2.28%.

**67.** $x = 105$, $x = 95$, $\mu = 100$, $\sigma = 2$

$$z \text{ (for } x = 105) = \frac{105 - 100}{2} = 2.5$$

$$z \text{ (for } x = 95) = \frac{95 - 100}{2} = -2.5$$

Fraction of parts to be rejected $= \text{Area } A_1 + A_2$

$$= 1 - 2(\text{area corresponding to } z = 2.5)$$
$$= 1 - 2(0.4938)$$
$$= 0.0124$$

Thus, the percentage of parts to be rejected is 1.24%.

**69.** (A) We are given that

$$P(0 \leq X \leq 1) = 0.3 = F(1) - F(0) = 1 - e^{-1/\lambda} - (1 - e^0).$$

Thus,

$$1 - e^{-1/\lambda} = 0.3$$
$$e^{-1/\lambda} = 0.7$$
$$-\frac{1}{\lambda} = \ln(0.7)$$
$$\lambda = -\frac{1}{\ln(0.7)}$$

But $E(X) = \mu = \lambda = -\dfrac{1}{\ln(0.7)} \approx 2.8$ years.

(B) $$P(X \geq 2.8) = \int_{2.8}^{\infty} f(x)dx = \lim_{R \to \infty} \int_{2.8}^{R} \frac{1}{\lambda} e^{-x/\lambda} \, dx$$

$$= \lim_{R \to \infty} \int_{2.8}^{R} \frac{1}{2.8} e^{-x/2.8} \, dx$$

$$= \lim_{R \to \infty} (-e^{-R/2.8} + e^{-2.8/2.8})$$

$$= e^{-1} \approx 0.368$$

**71.** $\mu = 240$, $\sigma = 20$

8 days = 192 hours = $x$

$z$ (for $x = 192$) $= \dfrac{192 - 240}{20} = -2.4$

Fraction of people having this incision who would heal in 192 hours or less = Area $A_1$

= 0.5 - (area corresponding to $z = 2.4$)

= 0.5 - 0.4918

= 0.0082

Thus, the percentage of people who would heal in 8 days or less is 0.82%.

**73.** $X$ is an exponential random variable with mean $\mu = 2$ (minutes). Thus, $\lambda = 2$ and

$$P(X \ge 5) = \int_5^\infty f(x)\,dx = \int_5^\infty \frac{1}{2}e^{-x/2}\,dx = \lim_{R \to \infty}\int_5^R \frac{1}{2}e^{-x/2}\,dx = \lim_{R \to \infty}(-e^{-x/2})\Big|_5^R$$

$$= \lim_{R \to \infty}(-e^{-R/5} + e^{-5/2}) = e^{-5/2} = e^{-2.5} \approx 0.082.$$

**75.** $\mu = 70$, $\sigma = 8$

We compute $x_1$, $x_2$, $x_3$, and $x_4$ corresponding to $z_1$, $z_2$, $z_3$, and $z_4$, respectively. The area between $\mu$ and $x_3$ is 0.2.

Hence, from the table, $z_3 = 0.52$ (approximately). Thus, we have:

$$0.52 = \frac{x_3 - 70}{8}$$

$x_3 - 70 = 4.16 \left[ \underline{\text{Note}}\text{: } z = \dfrac{x - \mu}{\sigma}. \right] \approx 4.2$

and $x_3 = 74.2$.

Also, $x_2 = 70 - 4.2 = 65.8$

The area between $\mu$ and $x_4$ is 0.4. Hence, from the table, $z_4 = 1.28$ (approximately). Therefore:

$$1.28 = \frac{x_4 - 70}{8}$$

$x_4 - 70 = 10.24 \approx 10.2$

and $x_4 = 70 + 10.2 = 80.2$.

Also, $x_1 = 70 - 10.2 = 59.8$.

Thus, we have $x_1 = 59.8$, $x_2 = 65.8$, $x_3 = 74.2$, $x_4 = 80.2$.

So,

$A$'s = 80.2 or greater, $B$'s = 74.2 to 80.2, $C$'s = 65.8 to 74.2,

$D$'s = 59.8 to 65.8, and $F$'s = 59.8 or lower.

**1.** $\displaystyle\int_0^\infty e^{-2x}\,dx = \lim_{b\to\infty}\int_0^b e^{-2x}\,dx = \lim_{b\to\infty}\frac{e^{-2x}}{-2}\Big|_0^b = \lim_{b\to\infty}\left(\frac{-e^{-2b}}{2}+\frac{1}{2}\right) = \frac{1}{2}$

Thus, the improper integral converges. (3-1)

**2.** $\displaystyle\int_0^\infty \frac{1}{x+1}\,dx = \lim_{b\to\infty}\int_0^b \frac{1}{x+1}\,dx$

$\displaystyle\qquad = \lim_{b\to\infty}\int_1^{b+1}\frac{1}{u}\,du = \lim_{b\to\infty}\ln u\Big|_1^{b+1}$

$u = x+1$
$du = dx$
$u = b+1$ when $x = b$
$u = 1$ when $b = 0$

$\displaystyle\qquad = \lim_{b\to\infty}\ln|b+1|$

This limit does not exist. Thus, the improper integral diverges. (3-1)

**3.** $\displaystyle\int_1^\infty \frac{16\,dx}{x^3} = 16\lim_{b\to\infty}\int_1^b x^{-3}\,dx = 16\lim_{b\to\infty}\frac{x^{-2}}{-2}\Big|_1^b = 16\lim_{b\to\infty}\left(\frac{-b^{-2}}{2}+\frac{1}{2}\right) = 16\left(\frac{1}{2}\right) = 8$

(3-1)

**4.** $\displaystyle P(0 \le X \le 1) = \int_0^1\left(1-\frac{1}{2}x\right)dx$

$\displaystyle\qquad = \left(x - \frac{1}{4}x^2\right)\Big|_0^1$

$\displaystyle\qquad = 1 - \frac{1}{4} = 0.75$

The graph of the function is given at the right.

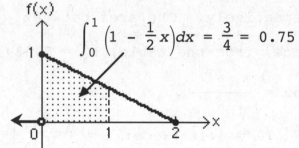

$\displaystyle\int_0^1\left(1-\frac{1}{2}x\right)dx = \frac{3}{4} = 0.75$

(3-2)

**5.** $\displaystyle \mu = E(X) = \int_{-\infty}^\infty x f(x)\,dx = \int_0^2 x\left(1-\frac{1}{2}x\right)dx = \int_0^2\left(x-\frac{1}{2}x^2\right)dx$

$\displaystyle\qquad = \left(\frac{x^2}{2}-\frac{1}{6}x^3\right)\Big|_0^2 = 2 - \frac{8}{6} = \frac{2}{3} \approx 0.6667$

$\displaystyle V(X) = \int_{-\infty}^\infty x^2 f(x)\,dx - \mu^2 = \int_0^2 x^2\left(1-\frac{1}{2}x\right)dx - \left(\frac{2}{3}\right)^2$

$\displaystyle\qquad = \int_0^2\left(x^2-\frac{1}{2}x^3\right)dx - \frac{4}{9} = \left(\frac{x^3}{3}-\frac{1}{8}x^4\right)\Big|_0^2 - \frac{4}{9}$

$\displaystyle\qquad = \frac{8}{3} - \frac{16}{8} - \frac{4}{9} = \frac{16}{72} = \frac{2}{9} \approx 0.2222$

$\displaystyle \sigma = \sqrt{V(X)} = \sqrt{\frac{2}{9}} = \frac{\sqrt{2}}{3} \approx 0.4714$

(3-3)

**6.** When $x < 0$, then $F(x) = 0$. When $0 \leq x \leq 2$,

$$F(x) = \int_{-\infty}^{x} f(t)\,dt = \int_{-\infty}^{0} f(t)\,dt + \int_{0}^{x} f(t)\,dt = 0 + \int_{0}^{x} \left(1 - \frac{1}{2}t\right) dt$$

$$= \left(t - \frac{1}{4}t^2\right)\Big|_{0}^{x} = x - \frac{1}{4}x^2.$$

When $x > 2$,

$$F(x) = \int_{-\infty}^{x} f(t)\,dt = \int_{-\infty}^{0} f(t)\,dt + \int_{0}^{2} f(t)\,dt + \int_{2}^{x} f(t)\,dt$$

$$= 0 + \int_{0}^{2} \left(1 - \frac{1}{2}t\right) dt + 0 = \left(t - \frac{1}{4}t^2\right)\Big|_{0}^{2} = 2 - \frac{1}{4} \cdot 4 = 1.$$

Thus,

$$F(x) = \begin{cases} 0 & x < 0 \\ x - \dfrac{1}{4}x^2 & 0 \leq x \leq 2 \\ 1 & x > 2 \end{cases}$$

The graph of $F(x)$ is shown at the right.

(3-2)

**7.** We must solve the following for $m$:

$$F(m) = P(X \leq m) = \frac{1}{2}$$

$$m - \frac{1}{4}m^2 = \frac{1}{2} \quad \text{(refer to Problem 6)}$$

$$m^2 - 4m + 2 = 0$$

$$m = \frac{4 \pm \sqrt{16 - 4(2)}}{2} = \frac{4 \pm \sqrt{8}}{2}$$

$$m = 2 \pm \sqrt{2}$$

Thus, the median is $x = 2 - \sqrt{2} \approx 0.5858$. (3-3)

**8.** $\mu = 100$, $\sigma = 10$, $x = 118$.

$$z = \frac{118 - 100}{10} = 1.8$$

$P(100 \leq X \leq 118) = P(0 \leq Z \leq 1.8) =$ area of region over the interval [0, 1.8]

$$= 0.4641 \text{ (from Table III)}. \qquad (3-4)$$

**9.** The uniform probability density function on [5, 15] is given by:

$$f(x) = \begin{cases} \dfrac{1}{15 - 5} & \text{if } 5 \leq x \leq 15 \\ 0 & \text{otherwise} \end{cases}$$

Thus,

$$f(x) = \begin{cases} \dfrac{1}{10} & \text{if } 5 \leq x \leq 15 \\ 0 & \text{otherwise} \end{cases}$$

The cumulative distribution function is given by:

$$F(x) = \begin{cases} 0 & \text{if } x < 5 \\ \dfrac{1}{10}(x - 5) & \text{if } 5 \leq x \leq 15 \\ 1 & \text{if } x > 15 \end{cases}$$

(3-4)

**10.** The exponential probability density function with $\lambda = \dfrac{1}{5} = 0.2$ is given by:

$$f(x) = \begin{cases} \dfrac{1}{0.2}e^{-x/0.2} & \text{if } x \geq 0 \\ 0 & \text{otherwise} \end{cases}$$

Thus,

$$f(x) = \begin{cases} 5e^{-5x} & \text{if } x \geq 0 \\ 0 & \text{otherwise} \end{cases}$$

The cumulative distribution function is given by:

$$F(x) = \begin{cases} 0 & \text{if } x < 0 \\ 1 - e^{-5x} & \text{if } x \geq 0 \end{cases}$$

$\hspace{12cm}$ (3-4)

**11.** $f(x) = e^{-x^2}$

$$\int_{-\infty}^{\infty} f(x)\,dx > \int_{-10}^{10} e^{-x^2}\,dx > 1.772;$$

$f$ is not a probability density function since

$$\int_{-\infty}^{\infty} f(x)\,dx \neq 1$$

$\hspace{12cm}$ (3-2)

**12.** $f(x) = \begin{cases} 0.75(x^2 - 4x + 3) & \text{if } 0 \leq x \leq 4 \\ 0 & \text{otherwise} \end{cases}$

$f$ is not a probability density function
since $f(x) < 0$ on $(1, 3)$ $\hspace{6cm}$ (3-2)

**13.** $P(1 \leq X \leq 4) = \int_1^4 \dfrac{5}{2}x^{-7/2}\,dx$

$\hspace{3cm} = \dfrac{5}{2}\left(-\dfrac{2}{5}\right)x^{-5/2}\Big|_1^4$

$\hspace{3cm} = -[(4)^{-5/2} - (1)^{-5/2}]$

$\hspace{3cm} = -\left(\dfrac{1}{32} - 1\right) = \dfrac{31}{32}$

$\hspace{3cm} \approx 0.9688$

The graph is shown at the right.

$$\int_1^4 \dfrac{5}{2}x^{-7/2}\,dx = \dfrac{31}{32} \approx 0.9688$$

$\hspace{12cm}$ (3-2)

**14.** $\mu = E(X) = \displaystyle\int_{-\infty}^{\infty} xf(x)\,dx = \int_1^{\infty} x \cdot \dfrac{5}{2}x^{-7/2}\,dx$

$\hspace{2cm} = \displaystyle\lim_{R \to \infty}\int_1^R \dfrac{5}{2}x^{-7/2}\,dx = \lim_{R \to \infty}\left(\dfrac{5}{2}\left(-\dfrac{2}{3}\right)x^{-3/2}\Big|_1^R\right)$

$\hspace{2cm} = -\dfrac{5}{3}\displaystyle\lim_{R \to \infty}(R^{-3/2} - 1^{-3/2}) = -\dfrac{5}{3}(-1) = \dfrac{5}{3} \approx 1.667$

$$V(X) = \int_{-\infty}^{\infty} x^2 f(x)\,dx - \mu^2 = \int_{1}^{\infty} x^2 \cdot \frac{5}{2} x^{-7/2}\,dx - \left(\frac{5}{3}\right)^2$$

$$= \lim_{R \to \infty} \int_{1}^{R} \frac{5}{2} x^{-3/2}\,dx - \frac{25}{9} = \lim_{R \to \infty} \left(\frac{5}{2}(-2)x^{-1/2}\,\Big|_{1}^{R}\right) - \frac{25}{9}$$

$$= -5 \lim_{R \to \infty} [R^{-1/2} - (1)^{-1/2}] - \frac{25}{9} = -5(-1) - \frac{25}{9} = 5 - \frac{25}{9} = \frac{20}{9}$$

$$\approx 2.2222$$

$$\sigma = \sqrt{V(X)} = \sqrt{\frac{20}{9}} = \frac{2}{3}\sqrt{5} \approx 1.4907 \qquad\qquad (3\text{-}3)$$

**15.** When $x < 1$, $F(x) = 0$.  When $x \geq 1$,

$$F(x) = \int_{-\infty}^{x} f(t)\,dt = \int_{-\infty}^{1} f(t)\,dt + \int_{1}^{x} f(t)\,dt = 0 + \int_{1}^{x} \frac{5}{2} t^{-7/2}\,dt$$

$$= \frac{5}{2}\left(-\frac{2}{5}\right) t^{-5/2}\,\Big|_{1}^{x} = -(x^{-5/2} - 1^{-5/2}) = 1 - x^{-5/2}.$$

Thus,

$$F(x) = \begin{cases} 1 - x^{-5/2} & x \geq 1 \\ 0 & \text{otherwise} \end{cases}$$

The graph of $F(x)$ is shown at the right.

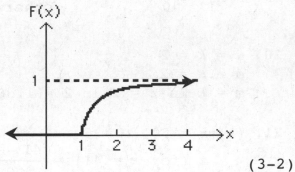

$$(3\text{-}2)$$

**16.** We must solve the following for $m$:

$$F(m) = P(X \leq m) = \frac{1}{2}$$

$$1 - m^{-5/2} = \frac{1}{2} \qquad \text{(refer to Problem 15)}$$

$$m^{-5/2} = \frac{1}{2}$$

$$\frac{1}{m^{5/2}} = \frac{1}{2}$$

$$m^{5/2} = 2 \qquad \text{(square both sides)}$$

$$m^5 = 4$$

$$m = (4)^{1/5}$$

$$m = 2^{2/5} \approx 1.32 \qquad\qquad (3\text{-}3)$$

**17.** $P(4 \leq X) = \int_{4}^{\infty} f(x)\,dx = e^{-2} \qquad$ (Solve for $\lambda$.)

$$= \int_{4}^{\infty} \frac{1}{\lambda} e^{-x/\lambda}\,dx = \lim_{R \to \infty} \frac{1}{\lambda} \int_{4}^{R} e^{-x/\lambda}\,dx = e^{-2}$$

$$= \lim_{R \to \infty} \frac{1}{\lambda} - \lambda e^{-x/\lambda}\,\Big|_{4}^{R} = -\lim_{R \oslash \infty} (e^{-R/\lambda} - e^{-4/\lambda}) = e^{-2}$$

$$= e^{-4/\lambda} = e^{-2}$$

Thus, we have $-\dfrac{4}{\lambda} = -2$

$\qquad\qquad\quad \lambda = 2$.

The probability density function, with $\lambda = 2$, is:

$$f(x) = \begin{cases} \dfrac{1}{\lambda}\, e^{-x/\lambda} & x \ge 0 \\ 0 & \text{otherwise} \end{cases} = \begin{cases} \dfrac{1}{2}\, e^{-x/2} & x \ge 0 \\ 0 & \text{otherwise} \end{cases} \qquad (3\text{-}4)$$

**18.** $P(0 \le X \le 2) = \displaystyle\int_0^2 f(x)\,dx = \int_0^2 \dfrac{1}{2}\, e^{-x/2}\, dx$ (refer to Problem 17)

$$= \dfrac{1}{2}\,(-2e^{-x/2})\,\Big|_0^2 = -(e^{-2/2} - e^0)$$

$$= -(e^{-1} - 1) \text{ or } (1 - e^{-1}) \approx 0.6321 \qquad (3\text{-}2)$$

**19.** $F(x) = \begin{cases} 1 - e^{-x/\lambda} & x \ge 0 \\ 0 & \text{otherwise} \end{cases} = \begin{cases} 1 - e^{-x/2} & x \ge 0 \\ 0 & \text{otherwise} \end{cases}$ [<u>Note</u>: $\lambda = 2$.]

$$\qquad\qquad (3\text{-}4)$$

**20.** $\mu = \lambda = 2$

$\sigma = \lambda = 2$

$m = \lambda \ln 2 = 2 \ln 2 \approx 1.3863 \qquad (3\text{-}4)$

**21.** (A) $\mu = 50$, $\sigma = 6$

$\qquad z$ (for $x = 41$) $= \dfrac{41 - 50}{6} = -1.5$

$\qquad z$ (for $x = 62$) $= \dfrac{62 - 50}{6} = 2.0$

$\qquad$ Required area $= A_1 + A_2$

$\qquad\qquad\qquad$ = (area corresponding to $z = 1.5$)
$\qquad\qquad\qquad$ + (area corresponding to $z = 2$)
$\qquad\qquad\qquad$ = $0.4332 + 0.4772$
$\qquad\qquad\qquad$ = $0.9104$

$\qquad$ (B) $z$ (for $x = 59$) $= \dfrac{59 - 50}{6} = 1.5$

$\qquad\qquad$ Required area $= 0.5 - \begin{smallmatrix}\text{(area corresponding}\\ \text{to } z = 1.5)\end{smallmatrix}$

$\qquad\qquad\qquad$ = $0.5 - 0.4332$
$\qquad\qquad\qquad$ = $0.0668$

$\qquad\qquad\qquad\qquad\qquad\qquad\qquad\qquad\qquad (3\text{-}4)$

**22.** Given $\mu = 82$ and $\sigma = 8$.

(A) We first find the number of standard deviations that 84 and 94 are from the mean.

For $x = 84$: $z = \dfrac{84 - 82}{8} = \dfrac{2}{8} = 0.25$

For $x = 94$: $z = \dfrac{94 - 82}{8} = \dfrac{12}{8} = 1.5$

Now, $P(84 \leq X \leq 94) = P(0.25 \leq z \leq 1.5)$
$$= 0.4332 - 0.0987$$
$$= 0.3345$$

(B) For $x = 60$: $z = \dfrac{60 - 82}{8} = -\dfrac{22}{8} = -2.75$

$P(X \geq 60) = P(z \geq -2.75) = 0.4970 + 0.5000$
$$= 0.9970 \qquad (3\text{-}4)$$

23. $\displaystyle\int_{-\infty}^{0} e^x \, dx = \lim_{a \to -\infty} \int_{a}^{0} e^x \, dx = \lim_{a \to -\infty} e^x \Big|_{a}^{0} = \lim_{a \to -\infty} (1 - e^a) = 1,$

since $e^a \to 0$ as $a \to -\infty$. $\qquad (3\text{-}1)$

24. $\displaystyle\int_{0}^{\infty} \frac{1}{(x + 3)^2} \, dx = \lim_{b \to \infty} \int_{0}^{b} \frac{1}{(x + 3)^2} \, dx$

   $\qquad\qquad$ Substitution: $u = x + 3$
   $\qquad\qquad\qquad\qquad\qquad du = dx$

   $\displaystyle = \lim_{b \to \infty} \frac{-1}{(x + 3)} \Big|_{0}^{b} = \lim_{b \to \infty} \left( \frac{-1}{b + 3} + \frac{1}{3} \right)$

   $\displaystyle = \frac{1}{3}$

Thus, the improper integral converges. $\qquad (3\text{-}1)$

25. Yes. Since $\displaystyle\int_{-1}^{\infty} f(x)\,dx = \int_{-1}^{1} f(x)\,dx + \int_{1}^{\infty} f(x)\,dx$ and $\displaystyle\int_{-1}^{\infty} f(x)\,dx = L$

exists, it follows that $\displaystyle\int_{1}^{\infty} f(x)\,dx = \int_{1}^{\infty} f(x)\,dx - \int_{-1}^{\infty} f(x)\,dx - \int_{-1}^{1} f(x)\,dx$

$\displaystyle = L - \int_{-1}^{1} f(x)\,dx$ exists. $\qquad (3\text{-}1)$

26. $f(x) = \begin{cases} e^{-10x} & \text{if } x \geq 0 \\ 0 & \text{otherwise} \end{cases}$

   $f(x) \geq 0$ on $(-\infty, \infty)$ and

   $\displaystyle\int_{-\infty}^{\infty} f(x)\,dx = \int_{-\infty}^{0} f(x)\,dx + \int_{0}^{\infty} f(x)\,dx$

   $\displaystyle = \int_{0}^{\infty} e^{-10x}\,dx = \lim_{b \to \infty} \int_{0}^{b} e^{-10x}\,dx$

   $\displaystyle = \lim_{b \to \infty} \left[ -\frac{1}{10} e^{-10x} \right]_{0}^{b}$

   $\displaystyle = \lim_{b \to \infty} \left[ \frac{1}{10} - \frac{1}{10} e^{-10b} \right] = \frac{1}{10}$

Therefore, let $k = 10$; $f(x) = 10e^{-10x}$ is a probability density function. $\qquad (3\text{-}2)$

**27.** $f(x) = \begin{cases} e^{10x} & \text{if } x \geq 0 \\ 0 & \text{otherwise} \end{cases}$

Since $\displaystyle\int_{-\infty}^{\infty} f(x)\,dx = \int_{-\infty}^{0} f(x)\,dx + \int_{0}^{\infty} f(x)\,dx$

$$= \int_{0}^{\infty} e^{10x}\,dx = \lim_{b \to \infty} \int_{0}^{b} e^{10x}\,dx$$

$$= \lim_{b \to \infty} \left[ \frac{1}{10} e^{10x} \right]_{0}^{b}$$

$$= \lim_{b \to \infty} \left[ \frac{1}{10} e^{10b} - \frac{1}{10} \right] \text{ diverges,}$$

no constant $k$ exists.    (3-2)

**28.** $X$ is an exponentially distributed random variable with median $m = 3 \ln 2$. It follows that the mean $\mu = 3$ and the probability density function is:

$$f(x) = \begin{cases} (1/3)e^{-x/3} & \text{if } x \geq 0 \\ 0 & \text{otherwise} \end{cases}$$

Now, $P(\overline{X}) \leq 3 = \displaystyle\int_{-\infty}^{3} f(x)\,dx = \int_{-\infty}^{0} f(x)\,dx + \int_{0}^{3} f(x)\,dx$

$$= \int_{0}^{3} \frac{1}{3} e^{-x/3}\,dx$$

$$= \frac{1}{3}\left[ -3e^{-x/3} \right]_{0}^{3} = 1 - e^{-1} \approx 0.6321 \qquad (3-4)$$

**29.** $\mu = \displaystyle\int_{0}^{\infty} x f(x)\,dx = \int_{0}^{\infty} \frac{50x}{(x+5)^3}\,dx$

$$= \lim_{R \to \infty} \int_{0}^{R} \frac{50x}{(x+5)^3}\,dx = 50 \lim_{R \to \infty} \left[ \frac{1}{(x+5)^2} - \frac{5}{(x+5)^3} \right] dx$$

$$= 50 \lim_{R \to \infty} \left( -\frac{1}{(x+5)} + \frac{5}{2} \cdot \frac{1}{(x+5)^2} \right)\Bigg|_{0}^{R}$$

$$= 50 \lim_{R \to \infty} \left[ \left( -\frac{1}{(R+5)} + \frac{5}{2} \cdot \frac{1}{(R+5)^2} \right) - \left( -\frac{1}{5} + \frac{5}{2} \cdot \frac{1}{25} \right) \right]$$

$$= 50 \left( \frac{1}{5} - \frac{1}{10} \right) = 50 \left( \frac{1}{10} \right) = 5$$

Now find the cumulative probability distribution function. When $x < 0$, $F(x) = 0$. When $x \geq 0$, we have:

$$F(x) = \int_{-\infty}^{x} f(t)\,dt = \int_{-\infty}^{0} f(t)\,dt + \int_{0}^{x} f(t)\,dt = 0 + \int_{0}^{x} \frac{50x}{(t+5)^3}\,dt$$

$$= -\frac{50}{2}\left(\frac{1}{(t+5)^2}\right)\Bigg|_0^x = -25\left(\frac{1}{(x+5)^2} - \frac{1}{25}\right) = 1 - \frac{25}{(x+5)^2}$$

Thus, $F(x) = \begin{cases} 1 - \dfrac{25}{(x+5)^2} & x \geq 0 \\ 0 & \text{otherwise} \end{cases}$

Next, to find the median, $m$, we must solve the following for $m$:

$$F(m) = P(X \leq m) = \frac{1}{2}$$

$$1 - \frac{25}{(m+5)^2} = \frac{1}{2}$$

$$\frac{25}{(m+5)^2} = \frac{1}{2}$$

$$(m+5)^2 = 50$$

$$m + 5 = \sqrt{50}$$

$$m + 5 = 5\sqrt{2}$$

Therefore, the median, $m$, equals $5\sqrt{2} - 5 \approx 2.071$. (3-3)

**30.** $f(x) = \begin{cases} \dfrac{0.8}{x^2} + \dfrac{0.8}{x^5} & \text{if } x \geq 1 \\ 0 & \text{otherwise} \end{cases}$

The cumulative distribution function is:

$$F(x) = \begin{cases} 0 & \text{if } x < 1 \\ 1 - \dfrac{0.8}{x} - \dfrac{0.2}{x^4} & \text{if } x \geq 1 \end{cases}$$

To find the median $m$, we solve

$$F(m) = P(X \leq m) = \frac{1}{2} \text{ for } m:$$

$$1 - \frac{0.8}{m} - \frac{0.2}{m^4} = \frac{1}{2}$$

Using a graphing utility, we find that $m \approx 1.68$. (3-3)

**31.** Consider the integral $\displaystyle\int \frac{e^x}{(1+e^x)^2}\, dx$.

If we let $u = 1 + e^x$, then $du = e^x\, dx$, and $\displaystyle\lim_{x\to\infty} u = \infty$, $\displaystyle\lim_{x\to-\infty} u = 1$.

Thus,

$$\int_{-\infty}^{\infty} \frac{e^x}{(1+e^x)^2}\, dx = \int_1^{\infty} \frac{1}{u^2}\, du = \lim_{b\to\infty} \int_1^b \frac{1}{u^2}\, du$$

$$= \lim_{b\to\infty}\left[-\frac{1}{u}\Bigg|_1^b\right]$$

$$= \lim_{b\to\infty}\left[1 - \frac{1}{b}\right] = 1 \qquad (3-1)$$

**32.** $\int_{-\infty}^{\infty} (ax^2 + bx + c)f(x) = a\int_{-\infty}^{\infty} x^2 f(x)\,dx + b\int_{-\infty}^{\infty} xf(x)\,dx + c\int_{-\infty}^{\infty} f(x)\,dx$

$$= a(\sigma^2 + \mu^2) + b\mu + c,$$

since $\int_{-\infty}^{\infty} x^2 f(x)\,dx = \sigma^2 + \mu^2$, $\int_{-\infty}^{\infty} xf(x)\,dx = \mu$, and $\int_{-\infty}^{\infty} f(x)\,dx = 1$.   (3-3)

**33.** $f_1(x) = \begin{cases} 0.25xe^{-x/2} & \text{if } x \geq 0 \\ 0 & \text{otherwise} \end{cases}$

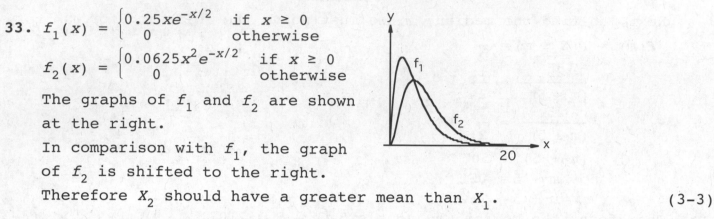

$f_2(x) = \begin{cases} 0.0625x^2 e^{-x/2} & \text{if } x \geq 0 \\ 0 & \text{otherwise} \end{cases}$

The graphs of $f_1$ and $f_2$ are shown at the right.

In comparison with $f_1$, the graph of $f_2$ is shifted to the right.

Therefore $X_2$ should have a greater mean than $X_1$.   (3-3)

**34.** Refer to the graphs in Problem 33. In comparison with $f_1$, the area under the graph of $f_2$ is more spread out. Therefore, the variance of $X_2$ should be greater than the variance of $X_1$.   (3-3)

**35.** $\mu_1 = \int_{-\infty}^{\infty} xf_1(x)\,dx = \int_0^{\infty} x(0.25xe^{-x/2})\,dx$

$$\approx \int_0^{30} 0.25x^2 e^{-x/2}\,dx \approx 4$$

$\mu_2 = \int_{-\infty}^{\infty} xf_2(x)\,dx = \int_0^{\infty} x(0.0625x^2 e^{-x/2})\,dx$

$$\approx \int_0^{30} 0.0625x^3 e^{-x/2}\,dx \approx 6 \qquad (3\text{-}3)$$

**36.** $V_1(x) = \int_{-\infty}^{\infty} x^2 f_1(x)\,dx - 4^2 = \int_0^{\infty} x^2(0.25xe^{-x/2})\,dx - 16$

$$\approx \int_0^{30} 0.25x^3 e^{-x/2}\,dx - 16 \approx 24 - 16 = 8$$

$V_2(x) = \int_{-\infty}^{\infty} x^2 f_2(x)\,dx - 6^2 = \int_0^{\infty} x^2(0.0625x^2 e^{-x/2})\,dx - 36$

$$\approx \int_0^{30} 0.0625x^4 e^{-x/2}\,dx - 36 \approx 48 - 36 = 12 \qquad (3\text{-}3)$$

**37.** (A) The total production is given by:

$\int_0^{\infty} R(t)\,dt = \lim_{T \to \infty} \int_0^T R(t)\,dt$

$$= \lim_{T \to \infty} \int_0^T (12e^{-0.3t} - 12e^{-0.6t})\,dt$$

$$= \lim_{T \to \infty} \left[ (-40e^{-0.3t} + 20e^{-0.6t}) \Big|_0^T \right]$$

$$= \lim_{T \to \infty} (-40e^{-0.3T} + 20e^{-0.6T} + 20) = 20$$

Thus, the total production is 20 million barrels.

(B) To find when the well will reach 50% of the total production, we must solve

$$\int_0^T R(t)\,dt = 10$$

for $T$. Now,

$$\int_0^T R(t)\,dt = \int_0^T (12e^{-0.3t} - 12e^{-0.6t})\,dt$$

$$= \left[ (-40e^{-0.3t} + 20e^{-0.6t}) \Big|_0^T \right]$$

$$= -40e^{-0.3T} + 20e^{-0.6T} + 20$$

Thus, we have

$$-40e^{-0.3T} + 20e^{-0.6T} + 20 = 10$$

or

$$-40e^{-0.3T} + 20e^{-0.6T} = -10$$

Using a graphing utility to solve this equation, we find that $T \approx 4.09$ years.

(3-1)

**38.** $f(x) = \begin{cases} 0.02(1 - 0.01x) & \text{if } 0 \le x \le 100 \\ 0 & \text{otherwise} \end{cases}$

(A) $\displaystyle\int_{40}^{100} f(x)\,dx = \int_{40}^{100} 0.02(1 - 0.01x)\,dx$

$$= 0.02\left[ (x - 0.005x^2) \Big|_{40}^{100} \right]$$

$$= 0.02[(100 - 50) - (40 - 8)] = 0.36$$

The probability that the weekly demand for popcorn is between 40 and 100 pounds is 0.36.

(B) $P(X \le 50) = \displaystyle\int_0^{50} f(x)\,dx = \frac{1}{50}\int_0^{50} (1 - 0.01x)\,dx = \frac{1}{50}(x - 0.005x^2) \Big|_0^{50}$

$$= \frac{1}{50}(50 - 0.005 \cdot 50^2) = 1 - 0.25 = 0.75$$

(C) Solve the following for $x$:

$$\int_0^x f(t)\,dt = 0.96 \qquad (x = \text{number of pounds of popcorn})$$

$$\frac{1}{50}\int_0^x (1 - 0.01t)\,dt = 0.96$$

$$\frac{1}{50}(t - 0.005t^2) \Big|_0^x = 0.96$$

$$\frac{1}{50}(x - 0.005x^2) = 0.96$$

$$x - 0.005x^2 = 48$$

$$5x^2 - 1000x + 48{,}000 = 0$$

$$x^2 - 200x + 9600 = 0$$

$$(x - 80)(x - 120) = 0$$

$$x = 80 \quad \text{or} \quad x = 120$$

Thus, 80 pounds of popcorn must be on hand at the beginning of the week.

(3-2)

**39.** Capital Value: $CV = \int_0^\infty 2400e^{-0.06t}\,dt = \lim_{T \to \infty} \int_0^T 2400e^{-0.06t}\,dt$

$$= \lim_{T \to \infty} 2400 \frac{e^{-0.06t}}{-0.06}\Big|_0^T$$

$$= \lim_{T \to \infty} -40{,}000(e^{-0.06T} - 1)$$

$$= \$40{,}000 \tag{3-1}$$

**40.** $f(x) = \begin{cases} 6x(1 - x) & \text{if } 0 \le x \le 1 \\ 0 & \text{otherwise} \end{cases}$

(A) $P(X \ge 0.2) = 1 - P(X < 0.2)$

$$= 1 - \int_{-\infty}^{0.2} f(x)\,dx$$

$$= 1 - \int_0^{0.2} f(x)\,dx$$

$$= 1 - \int_0^{0.2} 6x(1 - x)\,dx$$

$$= 1 - \left[ (3x^2 - 2x^3)\Big|_0^{0.2} \right]$$

$$= 1 - (0.12 - 0.016) = 0.896$$

(B) The expected value (mean) is given by:

$$\mu = \int_{-\infty}^\infty xf(x)\,dx = \int_0^1 (6x^2 - 6x^3)\,dx$$

$$= (2x^3 - \frac{3}{2}x^4)\Big|_0^1 = 2 - \frac{3}{2} = 0.5$$

The expected percentage is 50%.

(C) The cumulative distribution function is:

$$F(x) = \begin{cases} 0 & \text{if } x < 0 \\ 3x^2 - 2x^3 & \text{if } 0 \le x \le 1 \\ 1 & \text{if } x > 1 \end{cases}$$

To find the median $m$, we solve

$$F(m) = P(X \le m) = \frac{1}{2}$$

or $\quad 3m^2 - 2m^3 = \dfrac{1}{2}$

for $m$. Using a graphing utility, we find that $m = 0.5$.
The median percentage is 50%. $\qquad\qquad$ (3-2, 3-3)

**41.** Mean failure time $= \mu = 4000$. As an exponential density function, it is expressed by $\lambda = \mu = 4000$. Thus,

$$f(x) = \begin{cases} \dfrac{1}{\lambda}e^{-x/\lambda} = \dfrac{1}{4000}e^{-x/4000} & x \ge 0 \\ 0 & \text{otherwise} \end{cases}$$

The cumulative distribution function is given by:

$$F(x) = \begin{cases} 1 - e^{-x/\lambda} \\ 0 \end{cases} = \begin{cases} 1 - e^{-x/4000} & x \ge 0 \\ 0 & \text{otherwise} \end{cases}$$

(A) $P(X \geq 4000) = \displaystyle\int_{4000}^{\infty} f(x)\,dx = 1 - F(4000)$

$$= 1 - (1 - e^{-4000/4000}) = e^{-1} = 0.3679$$

(B) $P(0 \leq X \leq 1000) = \displaystyle\int_{0}^{1000} f(x)\,dx = F(1000) - F(0)$

$$= (1 - e^{-1000/4000}) - (1 - e^0) = 1 - e^{-.25} \approx 0.2212$$

(3-4)

42. $\mu = 35{,}000, \ \sigma = 5{,}000$

$z$ (for $x = 25{,}000$) $= \dfrac{25{,}000 - 35{,}000}{5{,}000} = -2$

Required probability = area $A$

$\qquad = 0.5 - \text{area } A_1$

$\qquad = 0.5 - 0.4772$

$\qquad = 0.0228$

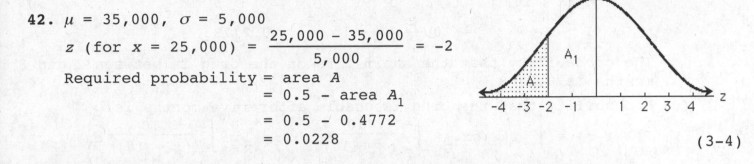

(3-4)

43. $\mu = 100, \ \sigma = 10$

(A) $z$ (for $x = 91.5$) $= \dfrac{91.5 - 100}{10} = -0.85$

$z$ (for $x = 108$) $= \dfrac{108.5 - 100}{10} = 0.85$

The probability of an applicant scoring between 92 and 108

$\qquad = \text{area } A$

$\qquad = 2 \cdot \text{area } A_1$

$\qquad = 2(\text{area corresponding to } z = 0.85)$

$\qquad = 2(0.3023) = 0.6046$

Thus, the percentage of applicants
scoring between 92 and 108 is 60.46%.

(3-4)

(B) $z$ (for $x = 114.5$) $= \dfrac{114.5 - 100}{10} = 1.45$

The probability of an applicant scoring 115 or higher

$\qquad = \text{area } A$

$\qquad = 0.5 - (\text{area corresponding to } z = 1.45)$

$\qquad = 0.5 - 0.4265$

$\qquad = 0.0735$

Thus, the percentage of applicants
scoring 115 or higher is 7.35%.

(3-4)

**44.** $f(x) = \begin{cases} \dfrac{10}{(x + 10)^2} & \text{if } x \geq 0 \\ 0 & \text{otherwise} \end{cases}$

(A) $\displaystyle\int_2^8 f(x)\,dx = \int_2^8 \dfrac{10}{(x + 10)^2}\,dx$

Let $u = x + 10$, then $du = dx$; $u = 12$ when $x = 2$, $u = 18$ when $x = 8$. Thus,

$$\int_2^8 \dfrac{10}{(x + 10)^2}\,dx = 10\int_{12}^{18} \dfrac{1}{u^2}\,du = 10\left[-\dfrac{1}{u}\Big|_{12}^{18}\right]$$

$$= 10\left(\dfrac{1}{12} - \dfrac{1}{18}\right) = \dfrac{5}{18} \approx 0.2778$$

The probability that the shelf-life of the drug is between 2 and 8 months is 0.2778.

(B) Probability that the drug is usable after five months is:

$$P(X > 5) = \int_5^\infty f(x)\,dx = \int_5^\infty \dfrac{10}{(x + 10)^2}\,dx = \lim_{R \to \infty} \int_5^R \dfrac{10}{(x + 10)^2}\,dx$$

$$= \lim_{R \to \infty} 10\left(-\dfrac{1}{x + 10}\right)\Big|_5^R = -10 \lim_{R \to \infty} \left(\dfrac{1}{R + 10} - \dfrac{1}{5 + 10}\right)$$

$$= -10\left(-\dfrac{1}{15}\right) = \dfrac{2}{3} \approx 0.6667$$

(C) In order to find the median, $m$, we must solve the following for $m$:

$$P(X \leq m) = \int_0^m f(x)\,dx = \dfrac{1}{2}$$

$$\int_0^m \dfrac{10}{(x + 10)^2}\,dx = \dfrac{1}{2}$$

$$-\dfrac{10}{(x + 10)}\Big|_0^m = \dfrac{1}{2}$$

$$-\left(\dfrac{10}{m + 10} - 1\right) = \dfrac{1}{2}$$

$$-\dfrac{10}{m + 10} + 1 = \dfrac{1}{2}$$

$$-\dfrac{10}{m + 10} = -\dfrac{1}{2}$$

$$m + 10 = 20$$

$$m = 10 \text{ months}$$

(3-2, 3-3)

**45.** $f(x) = \begin{cases} \dfrac{1}{\lambda}\, e^{-x/\lambda} & x \geq 0 \\ 0 & \text{otherwise} \end{cases}$

$P(X > 1) = \dfrac{1}{\lambda} \displaystyle\int_0^{\infty} e^{-x/\lambda}\, dx = e^{-2}$  (Given)

Thus, $\quad \dfrac{1}{\lambda} \displaystyle\lim_{R \to \infty} \int_1^{R} e^{-x/\lambda}\, dx = e^{-2}$

$$\dfrac{1}{\lambda}(-\lambda) \lim_{R \to \infty} (e^{-x/\lambda}) \Big|_1^{R} = e^{-2}$$

$$-1 \cdot \lim_{R \to \infty} (e^{-R/\lambda} - e^{-1/\lambda}) = e^{-2}$$

$$e^{-1/\lambda} = e^{-2}$$

Thus, $\qquad\qquad -\dfrac{1}{\lambda} = -2$

$$\lambda = \dfrac{1}{2}$$

Therefore, $f(x)$, with $\lambda = \dfrac{1}{2}$, is given by $f(x) = \begin{cases} 2e^{-2x} & x \geq 0 \\ 0 & \text{otherwise} \end{cases}$

(A) $P(X > 2) = \displaystyle\int_2^{\infty} f(x)\, dx = \int_2^{\infty} 2e^{-2x}\, dx = 2 \lim_{R \to \infty} \int_2^{R} e^{-2x}\, dx$

$\qquad\qquad = 2 \displaystyle\lim_{R \to \infty} \left( -\dfrac{1}{2} e^{-2x} \Big|_2^{R} \right) = -\lim_{R \to \infty} (e^{-2R} - e^{-4}) = -(-e^{-4}) = e^{-4}$

$\qquad\qquad \approx 0.0183$

(B) Mean life expectancy:

$\mu = \displaystyle\int_0^{\infty} x f(x)\, dx = \int_0^{\infty} x \cdot 2e^{-2x}\, dx$

$\quad = 2 \displaystyle\lim_{R \to \infty} \int_0^{R} x e^{-2x}\, dx$  (integration by parts; $u = x$, $dv = e^{-2x}\, dx$)

$\quad = 2 \displaystyle\lim_{R \to \infty} \left[ x\left(-\dfrac{1}{2} e^{-2x}\right) \Big|_0^{R} + \dfrac{1}{2} \int_0^{R} e^{-2x}\, dx \right] = 2 \lim_{R \to \infty} \left( -\dfrac{1}{2} x e^{-2x} - \dfrac{1}{4} e^{-2x} \right) \Big|_0^{R}$

$\quad = -2 \displaystyle\lim_{R \to \infty} \left[ \left( \dfrac{1}{2} R e^{-2R} + \dfrac{1}{4} e^{-2R} \right) - \left( 0 + \dfrac{1}{4} \right) \right]$

$\quad = -2 \left( -\dfrac{1}{4} \right) = \dfrac{1}{2}$ or 0.5 month *or* $\mu = \lambda = \dfrac{1}{2}$  $\qquad\qquad$ (3-2, 3-3)

**46.** $R(t) = 15e^{-0.2t} - 15e^{-0.3t}$

(A) The total amount of the drug that is eliminated by the body is given by:

$$\int_0^\infty R(t)\,dt = \lim_{T\to\infty}\int_0^T R(t)\,dt$$

$$= \lim_{T\to\infty}\int_0^T (15e^{-0.2t} - 15e^{-0.3t})\,dt$$

$$= \lim_{T\to\infty}\left[(-75e^{-0.2t} + 50e^{-0.3t})\Big|_0^T\right]$$

$$= \lim_{T\to\infty}\left[-75e^{-0.2T} + 50e^{-0.3T} + 25\right]$$

$$= 25 \text{ milliliters}$$

(B) To find how long it will take for 50% of the drug to be eliminated, we solve

$$\int_0^T R(t)\,dt = 0.5(25) = 12.5$$

for $T$:

$$\int_0^T (15e^{-0.2t} - 15e^{-0.3t})\,dt = 12.5$$

$$\left[(-75e^{-0.2t} + 50e^{-0.3t})\Big|_0^T\right] = 12.5$$

$$-75e^{-0.2T} + 50e^{-0.3T} + 25 = 12.5$$

$$-75e^{-0.2T} + 50e^{-0.3T} + 12.5 = 0$$

Using a graphing utility, we find that $T \approx 6.93$ hours.　　　(3-1)

**47.** $\mu = 108$, $\sigma = 12$

$$z = \frac{135 - 108}{12} = \frac{27}{12} = 2.25$$

$$P(X \geq 135) = P(Z \geq 2.25) = 0.5 - P(Z < 2.25)$$
$$= 0.5 - 0.4878$$
$$= 0.0122;$$

1.22% of the children can be expected to have IQ scores of 135 or more.　　　(3-4)

**48.** $N'(t) = \dfrac{100t}{(1 + t^2)^2}$

Therefore, $N(t) = \displaystyle\int \dfrac{100t}{(1 + t^2)^2}\,dt = 50\displaystyle\int \dfrac{1}{u^2}\,du$      Substitution: $u = 1 + t^2$
                                                                     $du = 2t\,dt$

$$= \dfrac{50u^{-1}}{-1} + C = \dfrac{-50}{1 + t^2} + C$$

Now, $N(0) = \dfrac{-50}{1 + 0} + C.$ Thus, $C = N(0) + 50$ and

$$N(t) = \dfrac{-50}{1 + t^2} + N(0) + 50$$

$$N(3) = \dfrac{-50}{1 + 3^2} + N(0) + 50 = N(0) + 45$$

Therefore, the voting population will increase by 45 thousand in 3 years. If the population grows indefinitely at this rate, then

$$\lim_{t \to \infty} N(t) = \lim_{t \to \infty}\left(\dfrac{-50}{1 + t^2} + N(0) + 50\right) = N(0) + 50$$

Therefore, the total increase in voting population is 50 thousand.

(3-1)

# APPENDIX D  SPECIAL TOPIC

## EXERCISE D-1

**1.** (A) $p(x) = a_0 + a_1(x - 1) + a_2(x - 1)(x - 3)$

(B) $2 = p(1) = a_0$

$6 = p(3) = a_0 + 2a_1$

$11 = p(4) = a_0 + 3a_1 + 3a_2$

(C) $a_0 = 2$

$a_1 = \dfrac{1}{2}(6 - a_0) = \dfrac{1}{2}(6 - 2) = 2$

$a_2 = \dfrac{1}{3}(11 - a_0 - 3a_1) = \dfrac{1}{3}(11 - 2 - 6) = 1$

The interpolating polynomial is:
$$p(x) = 2 + 2(x - 1) + (x - 1)(x - 3)$$

**3.** (A) $p(x) = a_0 + a_1(x + 1) + a_2(x + 1)x + a_3(x + 1)x(x - 2)$

(B) $6 = p(-1) = a_0$

$5 = p(0) = a_0 + a_1$

$15 = p(2) = a_0 + 3a_1 + 6a_2$

$-39 = p(4) = a_0 + 5a_1 + 20a_2 + 40a_3$

(C) $a_0 = 6$

$a_1 = 5 - a_0 = -1$

$a_2 = \dfrac{1}{6}(15 - a_0 - 3a_1) = \dfrac{1}{6}(15 - 6 + 3) = 2$

$a_3 = \dfrac{1}{40}(-39 - a_0 - 5a_1 - 20a_2) = \dfrac{1}{40}(-39 - 6 + 5 - 40) = -2$

The interpolating polynomial is:
$$p(x) = 6 - (x + 1) + 2(x + 1)x - 2(x + 1)x(x - 2)$$

**5.**

| $x$ | 1 | 2 | 3 |
|------|---|---|----|
| $f(x)$ | 4 | 8 | 14 |

| $x_k$ | $y_k$ | FIRST DIVIDED DIFFERENCE | SECOND DIVIDED DIFFERENCE |
|-------|-------|---------------------------|----------------------------|
| 1 | 4 | | |
| | | $\dfrac{8 - 4}{2 - 1} = 4$ | |
| 2 | 8 | | $\dfrac{6 - 4}{3 - 1} = 1$ |
| | | $\dfrac{14 - 8}{3 - 2} = 6$ | |
| 3 | 14 | | |

Interpolating polynomial: $p(x) = 4 + 4(x - 1) + (x - 1)(x - 2)$.

**7.**

| $x$ | -1 | 0 | 1 | 2 |
|---|---|---|---|---|
| $f(x)$ | -3 | 1 | 3 | 9 |

| $x_k$ | $y_k$ | FIRST DIVIDED DIFFERENCE | SECOND DIVIDED DIFFERENCE | THIRD DIVIDED DIFFERENCE |
|---|---|---|---|---|
| -1 | -3 | | | |
| | | $\dfrac{1-(-3)}{0-(-1)}=4$ | | |
| 0 | 1 | | $\dfrac{2-4}{1-(-1)}=-1$ | |
| | | $\dfrac{3-1}{1-0}=2$ | | $\dfrac{2-(-1)}{2-(-1)}=1$ |
| 1 | 3 | | $\dfrac{6-2}{2-0}=2$ | |
| | | $\dfrac{9-3}{2-1}=6$ | | |
| 2 | 9 | | | |

Interpolating polynomial:
$$p(x) = -3 + 4(x+1) - (x+1)x + (x+1)x(x-1).$$

**9.**

| $x$ | -2 | 1 | 2 | 4 |
|---|---|---|---|---|
| $f(x)$ | 25 | 10 | 17 | 13 |

| $x_k$ | $y_k$ | FIRST DIVIDED DIFFERENCE | SECOND DIVIDED DIFFERENCE | THIRD DIVIDED DIFFERENCE |
|---|---|---|---|---|
| -2 | 25 | | | |
| | | $\dfrac{10-25}{1-(-2)}=-5$ | | |
| 1 | 10 | | $\dfrac{7-(-5)}{2-(-2)}=3$ | |
| | | $\dfrac{17-10}{2-1}=7$ | | $\dfrac{-3-3}{4-(-2)}=-1$ |
| 2 | 17 | | $\dfrac{-2-7}{4-1}=-3$ | |
| | | $\dfrac{13-17}{4-2}=-2$ | | |
| 4 | 13 | | | |

Interpolating polynomial:
$$25 - 5(x+2) + 3(x+2)(x-1) - (x+2)(x-1)(x-2).$$

**11.** Given a table with 3 points, it will have a linear interpolating polynomial if the 2 first divided differences are equal. In general, it will have a quadratic polynomial. It will not have a cubic interpolating polynomial because 3 points do not determine a unique cubic.

**13.**

| $x$ | -4 | 0 | 4 | 8 |
|---|---|---|---|---|
| $f(x)$ | -64 | 32 | 0 | 224 |

| $x_k$ | $y_k$ | FIRST DIVIDED DIFFERENCE | SECOND DIVIDED DIFFERENCE | THIRD DIVIDED DIFFERENCE |
|---|---|---|---|---|
| -4 | -64 | $\dfrac{32-(-64)}{0-(-4)}=24$ | | |
| 0 | 32 | | $\dfrac{-8-24}{4-(-4)}=-4$ | |
| 4 | 0 | $\dfrac{0-32}{4-0}=-8$ | | $\dfrac{8-(-4)}{8-(-4)}=1$ |
| 8 | 224 | $\dfrac{224-0}{8-4}=56$ | $\dfrac{56-(-8)}{8-0}=8$ | |

Interpolating polynomial:
$$p(x)=-64+24(x+4)-4(x+4)x+(x+4)x(x-4).$$

(A) $f(2) \approx p(2) = -64 + 24(6) - 4(6)(2) + (6)(2)(-2) = 8$

(B) $f(6) \approx p(6) = -64 + 24(10) - 4(10)(6) + 10(6)(2) = 56$

**15.**

| $x$ | -1 | 0 | 1 | 4 |
|---|---|---|---|---|
| $f(x)$ | 0 | 0 | 0 | 15 |

| $x_k$ | $y_k$ | FIRST DIVIDED DIFFERENCE | SECOND DIVIDED DIFFERENCE | THIRD DIVIDED DIFFERENCE |
|---|---|---|---|---|
| -1 | 0 | $\dfrac{0-0}{0-(-1)}=0$ | | |
| 0 | 0 | | $\dfrac{0-0}{1-(-1)}=0$ | |
| 1 | 0 | $\dfrac{0-0}{1-0}=0$ | | $\dfrac{\frac{5}{4}-0}{4-(-1)}=\dfrac{1}{4}$ |
| 4 | 15 | $\dfrac{15-0}{4-1}=5$ | $\dfrac{5-0}{4-0}=\dfrac{5}{4}$ | |

Interpolating polynomial: $\dfrac{1}{4}(x+1)x(x-1).$

(A) $f(2) \approx p(2) = \dfrac{1}{4}(3)(2)(1) = \dfrac{3}{2}$

(B) $f(3) \approx p(3) = \dfrac{1}{4}(4)(3)(2) = 6$

**17.**

| $x$ | -4 | -2 | 0 | 2 | 4 |
|---|---|---|---|---|---|
| $f(x)$ | 24 | 2 | 0 | -6 | 8 |

| $x_k$ | $y_k$ | FIRST DIVIDED DIFFERENCE | SECOND DIVIDED DIFFERENCE | THIRD DIVIDED DIFFERENCE | FOURTH DIVIDED DIFFERENCE |
|---|---|---|---|---|---|
| -4 | 24 | $\dfrac{2-24}{-2-(-4)}=-11$ | | | |
| -2 | 2 | | $\dfrac{-1-(-11)}{0-(-4)}=\dfrac{5}{2}$ | | |
| 0 | 0 | $\dfrac{0-2}{0-(-2)}=-1$ | | $\dfrac{-\frac{1}{2}-\frac{5}{2}}{2-(-4)}=-\dfrac{1}{2}$ | |
| 2 | -6 | $\dfrac{-6-0}{2-0}=-3$ | $\dfrac{-3-(-1)}{2-(-2)}=-\dfrac{1}{2}$ | | $\dfrac{\frac{1}{2}-(-\frac{1}{2})}{4-(-4)}=\dfrac{1}{8}$ |
| 4 | 8 | $\dfrac{8-(-6)}{4-2}=7$ | $\dfrac{7-(-3)}{4-0}=\dfrac{5}{2}$ | $\dfrac{\frac{5}{2}-(-\frac{1}{2})}{4-(-2)}=\dfrac{1}{2}$ | |

Interpolating polynomial:

$$p(x) = 24 - 11(x + 4) + \frac{5}{2}(x + 4)(x + 2) - \frac{1}{2}(x + 4)(x + 2)x$$
$$+ \frac{1}{8}(x + 4)(x + 2)x(x - 2)$$

(A) $f(-3) \approx p(-3) = 24 - 11(1) + \frac{5}{2}(1)(-1) - \frac{1}{2}(1)(-1)(-3)$

$$+ \frac{1}{8}(1)(-1)(-3)(-5) = \frac{57}{8} = 7.125$$

(B) $f(1) \approx p(1) = 24 - 11(5) + \frac{5}{2}(5)(3) - \frac{1}{2}(5)(3)(1) + \frac{1}{8}(5)(3)(1)(-1)$

$$= -\frac{23}{8} = -2.875$$

**19.**

| $x$ | -3 | -2 | -1 | 1 | 2 | 3 |
|---|---|---|---|---|---|---|
| $f(x)$ | -24 | -6 | 0 | 0 | 6 | 24 |

| $x_k$ | $y_k$ | FIRST DIVIDED DIFFERENCE | SECOND DIVIDED DIFFERENCE | THIRD DIVIDED DIFFERENCE | FOURTH DIVIDED DIFFERENCE | FIFTH DIVIDED DIFFERENCE |
|---|---|---|---|---|---|---|
| -3 | - | | | | | |
| | | $\frac{-6 - (-24)}{-2 - (-3)} = 18$ | | | | |
| -2 | 24 | | $\frac{6 - 18}{-1 - (-3)} = -6$ | | | |
| | | $\frac{0 - (-6)}{-1 - (-2)} = 6$ | | $\frac{-2 - (-6)}{1 - (-3)} = 1$ | | |
| -1 | -6 | | $\frac{0 - 6}{1 - (-2)} = -2$ | | | |
| | | $\frac{0 - 0}{1 - (-1)} = 0$ | | $\frac{2 - (-2)}{2 - (-2)} = 1$ | $\frac{1 - 1}{2 - (-3)} = 0$ | |
| 1 | 0 | | $\frac{6 - 0}{2 - (-1)} = 2$ | | | 0 |
| | | $\frac{6 - 0}{2 - 1} = 6$ | | $\frac{6 - 2}{3 - (-1)} = 1$ | $\frac{1 - 1}{3 - (-2)} = 0$ | |
| 2 | 0 | | $\frac{18 - 6}{3 - 1} = 6$ | | | |
| | | $\frac{24 - 6}{3 - 2} = 18$ | | | | |
| 3 | 6 | | | | | |
| 24 | | | | | | |

Interpolating polynomial:

$$p(x) = -24 + 18(x + 3) - 6(x + 3)(x + 2) + (x + 3)(x + 2)(x + 1).$$

(A) $f(-0.5) \approx p\left(-\frac{1}{2}\right) = -24 + 18\left(\frac{5}{2}\right) - 6\left(\frac{5}{2}\right)\left(\frac{3}{2}\right) + \left(\frac{5}{2}\right)\left(\frac{3}{2}\right)\left(\frac{1}{2}\right) = \frac{3}{8} = 0.375$

(B) $f(2.5) \approx p\left(\frac{5}{2}\right) = -24 + 18\left(\frac{11}{2}\right) - 6\left(\frac{11}{2}\right)\left(\frac{9}{2}\right) + \left(\frac{11}{2}\right)\left(\frac{9}{2}\right)\left(\frac{7}{2}\right) = \frac{105}{8} = 13.125$

**21.**

| $x$ | -2 | 0 | 2 |
|---|---|---|---|
| $f(x)$ | 2 | 0 | 2 |

| $x_k$ | $y_k$ | FIRST DIVIDED DIFFERENCE | SECOND DIVIDED DIFFERENCE |
|---|---|---|---|
| -2 | 2 | | |
| | | $\frac{0 - 2}{0 - (-2)} = -1$ | |
| 0 | 0 | | $\frac{1 - (-1)}{2 - (-2)} = \frac{1}{2}$ |
| | | $\frac{2 - 0}{2 - 0} = 1$ | |
| 2 | 2 | | |

Interpolating polynomial:

$$p(x) = 2 - (x + 2) + \frac{1}{2}(x + 2)x = \frac{1}{2}x^2$$

**23.**

| $x$ | 0 | 1 | 2 |
|-----|---|---|---|
| $f(x)$ | -4 | -2 | 0 |

| $x_k$ | $y_k$ | FIRST DIVIDED DIFFERENCE | SECOND DIVIDED DIFFERENCE |
|-------|-------|-------------------------|---------------------------|
| 0 | -4 | | |
| | | $\dfrac{-2 - (-4)}{1 - 0} = 2$ | |
| 1 | -2 | | 0 |
| | | $\dfrac{0 - (-2)}{2 - 1} = 2$ | |
| 2 | 0 | | |

Interpolating polynomial:

$$p(x) = -4 + 2x$$

**25.**

| $x$ | -1 | 0 | 2 | 3 |
|-----|----|----|----|----|
| $f(x)$ | 0 | 2 | 0 | -4 |

| $x_k$ | $y_k$ | FIRST DIVIDED DIFFERENCE | SECOND DIVIDED DIFFERENCE | THIRD DIVIDED DIFFERENCE |
|-------|-------|--------------------------|---------------------------|--------------------------|
| -1 | 0 | | | |
| | | $\dfrac{2 - 0}{0 - (-1)} = 2$ | | |
| 0 | 2 | | $\dfrac{-1 - 2}{2 - (-1)} = -1$ | |
| | | $\dfrac{0 - 2}{2 - 0} = -1$ | | |
| 2 | 0 | | $\dfrac{-4 - (-1)}{3 - 0} = -1$ | 0 |
| | | $\dfrac{-4 - 0}{3 - 2} = -4$ | | |
| 3 | -4 | | | |

Interpolating polynomial:

$$p(x) = 2(x + 1) - 1(x + 1)x = 2 + x - x^2$$

**27.**

| $x$ | -2 | -1 | 0 | 1 | 2 |
|-----|----|----|----|----|----|
| $f(x)$ | 1 | 5 | 3 | 1 | 5 |

| $x_k$ | $y_k$ | FIRST DIVIDED DIFFERENCE | SECOND DIVIDED DIFFERENCE | THIRD DIVIDED DIFFERENCE | FOURTH DIVIDED DIFFERENCE |
|-------|-------|--------------------------|---------------------------|--------------------------|---------------------------|
| -2 | 1 | | | | |
| | | $\dfrac{5 - 1}{-1 - (-2)} = 4$ | | | |
| -1 | 5 | | $\dfrac{-2 - 4}{0 - (-2)} = -3$ | | |
| | | $\dfrac{3 - 5}{0 - (-1)} = -2$ | | $\dfrac{0 - (-3)}{1 - (-2)} = 1$ | |
| 0 | 3 | | $\dfrac{-2 - (-2)}{1 - (-1)} = 0$ | | 0 |
| | | $\dfrac{1 - 3}{1 - 0} = -2$ | | $\dfrac{3 - 0}{2 - (-1)} = 1$ | |
| 1 | 1 | | $\dfrac{4 - (-2)}{2 - 0} = 3$ | | |
| | | $\dfrac{5 - 1}{2 - 1} = 4$ | | | |
| 2 | 5 | | | | |

Interpolating polynomial:

$$p(x) = 1 + 4(x + 2) - 3(x + 2)(x + 1) + (x + 2)(x + 1)x$$
$$= x^3 - 3x + 3$$

**29.**

| $x$ | -2 | -1 | 0 | 1 | 2 |
|---|---|---|---|---|---|
| $f(x)$ | -3 | 0 | 5 | 0 | -3 |

| $x_k$ | $y_k$ | FIRST D.D. | SECOND D.D. | THIRD D.D. | FOURTH D.D. |
|---|---|---|---|---|---|
| -2 | -3 | $\frac{0-(-3)}{-1-(-2)}=3$ | | | |
| -1 | 0 | | $\frac{5-3}{0-(-2)}=1$ | | |
| | | $\frac{5-0}{0-(-1)}=5$ | | $\frac{-5-1}{1-(-2)}=-2$ | |
| 0 | 5 | | $\frac{-5-5}{1-(-1)}=-5$ | | $\frac{2-(-2)}{2-(-2)}=1$ |
| | | $\frac{0-5}{1-0}=-5$ | | $\frac{1-(-5)}{2-(-1)}=2$ | |
| 1 | 0 | | $\frac{-3-(-5)}{2-0}=1$ | | |
| | | $\frac{-3-0}{2-1}=-3$ | | | |
| 2 | -3 | | | | |

Interpolating polynomial:

$$p(x) = -3 + 3(x+2) + (x+2)(x+1) \quad\quad -2(x+2)(x+1)x$$
$$+ (x+2)(x+1)x(x-1)$$

$$= 5 - 6x^2 + x^4$$

**31.** $p(x) = x^3 - 3x + 3$; the two polynomials are identical.

**33.** $p(x) = x^4 - 6x^2 + 5$; the two polynomials are identical.

**35.**

| $x$ | 1 | 4 | 9 |
|---|---|---|---|
| $f(x)$ | 1 | 2 | 3 |

| $x_k$ | $y_k$ | FIRST DIVIDED DIFFERENCE | SECOND DIVIDED DIFFERENCE |
|---|---|---|---|
| 1 | 1 | | |
| | | $\frac{2-1}{4-1}=\frac{1}{3}$ | |
| 4 | 2 | | $\frac{\frac{1}{5}-\frac{1}{3}}{9-1}=-\frac{1}{60}$ |
| | | $\frac{3-2}{9-4}=\frac{1}{5}$ | |
| 9 | 3 | | |

Interpolating polynomial:

$$p(x) = 1 + \frac{1}{3}(x-1) - \frac{1}{60}(x-1)(x-4)$$

| $x$ | 1 | 2 | 3 | 4 | 5 | 6 | 7 | 8 | 9 |
|---|---|---|---|---|---|---|---|---|---|
| $p(x)$ | 1 | 1.4 | 1.7 | 2 | 2.2 | 2.5 | 2.7 | 2.9 | 3 |
| $\sqrt{x}$ | 1 | 1.4 | 1.7 | 2 | 2.2 | 2.4 | 2.6 | 2.8 | 3 |

**37.**

| $x$ | -2 | -1 | 0 | 1 | 2 |
|------|----|----|---|---|---|
| $f(x)$ | -4 | -5 | 0 | 5 | 4 |

| $x_k$ | $y_k$ | FIRST DIVIDED DIFFERENCE | SECOND DIVIDED DIFFERENCE | THIRD DIVIDED DIFFERENCE | FOURTH DIVIDED DIFFERENCE |
|-------|-------|--------------------------|---------------------------|--------------------------|----------------------------|
| -2 | -4 | $\dfrac{-5-(-4)}{-1-(-2)}=-1$ | | | |
| -1 | -5 | $\dfrac{0-(-5)}{0-(-1)}=5$ | $\dfrac{5-(-1)}{0-(-2)}=3$ | $\dfrac{0-3}{1-(-2)}=-1$ | |
| 0 | 0 | $\dfrac{5-0}{1-0}=5$ | 0 | $\dfrac{-3-0}{2-(-1)}=-1$ | 0 |
| 1 | 5 | $\dfrac{4-5}{2-1}=-1$ | $\dfrac{-1-5}{2-0}=-3$ | | |
| 2 | 4 | | | | |

Interpolating polynomial:

$$p(x) = -4 - (x+2) + 3(x+2)(x+1) - (x+2)(x+1)x$$
$$= 6x - x^3$$

---

**39.**

| $x$ | $-x_1$ | 0 | $x_1$ |
|------|--------|---|-------|
| $f(x)$ | $Y_1$ | $Y_2$ | $Y_1$ |

| $x_k$ | $y_k$ | FIRST DIVIDED DIFFERENCE | SECOND DIVIDED DIFFERENCE |
|-------|-------|--------------------------|----------------------------|
| $-x_1$ | $Y_1$ | $\dfrac{Y_2-Y_1}{0-(-x_1)}=\dfrac{Y_2-Y_1}{x_1}$ | $\dfrac{\dfrac{Y_1-Y_2}{x_1}-\dfrac{Y_2-Y_1}{x_1}}{x_1-(-x_1)}=\dfrac{Y_1-Y_2}{x_1^2}$ |
| 0 | $Y_2$ | | |
| $x_1$ | $Y_1$ | $\dfrac{Y_1-Y_2}{x_1-0}=\dfrac{Y_1-Y_2}{x_1}$ | |

Interpolating polynomial:

$$p(x) = Y_1 + \frac{Y_2-Y_1}{x_1}(x+x_1) + \frac{Y_1-Y_2}{x_1^2}(x+x_1)x$$

$$= Y_1 + \frac{(Y_2-Y_1)}{x_1}x + Y_2 - Y_1 + \frac{Y_1-Y_2}{x_1^2}x^2 + \frac{(Y_1-Y_2)x}{x_1}$$

$$= Y_2 + \frac{Y_1-Y_2}{x_1^2}x^2$$

**41.**

| $t$ | 0 | 4 | 8 | 12 |
|---|---|---|---|---|
| $C(t)$ | 2 | 32 | 38 | 20 |

(A)

| $t_k$ | $y_k$ | FIRST DIVIDED DIFFERENCE | SECOND DIVIDED DIFFERENCE | THIRD DIVIDED DIFFERENCE |
|---|---|---|---|---|
| 0 | 2 | $\dfrac{32 - 2}{4 - 0} = 7.5$ | | |
| 4 | 32 | $\dfrac{38 - 32}{8 - 4} = 1.5$ | $\dfrac{1.5 - 7.5}{8 - 0} = -0.75$ | |
| 8 | 38 | $\dfrac{20 - 38}{12 - 8} = -4.5$ | $\dfrac{-4.5 - 1.5}{12 - 4} = -0.75$ | 0 |
| 12 | 20 | | | |

Interpolating polynomial:

$p(t) = 2 + 7.5t - 0.75t(t - 4)$

$\quad\quad = 2 + 10.5t - 0.75t^2$

(B) $C(6) \approx p(6) = 2 + 10.5(6) - 0.75(6^2) = 38$ or $\$38{,}000$

(C) Average cash reserves for the first quarter:

$\dfrac{1}{3}\displaystyle\int_0^3 p(t)\,dt = \dfrac{1}{3}\int_0^3 (2 + 10.5t - 0.75t^2)\,dt$

$\quad\quad\quad\quad\quad = \dfrac{1}{3}\left[2t + 5.25t^2 - 0.25t^3\right]_0^3 = 15.5$

The average cash reserves are approximately $\$15{,}500$.

**43.**

| $x$ | 0 | 0.2 | 0.8 | 1 |
|---|---|---|---|---|
| $f(x)$ | 0 | 0.04 | 0.52 | 1 |

| $x_k$ | $y_k$ | FIRST DIVIDED DIFFERENCE | SECOND DIVIDED DIFFERENCE | THIRD DIVIDED DIFFERENCE |
|---|---|---|---|---|
| 0 | 0 | $\dfrac{0.04 - 0}{0.2 - 0} = 0.2$ | | |
| 0.2 | 0.04 | $\dfrac{0.52 - 0.04}{0.8 - 0.2} = 0.8$ | $\dfrac{0.8 - 0.2}{0.8 - 0} = 0.75$ | |
| 0.8 | 0.52 | $\dfrac{1 - 0.52}{1 - 0.8} = 2.4$ | $\dfrac{2.4 - 0.8}{1 - 0.2} = 2$ | $\dfrac{2 - 0.75}{1} = 1.25$ |
| 1 | 1 | | | |

(A) Interpolating polynomial:

$p(t) = 0.2x + 0.75x(x - 0.2) + 1.25x(x - 0.2)(x - 0.8)$

$\quad\quad = 0.25x - 0.5x^2 + 1.25x^3$

(B) Index of income concentration:

$C \approx 2\displaystyle\int_0^1 [x - p(x)]\,dx$

$\quad = 2\displaystyle\int_0^1 [x - (0.25x - 0.5x^2 + 1.25x^3)]\,dx$

$\quad = 2\displaystyle\int_0^1 (0.75x + 0.5x^2 - 1.25x^3)\,dx$

$\quad = 2\left[0.375x^2 + 0.1667x^3 - 0.3125x^4\right]_0^1$

$\quad = 2(0.2292) = 0.4584$

**45.**

| $x$ | 2 | 4 | 6 |
|---|---|---|---|
| $R(x)$ | 24.4 | 36 | 34.8 |

| $x_k$ | $y_k$ | FIRST DIVIDED DIFFERENCE | SECOND DIVIDED DIFFERENCE |
|---|---|---|---|
| 2 | 24.4 | $\dfrac{36-24.4}{4-2}=5.8$ | |
| 4 | 36 | $\dfrac{34.8-36}{6-4}=-0.6$ | $\dfrac{-0.6-5.8}{6-2}=-1.6$ |
| 6 | 34.8 | | |

(A) Interpolating polynomial:
$$p(x) = 24.4 + 5.8(x-2) - 1.6(x-2)(x-4)$$
$$= 15.4x - 1.6x^2$$

(B) $R(5) \approx p(5) = 15.4(5) - 1.6(5^2) = 37$ or $\$37,000$

(C) Production level to maximize revenue:
$$P'(x) = 15.4 - 3.2x$$
$$p'(x) = 0: \quad 15.4 - 3.2x = 0$$
$$x = 4.8125 \quad \text{(thousand lamps)}$$

To the nearest integer, the production level that will maximize the revenue is 4,813 lamps.

**47.**

| $t$ | 0 | 1 | 2 | 3 | 4 |
|---|---|---|---|---|---|
| $C(t)$ | 14 | 13 | 16 | 17 | 10 |

| $t_k$ | $y_k$ | FIRST DIVIDED DIFFERENCE | SECOND DIVIDED DIFFERENCE | THIRD DIVIDED DIFFERENCE | FOURTH DIVIDED DIFFERENCE |
|---|---|---|---|---|---|
| 0 | 14 | $\dfrac{13-14}{1-0}=-1$ | $\dfrac{3-(-1)}{2-0}=2$ | | |
| 1 | 13 | $\dfrac{16-13}{2-1}=3$ | $\dfrac{1-3}{3-1}=-1$ | $\dfrac{-1-2}{3}=-1$ | |
| 2 | 16 | $\dfrac{17-16}{3-2}=1$ | $\dfrac{-7-1}{4-2}=-4$ | $\dfrac{-4-(-1)}{4-1}=-1$ | 0 |
| 3 | 17 | $\dfrac{10-17}{4-3}=-7$ | | | |
| 4 | 10 | | | | |

(A) Interpolating polynomial:
$$p(t) = 14 - t + 2t(t-1) - t(t-1)(t-2)$$
$$= 14 - 5t + 5t^2 - t^3$$

(B) Average temperature:
$$\frac{1}{4}\int_0^4 p(t)\,dt = \frac{1}{4}\int_0^4 (14 - 5t + 5t^2 - t^3)\,dt$$

$$= \frac{1}{4}\left[14t - \frac{5}{2}t^2 + \frac{5}{3}t^3 - \frac{1}{4}t^4\right]_0^4$$

$$= \frac{1}{4}\left(\frac{176}{3}\right) \approx 14.7^{\circ}C$$

**49.**

| $t$ | 0 | 2 | 4 | 6 |
|---|---|---|---|---|
| $C(t)$ | 450 | 190 | 90 | 150 |

| $t_k$ | $Y_k$ | FIRST DIVIDED DIFFERENCE | SECOND DIVIDED DIFFERENCE | THIRD DIVIDED DIFFERENCE |
|---|---|---|---|---|
| 0 | 450 | $\dfrac{190 - 450}{2 - 0} = -130$ | | |
| 2 | 190 | $\dfrac{90 - 190}{4 - 2} = -50$ | $\dfrac{-50 - (-130)}{4 - 0} = 20$ | 0 |
| 4 | 90 | $\dfrac{150 - 90}{6 - 4} = 30$ | $\dfrac{30 - (-50)}{4} = 20$ | |
| 6 | 150 | | | |

(A) Interpolating polynomial:
$$p(t) = 450 - 130t + 20t(t - 2) = 450 - 170t + 20t^2$$

(B) Minimum concentration level:
$$p'(t) = -170 + 40t$$
$$p'(t) = 0: \quad -170 + 40t = 0$$
$$t = \frac{170}{40} = 4.25$$

The minimum concentration level will be reached in 4.25 days (approximately).

**51.**

| $t$ | 0 | 10 | 20 | 30 |
|---|---|---|---|---|
| $N(t)$ | 10,000 | 13,500 | 20,000 | 23,500 |

| $t_k$ | $Y_k$ | FIRST DIVIDED DIFFERENCE | SECOND DIVIDED DIFFERENCE | THIRD DIVIDED DIFFERENCE |
|---|---|---|---|---|
| 0 | 10,000 | $\dfrac{13,500 - 10,000}{10 - 0} = 350$ | | |
| 10 | 13,500 | $\dfrac{20,000 - 13,500}{20 - 10} = 650$ | $\dfrac{650 - 350}{20 - 0} = 15$ | $\dfrac{-15 - 15}{30 - 0} = -1$ |
| 20 | 20,000 | $\dfrac{23,500 - 20,000}{30 - 20} = 350$ | $\dfrac{350 - 650}{30 - 10} = -15$ | |
| 30 | 23,500 | | | |

(A) Interpolating polynomial:
$$p(t) = 10,000 + 350t + 15t(t - 10) - t(t - 10)(t - 20)$$
$$= 10,000 + 45t^2 - t^3$$

(B) Average number of registered voters
$$\frac{1}{20}\int_0^{20} p(t)\,dt = \frac{1}{20}\int_0^{20} (10,000 + 45t^2 - t^3)\,dt$$
$$= \frac{1}{20}\left[10,000t + 15t^3 - \frac{1}{4}t^4\right]_0^{20}$$
$$= \frac{1}{20}(280,000) = 14,000$$

The average number of registered voters over the first 20 years is approximately 14,000.

# INDEX

## A

Addition, Taylor series, 64–65
Advertising, 22
Agriculture, 127
Alternating series, 76–77, 84
Approximation:
    piecewise linear, 136
    using Taylor polynomials, 45–46
    using Taylor series, 73–82, 83
Arrival rates, 121
Average cost, 147–148
Average price, 49, 51, 52, 82, 85

## B

Bacteria control, 148
Basic Taylor series, 64

## C

Capital value, 92–93, 94, 128, 131
    of perpetual income stream, 93
Cash reserves, 147
Communications, 127
Component failure, 127
Computer failure, 131
Continuous compound interest, 11, 22, 28–29,
    33, 38
Continuous income stream, 92–93, 128
Continuous random variables, 95–96, 128
Convergence:
    improper integrals, 89–90, 128
    interval of, 56–57, 60, 83, 128
        finding, 58–59
    Taylor series, 55–62
Corporate profits, 22
Credit applications, 131
Crime scene investigation, 23
Crop yield, 38
Cumulative distribution functions, 101–102,
    123, 128
    defined, 101
    properties, 102
    using, 102–103

## D

Definite integrals, using Taylor series to
    approximate, 77
Depreciation, 37
Derivatives:
    higher-order, 40–41
    $n$th, 40–41
Differential equations, 1–38, 35
    defined, 2
    explicit solutions, 6–7
    family of solutions, 3, 35
    first-order, 2
    general solution, 3, 35
    implicit solutions, 5
    initial condition, 4, 35
    order of, 2
    particular solution, 3, 35
    separation of variables, 12–15, 35

singular solutions, 14, 35
slope field, 2
solutions of, 2–7
    finding, 4
    verifying, 3–4
Differentiation, Taylor series, 66–67
Discrete random variables, 95–96
Divergence:
    improper integrals, 89
    Taylor series, 55–62
Divided difference tables, 140–143
    application, 143–145
    final form, 141
    first divided differences, 140
    and interpolating polynomials, 141
    second divided differences, 140
    third divided differences, 140
    using, 142–143
Drugs:
    assimilation, rate of, 95
    dosage and concentration, 148
Dynamic price stability, 7

## E

Ecology, 23, 38
Electrical current, 119
Electricity consumption, 106, 116
Epidemic, 23
Equilibrium price, 7, 29–30
    long-range, 29
    at time $t$, 29
Error estimation for alternating series, 76–77, 84
Error in approximation, 74–77
Estes, William K., 34
Estes learning model, 34
Expanded notation, 45
Expected value, 108–110, 128
    computing, 109–110, 112
Explicit solutions, differential equations, 6–7
Exponential decay, 20, 35
Exponential distributions, 120–121, 129
Exponential growth, 15–16
    product analysis, 15–16
Exponential growth model, 15
Exponential growth phenomena,
    comparison of, 20
Exponential probability density
    function, 120, 129
    mean, 108, 128
    median, 112–114, 128
    standard deviation, 108, 109–110, 112,
        122, 128
    summary, 121
Exponential random variable, 120

## F

First-order differential equations, 2, 35
First-order linear differential equations, 24–36
    applications, 28–32
    defined, 24
    integrating factor, 25–26
        using, 27–28

natural logarithm function, basic formulas
    involving, 27
    product rule, 24
    solutions, 24–26
        procedure, 25, 36
    standard form, 25
Food preparation, 23
Forward substitution, 138

## G

Gasoline consumption, 106, 117
Gompertz growth model, 11
Grading, on a curve, 127
Guarantees, 127

## H

Higher-order derivatives, Taylor
    polynomials, 40–41

## I

Immigration, 95
Implicit solutions, differential equations, 5
Improper integrals, 88–95, 128
    converging, 89
    defined, 89
    diverging, 89
    evaluating, 89–91
Income distribution, 78–80, 82, 85, 147
Initial condition, 4, 35
Integral, improper, 88–95, 128
Integrating factor, 25–26
    defined, 25
    using, 27–28
Integration, 139
    constant of, 14, 26–28, 36
    of Taylor series, 67–68
Interpolating polynomials, 135–148
    and computer graphics, 135
    defined, 135, 137–138
    finding, 138–139
    forward substitution, 138
    Newton's form for, 138
    Newton's form for, 139
    steps for finding, 138
Interval of convergence, 56–57, 60, 83, 128
    finding, 58–59
Inventory, 143–145, 147

## L

Learning, 23–24, 52, 82, 107, 117, 127
Learning theory, 34–35
Life expectancy, 107, 110, 113–114, 117, 131
Limited growth, 16–18, 20, 35
    sales growth, 17–18
Limited growth model, 16
Logistic growth, 18–20, 20, 35
    population growth (example), 18–20
Logistic growth model, 11, 18
Long-range equilibrium price, 29
Lower triangular system, 138

**I-1**

## M

Maclaurin polynomials, 44fn
Manufacturing, 23
Marketing, 82, 85
Maximum revenue, 147
Mean, 108, 128
Measure of central tendency, 110
Median, 112–114, 128
    defined, 112
    examples, 113–114
    finding, 112–113
Medicine, 51–52, 82, 85, 86, 127, 131
Minimum average cost, 147–148
Multiplication, Taylor series, 65–66

## N

Newton's law of cooling, 23
Normal curve, 123, 129
Normal distributions, 121–125, 129
    area under normal curves, 123, 129
    finding probabilities for, 124–125
    mean, 108, 128
    median, 112–114, 128
    standard deviation, 108, 122, 128
Normal probability density function, 121, 129
    summary, 123
Normal random variable, 121
$n$th derivatives, 40–41
$n$th-degree Taylor polynomial
    at $a$, 48–49
    for $f$ at 0, 44

## O

Oil production, 91–92
Order of a differential equation, 2

## P

Pareto, Vilfredo, 126fn
Pareto random variable, 126
Particular solution, 3, 35
Perpetual income stream, 128
    capital value of, 93
Personal income, 22
Personnel screening, 131
Piecewise linear approximation, 136
Politics, 86, 131
Pollution, 34, 38, 52, 95
Pollution control, 30–32
Polynomials, approximating $e^x$ with, 41–43
Population growth, 11, 18–20
Price stability, 10–11, 38
Probability density functions, 96–97, 128
    defined, 97
    quartile points for, 116
    using, 97–100
Probability distribution, 96
Probability theory, 95
Product analysis, 15–16, 22
Product life, 117

Production, 51, 52, 85, 95, 130
Profit, 116
Psychology, 127

## Q

Quality control, 127
Quartile points, 116

## R

Radial tire failure, 131
Random variables, continuous, 95–96, 128
    alternative formula for variance, 111–112
    cumulative distribution functions, 101–102
    defined, 111
    expected values, 108–109
    mean, 108–109
    median, 109
    probability density functions, 96–97, 128
        defined, 97
        exponential distributions, 120–121, 129
        normal distributions, 121–125, 129
        quartile points for, 116
        using, 97–100
    quartile points for, 116
    standard deviation, 108–109
    uniform distributions, 117–119, 129
    variance, 108–109
Random variables, discrete, 95–96
Remainder, 73–74
    formula for, 74–76
Respiration, 148
Rumor spread, 11, 24, 38

## S

Sales analysis, 22–23
Sales growth, 17–18
Segal, Arthur, 34
Sensory perception, 23
Separation of variables, 12–15, 35
Series:
    alternating, 76–77, 84
    Taylor, 56–64
Service time, 127
Shelf life, 100, 103–104, 107, 117
Singular solutions, 14, 35
Slope field, 2, 35
Special calculus topics, 135–148
Special probability distributions, 117–127, 129
    exponential distributions, 120–121
    normal distributions, 121–125
    uniform distributions, 117–119
Standard deviation, 108, 128
    computing, 109–110, 112
Standard form first-order linear differential
    equations, 25
Standard normal curve, 123, 129
    area under (table), 133
Summation notation, 45
Survival time, 127

## T

Taylor, Brook, 39
Taylor polynomials, 39, 83
    at $a$, 47–48
    at 0, 44–45, 46–47
        concise form, 45
    applications, 49–50
    approximation using, 45–46
    defined, 39, 44
    expanded notation, 45
    higher-order derivatives, 40–41
    summation notation, 45
Taylor series:
    at $a$, 56–64, 83
    addition of, 64–65
    with alternating terms, 76–77
    approximations using, 73–82, 83
    basic, 64
    convergence, 55–62
    defined, 56
    differentiation, 66–67
    divergence, 55–62
    integration of, 67–68
    interval of convergence, 56–57, 60
    multiplication of, 65–66
    operations on, 64–73, 83
    remainder, 73–74
        formula for, 74–76
    representation of functions by, 59–62
    substitution, using to find, 68–71
    using to approximate definite
        integrals, 77–78
Taylor series for f at a, 83
Taylor's formula for the remainder, 84
Temperature, 82, 148
Testing, 131

## U

Uniform distributions, 117–119, 129
Uniform probability density function,
    118–119, 129
    mean, 119
    median, 119
    standard deviation, 119
Unlimited growth, 20, 35
Useful life, 82

## V

Variance, 108, 128
    computing, 109–110, 111–112
Verhulst growth model, 11
Voter registration, 148
Voter turnout, 117

## W

Waiting time, 107, 116, 118, 127
Water consumption, 117
Weight loss, 34

# APPLICATIONS INDEX

## Business & Economics

Advertising, 22
Approximating revenue, 135–136
Arrival rates, 121
Average cost, 147–148
Average price, 49, 51, 52, 82, 85
Capital value, 92–93, 94, 128, 131
Cash reserves, 147
Communications, 127
Component failure, 127
Computer failure, 131
Continuous compound interest, 11, 22, 28–29, 33, 38
Corporate profits, 22
Credit applications, 131
Demand, 107, 130
Depreciation, 37
Electrical current, 119
Electricity consumption, 106, 116
Equilibrium price, 29–30
Gasoline consumption, 106, 117
Guarantees, 127
Inventory, 143–145, 147
Long-range equilibrium price, 29
Manufacturing, 23
Marketing, 82, 85
Maximum revenue, 147
Oil production, 91–92
Personal income, 22
Personnel screening, 131
Price stability, 10–11, 38
Product analysis, 22
Product life, 117
Production, 51, 52, 85, 95, 130
Profit, 116
Quality control, 127
Radial tire failure, 131
Sales, 127
Sales analysis, 22–23
Sales growth, 17–18, 37–38
Service time, 127
Shelf life, 100, 103–104, 107, 117
Supply-demand, 34
Time-sharing, 106
Useful life, 82
Waiting time, 107, 116, 118, 127
Water consumption, 117

## Life Sciences

Agriculture, 127
Bacteria control, 148
Crime scene investigation, 23
Crop yield, 38
Drugs, assimilation, rate of, 95, 131
Drugs, dosage and concentration, 148
Ecology, 23, 38
Food preparation, 23
Life expectancy, 107, 110, 113–114, 117
Medicine, 51–52, 82, 85, 86, 127, 131
Pollution, 34, 38, 52, 95
Pollution control, 30–32
Population growth, 23
    logistic growth model, 11
    Verhulst growth model, 11
Respiration, 148
Simple epidemic, 23
Survival time, 127
Temperature, 82, 148
Weight loss, 34

## Social Sciences

Estes learning model, 34
Gompertz growth model, 11
Grading on a curve, 127
Immigration, 95
Income distribution, 78–80, 82, 85, 147
Learning, 23–24, 52, 82, 107, 117, 127
Learning theory, 34–35
Politics, 86, 131
Psychology, 127
Rumor spread, 24, 38–39
    Gompertz growth model, 11
Sensory perception, 23
Testing, 131
Voter registration, 148
Voter turnout, 117